From Photon to Neuron

Book cover: [Confocal fluorescence micrograph.] The vertebrate retina is made up of layers of different cell types. To create this image, the rod photoreceptor cells of a mouse (in the outer nuclear layer, *far right*) were labeled in red. Moving leftward, the next layers constitute the inner nuclear layer (*center*): bipolar cells were labeled in green; the amacrine and horizontal cells were labeled in blue. Next comes the inner plexiform layer (*green filaments*), including interconnections, and finally a layer containing ganglion cell bodies (*far left, blue*). These cell types and their functions are described in Chapter 11. The magnification is such that the cell diameters are about $5\,\mu$m.

In greater detail, the experimenters generated DNA plasmids that drove fluorescent protein expression, either enhanced GFP or a red emitter named mCherry, in bipolar cells or rods respectively. They also used short DNA segments (enhancers), that made the fluorescent proteins express in a cell type-specific manner (similarly to methods discussed in Chapter 2). Other cell types in the inner nuclear layer (INL, *center*) were tagged with fluorescent antibodies. (Yet other cell types that make up the retina were not labeled, including the cone photoreceptor cells and the majority of Müller glia cells. Rod outer segments are only faintly visible on *far right* because most of the fluorescent protein remained confined to the inner segment and cell body.) [Courtesy of Dr. Sui Wang and Dr. Constance Cepko; see also Wang et al., 2014.]

The basic structural organization of the retina shown here is conserved across vertebrates and is responsible for detecting, preprocessing, and conveying all of the different kinds of visual information about the outside world to the brain.

This image also demonstrates that specific cell types can be genetically manipulated in vivo, with potential applications to therapy and basic science. For example, an optogenetic protein can be introduced only in bipolar cells, rendering them photosensitive, an approach being taken to compensate for loss of photoreceptors in disease (see Section 11.4.3).

Inset: Fragment from Albert Einstein's 1905 article introducing the photon concept. The formula appears in the text, in modern notation, as Equation 1.6 (page 34). [From Einstein, 1905.]

Facing page: Protein structures shown in Figure 10.13 (page 338). *Vesicle:* A1 - synaptobrevin; A2 - synaptotagmin; A3 - Rab; A4 - synaptophysin; A5 - vGlut; A6 - vesicular ATPase. *Presynaptic cytoplasm:* B1 - dystrophin; B2 - actin; B3 - NSF; B4 - munc13 (inactive); B5 - GDI (guanine nucleotide dissociation inhibitor); B6 - munc13/munc18; B7 - bassoon; B8 - RIM, CAST, etc.; B9 - ribeye; B10 - PRA1 (connection between bassoon and Rab); B11 - speculative bridging protein. *Presynaptic membrane:* C1 - EAAT (excitatory amino acid transporter); C2 - SNAP25; C3 - syntaxin; C4 - sidekick/DSCAM (speculative); C5 - PSD-95; C6 - voltage-gated calcium channel; C7 - dystroglycan; C8 - src; C9 - LAR (receptor-type tyrosine-protein phosphatase); C10 - sodium/potassium ATPase. *Cleft:* D1 - laminin; D2 - pikachurin. *Postsynaptic membrane:* E1 - iGluR; E2 - mGluR; E3 - G protein; E4 - adenyate cyclase; E5 - TRPM1; E6 - nyctalopin; E7 - PSD-95/MAGI (speculative); E8 - Fyn; E9 - potassium channel; E10 - ErbB (as a generic receptor tyrosine kinase). *Postsynaptic cytoplasm:* F1 - GKAP; F2 - SHANK; F3 - Homer. [Art by David S Goodsell.]

From Photon to Neuron

Light, Imaging, Vision

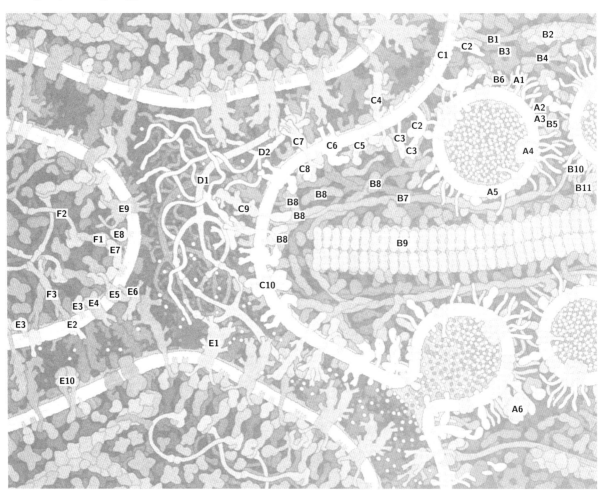

Philip Nelson

With the assistance of Sarina Bromberg,
Ann M. Hermundstad, and Jesse M. Kinder

Princeton University Press
Princeton and Oxford

Published by Princeton University Press, 41 William Street, Princeton, New Jersey 08540
In the United Kingdom: Princeton University Press, 6 Oxford Street, Woodstock, Oxfordshire OX20 1TR

press.princeton.edu

Cover: The retina is responsible for detecting and preprocessing visual information about the outside world and conveying that information to the brain. [Courtesy Constance Cepko, Harvard Medical School Department of Genetics and Ophthalmology and Howard Hughes Medical Institute, and Sui Wang, Stanford Medical School Department of Ophthalmology.] The inset shows a fragment from Albert Einstein's 1905 article introducing the photon concept.

ISBN 978-0-691-17518-8
ISBN (pbk.) 978-0-691-17519-5

British Library Cataloging-in-Publication Data is available.

Publication of this book has been aided by the United States National Science Foundation.

This book was composed using the LaTeX typesetting system.

The publisher would like to acknowledge the author of this volume for providing the print-ready files from which this book was printed.

Printed on acid-free paper. ∞

Printed in the United States of America.

1   3   5   7   9   10   8   6   4   2

For Scott Weinstein and William Berner

And yet, protest it if we will,
Some corner of the mind retains
The medieval man, who still
Keeps watch upon those starry skeins
And drives us out of doors at night
To gaze at anagrams of light.

— Adrienne Rich

Brief contents

Detailed contents

Chapter 2 | Photons and Life   61

Chapter 3 │ Color Vision 107

Chapter 4 | How Photons Know Where to Go   145

Chapter 5 | Optical Phenomena and Life   180

PART II   Human and Superhuman Vision

Chapter 10 │ The Mechanism of Visual Transduction   318

Chapter 11 | The First Synapse and Beyond   352

PART III   Advanced Topics

Chapter 12 | Electrons, Photons, and the Feynman Principle   381

Chapter 13 | Field Quantization, Polarization, and the Orientation of a Single Molecule   398

Chapter 14 | Quantum-Mechanical Theory of FRET   415

Web resources

The book's Web site (`http://press.princeton.edu/titles/11051.html`) contains links to the following resources:[1]

- *Datasets* contains datasets that are used in the problems, along with their descriptions. In the text, these are cited like this: Dataset 1, with numbers keyed to the list on the Web site.
- *Media* gives links to external media (graphics, audio, and video). In the text, these are cited like this: Media 1, with numbers keyed to the list on the Web site.
- Finally, *Errata* is self-explanatory.

---

[1]Secondary backup sources for these resources include [Not ready yet.] and `http://www.physics.upenn.edu/biophys/PtN/Student`.

To the student

> The man who cannot wonder is but a pair of spectacles
> behind which there is no eye.
> — *Thomas Carlyle, 1795–1881*

This is a book about the physical nature of light, and how it was transformed in the 20th century. This is also a book about the many uses living organisms have found for light, and the strategies they have evolved, especially the formation of visual images representing the world and their transmission to the brain. Finally, this is a book about extensions of light-based imaging, technologies invented from earliest times right up to the present, each of which brought revolutionary improvements in our understanding of the microworld.

That's a lot of territory, but I believe that every student, both in the life sciences and physical sciences, needs to know the basics of this field. For one thing, experiments keep demonstrating that Life makes specific use of the weird, yet manageable, quantum character of light: To understand your own vision, photosynthesis, or a host of other topics, you need to appreciate this aspect of Nature. Moreover, the recent explosion of superresolution and other advanced imaging techniques is also mostly inaccessible without that viewpoint.

The good news, which came as a surprise to me, is that *many important topics are no harder to understand from the modern viewpoint* than they are from the older one rooted in the classical, 19th-century model of light ("Maxwell's equations"). In fact, because so many biophysical topics seem to rely on quantum behavior, and so few on details of the classical model, the main part of this book will never introduce the heavy mathematical apparatus of Maxwell's equations at all. It's important for some rather specialized topics (for example, birefringence), but this book regards classical electrodynamics as a *more advanced topic* that you should learn *later*—it's an approximation that, when applicable, makes some detailed calculations more tractable. In this way, we will avoid the inevitable, uncomfortable moment when we must say, "Actually, everyone agrees that this (classical) model is wrong."

If you have already been indoctrinated in the older view, don't worry. Please keep an open mind to what the experiments described in the next chapters are telling us about light, and how the framework developed here can make sense of those experiments. With your background, you'll be able to delve into Chapter 13 and see the connection to what you've learned before.

Features of this book

- Most chapters end with an appendix labeled "Track 2." Some of these give extra details for advanced students, including literature citations. There are also Track 2 footnotes and problems, marked with the symbol $\boxed{T_2}$.
- Appendix A summarizes graphical and mathematical notation, then lists key symbols that are used consistently throughout the book. Appendix B discusses

some useful tools for solving problems. Appendix C gathers some numerical constants for reference. Appendix D gives a refresher on complex numbers.

- The notations "Equation x.y" and "Idea x.y" both refer to the same numbered series.
- When a distant figure gets cited, often its reference includes a small version of the image in the margin. You can look at this icon and decide whether you want to flip back to see the full figure and its caption.

Skills and habits

> You cannot become a ship's captain, nor any kind of craftsman, from reading a book.
> — *Galen of Pergamum, second century CE*

Science is not just a heap of facts for you to repeat on demand. It is a collection of *skills and habits* that over time have proven successful at creating *new knowledge*. To be creative in that sense, you must begin by gaining fluency with the building blocks of scientific ideas. And reading this book, or any book, is only one part of a conscious strategy to gain that fluency. When this book poses a question, or when you hit a stumbling block, don't immediately look to see who has already answered that question on the Web. What will make you grow as a scientist is to attempt that problem yourself with the tools at your disposal. Later, when you attack problems that nobody has answered yet, this discipline will pay off.

Again: science is about doing things that *nobody has ever done before*. This book will tell you stories of how it has happened in the past, and offer you many opportunities to develop the skills you'll need when it's your turn. In fact, some are literally flagged with the tag "Your Turn." You'll find others as you read; take the time to derive each formula in the text, hit obstacles, and overcome them. You may need help from your instructor or a classmate for that. Use them.

One major skill emphasized in this book is writing short computer programs. Today essentially all science is done with the help of computers, so the sooner you acquire this skill the better. Several excellent software platforms now exist to help you do the everyday tasks that arise in the lab, and also when learning any subject. Some of these are free and open-source, for example Python, R, or Octave. You also can find a wealth of free help online, but be warned: You'll need a lot of daily experience before this unnatural activity begins to feel natural. Some available resources are written with the specific goal of helping you learn what you need in order to manipulate and visualize both experimental data and theoretical models, for example, the two short guides Nelson & Dodson, 2015, and Kinder & Nelson, 2015.

About you

Parts I and II of this book are a one-semester course intended for anyone who has completed a year of university-level physics and calculus. (I have found, however, that even students with much more background than that find many topics and ideas here that did not come up in their earlier courses.) In addition, you may want to read some or all of the "Track-2" sections at the end of each chapter. The chapters in Part III cover advanced topics; here you will need background from other courses on quantum mechanics or electromagnetism.

Very few biological or chemical prerequisites are assumed, though from time to time you may need to use an external resource to fill in some of the background to the story.

Mostly, this book assumes a great deal of curiosity on your part about how the world works, including things you see around you every day. In fact, it's useful to visualize science as a form of espionage: We have some distant, complex adversary (perhaps cancer, or blindness). We belong to a far-flung network of people who are trying to find useful things. Some have obviously relevant missions. That's called applied research.

But other agents are out there trying to understand the world by looking for things that *don't fit*. That's called pure research. It could be something that other people have seen without realizing that it doesn't fit. It could be pointing to an important nugget of intelligence, as we will see in a number of case studies. We must integrate, think laterally, maybe discard preconceptions. Sometimes there is a high-tech gadget that we can invent to get the key datum, perhaps originally designed for a very different purpose.

Like real spycraft, the work is often mundane, and sometimes lonely. Often it ends up not directly benefiting anyone. But sometimes there's a valuable insight, maybe not useful for the purpose we had in mind, but someone else can see the connection to something important. Sometimes lives get saved. Let's get started.

To the instructor

The present generation has no right to complain of the great discoveries already made, as if they left no room for further enterprise. They have only given science a wider boundary.
— *James Clerk Maxwell*

This book embodies a course that I have taught for several years at the University of Pennsylvania. The students who enroll are mainly second- through fourth-year science and engineering undergraduates who have had a year of introductory physics and calculus. Many have heard rumors about new imaging techniques, quantum phenomena, or something else, and want to know more. Their interest led me to present the radical notion that physics is interesting and important for its own sake, even to non-majors, because over and over we find discoveries being made by people who know in detail how their apparatus works, and so can extend its reach. We find discoveries made years before they "ought" to have been possible, by people who were able to carry out certain kinds of indirect reasoning. And so on.

Thus, I have attempted to "keep the physics in biophysics," not only to enrich students' understanding of the underpinnings of biophysical applications, but also to make basic physics ideas more vivid and concrete by reference to those applications. My goal is not to be comprehensive, but to tell one story that touches on a huge body of ongoing work, reaching all the way into the foothills of neuroscience, a topic of intense interest among my students. Students who know about light, imaging, and vision will find themselves well positioned to learn the many other topics not covered in this finite book.

I have also chosen topics for which I felt I could bring students to the point where they could do the key calculations for themselves (sometimes with a computer). Certainly bringing myself to that point has been a struggle! I hope that students can arrive at a working knowledge of the field more readily with some more guidance than is found in the primary literature.

I've also come to believe that

- Whenever possible, we should try to relate abstract concepts to familiar experience.
- It is possible, and even desirable, to tell students about the nature of light as we currently understand it. It is true that Maxwell's theory is an excellent approximation for some purposes—but unfortunately not for some of the most basic processes of Life. Instead, a single unified framework is now understood to cover all light phenomena. Remarkably, for some applications that framework is no more difficult than the old one.
- The study of basic science is fundamentally intertwined with the development of *instrumentation*. Understanding current instrumentation requires a more sophisticated model of light than what we present in first-year physics. This book positions the student to participate in the ongoing revolution in optical techniques.

- The traditional lead-in to quantum physics, via the energy levels of the hydrogen atom, has some drawbacks. For one thing, the student needs quite a lot of mathematical apparatus to arrive at a result that life-science students, at least, will not see as central to their concerns. Also, the nonrelativistic (Schrödinger equation) approach leaves us unable to say anything about photons, and their many applications to imaging technology. This book instead uses Feynman's viewpoint, which, besides handling photons, has some conceptual advantages.[2]

Another reason to study light has to do with a key skill that every scientist has to exercise, that of *abandoning long-held assumptions*. Of course, we all contend daily with our own mistakes, but not always in a mindful way. So this book pays some attention to the discovery that the wave theory of light could not explain most of how light interacts with individual molecules. Changing our model is not as simple as just walking away, however: Most entrenched, wrong models became entrenched because they succeeded brilliantly at *something*. Any successor must walk the tightrope of preserving those successes, while avoiding the failures. Chapters 1–4 tell the light story, with an eye to the more general situation of having to revise any partially successful physical model.

Rather than organize the material by organism type or by length scale, I have tried to arrange the plot line in a way that builds up the framework needed to understand one important system, the vertebrate visual system (Chapters 9–11). Mathematical ideas such as complex numbers are developed as needed. Sometimes concepts are introduced before they can be fully explained, for example, fluorescence microscopy. In those cases, I have given enough detail at the first appearance to support the point being made, with a forward reference to the chapter that gives more details.

This book is independent of my earlier ones (Nelson, 2014; Nelson, 2015); that is, they are not prerequisites for reading this one. Nor is there much overlap between the coverage of these three books. For example, both prior books intentionally omitted almost any mention of quantum physics, which gets star billing here.

The basic quantum ideas are central to some important biological phenomena (phototransduction, photosynthesis), as well as a host of experimental methods (fluorescence imaging, including superresolution, two-photon, and FRET). Extending that point, other cutting-edge laboratory techniques also rest on physical principles, which are often poorly understood by their users. Understanding some physics not only can improve lab practice, but also prepares students to invent new techniques (or adapt old ones).

Ways to use this book

Undergraduate course: Parts I–II of this book could serve as the basis of a course on the science underpinning contemporary biological physics. Alternatively, that core has enough overlap with the traditional "Modern Physics" course that it could be used for a version of that course, for students with particular interest in life science. Or it can be used as a supplement in more specialized courses on physics, biophysics, nanoscience, or several kinds of engineering or applied math.

Depending on how much you time you need to devote to background, you may find that even Parts I–II cover more than one semester's worth of material. In that case, you may want to consider skipping, or treading lightly on, any of Chapters 3 or

---

[2] One *dis*advantage is that it's hard to find the levels of the hydrogen atom! However, Chapter 12 makes a start via a simpler problem.

6–8, none of which are essential for Chapters 9–11. Conversely, if vision is not your goal, you may instead wish to drop some or all of Chapters 9–11.

This book assumes that, in addition to first-year physics, the student has had some introduction to the basics of probability. If that's not true for your students, you can cover the Prologue material very carefully, perhaps assigning extra problems from another source (such as Nelson, 2015). Otherwise, you can skip the Prologue, reminding students that it's there to set notational conventions. In later chapters, "Background" sections summarize other foundations in the same terse style.

Most chapters end with "Track 2" sections. Some of these contain material appropriate for students with more advanced backgrounds in physical or life science. Others discuss topics that, although at the undergraduate level, will not be needed later in the book. They can be discussed à la carte, based on your and the students' interests. The main, "Track 1," sections do not rely on any of this material. Also, the *Instructor's Guide* contains many additional bibliographic references, some of which could be helpful for starting projects based on primary literature.

Graduate course: Although Track 1 is meant as an undergraduate course, it contains plenty of material not generally included in undergraduate physics curricula. Thus, it could easily form the basis of a graduate course, if you add all or part of Track 2, Part III, and/or some reading from your own specialty (or work cited in the *Instructor's Guide*). The chapters in Part III assume that the reader has some more advanced background than the main text; see the introductions to each chapter.

Numerical work

To do research, students need skills including graphical presentation of data and model results, numerical math, and handling of datasets.[3] But few people enjoy studying a computer math package (nor math itself) in an antiseptic, context-free way. That's what makes computers and math so boring to some people.[4] My students get motivated when they have a concrete problem, perhaps one involved in obtaining a classic result, driving them to build up the skills to solve it. Specifically, many students find biological problems to be a compelling starting point.

In my own course, many students arrive with no programming experience. Two separate *Student's Guides* give them some computer laboratory exercises and other suggestions for how to get started using MATLAB® or Python (Nelson & Dodson, 2015; Kinder & Nelson, 2015). Several other general-purpose programming environments would also work for the exercises, depending on your own preference, for example, *Mathematica*®, Octave, R, or Sage. Some of them are free and open source.

The *Instructor's Guide* gives solutions to the Problems and Your Turn questions in this book, including code. You can request it by following the instructions at `http://press.princeton.edu/titles/11051.html`.

Classroom demonstrations

One of the most powerful teaching techniques involves bringing a piece of apparatus into the class and showing the students something weird and *real*—not a simulation, nor a metaphor. The optical part of the course provides many opportunities for such

---

[3]The book's companion Web site features a collection of real experimental datasets to accompany the homework problems.

[4]It's also what makes them so exciting to other people!

experiences, partly justifying its prominent place in this book. The *Instructor's Guide* offers some suggestions.

Standard disclaimers

This is a textbook, not an encyclopedia. Many finer points have been intentionally suppressed in an attempt to give an approachable *first look* at this subject matter. Some appear in Track 2 or Part III; many more appear in the *Instructor's guide.*

No claim is made that anything in this book is original. No attempt at historical completeness is implied. The experiments described here were chosen simply because they illustrated points I needed to make. The citation of original works reflects books and articles I think would interest the reader, and/or sources where I learned things myself.

Other books

This book's goal is to help your students acquire some skills and frameworks, in the context of light, imaging, and vision. Two earlier books introduce different slices through biological physics: Nelson, 2014, discusses mechanics and fluid mechanics, entropy and entropic forces, bioelectricity and neural impulses, and mechanochemical energy transduction. Nelson, 2015, discusses probabilistic modeling, feedback control, and their union in synthetic biology.

Many other books instead offer a more complete coverage of the field of biophysics, and would make excellent complements to this one. A few recent examples include
General: Ahlborn, 2004; Bialek, 2012; Franklin et al., 2010; Nordlund, 2011.
Mathematical background: Bodine et al., 2014; Otto & Day, 2007; Shankar, 1995.
Cell and molecular biophysics: Boal, 2012; Milo & Phillips, 2016; Phillips et al., 2012.
Cell biology/biochemistry background: Alberts et al., 2014; Berg et al., 2015; Karp et al., 2016; Lodish et al., 2016; Steven et al., 2016.
Medicine/physiology: Amador Kane, 2009; Herman, 2016; Hobbie & Roth, 2015.
Biophysical chemistry: Atkins & de Paula, 2011; Dill & Bromberg, 2010.
Optics: Cox, 2012; James, 2014; Peatross & Ware, 2015; Pedrotti et al., 2007.
Computer methods: Hill, 2015; Kinder & Nelson, 2015. Computation: Landau et al., 2015; Newman, 2013. Other computer skills: Haddock & Dunn, 2011.
Visual neuroscience: Byrne et al., 2014; Cronin et al., 2014; Nicholls et al., 2012; Purves et al., 2012.
Many other books are cited at the ends of chapters.

Finally, you may want to consult `http://bionumbers.hms.harvard.edu/` for specific numerical values, so often needed when constructing physical models of living systems.

Last

Please remember that your work is important. Long ago, a stranger came to one of my classes. He only came twice. His name was Roger Dashen, and he discussed some material now echoed in Chapters 4 and 12 of this book. I scarcely saw him again after that. But in those three hours, he changed my life.

From Photon to Neuron

Prologue: Preliminaries

> The beauty of Nature lies in detail; the message, in
> generality. Optimal appreciation demands both.
> — *Stephen Jay Gould*

Our story begins in Chapter 1. This Prologue briefly reviews some ideas from probability, in part to establish notation. If these ideas look unfamiliar, you may also want to refer to one of the sources listed at the end of the chapter. Before doing that, however, just try to derive some of the listed results; many can be obtained in a few lines from the clues given here.

Later chapters will have similarly terse "Background" sections extending the ideas introduced here or reviewing other foundational material.

0.1 SIGNPOST: *UNCERTAINTY*

In everyday life, we make countless decisions about how to act in order to optimize something, for example, crossing a street quickly but safely. In science, we try to understand how things work, either for its own sake or in the service of some bigger goal, for example, finding a treatment for some illness. In each case, our efforts are hampered by *uncertainty:* Repeating the exact same action (or experiment), under what appear to be the exact same circumstances (or experimental protocol), will not necessarily yield the same results each time.

The uncertainty may be due to our limited knowledge of relevant facts that could in principle be better known:

- The general personality, and momentary mental state, of a car's driver will affect how much caution we should use when crossing a street when that car is present.
- A patient's personal history, family history, and genotype may condition her response to a particular treatment.

In other cases, however, uncertainty reflects some kind of intrinsic **randomness** (or "noise") in the world:

- Sudden gusts of wind and other weather phenomena are impossible to predict, due to the high complexity of the Earth system.
- Gene mutations from absorption of cosmic rays are unpredictable, too.

In science, we generally attempt to make a distinction between the system under study and the apparatus used to study it. Uncertainty can arise in either of these domains:

- When a cell divides, some regulatory molecules present in it will end up in one daughter and some in the other one, but the exact numbers will be unpredictable.

- Repeating the same measurement, say, of the period of a pendulum, will yield slightly different results on every occasion, due to imprecision in our measuring apparatus and procedure.

We'll now begin to make a framework to quantify uncertainty, and to see what conclusions can be drawn from random events.

0.2 DISCRETE PROBABILITY DISTRIBUTIONS

Section 0.1 emphasized that any physical system has at least some randomness, and (almost) any measurement apparatus injects more. Nevertheless, there are some rules at work in the Universe; to find them, we need to develop tools to describe and manage our uncertainty.

Even in uncertain situations, we are never *completely* unable to make predictions. For example, when crossing a street, stepping out one meter away from a fast-moving car is always a bad idea. Thus, to every proposition about a measurement that hasn't been made yet we ascribe a *degree of belief,* or **probability**, based on all the partial information we currently possess. We represent probabilities by numbers lying between 0 (proposition is certainly false) to 1 (proposition is certainly true).

0.2.1 A probability distribution summarizes our knowledge about an uncertain situation

To start thinking about how to quantify probability, consider a situation that, while rather artificial in everyday life, arises often in the lab:

- We imagine an experiment or measurement with a discrete list of possible outcomes, like counting the number of molecules of some type in a cell. If we know that there were M molecules of this type just prior to cell division, and none are created or destroyed during the division, then the number ℓ in one chosen daughter cell will be an integer between 0 and M inclusive.
- We suppose that the experiment can be performed many times ("repeated trials") and that every relevant aspect of the situation has been duplicated exactly on each trial. The situation just described is sometimes called a **replicable experiment**.
- We also suppose that we have no relevant prior information other than the actual values ℓ_i observed on many previous trials, that is, for $i = 1, \ldots, N_{\text{tot}}$.

In such a situation, it makes sense to count the number of times, N_ℓ, that each outcome was observed (also called the "frequency" of outcome ℓ), and to ascribe a degree of belief $\mathcal{P}(\ell)$ to each allowed value of ℓ by the formula

$$\mathcal{P}(\ell) = \lim_{N_{\text{tot}} \to \infty} N_\ell/N_{\text{tot}}. \qquad \begin{array}{l} \text{empirical probability for} \\ \text{a replicable experiment} \end{array} \qquad (0.1)$$

This formula may not be very practical (we can't make an infinite number of observations), but in principle it does define a function of ℓ, an example of a **discrete**

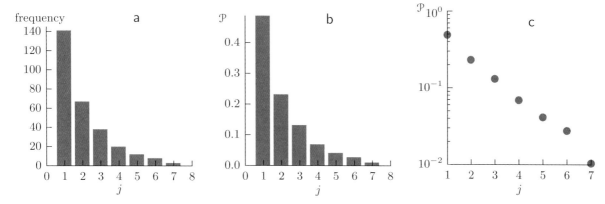

Figure 0.1: [Data summaries.] **Empirically estimating a discrete probability distribution.** (a) A fair coin was flipped 600 times, and the intervals j between successive "heads" were tabulated. This panel is a histogram, or bar chart of the frequencies of various outcomes. No values of j greater than 7 were observed in this particular experiment and there were a total of $N_{\text{tot}} = 289$ intervals. (b) Each frequency was divided by N_{tot} to estimate the probability distribution, using Equation 0.1, and presented as another bar graph. (c) The same estimated distribution in (b), this time drawn on semilogarithmic axes to bring out its roughly Exponential form. You'll explain this empirical observation in Problem 0.3.

probability distribution.[1] It's worth noting that

> *The values of any discrete probability distribution are dimensionless numbers lying between 0 and 1.*

We say that our measurements are "**draws** from the distribution $\mathcal{P}$." Figure 0.1 shows an example of estimating a distribution from a finite number of draws by making a histogram.

Because we assumed that every trial must lead to exactly one of the allowed outcomes, it follows that the sum of all the N_ℓ must equal N_{tot}, or[2]

$$\sum_\ell \mathcal{P}(\ell) = 1. \quad \text{normalization condition, discrete case} \qquad (0.2)$$

A closely related formula follows when we define the **sample space** as the list of all allowed outcomes, and **events** as subsets of the sample space.[3] In the context of games, for example, the sample space could be every distinct collection of five cards from a full 52-card deck. An example of an event in that sample space is the subset E_{fh} of five-card hands called "full house." Equivalently, we call E_{fh} the event "that a full house is drawn." A meaningful question could be to find the probability for this event under the assumption that five cards are picked from a well-shuffled, standard deck.

---

[1]Some authors call $\mathcal{P}$ a "probability mass function."

[2]When a variable appears under a summation symbol without any limits, we mean the sum over all values of that variable. Context will determine what are the relevant values, for example, all integers, all nonnegative integers, or whatever is appropriate.

[3]This standard terminology differs from many physics books, in which "event" is a synonym for "point in space–time," with no connection to probability.

If two events, E_1 and E_2, have no outcomes in common, we say they are "mutually exclusive"; then Equation 0.1 implies that

$$\mathcal{P}(E_1 \text{ or } E_2) = \mathcal{P}(E_1) + \mathcal{P}(E_2). \qquad \textbf{addition rule } \text{for mutually exclusive events} \qquad (0.3)$$

More generally,

$$\mathcal{P}(E_1 \text{ or } E_2) = \mathcal{P}(E_1) + \mathcal{P}(E_2) - \mathcal{P}(E_1 \text{ and } E_2). \qquad (0.4)$$

In both of these formulas, the word **or** denotes the union of sets, in this case, the event that an outcome lies in E_1 or E_2 (or both); the word **and** denotes the intersection. For mutually exclusive events, the intersection is empty and we recover Equation 0.3.

Because every outcome must lie either in E or outside it, we also have a "negation rule":

$$\mathcal{P}(\textbf{not-}E) = 1 - \mathcal{P}(E). \qquad (0.5)$$

0.2.2 Conditional probability quantifies the degree to which events are correlated

We may have knowledge about the probabilities of multiple events and their combinations. For example, E could represent the event that an individual has some illness and E' could be the event that a particular test for that illness comes out positive. If we then perform a measurement, determining that E' is true, then we may learn something about the probability of E. To express this intuition precisely, we introduce the **conditional probability**, defined as

$$\mathcal{P}(E \mid E') = \frac{\mathcal{P}(E \text{ and } E')}{\mathcal{P}(E')}. \qquad (0.6)$$

When speaking, the left-hand side of this formula is pronounced "the probability of E given E-prime." Rearranging the formula gives

$$\mathcal{P}(E \text{ and } E') = \mathcal{P}(E \mid E') \times \mathcal{P}(E'). \quad \text{general } \textbf{product rule} \qquad (0.7)$$

Sometimes establishing that E' is true tells us *nothing* relevant for predicting E, or in other words $\mathcal{P}(E \mid E') = \mathcal{P}(E)$. In that case, we say that the events are **statistically independent**, and the product rule simplifies to

$$\mathcal{P}(E \text{ and } E') = \mathcal{P}(E) \times \mathcal{P}(E'). \qquad \text{independent events} \qquad (0.8)$$

Two events that are not statistically independent are called **correlated**.

0.2.3 A random variable can be partially described by its expectation and variance

Often, the events that interest us involve the measured values of a numerical quantity, called a **random variable**. If that quantity is always an integer (such as the number of molecules of some type), then we get a discrete distribution: Let E_{ℓ_0} denote the event in which the measured quantity ℓ takes the particular value ℓ_0. Then we abbreviate $\mathcal{P}(\mathsf{E}_{\ell_0})$, writing it as $\mathcal{P}_\ell(\ell_0)$ or simply $\mathcal{P}(\ell_0)$.

Although the distribution $\mathcal{P}(\ell_0)$ is a function of ℓ_0, often we can summarize it adequately by stating just two numbers, the **expectation** of ℓ,

$$\langle \ell \rangle = \sum_{\ell_0} \ell_0 \mathcal{P}(\ell_0), \tag{0.9}$$

and its **variance**:

$$\operatorname{var} \ell = \left\langle \left(\ell - \langle \ell \rangle \right)^2 \right\rangle. \tag{0.10}$$

Note that, despite appearances, neither $\langle \ell \rangle$ nor $\operatorname{var} \ell$ is a function of the variable ℓ. The symbol ℓ appears in these expressions only to tell us which variable's expectation and variance is being discussed; the expressions themselves each represent a single number. Both quantities, however, do depend on the *distribution* $\mathcal{P}(\ell)$ that describes what we know about ℓ.

Any function of ℓ can be used to generate a new random variable. If f is such a function, we will use the same letter f to represent the corresponding random variable,[4] which is defined by making draws of ℓ and feeding them into f. Then we extend the preceding formulas to define

$$\langle f \rangle = \sum_{\ell_0} f(\ell_0) \mathcal{P}(\ell_0) \ \text{ and } \ \operatorname{var} f = \left\langle \left(f - \langle f \rangle \right)^2 \right\rangle. \tag{0.11}$$

Other books use the symbols and phrases $\mathbb{E}(f)$, μ_f, "expected value of f," and "expectation value of f" as synonyms for what we will call "the expectation of f" and write as $\langle f \rangle$. Equation 0.9 says that any of these notations refers to the mean (average) of an infinitely replicated set of measurements of a random variable. Also, the **standard deviation** of a distribution is defined as the square root of the variance,[5] and the **relative standard deviation** as the standard deviation divided by the expectation:[6]

$$\mathrm{RSD} = \frac{\sqrt{\operatorname{var} \ell}}{|\langle \ell \rangle|}. \tag{0.12}$$

The expectation is different from "the mean of a particular, finite set of measurements," which we will call the **sample mean** and denote by $\overline{f}$. The sample mean is itself a random variable: If we took another finite set of measurements, we'd get a different value. In contrast, the expectation is a property of the distribution of f itself.

The expectation has the key property that it is *linear:* That is, if f and g are any two random variables, and a and b any two constants,

$$\langle af + bg \rangle = a\langle f \rangle + b\langle g \rangle. \tag{0.13}$$

---

[4]We rely on context to show which meaning of f is meant. Mathematical books on probability use a more elaborate notation to avoid any possibility of confusion.

[5]Some authors use "root-mean-square deviation" (RMSD) as a synonym for the standard deviation estimated from a finite sample of data (the square root of the sample variance). The term RMSD can also apply to more complex objects, like random vectors.

[6]The term "coefficient of variation" is a synonym for RSD.

The variance is not so simple. For example, $\mathrm{var}(af) = a^2\,\mathrm{var}(f)$, and the variance of a sum isn't necessarily the sum of the individual variances (see the next section). However, the variance does have a useful alternative form:

$$\mathrm{var}\,f = \langle f^2 \rangle - \langle f \rangle^2. \tag{0.14}$$

0.2.4 Joint distributions

Sometimes we measure more than one quantity, giving rise to a **joint distribution**: Let E_{ℓ_0} be as before, and E'_{s_0} the event that a second quantity, observed on the same trial, has the value s_0. Then we will write $\mathcal{P}(\ell_0, s_0)$ as an abbreviation for the probability of (E_{ℓ_0} **and** E'_{s_0}).

We can partition the sample space (the possible outcomes) into a lot of nonoverlapping subsets (classes of outcomes) by considering the various compound events (E_{ℓ_0} **and** E'_{s_0}) as ℓ_0 and s_0 range over all their allowed values. That's useful if we wish to calculate the expectation of some function of ℓ_0 and s_0 because, rather than summing over every observed outcome, we may instead sum over the compound events. For example, we may want the expectation of the product ℓs:

$$\langle \ell s \rangle = \sum_{\substack{\text{classes of} \\ \text{outcomes}}} \mathcal{P}(\text{outcome})\ell_0 s_0 = \sum_{\ell_0, s_0} \mathcal{P}(\ell_0, s_0)\ell_0 s_0. \tag{0.15}$$

Suppose now that two random variables are statistically independent. Equation 0.8 then says that

$$\mathcal{P}(\ell, s) = \mathcal{P}_\ell(\ell)\mathcal{P}_\mathrm{s}(s). \qquad \text{for independent variables}$$

In this formula, $\mathcal{P}_\ell$ is the probability distribution for ℓ to have a particular value regardless of the value of s, and similarly for $\mathcal{P}_\mathrm{s}$. The expectation of the product, Equation 0.15, then becomes simpler:

$$\langle \ell s \rangle = \sum_{\ell_0, s_0} \mathcal{P}_\ell(\ell_0)\mathcal{P}_\mathrm{s}(s_0)\ell_0 s_0 = \sum_{\ell_0}\sum_{s_0} \mathcal{P}_\ell(\ell_0)\ell_0 \mathcal{P}_\mathrm{s}(s_0)s_0$$

$$= \left(\sum_{\ell_0} \mathcal{P}_\ell(\ell_0)\ell_0 \right)\left(\sum_{s_0} \mathcal{P}_\mathrm{s}(s_0)s_0 \right) = \langle \ell \rangle\langle s \rangle. \qquad \text{for independent variables}$$

$$\tag{0.16}$$

One immediate consequence of this result and Equation 0.14 is that

$$\mathrm{var}(\ell + s) = \big\langle (\ell + s)^2 \big\rangle - \big(\langle \ell + s \rangle \big)^2 = \langle \ell^2 \rangle + \langle s^2 \rangle - \big(\langle \ell \rangle^2 \big) - \big(\langle s \rangle^2 \big)$$

$$= \mathrm{var}(\ell) + \mathrm{var}(s). \qquad \text{for independent variables} \tag{0.17}$$

You can try the same thing with $\mathrm{var}(\ell - s)$, to find that

The variance of the sum or difference of two independent random variables equals the sum of their individual variances. $\tag{0.18}$

One measure of the degree to which two random variables are correlated is called the **covariance**, defined by the formula

$$\mathrm{cov}(\ell, s) = \big\langle\, \big(\ell - \langle \ell \rangle \big)\big(s - \langle s \rangle \big)\, \big\rangle = \langle \ell s \rangle - \langle \ell \rangle\langle s \rangle. \tag{0.19}$$

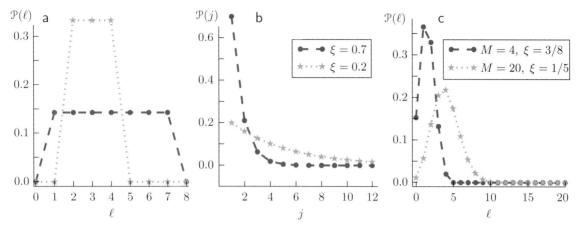

Figure 0.2: [Mathematical functions.] **Examples of discrete probability distributions.** These functions are defined only at integer values; the lines merely join the points. (a) Two examples of Uniform discrete probability distributions, on the ranges $2 \le \ell \le 4$ (*stars*) and $1 \le \ell \le 7$ (*circles*). (b) Two examples of Geometric distributions, with different values of ξ. In each case, the range of j is from 1 to infinity. (c) Two examples of Binomial distributions. The one shown with *stars* also closely approximates the Poisson distribution with $\mu = 4$, because $M = 20$ is much larger than 1. Unlike the Poisson distribution, however, it is exactly zero for $\ell > 20$.

Equation 0.16 implies that this quantity equals zero if ℓ and s are statistically independent. Another useful measure of correlation is called the **correlation coefficient**, a dimensionless quantity defined by

$$\mathrm{corr}(\ell, s) = \mathrm{cov}(\ell, s)/\sqrt{(\mathrm{var}\,\ell)(\mathrm{var}\,s)}. \tag{0.20}$$

Although two uncorrelated variables will have correlation coefficient equal to zero, the converse is not necessarily true: The correlation coefficient can detect a *linear* relation between two random variables, but other kinds of nonindependence are possible.

0.2.5 Some explicit discrete distributions

Given a replicable experiment, we can make a large number of measurements and try to estimate the probability distribution of outcomes by using Equation 0.1. But in many cases, we can instead reason from what we know of the system, and the mechanism by which it generates its outcomes, to predict the overall form of the distribution. That is, we construct a **probabilistic model** of the system; then, to the extent that we believe the model, it assigns probabilities to all the outcomes in the sample space.

In other words, a probabilistic model assigns an explicit mathematical function as $\mathcal{P}(\text{event})$. Remarkably, a small set of such functions suffices to cover a wide range of situations that arise in physical and life sciences. Each of these model distribution functions has one or more unknown parameters. But it is far easier to estimate a few parameters empirically from data,[7] or calculate them from a detailed description of the system, than to find the full distribution directly.

In this book, the names of the explicit model distributions are capitalized, for example, the Uniform distribution.

---

[7]Section 7.2 (page 248) will outline how to do this.

Uniform discrete distribution

The simplest model distribution is **Uniform**: It attributes equal probability to each of L possible outcomes (Figure 0.2a). That is, its distribution is a constant function over some range. The normalization condition implies that

$$\mathcal{P}_{\mathrm{unif}}(\ell_0) = 1/L.$$

If the list of allowed values consists of all the integers from $\ell_{\min}$ to $\ell_{\max}$ inclusive, then $L = \ell_{\max} - \ell_{\min} + 1$ and the expectation is just the average of the extreme values:

$$\langle \ell \rangle = (\ell_{\max} + \ell_{\min})/2. \qquad \text{Uniform distribution} \qquad (0.21)$$

Also, defining $\Delta = \ell_{\max} - \ell_{\min}$, we have

$$\mathrm{var}\,\ell = \frac{\Delta}{12}(\Delta + 2). \qquad \text{Uniform distribution} \qquad (0.22)$$

Bernoulli trial

The next simplest example is applicable to a system with only a single, two-valued outcome (like a coin flip), which we represent by a random variable s with two allowed values, 0 and 1. The **Bernoulli trial distribution** is then just

$$\mathcal{P}_{\mathrm{bern}}(1;\xi) = \xi; \quad \mathcal{P}_{\mathrm{bern}}(0;\xi) = 1 - \xi.$$

In the notation $\mathcal{P}_{\mathrm{bern}}(s;\xi)$, the symbol(s) to the left of the semicolon, in this case just s, is the random variable described by this distribution. The symbol(s) to the right of the semicolon, in this case just ξ, is a **parameter** whose value distinguishes different distributions in the family. For a Bernoulli trial, the parameter ξ is a constant with value between 0 and 1 describing whether the coin is fair (the case $\xi = 1/2$), and if not, how unfair it is. For a fair coin, the expectation $\langle s \rangle$ equals 1/2; more generally,

$$\langle s \rangle = \xi, \qquad \mathrm{var}\,s = \xi(1 - \xi). \qquad \text{Bernoulli trial} \qquad (0.23)$$

Geometric distribution

Another distribution arises when we consider sequences of many independent Bernoulli trials, each with the same probability ξ of "success." We look at any point in this sequence and count how many trials, j, are needed before we come to the first "success."[8] Thus, if the very next trial yields $s = 1$ we record $j = 1$; if instead the next trial is a "failure" but the one following is "success" we record $j = 2$; and so on. The discrete distribution of the random variable j is called the **Geometric distribution** (Figure 0.2b):

$$\mathcal{P}_{\mathrm{geom}}(j;\xi) = \xi(1 - \xi)^{j-1}, \quad \text{for } j = 1, 2, \dots. \qquad (0.24)$$

Note that, although j is an integer, there is no upper limit to its value. Nevertheless, the sum of $\mathcal{P}_{\mathrm{geom}}(j;\xi)$ over all (infinitely many) values of j equals 1 for any value of ξ.[9] Similarly, the infinite sums that give the expectation and variance also have finite values:

$$\langle j \rangle = 1/\xi, \qquad \mathrm{var}\,j = (1 - \xi)/\xi^2. \qquad \text{Geometric distribution} \qquad (0.25)$$

---

[8]Some authors instead describe this distribution with a variable that equals *zero* if success is immediate; it is related to our j as $j - 1$.
[9]You'll show this in Problem 0.4.

For our purposes, the main interest of the Geometric distribution will be that it has a useful continuous-time limit (Section 0.4.2 and Problem 0.6).

Binomial and Poisson distributions

Another common situation arises when we study batches of M independent Bernoulli trials, all with the same ξ, but we only care about the total number ℓ of "successes" in each batch.[10] Then ℓ is an integer between 0 and M, and it follows the **Binomial distribution** (Figure 0.2c):

$$\mathcal{P}_{\text{binom}}(\ell; \xi, M) = \frac{M!}{\ell!(M-\ell)!}\, \xi^{\ell}(1-\xi)^{M-\ell}. \tag{0.26}$$

Note that the factorial of zero is considered to be equal to 1. This time we find

$$\langle \ell \rangle = M\xi; \qquad \text{var}\,\ell = M\xi(1-\xi). \qquad \text{Binomial distribution} \tag{0.27}$$

An important limiting case of this distribution arises frequently. If M is much larger than 1, the Binomial distribution can be approximated by a simpler version, called the **Poisson distribution**:

$$\mathcal{P}_{\text{pois}}(\ell; \mu) = \frac{1}{\ell!}\, \mu^{\ell}\mathrm{e}^{-\mu}. \tag{0.28}$$

The two parameters M and ξ specifying the Binomial distribution (Equation 0.26) appear here in just a single combination, their product $\mu = M\xi$, which is held fixed as $M \to \infty$. In terms of that parameter, Equation 0.27 reduces in the limit to

$$\langle \ell \rangle = \mu; \qquad \text{var}\,\ell = \mu. \qquad \text{Poisson distribution} \tag{0.29}$$

Thus, every Poisson distribution has the property that its expectation equals its variance. The relative standard deviation is then[11]

$$\text{RSD} = \frac{1}{\sqrt{\mu}}. \qquad \text{Poisson distribution} \tag{0.30}$$

The Poisson distribution will arise many times, for example, when we study psychophysical experiments in Chapter 9. Also, we will later generalize the underlying Binomial distribution to cover the case of M flips of a k-sided "coin."[12]

Section 0.2 has introduced the most fundamental kind of probability distributions, those involving discrete alternatives. We also saw several families of idealized distributions that apply to many situations arising in physics and biology.

0.3 DIMENSIONAL ANALYSIS

Before we generalize the preceding ideas to continuous variables, take a look at Appendix B. That appendix summarizes some key ideas that we will need, both as tools for doing more accurate work and as a way to organize our thoughts about any new situation we may face, and even discover new physical laws.

---

[10]The cell-division example mentioned on page 2 is a situation of this type. Here "success on trial number i" is interpreted as "molecule number i ended up in the first daughter cell."
[11]RSD is defined in Equation 0.12.
[12]See Problem 11.2.

In this book, the names of units are set in a special typeface, to help you distinguish them from named quantities. Thus, km denotes "kilometers," whereas km could denote the product of a rate constant times a mass, and "km" could be an abbreviation for some ordinary word or special function. Units are usually indispensable for communicating quantitative ideas to someone else, but sometimes only relative numbers are needed, for example, when labeling a graph. Appendix B introduces the phrase "arbitrary units" (a.u.) for such situations.

Symbols like $\mathbb{T}$ denote more abstract **dimensions** (in this case, time); again see Appendix B.

Named quantities are generally single italicized letters. We can assign them arbitrarily, but we must use them consistently, so that others know what we mean. Appendix A collects definitions of many of the named quantities, and other symbols, used in this book, along with their dimensions.

0.4 CONTINUOUS PROBABILITY DISTRIBUTIONS

0.4.1 Probability density functions

Most experimentally measured quantities are not discrete. For example, distances, times, electric potentials, and indeed any measured quantities that carry dimensions are continuously distributed. To describe the distribution of a continuous random variable x, we first imagine dividing its range of allowed values into **bins** of width Δx. Then we define the event $\mathsf{E}_{x_0, \Delta x}$ by the proposition that a measurement of x yielded a value within a range Δx around a particular value x_0. The **probability density function** (or **PDF**) for x can then be thought of as a second limit[13]

$$\wp_{\mathrm{x}}(x_0) = \lim_{\Delta x \to 0} \frac{\mathcal{P}(\mathsf{E}_{x_0, \Delta x})}{\Delta x}. \tag{0.31}$$

We can omit the subscript x on the function $\wp$ if that does not lead to ambiguity. A key feature of this definition is that *a probability density function always has dimensions inverse to those of its variable.*

In practice, given a finite number of observations, we cannot actually send Δx to zero, because then almost all of the bins would contain *zero* observations. Nevertheless, if the number N_{tot} of observations is large enough, then we can get an estimate of the probability density function by combining Equation 0.31 with Equation 0.1 (Figure 0.3c):

> Given N_{tot} observations of a continuous random variable x, choose a set of bins that cover the range of x and are narrow, though wide enough that each contains many observations. Find the frequencies N_j for each bin centered on x_j. Then the estimated PDF at x_j is $\wp_{\mathrm{x,est}}(x_j) = N_j/(N_{\mathrm{tot}}\Delta x)$. $\tag{0.32}$

Because the probability of $\mathsf{E}_{x_0, \Delta x}$ is $\wp(x_0)\Delta x$, Equation 0.2 becomes[14]

$$\int \mathrm{d}x \, \wp(x) = 1. \qquad \text{normalization condition, continuous case} \tag{0.33}$$

---

[13]The first limit is the one in Equation 0.1, which defines the numerator of Equation 0.31. We will sometimes use the phrase "probability distribution" to mean either a discrete distribution or a probability density function, if the desired meaning is clear from context.

[14]Similarly to summations, here the integral sign with no limits means the definite integral over "all values" of x, that is, all values appropriate in the problem under consideration.

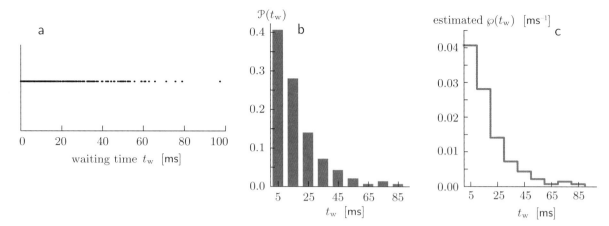

Figure 0.3: [Experimental data.] **Three ways to visualize an empirical probability density function.** (a) A graph showing the waiting times between 290 successive light detector blips as a "cloud" of 289 dots. The dot density is higher at the left of the diagram, implying a higher probability of shorter waiting times, but it is not easy in this representation to see anything quantitative. (b) The same data, presented as a scaled histogram. The range of the data has been subdivided into 10 "bins." The bar graph shows the probability distribution of the discrete random variable that indicates the bin containing each observed waiting time. Thus, taller bars correspond to greater density of dots in (a). (c) The same data, presented as an estimated PDF. Each value in (b) has been further scaled by the bin width, in this case 10 ms, so that the area under the curve equals 1 (see Equation 0.31). [Data courtesy John F Beausang, available in Dataset 1.]

We can also define an event E by a finite range of x values, for example, from x_0 to x_1. Then the probability of E is the corresponding area under the graph of the probability density function:

$$\mathcal{P}(\mathsf{E}) = \int_{\mathsf{E}} \mathrm{d}x\, \wp(x) = \int_{x_0}^{x_1} \mathrm{d}x\, \wp(x). \qquad (0.34)$$

In this expression, the dimensions of $\mathrm{d}x$ cancel those of $\wp$, giving a dimensionless answer for the probability of the event.

We also define joint PDFs by

$$\wp_{\mathrm{x,y}}(x_0, y_0) = \lim_{\Delta x, \Delta y \to 0} \frac{\mathcal{P}(\mathsf{E}_{x_0, \Delta x}\ \textbf{and}\ \mathsf{E}_{y_0, \Delta y})}{\Delta x\, \Delta y}. \qquad (0.35)$$

The function $\wp_{\mathrm{x,y}}$ carries dimensions inverse to those of xy, because of the factors in the denominator of Equation 0.35. Other formulas also have natural generalizations to continuous distributions. For example, the definition of conditional probability (Equation 0.6) becomes

$$\wp(x \mid y) = \wp(x, y)/\wp(y). \qquad (0.36)$$

Note that $\wp(x \mid y)$ has dimensions inverse to those of x; the dimensions of y cancel on the right-hand side of Equation 0.36. Equation 0.36 is the foundation from which we'll get the Bayes formula in Chapter 7.

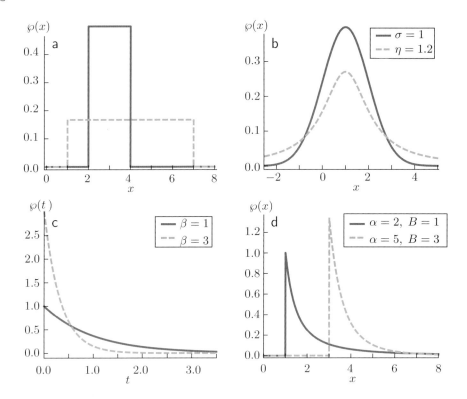

Figure 0.4: [Mathematical functions.] **Examples of probability density functions.** (a) Two examples of Uniform continuous probability distributions, on the ranges $2 \leq x \leq 4$ (*solid line*) and $1 \leq x \leq 7$ (*dashed line*). (b) Examples of Gaussian (*solid*) and Cauchy (*dashed*) distributions, both defined on infinite ranges. Although both of the distributions shown have the same full width at half maximum, the Cauchy distribution assigns greater probability to extreme events (those far from the central peak). (c) Examples of Exponential distributions, defined for all nonnegative t. The discrete distribution in Figure 0.2b has this limiting form when the bin sizes are sent to zero. (d) Examples of the power-law distributions defined by Equation 0.43. Note that, unlike a discrete distribution, a probability density function can take values that exceed 1 [panels (c,d)].

0.4.2 Some explicit continuous distributions

Section 0.2.5 mentioned the difficulty of accurately estimating a discrete probability distribution from data. The additional limit appearing in Equation 0.31 makes even greater demands on the available data in the continuous case, so in this case it is even more helpful to know, or at least have reason to believe, that an observed random variable belongs to a specific family of distributions.

Uniform

Any computer math package includes a "random" function that returns values between 0 and 1. This function's results are not truly random (the computer carries out some algorithm to generate them), nor are they truly continuous (the computer represents numbers with limited precision). Nevertheless, for most practical purposes, the results returned by this function resemble draws from a probability density function with the property that the probability for x to lie in any range of values equals the width of that range.

More generally, the **Uniform continuous distribution** on the range $x_{\min}$ to $x_{\max}$

is defined by the probability density function (Figure 0.4a)

$$\wp_{\text{unif}}(x; x_{\min}, x_{\max}) = \begin{cases} (x_{\max} - x_{\min})^{-1} & \text{if } x_{\min} \leq x \leq x_{\max} \\ 0 & \text{otherwise.} \end{cases} \tag{0.37}$$

Your Turn 0A

Work out the expectation and variance of this distribution. They depend on the two parameters $x_{\min}$ and $x_{\max}$.

Gaussian

Many measured quantities are well described by a family of continuous probability density functions called **Gaussian distributions**, which are characterized by two parameters (Figure 0.4b):

$$\wp_{\text{gauss}}(x; \mu_{\text{x}}, \sigma) = \frac{1}{\sigma\sqrt{2\pi}} e^{-(x-\mu_{\text{x}})^2/(2\sigma^2)}. \tag{0.38}$$

The variable x may have any dimensions; the parameters μ_{x} and σ have the same dimensions as x. The relations

$$\langle x \rangle = \mu, \qquad \text{var } x = \sigma^2 \qquad \text{Gaussian distribution} \tag{0.39}$$

are consistent with this dimensional rule. Some authors instead refer to Gaussian distributions as "normal."

Cauchy

The Gaussian PDF has a bump at its central value μ_{x}, and falls continuously away from that point with a width controlled by its second parameter σ. Other mathematical functions have similar properties, and, if an appropriate normalization can be found, any such function could in principle describe a random variable. One example that arises in practice is the **Cauchy distribution** (Figure 0.4b), defined by

$$\wp_{\text{cauchy}}(x; \mu_{\text{x}}, \eta) = \frac{1/(\pi\eta)}{1 + (x-\mu_{\text{x}})^2/\eta^2}. \tag{0.40}$$

The parameter η determines how broad the distribution is, analogously to the variance parameter σ for a Gaussian distribution. Although its graph is again a smooth bump function centered on μ_{x}, surprisingly the Cauchy distribution's variance is *infinite*. Even its central value μ_{x} cannot be estimated by taking the sample mean of any finite number of draws. You'll explore that statement in Problem 7.1, and show that nevertheless, we can reliably estimate the central value μ_{x} by a different method.[15]

Exponential

Some random systems generate **time series**: Each draw consists of a sequence of numbers. These numbers may be the measured values of a fluctuating quantity at

---

[15] $\boxed{T_2}$ Besides teaching us some lessons about the behavior of variance, the Cauchy distribution also describes the shapes of atomic or nuclear spectral lines, and appears in fluorescence resonance energy transfer (see Section 14.3.1 and Problems 1.5 and 2.4). It is sometimes called the "Breit-Wigner" or "Lorentzian" distribution.

a series of predetermined times, or they may be the random actual arrival times of discrete "blips," like the passages of cosmic rays through a detector. In the case of cosmic rays, the time intervals between successive blips follow an **Exponential distribution**, that is, they are distributed according to the one-parameter family of probability density functions (Figure 0.4c)

$$\wp_{\mathrm{exp}}(t_{\mathrm{w}}; \beta) = \beta \mathrm{e}^{-\beta t_{\mathrm{w}}}. \tag{0.41}$$

The random variable t_{w} is called the **waiting time**. The parameter β is called the **mean rate** of the distribution, because its reciprocal is the expectation of t_{w}:

$$\langle t_{\mathrm{w}} \rangle = \beta^{-1}; \qquad \mathrm{var}\, t_{\mathrm{w}} = \beta^{-2}. \qquad \text{Exponential distribution} \tag{0.42}$$

Note, however, that β is *not* the expectation of $1/t_{\mathrm{w}}$; see Problem 0.5.

The form of the Exponential distribution resembles a continuous-time version of the Geometric distribution:[16] Each involves raising some constant to a power that involves the waiting time. For our purposes, the importance of these distributions will be that waiting times between blips of a light detector are often Exponentially distributed (as suggested by Figure 0.3).

Power-law

The Cauchy distribution has the property that, at very large values of x, its form simplifies to $\wp_{\mathrm{cauchy}}(x) \approx \mathrm{const} \times x^{-2}$. This behavior falls off with x much more slowly than the Gaussian distribution, though still fast enough for the normalization integral to converge (Equation 0.33). It is emblematic of a broader class of **power-law distributions**, including, for example, (Figure 0.4d)

$$\wp(x; \alpha, B) = \begin{cases} Ax^{-\alpha} & \text{if } x > B \\ 0 & \text{otherwise.} \end{cases} \tag{0.43}$$

In this expression, the parameter α is a number greater than 1 and B is a number greater than 0. The prefactor A is determined by the requirement that $\wp$ be properly normalized; you can work it out for yourself, along with the expectation and variance of x (and for what values of α they are defined).[17]

We will see later that the "blinking" of a fluorescent molecule is sometimes described by a power-law distribution (Problem 7.6).

Section 0.4 has outlined some tools to describe continuous random variables, and several examples of idealized distributions that describe many situations in physics and biology.

0.5 MORE PROPERTIES OF, AND OPERATIONS ON, PROBABILITY DISTRIBUTIONS

This section mentions a few more properties of random variables, and operations that can be performed on their distributions, in either discrete or continuous situations.

---

[16]You'll show this in Problem 0.6. Equation 0.24 (page 8) introduced this distribution.
[17]You'll do this in Problem 0.2.

0.5.1 Transformation of a probability density function

Section 0.2.3 described how a function f can be applied to a random variable x to yield a new random variable $y = f(x)$. In general, the new y will have units that differ from those of x. For example, we may be interested in the logarithm of a quantity that we have measured. If x is a length, and D is a constant with units of length, then the function $f(x) = \ln(x/D)$, is dimensionless, and so is the corresponding new random variable y.

The probability density function describing y therefore in general needs a factor that converts the units of $\wp_x$ to units inverse to y. The appropriate relation is

$$\wp_y(y_0) = \frac{\wp_x(x_0)}{\left|\mathrm{d}y/\mathrm{d}x\big|_{x_0}\right|}. \qquad \text{transformation of a PDF} \qquad (0.44)$$

In this formula, x_0 is an x value that gets mapped to y_0 under the function f. If more than one value of x maps to the given y_0, then the right-hand side of the formula becomes a sum over all such points.

We'll see later that the transformation property is useful when we need to switch between describing a light spectrum in terms of wavelength and describing it in terms of frequency.[18] The factor in the denominator of Equation 0.44 arises from the fact that dividing the spectrum into bins of equal wavelength interval is not equivalent to dividing into bins of equal frequency interval.

0.5.2 The sample mean of many independent, identically distributed random variables has lower variance than any one of its constituents

Often we repeat a measurement several times, leading to a batch of independent random variables, all with the same probability distribution. We may feel intuitively that taking the average of M measurements will give us a quantity that, while still subject to random variation, is nevertheless a better guide to the true expectation than any one of its components—the fluctuations "average out." To justify this intuition, first consider the case of just two independent measurements, and hence two independent random variables x_1 and x_2, each with the same probability distribution. Using Equation 0.17 (page 6), we find

$$\mathrm{var}(x_1) + \mathrm{var}(x_2) = 2\,\mathrm{var}(x_1). \qquad (0.45)$$

We can apply the same reasoning to the sum of M measurements divided by M:

$$\mathrm{var}(\bar{x}) = \mathrm{var}\Big(\frac{1}{M}\big(x_1 + \cdots + x_M\big)\Big) = \Big(\frac{1}{M^2}\Big)\mathrm{var}\big(x_1 + \cdots + x_M\big) \qquad (0.46)$$

$$= \Big(\frac{1}{M^2}\Big)\big(M\big)\big(\mathrm{var}\,x_1\big) = \frac{1}{M}\,\mathrm{var}\,x_1. \qquad (0.47)$$

Thus, the sample mean of repeated measurements indeed has lower variance than any one of its constituents. However, the standard deviation of the sample mean decreases rather slowly, as $1/\sqrt{M}$, as the number of measurements is increased.

---

[18] You'll work this out in Problem 3.2.

Your Turn 0B

a. How does the *relative* standard deviation depend on M?

b. Test Equation 0.47, and your answer to (a), by using a computer's random number generator.

0.5.3 Count data are typically Poisson distributed

Frequently we encounter discrete objects or events each characterized by a continuous random quantity. Equation 0.32 outlined a procedure to estimate a continuous probability density function of that quantity, but in practice we never have an infinite number of measurements. Thus, the count N_j within each bin is a random variable. If the bins are chosen finely enough so that every one of them contains only a small fraction of all the measurements, then we can consider the population of a particular bin, number j, as a Bernoulli trial: Each measurement has a small, fixed probability to land in bin j. The total population N_j is then the sum of many such trials, and so will be Poisson distributed, with mean equal to $N_{\text{tot}}\mathcal{P}(j)$. Equation 0.30 then tells us how accurate our estimate of the PDF based on this sample will be, for bin number j and the bin spacing we chose.

0.5.4 The difference of two noisy quantities can have greater relative standard deviation than either by itself

Suppose that you wanted to measure the width of the dot at the end of this sentence. You can imagine using a ruler to measure the distance d_{L} from the edge of the page to the left end of the dot, and the distance d_{R} from the same edge to the right end of the dot, and subtracting. But that approach would give a very inaccurate result.

The problem is that the independently measured values of d_{L} and d_{R} each have some randomness, and you measured them in two separate steps. If each has variance σ^2, then their difference has variance $2\sigma^2$ by Equation 0.17 (page 6). The relative standard deviation is then $\sqrt{2}\sigma/|\langle d_{\text{L}} - d_{\text{R}}\rangle|$, which can be large because the denominator can be comparable in magnitude to the numerator—even if $\sigma/\langle d_{\text{L}}\rangle$ and $\sigma/\langle d_{\text{R}}\rangle$ are both small quantities.

0.5.5 The convolution of two distributions describes the sum of their random variables

Section 0.2.4 discussed a random variable defined as the sum of two other, independent, variables. In that section, we were only interested in the case where the two constituents had the same distribution, and we only wanted the variance of the sum. Often, we face a more detailed question, which is to find the *full* distribution of the sum of two independent but *non*-identical variables.

Suppose that two independent random variables ℓ and j are both integer valued. In order for their sum to have a particular value n, we must have that $j = n - \ell$. Adding up the probabilities for each such situation gives the probability distribution of n as the **convolution**, defined by

$$(\mathcal{P}_\ell \star \mathcal{P}_{\text{j}})(n) = \sum_\ell \mathcal{P}_\ell(\ell)\mathcal{P}_{\text{j}}(n - \ell). \tag{0.48}$$

In this formula, the sum extends over all allowed values of ℓ for which $n - \ell$ is also an allowed value of j. A similar formula applies for continuous variables, with the sum replaced by an integral.

Convolution will arise later in several contexts, for example, the blurring of images by diffraction, x-ray crystallography, and image processing in the retina.[19]

Convolution property of Poisson distributions

Poisson random variables have the useful property that the sum of two such independent variables is again Poisson distributed. Because any Poisson distribution is completely characterized by its expectation, we can be more precise: Equation 0.13 tells us that

> *The sum of independent Poisson random variables with expectations μ_1 and μ_2 is itself a Poisson random variable with expectation $\mu_1 + \mu_2$.* (0.49)

Section 0.5 has rounded out our survey of probability with some observations that will be needed in later chapters.

0.6 THERMAL RANDOMNESS

One way in which randomness enters biology and physics is through the complex molecular motion we call "heat." Any situation with a subsystem of interest containing more than a few particles, in contact with the macroscopic world, is likely to have motions so complex that for practical purposes measurements made on it have no discernible, relevant structure. In such situations, it is often a good approximation to suppose that the kinetic energy of every independently moving molecule has expectation proportional to the temperature of the surrounding world. Specifically, that energy equals $(3/2)k_\mathrm{B}T$ per molecule, where T is absolute temperature (degrees above absolute zero) and k_B is the **Boltzmann constant**.[20] At room temperature, the energy scale $k_\mathrm{B}T_\mathrm{r}$ can be expressed as about 4.1 pN nm. Molecules also have thermal energy associated to their random *internal* flexing motions, although these can be more complicated.

$\boxed{T_2}$ *In quantum physics, the Boltzmann constant plays a similar role, although the influence of temperature cannot be quite so simply related to mean kinetic energy; see Section 1.3.3′c (page 52).*

THE BIG PICTURE

This Prologue has introduced some concepts involving the quantification of *uncertainty*, which will arise often later on. Most immediately, the next chapter will upgrade the Poisson distribution (Section 0.2.5) to apply to a wider context, that of random processes. The resulting "Poisson process" will turn out to be a good description of many forms of light.

---

[19]See Section 6.8.1, Section 8.3.3′, and Section 11.4.1′b.

[20]Chemists often find it convenient to use a related quantity, the **gas constant** $R = N_\mathrm{mole}k_\mathrm{B}$, where N_mole is Avogadro's number.

KEY FORMULAS

In later chapters, a section like this one will contain key formulas obtained in the main text. You can use those sections to review where each formula came from, what sort of situations it can help illuminate, and what each symbol represents. You'll understand the formulas better if you also check how the units work out in each one.

This entire Prologue has been largely a collection of formulas, however, so there's little point in repeating them here. Instead, here are some miscellaneous math results from your earlier classes, which will be needed later. More mathematical background, involving complex numbers, appears in Appendix D.

- *Series expansions:* Make sure you recall these series and how they follow from Taylor's theorem. Some are valid only when x is "small" in some sense.

$$\exp(x) = 1 + x + \cdots + \frac{1}{n!}x^n + \cdots$$

$$\cos(x) = 1 - \frac{1}{2!}x^2 + \cdots$$

$$\sin(x) = x - \frac{1}{3!}x^3 + \cdots$$

$$\tan(x) = x + \frac{1}{3}x^3 + \cdots$$

$$1/(1-x) = 1 + x + \cdots + x^n + \cdots \quad \text{(the geometric series formula)}$$

$$\sqrt{1+x} = 1 + \frac{1}{2}x - \frac{1}{8}x^2 + \cdots$$

- In addition, we'll need these formulas later:
The Gaussian integral: $\int_{-\infty}^{\infty} \mathrm{d}x \, \exp(-x^2) = \sqrt{\pi}$.
The compound interest formula:[21]

$$\lim_{M \to \infty} \left(1 + \frac{x}{M}\right)^M = \exp(x). \tag{0.50}$$

The law of cosines: Suppose that the edges of a triangle have lengths a, b, and c, and let γ be the angle opposite edge c. Then $c^2 = a^2 + b^2 - 2ab\cos\gamma$.

FURTHER READING

There is an infinite supply of excellent references in each of the categories below. The books listed are close to the viewpoint and applications used in the present book.

Semipopular:
Silver, 2012.

Intermediate:
All of the results in this Prologue are discussed in Nelson, 2015. Many can also be found in Blitzstein & Hwang, 2015; Dill & Bromberg, 2010; Otto & Day, 2007.

Technical:
Linden et al., 2014; Jaynes & Bretthorst, 2003.

---

[21]The left side of this formula is the factor multiplying an initial balance on a savings account after one year, if interest is compounded M times a year at an annual interest rate x.

PROBLEMS

0.1 *Burn it*

The Sun is proverbially bright. Each second it emits about 3.8×10^{26} J of energy, generated by nuclear fusion, another proverbially energetic process.

You're bright too, but in a different way. Your basal metabolic rate is roughly 1500 kilocalories per day, which you obtain from food and convert to heat, along with mechanical work, biosynthesis, and other useful things.

It would be unfair to compare you directly to the Sun, so compare your energy conversion rate *per body mass* to the Sun's. Use the ideas of Appendix B to make sure you do the conversions correctly.

0.2 *Probability foundations*

Some of the formulas quoted in this chapter are definitions. Many others, however, are results that you can prove, starting from those definitions. Prove these results analytically:

a. Equations 0.3, 0.4, and 0.5
b. Show that the conditional probability defined by Equation 0.6 is always properly normalized. For example, if ℓ and s are two random variables, show that $\sum_{\ell_0} \mathcal{P}(\ell_0 \mid s_0) = 1$, regardless of the value selected for s_0.
c. Equation 0.13
d. Equation 0.14
e. Equations 0.21 and 0.22
f. Equation 0.23
g. Equation 0.25
h. Equation 0.27 [*Hint:* The hard way is to start with Equation 0.26. The easy way is to start with Equations 0.23 and 0.18.]
i. Show that the Poisson distributions, Equation 0.28, are all properly normalized (for every value of μ). [*Hint:* An identity on page 18 is useful.]
j. Equation 0.29 [*Hint:* The hard way is to start with Equation 0.28. The easy way is to start with Equation 0.27 and take a limit of large M, holding μ fixed.]
k. Find the expectation and variance of the continuous Uniform distribution (Equation 0.37).
l. Confirm that the Cauchy distributions, Equation 0.40, are correctly normalized.
m. Equation 0.42
n. Confirm that the power-law distributions, Equation 0.43, can all be correctly normalized, and find the normalization factor for each. Then work out their expectations and variances.

0.3 *Lunch time*

Imagine a frog that periodically strikes at flies. Each attempt to catch a fly is an independent Bernoulli trial with some probability ξ reflecting the frog's skill (and the flies' skill at evasion). Let's make the simplifying assumption that ξ does not change throughout the experiment. We make a long list of the outcomes (success/failure) of repeated strikes.

For each successful strike, define j as the number of strikes before the *next* successful one. Thus if the very next strike is successful, we record $j = 1$, and so on. Find the probability distribution of the random variable j, and comment in the light of Figure 0.1 (page 3).

0.4 *Properties of* $\mathcal{P}_{\text{geom}}$

a. Confirm that Equation 0.24 (page 8) defines a properly normalized probability distribution, for any fixed value of the parameter ξ. [*Hint:* There is a useful formula on page 18; let $x = 1 - \xi$.]

b. Take the Taylor series expansion that you used in (a), multiply both sides by $(1 - x^K)$, and simplify the result. Use your answer to find the probability that, in the Geometric distribution, the first "success" occurs at *or before* the Kth attempt.

0.5 *Units are not the whole story*

The main text stated that in an Exponential distribution, β is not the expectation of $1/t_{\text{w}}$, even though both quantities have dimensions $\mathbb{T}^{-1}$. Explain.

0.6 *From Geometric to Exponential*

Section 0.4.2 introduced the Exponential family of probability density functions, and mentioned that they are related to the Geometric distribution. Consider a string of repeated independent Bernoulli trials, each with probability $\xi = \beta \Delta t$ of success. Here β is a constant with dimension $\mathbb{T}^{-1}$, and Δt is a small time interval. Suppose that the trials repeat every time interval Δt. If the first success is at trial $j = 1$, then we didn't have to wait at all for another success; if it is at $j = 2$ then the waiting time was Δt, and so on. Thus, the waiting time for success is $t_{\text{w}} = \Delta t(j - 1)$. The probability for the waiting time to land in a particular bin is then $(\beta \Delta t)\left(1 - (\beta \Delta t)\right)^{t_{\text{w}}/\Delta t}$.

Now consider a limit in which $\Delta t \to 0$ holding fixed β and t_{w}. The probability to be in any particular bin tends to zero, but in the limit t_{w} is becoming a continuous variable, so we expect that we should describe its distribution by a probability density function. Do that, and compare the result to Equation 1.10.

0.7 *Mechanical sensitivity*

One way to quantify sensitivity to touch is by finding the smallest discernible mechanical disturbance on a sensitive region of skin, for example, the back of one's hand. For many people, a single grain of salt dropped from a height of $10\,\text{cm}$ is close to this edge of conscious awareness. Estimate the kinetic energy of that grain by modeling it as a cube of volume $(0.2\,\text{mm})^3$ and mass density about $10^3\,\text{kg}\,\text{m}^{-3}$, released from rest. [*Hint:* That energy equals the change in gravitational potential energy, which you can reconstruct by using dimensional analysis to find a quantity with the appropriate units. The acceleration due to gravity is $g \approx 10\,\text{m}\,\text{s}^{-2}$.]

0.8 *Convolution property of Gaussian distributions*

There is a result analogous to Equation 0.49 (page 17) in the continuous domain:

> *The sum of independent Gaussian variables, with expectations μ_1 and μ_2 and variances $\sigma_1{}^2$ and $\sigma_2{}^2$, is itself Gaussian-distributed.*   (0.51)

a. What are the expectation and variance of the convolution? (You need not assume that Equation 0.51 is true.)

b. $\boxed{T_2}$ Now prove Equation 0.51, starting from the continuous version of Equation 0.48 (page 16), for the special case where the two original distributions are identical and centered on $\mu = 0$. [*Hint:* You'll need the Gaussian integral, page 18.]

c. $\boxed{T_2}$ And do the general case.

PART I

Doorways of Light

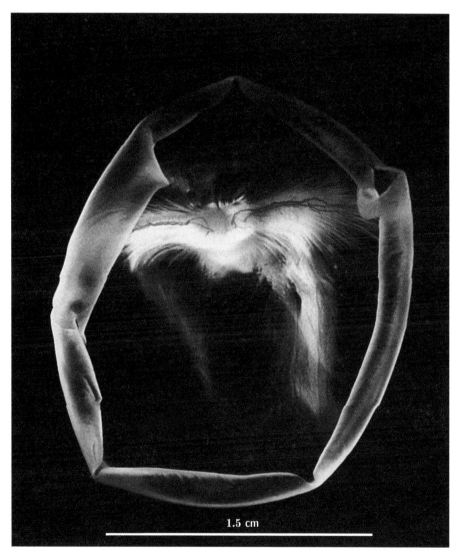

1.5 cm

[Photograph.] **Isolated rabbit retina**, a sheet of neural tissue about $100\,\mu$m thick. Photoreceptor cells in the retina transduce light into a neural signal, which is relayed by successive layers of neurons; ultimately the signal excites ganglion cells. Mostly what is visible in this picture are the output fibers of the ganglion cells (*white filaments*) that join into a bundle called the optic nerve. (Also some of the larger blood vessels can be seen.) [Courtesy Richard H Masland; see also Masland, 1986.]

CHAPTER 1

What Is Light?

> The fact that [Einstein] may occasionally have overshot the mark in his speculations, for example, in his hypothesis concerning light quanta, should not be counted against him too strongly. For without ever taking a risk, it is impossible to introduce real innovation. . . .
> — *Max Planck and others, in their nomination of Einstein for membership in the Prussian Academy of Sciences, 1913*

1.1 SIGNPOST: *PHOTONS*

Living organisms are physical systems that can obtain information about their environment and act on it. Certainly there are other hallmarks of Life: Organisms also obtain energy from external sources and transduce it; that is, they convert it into useful motion, use it to construct their own bodies, and even use it to perform the computations needed to integrate and act on the information they receive.

The nature of light is a thread that runs through the preceding ideas:

- Good vision can give one organism a decisive fitness advantage over another, so many animals have evolved remarkably sophisticated visual systems, and dedicate significant amounts of their energy budget to them.
- Solar energy—harnessed by photosynthesis—is the primary driver of nearly all life on Earth.

Scientists are a particular class of living organisms, also involved in obtaining certain kinds of information. In that vein, we may add to our list that

- Light-based imaging has enabled much of the progress in our understanding of Life over the past three and a half centuries, including new advances right up to the present.

Light also offers us a case study of how scientists were forced to abandon a cherished physical model that was long regarded as the paragon of a successful scientific theory. This chapter begins by reviewing some of the phenomena that led to this model (the "classical wave theory" of light). Later sections will describe *other* phenomena, however, that forced physicists to *reject* this model in the early 20th century, and replace it by something quite different.

This chapter and the next ones will argue that the newer model—"quantum physics"—is needed to understand many phenomena that are directly relevant to biology, and to biophysical instruments. Chapter 4 will then return to the task of reconciling quantum physics, and its picture of light as a stream of *photons*, with the phenomena that appeared to demand a wave interpretation.

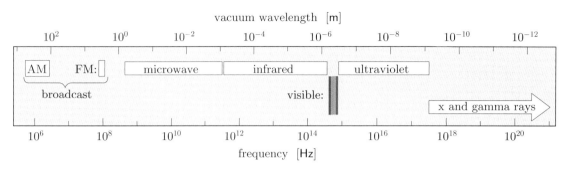

Figure 1.1: [Diagram.] **Part of the light spectrum,** showing different names traditionally given to various ranges. Because there is no fundamental difference between any of these, this book will refer to them all generically as "light"; the colored band will be called "visible light." Frequencies and their associated vacuum wavelengths are shown on log scales.

Along the way, we will see that light phenomena have an irreducibly random character, both in space and in time. Concepts developed in the Prologue will therefore be needed to describe light.

The Focus Question is

Biological question: Why do you need natural sunlight, not indoor artificial lighting, to generate vitamin D?

Physical idea: Sunlight contains an invisible component required for vitamin D synthesis in the skin.

1.2 LIGHT BEFORE 1905

Light can mediate interactions between two material objects. Among other effects, it can transfer energy from one object to another. Originally, "light" referred only to radiation that could stimulate our eyes, but eventually many other kinds were found—an entire **light spectrum.** That is, light from any source can be separated by physical means (for example, by sending it through a prism) into a continuum of different types (see Figure 1.1):

- Ordinary visible light occupies a narrow band in this spectrum, with red and orange at one end, green in the middle, and blue and violet at the other extreme.
- Adjacent to the visible on one side is the **infrared** (IR) region, and beyond that lie the microwave and radio regions.
- On the other side of the visible lies the **ultraviolet** (UV) region. Beyond it lie x rays and then gamma rays.

Within any of these ranges, light can be more finely subdivided. For example, within the visible range the various different types of light appear to our eyes to have different *colors.*[1] More precisely, we can construct a filter that transmits any desired part of the spectrum, while blocking other parts. If the transmitted part consists of one very narrow spectral band, we call that light **monochromatic.**

---

[1] There exist colors, such as magenta, that do not appear anywhere in the spectrum. Chapter 3 will discuss perceived color systematically; the present chapter considers only the physical characterization of light.

1.2.1 Basic light phenomena

Today, every part of the light spectrum is viewed as a different aspect of a single thing, because all share a suite of phenomena. We will refer to any of them as "light."

Light can connect two objects separated by vacuum, for example, the space between Earth and the Sun. Most kinds of light also pass readily through media of very low density, such as air. Some kinds even pass through certain dense media: Visible light can pass through water or glass; x rays pass through the soft tissue in our bodies.

Light *moves:* That is, it takes time to get from one place to another. Unlike everyday objects, however, light in vacuum always travels at a single speed, regardless of its type or how the emitting object was moving. Because it is universal, we call this speed a constant of Nature. It's denoted by the letter c; its value is $c \approx 3.0 \times 10^8$ m/s.

Light generally travels in straight-line paths.[2] Thus, a point source of light (or a large source that is very distant from us, like the Sun) can cast sharply defined shadows. When light encounters an opaque object, it can be partly absorbed; for example, it may warm the object. But some or all of the light can instead reflect (bounce) off the obstruction. If the object is smooth with a well-defined shape, the light reflects in a predictable way.

These phenomena inspired some early scientists with a metaphor from everyday life: Light appeared to be a stream of tiny material particles, like grains of sand but much smaller. They emerged from a luminous body, flew unimpeded through empty space (or partially impeded by a medium), bounced off certain kinds of obstructions, and ultimately delivered their energy by being absorbed. For example, those entering our eyes would give rise to the sensation of light, and so on.

1.2.2 Light displays wavelike behavior in many situations

However, light also exhibits some phenomena that are hard to reconcile with the material-particle metaphor, for example, **diffractive** effects, in which light deviates from straight-line motion. You can see such effects yourself, following an observation made by F. Hopkinson in 1785:[3] Hold a piece of finely woven fabric close to your eye, and look through it at a distant point source of light, perhaps a street light at night. In addition to the main image, you'll see other images, a "transmission grating" effect that we will study in later chapters. Hopkinson mentioned his observations to his friend D. Rittenhouse. Although it was too early to be able to explain the effect, Rittenhouse did document it in quantitative detail, constructing finely spaced, parallel arrays of human hairs in place of fabric and measuring the angular displacement of the images. These results are difficult or impossible to explain from a physical model of light as a stream of tiny material particles.

Shortly after Rittenhouse's experiments, Thomas Young championed a metaphor for light as a *wave* phenomenon, in part to explain diffractive effects. Although this **classical wave theory** of light was eventually found to be incomplete, it does explain phenomena that the particle model cannot. Later, when we propose a third theory of light (quantum physics), we'll need to ensure that it, too, can explain diffractive effects.

---

[2]Abu Ali al-Hasan ibn al-Hasan ibn al-Haitham made this observation, and reported many other experiments with lenses, mirrors, refraction, and reflection, in his *Kitab al-Manazir,* written from 1011 to 1021CE.

[3]Hopkinson was an American musician and public official; he signed the US Declaration of Independence and helped to design the first US flag in 1777.

Any wave—for example, the sound from a flute, or ripples on the surface of water—has properties called **frequency** and **amplitude**. Because the energy carried by a wave depends on its amplitude, Young also interpreted the intensity of a light source in terms of the amplitude of a corresponding wave. As for the frequency, Young proposed that it was related to where the light fell on the spectrum. Instead of frequency ν, one can equivalently characterize a wave by specifying the quantity $T = 1/\nu$ (called "the period"), or by stating the distance λ that the wave travels in vacuum during one period:

$$\lambda = c \times (1/\nu). \quad \textbf{vacuum wavelength} \text{ corresponding to frequency } \nu \qquad (1.1)$$

Fig. 4.2c (page 147)

Young then explained light diffraction by analogy to corresponding phenomena with sound or water waves: Similarly to light, each segment of a water wavefront moves on a straight-line path in open space, but will spread out after passing the edge of an obstruction. The interplay between λ and the physical dimensions of a grating then accounted for the patterns of light and darkness seen in diffraction patterns.[4] This connection also made sense of mysterious colors that appeared in Rittenhouse's diffraction experiments: A grating bends each part of the spectrum of visible light by a different angle, and so can separate it. Young also noted that water and sound waves bounce when they encounter a straight wall, with the same geometry as that followed by light hitting a mirror.

Another kind of deviation from straight-line motion occurs when light passes from air into water. C. Huygens had already noted that such **refraction** phenomena also follow naturally from the wave theory, if we extend it by adding the assumption that light waves move more slowly in water than they do in air. When it later became possible to measure the speed of light directly, this assumption was confirmed.

For reasons like these, most scientists eventually adopted the classical wave model of light. Its triumph seemed complete when J. C. Maxwell found equations of motion for electricity and magnetism, then showed that those equations have wavelike solutions that give quantitative explanations of diffraction, refraction, and many other light phenomena. The equations also showed how the waves could be created when electric charges are set in motion, and conversely, that they could *cause* initially motionless charges to shake; these two processes correspond to the generation and reception of light. Shortly later, H. Hertz and others performed experiments confirming that oscillating electric currents do indeed give rise to traveling waves with all the characteristics of light (including propagation through vacuum at speed c, diffraction, reflection, and refraction). Around the same time, W. Röntgen also discovered x rays, which also proved to be lightlike, though with shorter wavelengths than visible light. An enormous range of phenomena thus seemed to fit comfortably into a single framework.

Section 1.2 has surveyed a physical model of light that appeared to be satisfactory up until the early 20th century. This model not only explained many phenomena; it also had an appealingly intuitive foundation.

1.3 LIGHT IS LUMPY

The classical wave model supplanted the older model of light as a stream of material particles, in part because the material-particle model gave no clear explanation of

---

[4]Chapter 4 will discuss diffraction effects, for example Figure 4.2, in detail.

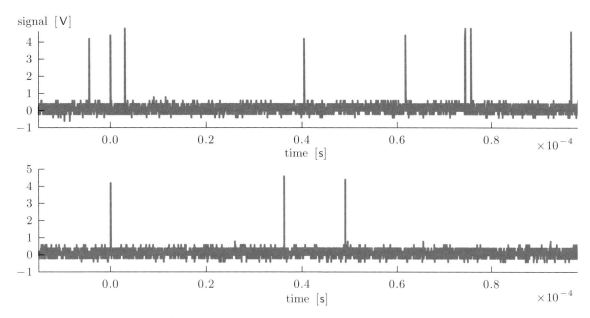

Figure 1.2: [Experimental data.] **Signals from a sensitive detector exposed to light.** *Top:* Time series of the output from the detector exposed to dim illumination. *Bottom:* Even dimmer illumination. In each case, the signal consists of similar blips superimposed on a background of instrumental noise. The difference between dim and dimmer light lies in the mean rate of the blips' appearance, not in the strength of individual blips. [Data courtesy John F Beausang, available in Dataset 2.]

diffraction phenomena. But today we can observe still other phenomena, for which the *wave* model gives no explanation. Thus, *neither* of these models is fully satisfactory.

1.3.1 The discrete aspect of light is most apparent at extremely low intensity

Many living organisms have evolved means to detect very dim light, for example starlight, or the greatly diminished sunlight available deep underwater. Light shows some surprising behavior in the regime of faint illumination.

Before we tackle the rather involved visual apparatus of living organisms, let's begin with a less sophisticated device: a laboratory light sensor. When exposed to steady illumination, it creates a measurable signal. Reducing the illumination reduces this signal, but it does so in a surprising way. We might expect to find that, as we gradually dim the light, the signal would gradually decrease in strength, eventually disappearing into background electrical noise from the equipment. This is what happens when a friend speaks too softly over a noisy telephone connection. But it's not what is observed with light.

A sensitive light detector in total darkness does show some noise in its output signal. But the effect of faint illumination is to superimpose distinct "blips" on the noise background (Figure 1.2). Changing the illumination level does not change the strength of each blip; what changes is the mean *rate* at which blips arrive.

The discrete, or "lumpy," character of the signals in Figure 1.2 does not automatically imply anything about light itself. Following the classical wave model, we might imagine that light really is a continuous stream of energy (symbolized by water in Figure 1.3), but that some mechanism in the detector triggers a blip only after enough of this energy has been captured. One difficulty with this metaphor is that it predicts

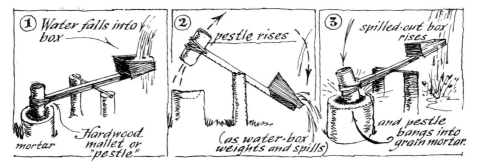

Figure 1.3: [Metaphor.] **Discrete response to continuous flow.** A "plumping mill," used in the early nineteenth century to convert continuous water flow to periodic bursts of mechanical work. The text discusses why this is *not* an adequate metaphor to describe the detection of light. [Art by Eric Sloane.]

that steady illumination would lead to blips at uniformly spaced times. Experiments show, however, that the blips instead form a *random process,* called **shot noise**.[5]

You can appreciate the difference between uniformly spaced blips and shot noise by listening to the audio clips in Media 1.[6] One of these contains clicks that are uniformly spaced in time. The other is an audio version of an experimental dataset (partially shown in Figure 1.2); it contains clicks at the arrival times of each of the blips. The two audio clips sound quite different, despite the fact that they have the same mean click rate.

We might at first suspect that the discrete, random character of the blips is a pathology of one particular light source or detector technology. But on the contrary, a variety of light sources (including the Sun, a candle, a hot filament, or a laser) all give random blips when reduced to extremely low intensity, regardless of how carefully we try to make them steady. Also, several different detector technologies (including an avalanche photodiode, a photomultiplier tube, or high-sensitivity photographic film) all give similar results. Nor is this randomness a byproduct of thermal motion:[7] Cooling the detector, even to nearly absolute zero temperature, doesn't affect it. Instead,

> *The discrete, random character of signals from a sensitive light detector is a property of* **light itself**. (1.2)

Experiments like the one just discussed also reveal another important property of light. Suppose that we construct a grid of millions of identical detectors, like the camera in a portable electronic device, then illuminate this array uniformly. The classical wave model would lead us to imagine a succession of "wavefronts" spreading from the source, arriving at a distant array of detectors, and inducing measurable signals in each of them, in unison. An analogy could be a spreading wave on water, impinging on an array of floating corks. The corks respond by all bobbing in synchrony, with similar amplitudes.

In contrast to that expectation, experiments instead show that each detector creates blips independently of the others, and that no two ever respond at exactly the same time. This behavior implies that light consists of a random stream of something that creates *highly localized* effects. In the wave metaphor with floating corks, the

---

[5]Section 1.4 will examine random processes in detail.
[6]See page xx for how to obtain the Media.
[7]Section 0.6 (page 17) introduced thermal motion. Cooling a detector certainly does reduce random signals *not* caused by light (the detector's "dark current").

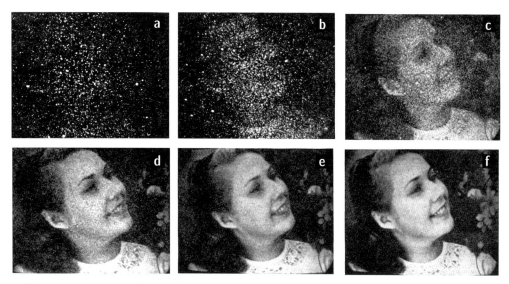

Figure 1.4: [Experimental data.] **Lumpy character of a visual scene.** Images reconstructed by scanning a visual scene with a photomultiplier tube, and recording the number of blips received in each pixel over a fixed time interval. Successive panels show the results at progressively longer exposures, with (f) collecting about 9000 times more photons overall than (a). [From Rose, 1953.]

observed behavior would correspond to all the corks remaining motionless, except for one, chosen at random, that somehow gets flung out of the water, at the same time stilling the wave everywhere!

Similar behavior is also observed with more complex visual scenes. Such a scene is often described as a continuously varying distribution of energy to different parts of our eye, from corresponding points in the external world. The preceding discussion, however, suggests a radically different view:

> *Each part of a visual scene should be regarded as supplying its own random time series of light blips to a detector, and the apparent modulation of intensity as we move over a scene is actually a variation of the mean rates of these random arrival processes.* (1.3)

To explore this distinction, A. Rose created the images in Figure 1.4, by using a scanning photomultiplier apparatus he had constructed in 1948. The figure shows an intelligible image emerging gradually out of the randomness of low-number counting statistics as the exposure time gets longer, or equivalently as the illumination goes from very dim to bright. At intermediate illumination, the brighter parts of the scene show clearer detail than the dark parts because the corresponding pixels contain larger numbers of blips.[8] The dim parts of the scene remain noisy until the illumination is increased still further. Images made under the microscope when observing single fluorescent molecules are also very dim, and here, too, the light is observed to arrive in discrete blips, building up gradually to form an image.

Nor is the "blip" interpretation (Idea 1.3) limited to camera images: Chapter 4 will show that even diffractive phenomena, long thought to be the decisive proof of

---

[8]Section 0.5.3 argued that blip counts are Poisson distributed, and Equation 0.30 (page 9) says that relative standard deviation decreases with increasing mean blip count.

the wave character of light, turn out to be discrete (lumpy) at low light intensity.[9] That is, their patterns of illumination, too, are found to be position-dependent mean rates for random, localized blip arrivals.

The classical wave theory cannot explain phenomena in which light displays a lumpy character. But this does not automatically imply that the older theory was right, and light is a stream of tiny material particles; after all, it remains true that light can also display diffractive (wavelike) phenomena. In fact, we now think of light via a third physical model, which is *neither* classical wave nor material particle. We will explore some of the details of this model in Chapter 4. Until then, we will regard "frequency" and "vacuum wavelength" in Figure 1.1 (page 24) merely as convenient labels for the various kinds of light, equivalent to stating a position within the spectrum.[10] Chapter 4 will explain the role of these quantities in the model.

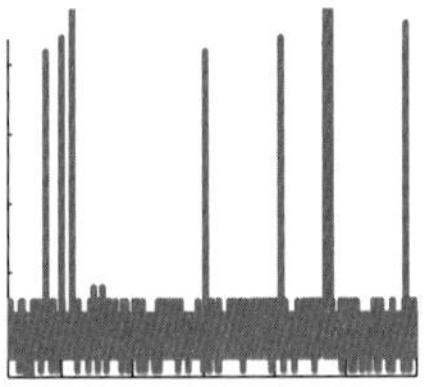

$\boxed{T_2}$ *Section 1.3.1′ (page 50) discusses other kinds of randomness in physics.*

1.3.2 The photoelectric effect

Modern instruments, like the one that yielded the shot noise data in Figures 1.2 and 0.3b, reveal the lumpy character of light directly. Remarkably, however, indirect experimental evidence for this proposition became available much earlier, in the late 19th century. Those early experiments also found a key detail, not visible in Figure 1.2, that proves to be essential for understanding the role of light in biology. Ironically, the first of these experiments was done by H. Hertz, in the course of experiments that (temporarily) seemed to assure the wave theory's supremacy! The story of Hertz's accidental discovery gives insight into the approach of a meticulous researcher when confronted with a totally unexpected result.

Hertz was studying the generation of light in the spectral range now called "radio," but he quickly realized that this aspect of his experiment was immaterial to the new phenomenon he had discovered. So he temporarily shelved his original research and set up an apparatus designed to focus attention only on the new phenomenon (Figure 1.5).

In Figure 1.5, a high-voltage power source is applied to two metal electrodes, generating a large (up to 10 cm) "primary" spark (electrical arc) across the gap between them. The power supply also applies high voltage to a smaller, "secondary" spark gap located a distance L from the primary. Holding the primary spark fixed, Hertz gradually reduced the secondary spark gap, to find the point at which sparks just began to be created there. He recorded this critical separation as a function of several variables in the apparatus, to see which changes affected it.

The observation that Hertz found surprising was that the secondary gap's propensity to spark seemed to depend on whether it was in a direct line of sight from the primary spark, as if some influence came out of the primary spark and traveled on straight lines. Interrupting this visual contact by an opaque screen—or even by a piece of clear window glass—greatly impeded sparking at the secondary gap. Eliminating the primary spark altogether had the same effect.

Hertz wanted to know how the primary spark was influencing the secondary. First, he found that the influence decreased in strength as the distance L was increased, qualitatively similar to the decrease in the intensity of light from a point source with

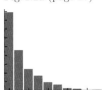

Fig. 1.2a (page 27)

Fig. 0.3b (page 11)

---

[9]See Figure 4.2 (page 147).

[10]When light passes into a transparent medium, such as water, its wavelength changes. Nevertheless, it is traditional to identify the kind of light by the wavelength that it *would have had* in vacuum (Equation 1.1, page 26). We will call this quantity vacuum wavelength, but many authors abbreviate it to simply "the wavelength."

Figure 1.5: [Hertz's original schematic.] **Hertz's apparatus.** After his accidental discovery of the photoelectric effect, Hertz built a simplified apparatus that did not involve radio transmission. A battery (*b*) energizes both primary (*d*) and secondary (*f*) spark gaps (*pink*). The experimenter adjusts the secondary gap, to find the maximum size beyond which no spark appears there. Hertz varied the distance *L* between the two spark gaps, and interposed various materials (*p*) between them. In one typical trial, the maximum length of the secondary spark was reduced to half its original value by blocking ultraviolet light (*wavy arrows*) from the primary spark. Hertz concluded that UV light from the primary enhanced the ability of the secondary to spark—a "photoelectric effect." (The diagram also shows coils of wire (*a,e*) and an interrupter (*c*), which generated the high-voltage surges needed to create sparks.) [From Hertz, 1893.]

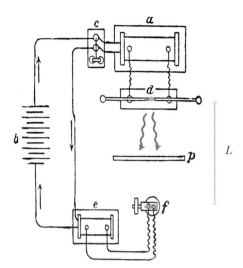

increasing distance. And whatever the cause of this influence, Hertz also found that it could be reflected by mirrors and bent by prisms, displaying the same reflection and refraction behavior as ordinary light. Crucially, the bending *angle* after passing through a prism was even larger than with visible blue light. Hertz also knew two more relevant facts: Sparks create both visible and ultraviolet light, and *glass blocks ultraviolet.*

For all these reasons, Hertz proposed that

> *Ultraviolet light somehow promotes spark generation when it lands on metal electrodes.*   (1.4)

This hypothesis made an experimentally testable prediction, namely, that an *independent* source of UV light could restore sparking, even if direct illumination from the primary was blocked. Hertz tested this prediction, and found that indeed with this arrangement the sparking was restored to the higher level seen in the original experiments. Visible light with no UV component yielded no such enhancement. Hertz documented his results carefully, but then returned to his main investigation.

In short, Hertz found that ultraviolet light falling on a metal surface can enhance the metal's ability to initiate an electric spark—a **photoelectric effect.** Followup experiments by several other scientists turned up additional clues:

1. When a piece of metal in air is negatively charged, exposure to UV light can cause it to lose its excess charge. No such effect occurs with a positively charged object.

2. The ability to discharge a negative object depends on the kind of light: Blue light is more effective than red, and UV light still more effective. Each particular kind of metal has a characteristic **photoelectric threshold** ν_*, a cutoff point on the light spectrum below which there is no photoelectric effect.[11]

---

[11]Most metals have a photoelectric threshold located in the ultraviolet range, though a few, such as sodium, have it in the visible.

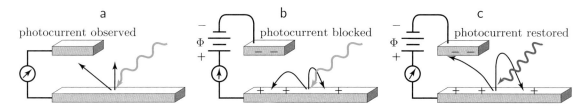

Figure 1.6: [Schematics.] **Generation of continuous photocurrents.** The primary spark gap in Hertz's apparatus (Figure 1.5) has been replaced by an ultraviolet light source (*not shown*). The secondary spark gap has been replaced by a pair of metal plates. The high-voltage secondary coil has been replaced by a current detector (*left*). (a) Light strikes the lower plate, ejecting electrons. Some of these land on a second, unilluminated plate and pass through wires, activating the detector. (b) In P. Lenard's apparatus, a battery is added to the circuit; now the upper plate has excess negative charge. Electrons can still be ejected from the lower plate, but now they are attracted back toward it (and repelled from the upper plate). No current is detected. (c) The plates are kept at the same electric potential difference Φ as in (b), but this time, light of even higher frequency is used. Electrons now emerge with greater kinetic energy than in (a,b). Some can now cross all the way to the upper plate, so a photocurrent is again detected. For any fixed value of the retarding potential Φ, there is a threshold frequency $\nu_*(\Phi)$ beyond which the photocurrent is restored. Conversely, if the light frequency is held fixed, then the stopping energy $U_{\text{stop}}(\nu)$ is defined as the product of the electron charge and the potential difference beyond which no photocurrent flows.

3. The photoelectric threshold ν_* does not depend on how bright the light is. That is, increasing the intensity of light at a fixed frequency below the threshold does not result in any discharge. (Above the threshold, however, the *rate* of discharge does depend on the illumination intensity.)

4. The mechanism that ejects negative charges does not depend on thermal motion: Cooling the metal makes no change to its photoelectric effect.

5. If we place a second metal plate near the first one, it can collect the charge ejected from the first (Figure 1.6a). If the two plates are joined by a wire, then a continuous "photocurrent" will flow indefinitely (as long as the illumination remains on), even when no high-voltage source is present.

6. The photocurrent always begins immediately upon turning on the illumination. In particular, although the strength of the photocurrent depends on the intensity of the light, there is no delay that increases at weaker illumination.

7. For a given type of metal, we can modify the photoelectric threshold by connecting a battery between the two plates described in point **5** (Figure 1.6b). P. Lenard found that, if the illuminated plate is at a positive electric potential relative to the collection plate (a "retarding potential"), then the photoelectric threshold ν_* always increases (Figure 1.6b,c).

Other experiments also showed that negatively charged *objects* were being ejected from the metal, and that they were the same "electrons" that had recently been found to stream across a vacuum in cathode-ray tubes. This interpretation explained why UV light facilitated the production of sparks in Hertz's experiment: It liberated electrons already present in the secondary electrode. These free electrons left the metal surface, ready to be driven by the electric field present at the secondary spark gap.

In Lenard's photocurrent experiments, introducing a battery created an *energy barrier,* related to the retarding electric potential Φ by the formula $U = -e\Phi$, where $-e$ is the charge on an electron. Such a barrier repels electrons from the second plate, and attracts them back to the illuminated plate. Both effects impede their passage, reducing or eliminating the photocurrent.

Physicists eventually interpreted Lenard's results by arguing that incident light liberates electrons by giving each one a "kick," enabling some of them to overcome the barrier. More precisely, for any given frequency ν of incident light, we define the **stopping energy** $U_{\text{stop}}(\nu)$ as the value of the potential energy barrier that is just large enough to eliminate the photocurrent when light of that frequency shines on the metal surface. Lenard found that

- The stopping energy depends on the type of metal. This result fits with point **2** above: The photoelectric threshold mentioned there is the value ν_* at which $U_{\text{stop}}(\nu_*) = 0$, and it depends on the metal.
- The stopping energy does not depend on the intensity of the light (compare point **3**); and
- The stopping energy is an increasing function of the frequency of the light (again see point **2**).

These observations support a physical model for the photoelectric effect, in which

> *The maximum "kick" that light can deliver to a single electron depends only on its frequency. Brighter light can kick more electrons per second, but it gives **each one** the same kick as dimmer light of the same frequency.*   (1.5)

Fig. 1.2 (page 27)

This idea also fits with the modern data shown in Figure 1.2: Higher light intensity means more electrons getting kicked (more blips from the detector per second), but not more energy delivered to each electron (each blip's strength is unchanged).

1.3.3 Einstein's proposal

The photoelectric effect posed a serious challenge to the model of light as a classical wave. For example, the wave picture implies that, for monochromatic light, increasing the intensity amounts to increasing the amplitude of the wave. If the photoelectric effect amounted to pulling electrons off atoms and giving them kinetic energy, surely a wave of greater amplitude would deliver more energy to each electron—but Lenard's results disproved this expectation. Moreover, the transfer from energy spread out in a wave to energy concentrated on one electron would require time, as the energy gradually built up.[12] Thus, the wave model predicted a *delay* before the first electrons acquired enough energy to break loose. But no such delay was observed experimentally, even for low illumination intensity.[13]

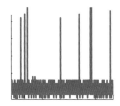

Fig. 1.3 (page 28)

Albert Einstein realized that the emerging concept of light delivering a definite kick to a single electron (Idea 1.5), although incomprehensible in the classical wave model, did fit with some recent experiments concerning the light given off by red-hot objects. Max Planck had proposed a mathematical formula that described the measured spectrum of this "thermal radiation." The formula included a new constant of Nature that we now call the "reduced **Planck constant**" and denote by the symbol $\hbar$.[14] Einstein found that Planck's formula could be obtained by starting from the

---

[12]Recall the metaphor in Figure 1.3 (page 28).

[13]This observation was point **6** on page 32.

[14]Pronounced "aitch-bar." Some authors introduce the symbol h to denote the quantity that we will call $2\pi\hbar$. But this notation runs the risk of confusion with other quantities also named h, so we will instead always express our formulas using the unique symbol $\hbar$. Other authors introduce the "angular frequency" $\omega = 2\pi\nu$, in terms of which Equation 1.6 says $E_{\text{photon}} = \hbar\omega$.

hypothesis that monochromatic light consists of lumps, now called **photons**, each carrying a packet of energy equal to

$$E_{\text{photon}} = 2\pi\hbar\nu. \quad \text{Einstein relation} \tag{1.6}$$

In this formula, ν refers to the frequency that we would have attributed to the light, under the classical wave model. For example, ν describes where in the spectrum the light in question would fall if sent through a prism or grating.

Einstein then predicted that the *same* numerical constant $\hbar$ would also control the photoelectric effect, by setting the maximum kick that any particular kind of light could give to an electron in a collision. This physical model led to a quantitative prediction for the stopping energy: Suppose that the minimum energy cost to remove an electron is some constant W, whose value is a characteristic of the particular metal used in the experiment. Therefore, the electron must expend some of its initial kick just exiting the metal, leaving at most $2\pi\hbar\nu - W$ to overcome any additional retarding potential imposed by the experimenter. If the kick is less than the energy required to eject an electron (that is, if $2\pi\hbar\nu < W$), then the electron can't exit the metal at all, and there's no photocurrent even with no applied retarding potential. Otherwise, Equation 1.6 predicts that the stopping energy is the difference between $2\pi\hbar\nu$ and W. Because W depends only on the metal, not on the kind of incoming light, Einstein obtained a detailed prediction for the dependence of stopping energy on frequency:

> *Every metal must have a linear relationship between stopping energy and frequency: $U_{\text{stop}}(\nu) = 2\pi\hbar\nu - W$. The slope, $2\pi\hbar$, of this relation must always equal the numerical value that Planck found when explaining the thermal radiation spectrum. The intercept, $-W$, depends only on the kind of metal (not the frequency nor intensity of the light).* (1.7)

This prediction was highly falsifiable: It was quantitative, and the key quantity $\hbar$ was not a fitting parameter—it was independently known from the spectrum of thermal radiation.[15]

Einstein was out on a limb, because at the time there were no experimental results that could confirm his prediction. Definitive confirmation took another 11 years, but eventually it was found that indeed the slope in Idea 1.7 is universal: It always agrees with Planck's value, regardless of the type of metal in the electrode, source and intensity of the light, temperature, and other adjustable parameters.

Einstein's proposal fits with the observation of discrete blips in today's sensitive light detectors (Figure 1.2), because they operate by a mechanism similar to the photoelectric effect. But it adds a crucial element, the dependence of the effect on frequency.

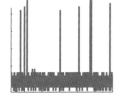

Fig. 1.2a (page 27)

$\boxed{T_2}$ *Section 1.3.3′a (page 50) discusses more evidence for the reality of photons. Section 1.3.3′b (page 51) discusses the momentum that they carry. Section 1.3.3′c (page 52) discusses the link between the Einstein relation and the thermal radiation spectrum. Section 1.3.3′d (page 53) discusses the concept of "frequency." Appendix B discusses Planck's discovery of his constant.*

---

[15]Section B.4 (page 442) explains how Planck determined $\hbar$.

1.3.4 Light-induced phenomena in biology qualitatively support the Einstein relation

Einstein's proposal meshes with some things we know about biology and light. A number of biologically and medically relevant processes require light, and indeed light with frequency greater than a certain threshold, in order to operate. For example,

- Visible light does not induce skin cancer. Only light with higher frequency (or shorter wavelength) than visible will have this biological effect.
- Similarly, UV light is required for the production of vitamin D in our skin; artificial indoor lighting, which lacks a UV component, won't suffice.
- Exposure to blue light (phototherapy) can clear excess bilirubin in the blood of a premature infant—but red light does not work.

Each of these examples highlights a threshold dependence on frequency, not intensity, of light, consistent with Einstein's proposal. Each will be discussed further in the following sections.

Section 1.3 has surveyed some phenomena that contradict the continuous wave model of light and seem to call for a "lumpy" model. We have also seen some hints that the lumpy aspect of light is relevant for biological phenomena. Before we propose a replacement for the wave model of light, the rest of this chapter and Chapter 2 will begin by documenting this aspect more systematically. To that end, we now pause to construct some ideas about probability that we'll need.

1.4 **BACKGROUND:** POISSON PROCESSES

"Background" sections, like this one, give brief reviews of foundational material.

Figure 1.2 illustrates how the arrival times of individual photon blips are unpredictable, even for a light source that is carefully constructed to be as stable as possible.[16] But the Prologue pointed out that no system is *completely* unpredictable: We characterize whatever we do know about it by stating a probability distribution. In fact, experiments showed that, for many kinds of light source, the arrival of blips on a sensitive light detector follows a simple, universal form, called a Poisson process. Many other apparently unrelated processes are also approximately Poisson in character, for example, radioactive decay, enzyme turnovers, release of neurotransmitter vesicles,[17] and even in some cases neural signaling itself.

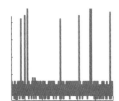

Fig. 1.2a (page 27)

1.4.1 A Poisson process can be defined as a continuous-time limit of repeated Bernoulli trials

A **random process** is a random system whose draws are time series. For example, the blips coming from the light detector in Figure 1.2 are all identical, so the only information we need to describe the outcome of a particular trial is the series of times $\{t_1, t_2 \ldots\}$ at which the blips occurred. If we repeat the experiment, we obtain another draw from the random process, represented by another such sequence of numbers.

One class of random processes, called **Poisson processes**, describes many biophysical phenomena. To define it, first imagine taking a time interval of interest and dividing

---

[16]You'll study such data in Problem 1.3.
[17]For vesicle release, see Figure 2.8 (page 72), Problem 2.1, and Figure 10.13 (page 338).

it into small, consecutive slices of duration Δt. Then we decide whether or not to place a blip in the middle of each time slice. A Poisson process is characterized by the properties that

- *The probability of a blip occurring in any small time slice equals $\beta\Delta t$, independent of what is happening in any other slice, and*   (1.8)
- *We take the continuous-time limit $\Delta t \to 0$ holding β fixed.*

Thus, a Poisson process is characterized by a single parameter β. Because β gives probability per unit time, it has dimension $\mathbb{T}^{-1}$.

1.4.2 Blip counts in a fixed interval are Poisson distributed

Each draw from a random process consists of many numbers, so it is not easy to give the complete probability density function on this large sample space. For Poisson processes, however, it often suffices to know about two kinds of reduced distributions.

The first of these reductions asks, "How many blips will we observe in a fixed time interval T_1?" The answer is not a single number. Rather, it is a discrete probability distribution for a random variable ℓ, and in fact the form of this distribution is familiar:

For a Poisson process with mean rate β, the probability of getting ℓ blips in a given time interval T_1 is Poisson distributed: $\mathcal{P}_{\mathrm{pois}}(\ell; \beta T_1) =$ (1.9) $(\ell\,!)^{-1}\mathrm{e}^{-\beta T_1}(\beta T_1)^\ell$.

Your Turn 1A

a. Use what you know about the Poisson distribution to find the expectation of the number of blips in a fixed time interval of duration T_1, then divide by T_1 to get blips per unit time. How does this quantity depend on β and T_1?

b. Normally, we don't notice the Poisson character of light, because the photons are arriving so rapidly that we don't register them as individual events. Show that, moreover, the relative standard deviation of the blip number ℓ in T_1 is small if T_1 is much greater than β^{-1}.

1.4.3 Waiting times are Exponentially distributed

The second reduced form of a Poisson process asks instead, "How long must I wait until the next blip arrives?" Here again, the answer is different each time; there is a continuous probability density function of waiting times. Section 0.4.2 already gave the answer:[18]

The waiting times in a Poisson process are distributed according to the Exponential distribution, $\wp_{\mathrm{exp}}(t_{\mathrm{w}}; \beta) = \beta\mathrm{e}^{-\beta t_{\mathrm{w}}}$, regardless of when any   (1.10) previous blips may have occurred.

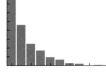

Fig. 0.3b (page 11)

Figure 0.3b illustrates that very dim light does follow this distribution, a hint of its underlying Poisson-process character. The intensity of the light source is related to the mean rate of the random process.

This background section has constructed a new kind of probability distribution (a random process), which will prove useful for describing light.

---

[18]See Equation 0.41 (page 14).

$\boxed{T_2}$ *For further details on Poisson processes, see the references at the end of this chapter. Section 1.4′ (page 54) says more about when we can and cannot expect photon arrivals to form a Poisson process.*

1.5 A NEW PHYSICAL MODEL OF LIGHT

Earlier sections described some results that seem to imply that light cannot be a stream of material particles ("grains of sand"), as well as others that seem to imply that light cannot be a classical wave:

- A monochromatic light source of uniform intensity activates a detector with a random series of discrete events.
- Uniform illumination over a broad area gives rise to independent and highly localized events. Thus, two different pixels in a sensitive camera never respond at exactly the same time. In contrast, a classical wave would spread its energy continuously over the illuminated region; if the wave were of uniform amplitude, any response would be expected to be synchronous across the region.
- In the photoelectric effect, at one fixed wavelength, fainter light ejects fewer electrons, but does not affect the stopping energy.

The rest of this chapter will focus on these three points. Chapter 4 will return to the wavelike phenomena (Section 1.2.2), and reconcile them with the discrete ones.

1.5.1 The Light Hypothesis, part 1

The phenomena described in previous sections motivate a series of related claims about light:[19]

Light Hypothesis, part 1:
- *Light travels through empty space at speed $c \approx 3 \times 10^8$ m/s.*
- *It comes in lumps (photons). The photons from a monochromatic light source each carry the same amount of energy, determined by the light's position in the spectrum.*
- *A photon interacts with matter by transferring all its energy to a single electron and disappearing (absorption). Or a photon can pass by the electron without disturbing it (or being itself changed) in any way. The choice of which option occurs is random (a Bernoulli trial).*   (1.11)
- *Conversely, a photon can be created from nothing by a single electron, which loses an amount of energy equal to that of the new photon. Photon creation (also called "emission") is also probabilistic; for example, a source like the Sun, while seemingly of constant intensity, actually gives off photons in a random process.*

We won't derive the Light Hypothesis from any deeper bedrock of truth. Instead, we'll regard it as just a compact statement of a few principles, into which a large

---

[19]Idea 1.11 is called "part 1" because Chapter 4 will extend it.

number of biophysically relevant phenomena will fit. Although physicists regard it as confirmed, we will nevertheless use the word "hypothesis," to emphasize that it's inherently implausible as it stands. We'll need to examine many more phenomena before concluding that it is even promising.

The Light Hypothesis asserts that every form of light has a discrete character. However, low-frequency forms, like radio, have such small E_{photon} that their discreteness is usually not observable. In contrast, high-frequency forms, like gamma rays, have such large E_{photon} that their effects appear discrete in nearly every situation. Intermediate forms, such as visible light, can appear either discrete or continuous, depending on the specific situation we study. Many biophysical applications involve the visible region of the spectrum, so we will need to come to grips with the dual character of light.

$\boxed{T_2}$ *Section 1.5.1′b (page 55) mentions some finer points concerning the Light Hypothesis.*

1.5.2 The spectrum of light can be regarded as a probability density times an overall rate

Section 1.3.3 noted a quantitative relation between E_{photon} and the apparent frequency of light: the Einstein relation $E_{\text{photon}} = 2\pi\hbar\nu$. Instead of including this observation in the Light Hypothesis, however, Chapter 4 will obtain it as a consequence of a more general principle.[20] Until then, we will continue to think of ν as simply an abbreviation for the quantity $E_{\text{photon}}/2\pi\hbar$; similarly, for now the vacuum wavelength is simply an abbreviation for the quantity c/ν.

We can find the probability density function for the vacuum wavelength of any photon in the stream, $\wp(\lambda)$, with dimensions $\mathbb{L}^{-1}$ appropriate to such a PDF.[21] Equivalently, we can let Φ_{p} denote the total mean rate for photon arrivals of all wavelengths, and introduce the combined function $\mathcal{I}(\lambda) = \Phi_{\text{p}}\wp(\lambda)$, called the **spectral photon arrival rate**, or simply the "spectrum" of the light.[22] A possible choice of units for $\mathcal{I}$ is $\text{s}^{-1}\,\text{nm}^{-1}$. The normalization condition on $\wp(\lambda)$ then says that the total mean photon arrival rate, Φ_{p}, is given by $\int \text{d}\lambda\, \mathcal{I}(\lambda)$.

Instead of giving the photon arrival rate per wavelength interval, $\mathcal{I}$, some authors introduce the function $(2\pi\hbar c/\lambda)\mathcal{I}(\lambda)$. This function gives the rate of *energy* transfer per wavelength interval, so it has units like $\text{W}\,\text{nm}^{-1}$. It, too, is often called "the spectrum," potentially leading to confusion, so this book will not use this description of light; instead we will always use the function $\mathcal{I}$ defined in the previous paragraph. When reading other books and publications, you can see which description is being used by looking at the units.[23]

---

[20] See Idea 4.5 (page 154).

[21] We could equally well describe the photons by their frequency or energy and find the corresponding transformed PDFs, but this book will use the more customary description, via vacuum wavelength. See Problem 3.2.

[22] We will use the same word "spectrum" to denote either the actual pattern of light cast on a screen by a prism, or the mathematical function $\mathcal{I}(\lambda)$ describing the content of a light beam. When there seems to be danger of ambiguity, the longer term "spectral photon arrival rate" can be used instead to denote $\mathcal{I}(\lambda)$.

[23] $\boxed{T_2}$ When writing your own work, you can if necessary refer to $\mathcal{I}$ as the "actinometric spectrum" and to $(2\pi\hbar c/\lambda)\mathcal{I}(\lambda)$ as the "radiometric spectrum."

1.5.3 Light can eject electrons from individual molecules, inducing photochemical reactions

A lump of matter consists of a lot of positively charged atomic nuclei that attract ("bind") electrons. Each electron may be associated to a single nucleus, or the electrons may be shared. Metals are an extreme case, in which some of the electrons wander at will throughout the entire sample.

In the photoelectric effect, a photon's energy is converted to kinetic energy of one electron (Section 1.3.2). That kinetic energy may then suffice to overcome the electron's attraction to a metal surface. Any energy left over after this ejection is then available to overcome an additional retarding potential in the surrounding space. Similarly, even a single atom can lose an electron ("ionize") upon exposure to light with high enough frequency.[24] Here are two related phenomena that further illustrate what the Light Hypothesis means when it says that a photon "transfers all its energy to a single electron."

- Most materials are electrical insulators: Their electrons are locked into a rigid state with no net flow, even when we apply an electric potential difference across the sample. However, in a special class of insulators, called photovoltaics, an incoming photon can deliver energy to an electron, promoting it into a higher-energy state, in which it is mobile. Even without actually leaving the material, such "excited" electrons can then travel through it, creating a net current. This mechanism forms the basis of solar cells. More sophisticated versions (for example, charge-coupled devices found in cameras) operate on a similar principle.

- Next, consider an electron that participates in a chemical bond in a single molecule. An incoming photon can knock the electron out of its molecule altogether or just move it internally, breaking the original bond and possibly allowing a different one to form—a **photochemical reaction**. If the original bond was part of the structure of a strand of DNA, changing it can cause a mutation, a first step to skin cancer (Figure 1.7). That's why ultraviolet radiation is dangerous, whereas visible light is not: Analogously to the photoelectric effect, disrupting a chemical bond requires a certain threshold energy, and therefore light of at least a threshold frequency. Similarly, many paints and inks fade in direct sunshine (they photobleach), but not in artificial light, because only sunlight contains a significant UV component.

The following sections will explore how the Light Hypothesis helps us to understand several more physical, chemical, and biological phenomena beyond the ones already discussed. These phenomena cannot be understood in the old model of light as a classical wave. We will also find some surprising behavior involving electrons: They, too, display both wavelike and particle-like behavior.

Section 1.5 has begun our exploration of a new physical model of light. Before completing our Light Hypothesis, we next look around for more phenomena that can lend it qualitative support.

$\boxed{T_2}$ *Section 1.5.3′ (page 55) discusses DNA photodamage in more detail.*

---

[24]In fact, Einstein's 1905 paper already noted that his proposal explained a related phenomenon: Exposure to light can ionize a gas (photoionization), but only if the light's wavelength is sufficiently short. Einstein also noted that the hypothesis could explain the Stokes shift phenomenon (Section 1.6.3).

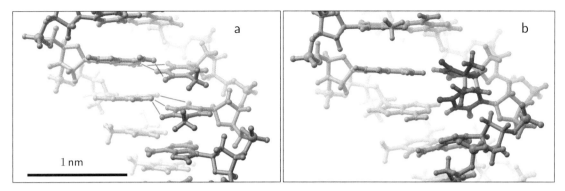

Figure 1.7: [Artist's reconstruction based on structural data.] **DNA photodamage.** (a) Normal segment of DNA. Each base lies in a plane roughly coming out of the picture; the bases are stacked to form two helical strands. Paired bases interact via hydrogen bonds (*gray lines*), holding the two strands together. (b) One kind of photodamage to DNA. Two adjacent thymine bases (*magenta*) have bonded covalently to each other (*cyan lines*). The new structure will subsequently cause an error during DNA replication, leading to a genetic mutation. [Art by David S Goodsell.]

1.6 FLUORESCENCE AND PHOTOISOMERIZATION CAN OCCUR AFTER PHOTON ABSORPTION

1.6.1 The Electron State Hypothesis

A photon need not eject an electron completely from an atom, nor break a bond in a molecule, to have an effect on it. To understand such phenomena qualitatively, we must combine the Light Hypothesis with an additional set of ideas, which may be familiar to you from chemistry:

> **Electron State Hypothesis:**
> - *The electrons bound in an isolated atom have a discrete set of allowed physical states. Each such state has a characteristic energy and spatial structure (size and shape).*
> - *The electrons in an isolated atom tend spontaneously to revert to their lowest-energy state (the* **ground state***). Those reversions are abrupt; they occur only after a randomly chosen waiting time.*
> - *When two or more atomic nuclei come close together, the allowed states of their electrons merge into a new set of shared, or "molecular," states. The electron energy in each allowed state depends on the relative positions of the nuclei.*

(1.12)

The electron energy mentioned in Idea 1.12 includes the kinetic energy of the electrons as well as the potential energies of their mutual repulsion and their attraction to the nuclei. We'll denote the nuclear positions collectively by a symbolic variable $\boldsymbol{y}$.

Because electrons are much less massive than nuclei, their motions are faster; they rearrange themselves to account for slow changes in the nuclear positions, maintaining a fixed excitation level (ground or "excited" state). So it makes sense to think of the electron state energy as simply a function of $\boldsymbol{y}$, effectively a kind of potential energy for the nuclear positions. Combining it with the nuclear mutual repulsion energy then gives a total effective potential energy $U_\alpha(\boldsymbol{y})$ that depends on nuclear positions and also on the discrete excitation level α. Thus, there are many such functions. We will give the special name $U_0(\boldsymbol{y})$ to the one associated with the electron ground state. For

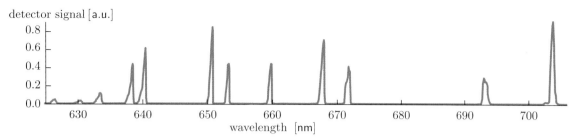

Figure 1.8: [Experimental data.] **Observed emission spectrum of neon gas.** An ordinary neon light bulb emits light with a spectrum showing multiple sharp peaks.

any excitation level, the nuclei tend to choose a spatial arrangement y that minimizes the effective potential energy.

$\boxed{T_2}$ *Section 1.6.1′ (page 56) mentions some fine points about the Electron State Hypothesis applied in a water solution. The discreteness of electron energy levels can actually be derived, starting from ideas similar to the ones describing light; see Chapter 12.*

1.6.2 Atoms have sharp spectral lines

The Electron State Hypothesis (Idea 1.12) fits with experimental facts about how a single atom interacts with light. Suppose that we inject energy into a few isolated atoms by filling a chamber with low-pressure neon gas and passing electricity through it. The electrons in each atom can then get knocked from their ground state to one of higher energy. For an atom to revert to its ground state, it must lose a sharply defined amount of energy (the difference between the two discrete levels). According to the Light Hypothesis, that energy can appear as a single photon, which must therefore also have a sharply defined frequency. Thus, if we sort the emerging photons by frequency (or by wavelength), then we expect their distribution to display sharp "spectral lines" (Figure 1.8). Those lines are characteristic of the type of atoms we placed in the chamber.

Another way to promote the electrons to an excited state is to shine light onto the gas. If an incoming photon has just the right energy to promote the electrons in an atom to an allowed state of higher energy, then the photon may be absorbed. Later, an identical photon may be reemitted, bringing the atom back to its ground state. But the new photon may not be traveling in the same direction as the original one; that is, the net effect of the absorption/emission process may be to "scatter" the incoming light.

This system also displays a crucial difference between ejection and rearrangement of electrons: Rather than just a *threshold* for excitation, as in the photoelectric effect, triggering a transition between two electron states requires a well-defined energy, and hence a *specific* wavelength of light. That is, a photon may be either "too red" (that is, not energetic enough) *or* "too blue" (too energetic) to participate in this process; in those cases, it will pass by the atom without interacting.

$\boxed{T_2}$ *Section 1.6.2′ (page 56) discusses some finer points about the Electron State Hypothesis.*

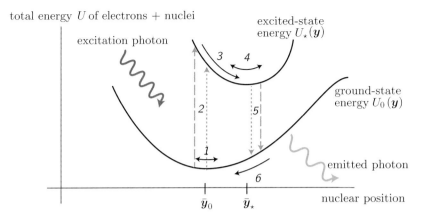

Figure 1.9: [Schematic energy diagram.] **Physical model for fluorescence.** *1*: The positions of the nuclei in a molecule fluctuate near $\bar{y}_0$, the minimum of the energy function $U_0(y)$ appropriate for electrons in their ground state. *2*: Absorption of a photon promotes the electrons to an excited state. The differing lengths of the *dashed* and *dotted* lines indicate that the required photon energy depends on the momentary value of the nuclear position coordinate y. *3*: The nuclei respond by moving toward the minimum of the new, excited-state energy function $U_\star(y)$. *4*: Then the nuclei fluctuate about the new minimum $\bar{y}_\star$. *5*: Eventually a new photon is emitted (*right*), sending the electrons back to their ground state. The lengths of the *dashed* and *dotted* lines again indicate that there is variation in the precise energy of the emitted photon. *6*: The nuclear positions then readjust back to the neighborhood of $\bar{y}_0$. The typical difference in length between the ascending and descending arrows represents the Stokes shift of this fluorescent molecule (fluorophore).

1.6.3 Molecules: fluorescence

Molecules, too, can absorb and emit light, but with some features not found in single atoms. This section will describe a particular form of the combined process of absorption followed by reemission of light by a molecule, called **fluorescence**.

A molecule differs from an isolated atom because it can rearrange the relative positions of its atomic nuclei. For example, a molecule of carbon dioxide has three nuclei. The nuclei are usually drawn as lying on a straight line, but they can move slightly, flexing the molecule. Similarly, the three nuclei of a water molecule normally lie in a bent configuration, but again the angle of that bend can change slightly, as can the distances between nuclei. The relative positions of the nuclei are a set of vectors that we can collectively call y (the "configuration" of the molecule). We can think of each configuration as a point in a complicated "configuration space" of allowed y values. Then a molecular transformation corresponds to a *trip* in configuration space from the starting to the ending conformation. There will be many possible routes to take on that trip, because y is a many-component quantity, but for our purposes we can restrict attention to just the one route that follows the easiest flexing motion of the molecule. Thus, we will regard y as a single variable (the "reaction coordinate"), describing where the molecule is currently sitting along that route.

Each allowed state of the electrons depends on the positions of the nuclei. Hence, the total energy of the molecule in each such state, including mutual repulsion of the nuclei, also depends on y. For example, the energy of the ground state is the function we called $U_0(y)$ in Section 1.6.1 (depicted as the lower curve in Figure 1.9). The molecule spends most of its life near the configuration $\bar{y}_0$ that minimizes this function.

When molecules are not isolated, for example, when they are dissolved in water (or

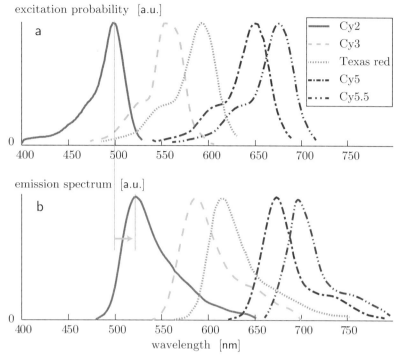

Figure 1.10: [Experimental data.] **Excitation and emission spectra.** (a) Excitation spectra for five popular fluorophores. The curves have been separately rescaled; each is proportional to the probability that an incoming photon will excite a fluorophore, as a function of that photon's vacuum wavelength. The abbreviation a.u. refers to "arbitrary units"; see Appendix B. (b) Emission spectra for the same fluorophores. This time, the vertical axis represents the probability density function for emission of photons of various wavelengths by the excited fluorophore. The spectra are much broader than those of single atoms (compare Figure 1.8). Also, the peak of each molecule's emission spectrum is Stokes-shifted toward red, relative to its peak wavelength for excitation (*arrow*; see Idea 1.13). [Data courtesy Yuval Garini.]

in the complex environment of a living cell), they constantly bump one another, due to their thermal motion. Each such collision imparts relative motion to a molecule's nuclei—that is, it changes y. Thus, the nuclei are likely to be close to, but not exactly in, the configuration $\bar{y}_0$: y fluctuates about $\bar{y}_0$. The curved, double-headed arrow in step *1* of Figure 1.9 represents this incessant flexing and stretching motion.

Once the molecule has been excited, its new electron state creates a *different* effective potential energy function for the nuclei (upper curve labeled $U_\star(y)$ in Figure 1.9). The amount of energy needed to change the electron state is represented by the distance between a point on the ground-state energy curve and the corresponding point on the excited-state curve. Because y was fluctuating prior to the transition, this "excitation energy" is variable. Thus, the energy required for excitation is not as sharply defined as it is for a single atom: We say that the molecule's **excitation spectrum** is "broadened," compared with the narrow spectral lines of a single atom (compare Figures 1.8 and 1.10a). The range of wavelengths at which the excitation spectrum is nonnegligible is called the molecule's **excitation band**. Similarly, after a molecule has been excited, it also gives off light in a broadened distribution (its

emission spectrum; see Figure 1.10b) as it transitions back to the ground state.[25]

These ideas lead to an important insight about fluorescence. The minimal-energy configuration for the excited-state potential energy function, called $\bar{y}_\star$ in Figure 1.9, may be different from the one appropriate to the ground state ($\bar{y}_0$). After the transition, the nuclei therefore begin to move from the neighborhood of $\bar{y}_0$ toward $\bar{y}_\star$, typically shedding their excess potential energy by colliding with the surrounding water molecules and heating them slightly. Figure 1.9 represents this process by the step labeled *3*. Thus, the energy released when the excited molecule reverts to its ground state is somewhat *less than* the energy required to enter the excited state, because the downhill "sliding" step *3* reduces the energy available for the reversion.[26] In short,

> The emitted light from molecular fluorescence has a broad spectrum, which peaks at a longer wavelength than the light required to excite that fluorescence. In contrast, light emitted from an isolated excited atom has a sharply defined wavelength and is unshifted.   (1.13)

A molecule, or group within a molecule, with this behavior is called a **fluorophore**. The difference between the excitation and reemission peak wavelengths is a characteristic of the fluorophore, called its **Stokes shift** (Figure 1.10). The Stokes shift may be familiar to you from the visible glow that many fabrics and brightly colored inks give off when lit by a "black light" (ultraviolet light source).

The excitation and emission spectra are signatures that can be used to identify which fluorophore we are seeing. Analysis of some phenomenon (here fluorescence) in terms of some spectrum (here those of the incoming and outgoing light) is generically called **spectroscopy**. Chapter 2 will discuss how the Stokes shift, a spectroscopic feature of fluorescence, led to a revolution in microscopy.

Good fluorophores are distinguished from generic molecules by having excited states that have a high probability of releasing their energy by photon emission, and that persist long enough (typically a nanosecond or so) for the nuclei to arrive at $\bar{y}_\star$ before reverting to the ground state.

A fluorophore is a distinct concept from a "chromophore," which is a molecule or group that absorbs light preferentially in a particular wavelength band but does not reemit a Stokes shifted photon.[27] That is, a chromophore creates color by selectively absorbing light; light that is not absorbed is transmitted or scattered without changing its wavelength.

Some other light-emission phenomena are closely related to fluorescence. If the energy source is a chemical reaction, not an incoming photon, the process is called "chemiluminescence" (as seen in chemical glow-wand toys). In the special case where the source is the metabolism of a living cell, we use the more specific term "bioluminescence" (as seen in the glow of fireflies, jellyfish, and even some bacteria; see Figure 1.11).

$\boxed{T_2}$ *Section 1.6.3′ (page 57) discusses why the electronic transitions are represented*

---

[25] Another common situation is that an excited molecule loses energy by some process not involving a photon, typically in a collision with some other molecule. The excited molecule does not fluoresce, but if its excitation spectrum has a peak in the visible range it may be useful as a paint pigment, because it preferentially removes light in that band.

[26] Another, similar energy loss mechanism is also depicted in step *6* of Figure 1.9: When the electron configuration reverts to its ground state, the nuclei are initially not at the minimum of $U_0(\boldsymbol{y})$. Thus, $\boldsymbol{y}$ must again slide down to $\bar{y}_0$, dissipating even more energy as heat.

[27] The term **pigment** can refer to a chromophore, or a large molecule containing a chromophore. The less specific word **dye** is often used to mean either chromophore or fluorophore.

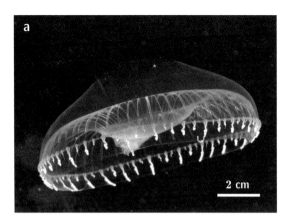
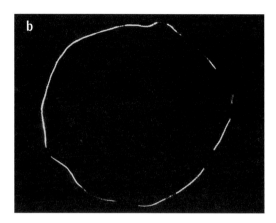

Figure 1.11: [Photos.] **Bioluminescent jellyfish.** (a) Side view of *Aequorea victoria* under external illumination. (b) End view of a related species, in darkness. When the animal is disturbed, a ring of bioluminescent organs around the base of the bell give off blue light, which is converted to green by a process to be discussed in Section 2.8.3 (page 87). [(a) Photo by Sierra Blakely. (b) Courtesy Steven Haddock; see also Media 3.]

by vertical lines in Figure 1.9, and makes more precise the meaning of energy functions like $U(\boldsymbol{y})$.

1.6.4 Molecules: photoisomerization

The ground-state potential energy for the nuclei in a molecule may have more than one local minimum, separated by an energy barrier (Figure 1.12). If the nuclei are arranged in a way corresponding to one of the local minima, then a small deformation will generate a restoring force pushing them back. Larger deformations, however, can "pop" the molecule over to the other stable state. If the barrier between the states is much larger than the typical thermal energy,[28] then collisions will only rarely suffice to excite such transitions; we then say that the molecule has multiple **geometric isomers**, or **conformations**. Once in the alternate conformation, the molecule stays there until we push on it again.

Figure 1.12 outlines a way in which light can change the conformation of a molecule, a process called **photoisomerization**. Extending the ideas of Section 1.6.3, we can think of the variable $\boldsymbol{y}$ in the figure as giving the position along the **critical path**. This is the "easiest" path in configuration space joining two conformations of the molecule, that is, the path for which the highest value of U_0 attained along the way is minimal.

Your Turn 1B

Compute the energy of a visible photon (choose any wavelength in the visible range). Next, compute the typical energy of thermal motion at room temperature (Section 0.6, page 17). Then compare those two numbers and make a comment on photoisomerization.

As before, the excited-state energy curve $U_\star(\boldsymbol{y})$ generally has a different minimum-energy conformation $\bar{\boldsymbol{y}}_\star$ from the original. If the excited-state equilibrium conformation

---

[28]Section 0.6 (page 17) introduced the energy of thermal motion.

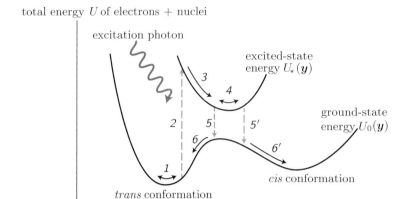

Figure 1.12: [Schematic energy diagram.] **Physical model for photoisomerization.** The first steps are analogous to those in Figure 1.9. Now, however, the ground-state energy function has two local minima, at $\bar{y}_{\text{trans}}$ and $\bar{y}_{\text{cis}}$. Promoting the molecule to its excited electron state sometimes enables it to fall back to a different conformation from the one in which it began (*5′–6′*), bypassing the ground-state energy barrier between the two conformations. Depending on the momentary configuration at the time of emission, however, the molecule may instead return to its initial conformation (*5–6*). Steps *5* and *5′* may or may not involve photon emission; for example, energy may instead be lost via collision.

lies between the alternate isomeric form and the original one, as in Figure 1.12, then the molecule may end up in the alternate conformation upon reversion to its electronic ground state. Thus, even if isomerization is extremely unlikely to happen spontaneously via thermal agitation, it may nevertheless happen readily in the presence of light of suitable wavelength.

Although the ideas just outlined are general, they have special significance in the context of biomolecules, whose detailed conformation profoundly affects their function. In its new conformation, the molecule may engage in different molecular recognition events than the original:

> *Light can isomerize some molecules from biologically inactive to active conformations, or vice versa.* $\qquad$ (1.14)

Chapter 10 will apply this insight to understanding the first events in human vision.

A simpler example concerns **phototherapy**. Section 1.3.4 mentioned that prematurely born infants sometimes have an excess of *trans*-bilirubin in the blood, which their immature livers cannot clear effectively. Shining blue light on their skin can isomerize this molecule to a water-soluble form called *cis*-bilirubin, minimizing the toxic effects of the *trans* form without the need for any drug or other invasive procedure.

Photoisomerization can also have a dark side! A normally fluorescent molecule can pop into a nonfluorescent conformation, destroying its fluorescence either temporarily (**blinking**) or permanently (**photobleaching**).[29]

Section 1.6 has shown how fluorescence and photoisomerization, two processes important throughout biophysics, are qualitatively understandable based on the Light

---

[29]You'll study blinking further in Problem 7.6.

Hypothesis and Electron State Hypothesis.

$\boxed{T_2}$ *Section 1.6.4′ (page 58) discusses a fine point about conformational change relevant for the very fast transitions that occur in visual pigments.*

1.7 TRANSPARENT MEDIA ARE UNCHANGED BY THE PASSAGE OF LIGHT, BUT SLOW IT DOWN

The ejection of an electron from its atomic home is a major change. The promotion of an electron to a new state is less violent, but still a discrete event (Section 1.6). Even milder forms of interaction are possible as well. An electron can absorb a photon, then immediately reemit it without lingering in any definite new state. Most phenomena involving reflection or scattering of light by matter work in this way, for example, the reflection of light from white paper. In other materials, the electrons are more likely to absorb a photon, then transfer the energy to the general thermal motion of all the atoms in a sample, heating it. Here an everyday example is the heating of black pavement in bright sunshine.

More subtly, some materials are transparent, at least to some kinds of light. For example, glass is transparent to visible light. When a stream of photons travels through such a medium, their frequencies do not change; however, the interactions with its electrons do have the net effect of *slowing the light down*.[30] We write the net speed for light of frequency ν as c/n, where n is called the medium's **index of refraction**, a dimensionless number larger than 1. More precisely, the index can depend on frequency, so the relation can better be written as $c/n(\nu)$.[31] We can measure the speed of light in each medium, and for each frequency determine n from these measurements. For example, visible light travels in water at roughly 3/4 its vacuum speed:[32] $n_w(\nu) \approx 1.33$–1.34 over this frequency range. A change in speed implies a change in Equation 1.1, which now says that the wavelength in the medium is

$$\lambda_{medium} = (\text{speed})(\text{period}) = \frac{c}{n(\nu)}\frac{1}{\nu}. \quad \text{wavelength in transparent medium}$$

(1.15)

Air is another common transparent medium. But the density of air is much smaller than that of water or glass, so its index of refraction is very close to 1; for most optical phenomena, we may treat air as essentially the same as vacuum.

As mentioned in Section 1.3.1, it is customary to describe monochromatic light by the wavelength that light of the same frequency *would have had,* were it propagating in vacuum. This "vacuum wavelength" is still given as a function of frequency by $\lambda = c/\nu$, even though the true wavelength is given by Equation 1.15.

---

[30] $\boxed{T_2}$ Section 12.3.3 (page 396) will discuss the origin of this slowdown.

[31] Chapter 4 will discuss how this phenomenon can give rise to rainbows when light passes through a prism (or through droplets of water suspended in the air after a rainstorm).

[32] In a sentence that defines notation, like this one, the underlined initial "w" in the word "water" is a hint about the meaning of the notation: The subscript on the symbol n_w refers to water.

THE BIG PICTURE

This chapter has presented some surprising phenomena involving light, and proposed that "light consists of *photons*." At the moment, that phrase is little more than a slogan, involving a word whose definition so far is "not exactly analogous to a water wave, nor to a stream of sand grains flying through space." This chapter has mostly pointed out phenomena that require a non-wave interpretation. Chapters 2 and 3 will show some more biophysical phenomena that are consistent with the minimal portrait of photons described here. Chapter 4 will then develop the physics of photons in greater detail, developing a physical model that embraces both particle and wave aspects. Chapter 9 will examine the question of whether the lumpy character of light is actually relevant to human vision.

There is a third aspect of light that you probably studied in first-year physics, namely its connection to electricity and magnetism. This book will not emphasize that aspect, because a detailed understanding is rarely essential for biophysics. A bigger theoretical framework, called field quantization, incorporates all three aspects of light.[33] The more limited framework we will use in this book, however, shows some lessons more simply than the elaborate complete theory, including many that are relevant to living systems and the instruments used to study them.

KEY FORMULAS

- *Photons:* Light consists of lumps called photons. A photon can be thought of as carrying energy E_{photon},[34] whose value determines where it falls in the light spectrum. In some situations, monochromatic light behaves analogously to a wave with frequency $\nu = E_{\text{photon}}/2\pi\hbar$, where the reduced Planck constant $\hbar \approx 1.05 \times 10^{-34}\,\text{J s}$. In vacuum or air, the wavelength λ equals c/ν, where $c \approx 3.0 \times 10^{8}\,\text{m/s}$ is the speed of light.
- *Electric potential energy versus electric potential:* The potential energy U of a charge q in a region with electric potential Φ is equal to $q\Phi$.
- *Photoelectric effect:* Light can eject electrons from the surface of a piece of metal in vacuum. The minimum potential energy barrier needed to stop the photocurrent (the stopping energy) is

$$U_{\text{stop}}(\nu) = 2\pi\hbar\nu - W. \qquad \text{[1.7, page 34]}$$

  The intercept, $-W$, depends only on the kind of metal. The stopping energy can be reexpressed in terms of an electric potential, by using the relation in the preceding point.
- *Poisson process:* A random process generates sequences of numbers ("blip times"). A Poisson process has the properties that any infinitesimal time slot from t to $t + \Delta t$ has probability $\beta\Delta t$ of containing a blip, and the number in one such slot is statistically independent of the number in any other (nonoverlapping) slot, where β is a positive number called the mean rate.

  The waiting times for this process have an Exponential distribution, with expectation β^{-1}.

---

[33] Chapter 13 will introduce field quantization.

[34] $\boxed{T2}$ Photons also carry momentum, but its magnitude is tied to the energy by $p = E/c$ (Section 1.3.3′b, page 51). There is also a property called "polarization," which is discussed in Chapter 13.

The probability of getting ℓ blips in a finite time interval of duration T_1 follows a Poisson distribution with expectation $\mu = \beta T_1$.

- *Molecular energy:* The total energy of the electrons in some state of an atom or molecule equals their potential energy in each others' fields (and those of the atomic nuclei), plus other contributions such as their kinetic energy. Section 1.6 introduced a viewpoint in which all forms of electron energy in a molecule were considered as effectively a contribution to the potential energy function for the atomic nuclei. When combined with the mutual electric repulsion energy of the nuclei, we called this function $U_\alpha(\boldsymbol{y})$; it depends on electron state α as well as on nuclear positions, collectively denoted $\boldsymbol{y}$.

- *Transparent medium:* In a medium like water, the speed of light is reduced to c/n, where n is the index of refraction, a dimensionless factor greater than 1. The wavelength formula is then $\lambda = c/(n\nu)$, but we continue to characterize light by its vacuum wavelength c/ν (or by its frequency ν).

FURTHER READING

Semipopular:
Feynman, 1985; Feynman, 1967; Breslin & Montwill, 2013. (See also Media 2.)
Historical: Stone, 2013; Pais, 1982.
Bioluminescence: Wilson & Hastings, 2013.

Intermediate:
Quantum character of light: Townsend, 2010; Greenstein & Zajonc, 2006.
Poisson processes: Nelson, 2015; Allen, 2011.
Photochemistry: Bialek, 2012; Atkins & Friedman, 2011.
Fluorescence, general: Jameson, 2014.
Fluorescent tagging: Nadeau, 2012.
Fluorescence microscopy: Cox, 2012, chapt. 3; Nadeau, 2012, §6.5.
DNA damage: Atkins & de Paula, 2011, chapt. 12; Nordlund, 2011, chapt. 11.

Technical:
Historic: Einstein's light-quantum article: Stachel, 1998.
Phototherapy for premature infants: Maisels & McDonagh, 2008.
Bioluminescence: Branchini et al., 2015; Haddock et al., 2010.
Single-molecule fluorescence methods: Hinterdorfer & van Oijen, 2009, chapts. 1, 2; Roy et al., 2008; Selvin & Ha, 2008.
Blinking of fluorophores: Stefani et al., 2009.
$\boxed{T_2}$ Optical tweezers and traps: Nelson, 2014, chapt. 6; van Mameren et al., 2011; Hinterdorfer & van Oijen, 2009, chapt. 12; Selvin & Ha, 2008; Appleyard et al., 2007.

$\boxed{T_2}$ **Track 2**

1.3.1′ Quantum randomness is distinct from classical chaos

1. Some classical dynamical systems can give effectively unpredictable results, a situation called deterministic chaos. Such a system's behavior may be "random" in the practical sense that many attempted repetitions of an experiment, with initial conditions set up identically (within the limits of the apparatus precision), can lead to final results with no discernible, relevant structure. Indeed, probability ideas are often helpful when modeling such systems; for example, Section 0.1 (page 1) used weather as an illustration of an effectively random system.

However, it's more correct to say that a chaotic dynamical system has a characteristic *time scale* beyond which we cannot make accurate predictions. If we measure the initial conditions with some level of accuracy, then we *can* predict the system's state into the future for times less than this "Lyapunov time." In contrast, quantum randomness is intrinsic and cannot be suppressed, or predicted, even over short times.

Fig. 1.3 (page 28)

2. The main text argued somewhat loosely that the steady flow of a continuous fluid will not give rise to random blips (Figure 1.3). In fact, a dripping faucet can behave chaotically, seeming to invalidate this argument. However, there is no obvious way to implement this idea with energy transport by waves.

3. Quantum physics does not say that "everything is uncertain." For example, if we prepare an initial state of an isolated system with definite energy, then that energy will remain definite (and unchanging) throughout its future development. Quantum physics does say that there are states in which the measured values of certain observable quantities are not predictable, and that such states are not only possible, but even unavoidable in some situations.

$\boxed{T_2}$ **Track 2**

1.3.3′a The reality of photons

Einstein was aware that the evidence for his model that was available in 1905 was not fully convincing. In fact, much later, it was found that the main features of the photoelectric effect can also be explained in a "semiclassical" model, in which light itself is not quantized (reviewed in Mandel & Wolf, 1995, chapt. 9).

An experiment by P. Grangier and coauthors finally provided compelling evidence for Einstein's intuition (Grangier et al., 1986). The experimenters prepared single-photon states by waiting for a single excited calcium atom to drop to its ground state. The resulting light was directed through a beam splitter. Detectors on each branch of the beam were then found never to respond simultaneously. In contrast, if light were a classical wave we could readily divert bits of it to two different detectors, where it would (at least sometimes) trigger simultaneous events in each branch. For more details, see Pearson & Jackson, 2010, and Greenstein & Zajonc, 2006, chapt. 2.

Actually, exotic materials do exist that can split a single photon into two, via an effect called spontaneous parametric down-conversion (SPDC).[35] The resulting two photons can then simultaneously arrive at, and excite, two detectors. But ordinary

---

[35] Section 9.4.3 (page 305) gives an application of SPDC to biology.

partial reflection does not produce this effect. Moreover, unlike in the classical model, the two emerging photons have frequencies different from the incoming one: By energy conservation and the Einstein relation, their frequencies must sum to the original photon's value. Finally, SPDC crystals typically split only one out of a million incident photons, leaving the rest unaffected.

1.3.3′b Light also carries momentum

Einstein proposed that light be regarded as consisting of lumps. We have seen that, in some ways, that description is reasonable:

- Light can travel through empty space at a huge but finite speed.
- Light interacts in a highly localized way with matter.
- Any projectile carries momentum as well as energy, and indeed, J. C. Maxwell's older theory of electromagnetic radiation already said something about this.

About that last point, Maxwell's classical wave theory quantitatively predicted that a beam of light transports momentum per time per area equal to its energy per time per area divided by c. This momentum transport gives rise to a "radiation pressure" effect; it had been experimentally observed prior to Einstein's work.[36]

But the notion of light as a "particle" had many unappealing aspects as well, including

    a. Unlike sand grains or electrons, light in vacuum always travels at a single universal speed.

    b. Nevertheless, the energy of a photon is variable, not the constant $\frac{1}{2}m_{\text{photon}}c^2$, as Newtonian particle mechanics might lead us to believe.

    c. Newton's formulas for energy, $E = \frac{1}{2}mv^2$, and momentum, $\boldsymbol{p} = m\boldsymbol{v}$, imply $E = vp/2$, which disagrees the electrodynamic result given above by a factor of 2.

Einstein was able to see through these paradoxes because at the same time as his light-quantum work, he was also creating the theory of relativity. He proposed that the Newtonian formulas were approximate, valid only in the realm of objects moving with speeds much slower than c, and that more generally they should be replaced by

$$\boldsymbol{p} = \frac{m\boldsymbol{v}}{\sqrt{1 - (v/c)^2}} \quad \text{and} \quad E = \frac{mc^2}{\sqrt{1 - (v/c)^2}}. \tag{1.16}$$

For a particle that moves with $v \ll c$, the first formula reduces to Newton's $m\boldsymbol{v}$, while the second reduces to a constant plus Newton's $\frac{1}{2}mv^2$. For velocities approaching that of light, both formulas seem to be divergent. Einstein realized, however, that there was an intriguing loophole: If the mass m is sent to *zero* while v approaches c, then E and $\boldsymbol{p}$ can both exist, with the relation

$$p_{\text{photon}} = E_{\text{photon}}/c, \tag{1.17}$$

in agreement with the electrodynamic result and hence addressing point **c** above. Indeed, a massless particle has *no choice* but to move at speed c: Otherwise, its energy and momentum would both vanish. That observation addresses point **a** above. Finally, we can take the limit in a way that gives E any value we like, addressing point **b**.

---

[36]Chapter 13 gives a route to this result. P. Debye later used the concept to explain why comet tails always point away from the Sun.

For other kinds of particles, eliminating v from Equations 1.16 gives

$$E = \sqrt{m^2 c^4 + p^2 c^2}.\qquad(1.18)$$

Experiments on the collisions of individual x-ray photons with electrons confirmed that momentum is conserved if we take the photon's and electron's momenta to be connected to energy via Equations 1.17 and 1.18, respectively.

The photon hypothesis, combined with momentum conservation, thus imply that, when a macroscopic object either absorbs or emits a lump of light, it must also change its momentum. If photons are absorbed or scattered at some mean rate, then the object will receive momentum at that mean rate times E_{photon}/c; in other words, it will experience a *force*. The biophysical relevance of these remarks comes when we seek ways to apply precisely controlled forces to single molecules (for example, a motor protein). A small, transparent object that bends a beam of light can be attached to the molecule in question by using an antibody linker. Typically this object is a micrometer-scale sphere of plastic, which acts as a lens. Deflecting the photons in a beam of light changes their momentum, because momentum is a vector. That change of photon momentum implies a compensating change in the plastic sphere; the resulting force (rate of momentum transfer) can be adjusted to the piconewton range. This **optical tweezers** apparatus has proven to be versatile, enabling delicate measurements on single biomolecules (Nelson, 2014, chapt. 6; van Mameren et al., 2011).

1.3.3′c The thermal radiation spectrum

A key motivation for the Light Hypothesis was the outstanding mystery in 1905 of the spectrum of light from a hot object (or more precisely, the light filling a cavity inside such an object).[37] This spectrum had recently been measured experimentally, as a function of frequency and the object's temperature, but by 1905 its high-frequency limiting behavior still defied theoretical explanation. Besides its fundamental importance, such **thermal radiation** also appears in the living world: Light from the Sun, with which we see, has a spectrum of roughly this form.

The experimental result can be expressed by stating the energy density du (energy per volume) in a cavity, carried by photons with frequencies in a certain range, $d\nu$. Max Planck found that the experimental data could be fit to a function of this form:

$$du = \frac{16\pi^2 \hbar \nu^3}{c^3}\left(e^{2\pi\hbar\nu/(k_{\mathrm{B}}T)} - 1\right)^{-1}d\nu.\qquad(1.19)$$

Here $k_{\mathrm{B}}T$ denotes the product of the absolute temperature times the Boltzmann constant.[38] Planck's formula cannot be understood in terms of classical electrodynamics combined with classical statistical physics; we will now see how it emerges as a consequence of Einstein's photon hypothesis.

We begin by focusing attention only on photons located in one region of space $d^3\boldsymbol{r}$, with momenta only in a range $d^3\boldsymbol{p}$. (Other photons not in these ranges are statistically independent of these ones, and make additive contributions to the mean energy density.) We need one crucial idea not yet known to Einstein in 1905, which is that all photons with a particular momentum, in a particular location in space, are *indistinguishable*.[39] That is, a state is completely specified as soon as we give the *numbers* of photons present in each such range $d^3\boldsymbol{r}d^3\boldsymbol{p}$.

---

[37]Section B.4 gives more details about thermal radiation. Some authors call it "blackbody radiation."
[38]Section 0.6 (page 17) introduced this constant.
[39]Einstein only arrived at this idea in 1924, starting from a suggestion by S. Bose. To see how

Photons can be emitted or absorbed by atoms in the walls of the surrounding chamber.[40] The energy E of those photons is a random variable; we want its expectation. We cannot make direct use of the classical result in Section 0.6 (page 17), because giving each of the infinitely many photon states the same energy would give the vacuum an unphysical, infinite energy content (and specific heat). However, a related statement from classical physics does make sense and continues to hold in the quantum world:

> At absolute temperature T, a system can occupy any of its allowed states. The probability of occupying a state of energy E depends on temperature   (1.20) as $e^{-E/k_B T}$ times a normalization constant.

Thus, the expectation we seek is the sum of terms corresponding to states with $n = 0$, 1, ... photons present. Each state has energy $E_n = (2\pi\hbar\nu)n$, where Equations 1.17 and 1.6 (page 34) give the frequency as $\nu = \|\boldsymbol{p}\|c/(2\pi\hbar)$. Each state has occupation probability given by the Boltzmann factor (Idea 1.20) divided by an overall constant (independent of n) for normalization:

$$\langle E\rangle_{\boldsymbol{r},\boldsymbol{p}} = \frac{E_0 e^{-E_0/k_B T} + E_1 e^{-E_1/k_B T} + \cdots}{e^{-E_0/k_B T} + e^{-E_1/k_B T} + \cdots}.$$

The denominator of this expression can be evaluated by using the geometric series formula (page 18). To evaluate the numerator, we can write it as a derivative:

$$\sum_{n=0}^{\infty} nE_1 e^{-nE_1/k_B T} = -\frac{\mathrm{d}}{\mathrm{d}\beta} \sum_{n=0}^{\infty} e^{-n\beta E_1}\Big|_{\beta=1/(k_B T)} = -\frac{\mathrm{d}}{\mathrm{d}\beta}\left(1 - e^{-\beta E_1}\right)^{-1}\Big|_{\beta=1/(k_B T)}$$

$$\langle E\rangle_{\boldsymbol{r},\boldsymbol{p}} = 2\pi\hbar\nu\left(e^{2\pi\hbar\nu/(k_B T)} - 1\right)^{-1}.$$

To arrive at the total energy density, we integrate this expression over $\mathrm{d}^3\boldsymbol{r}\,\mathrm{d}^3\boldsymbol{p}/(2\pi\hbar)^3$ (the $\hbar$ factors are needed to give a dimensionless quantity). Doing the integral over position yields a factor of the cavity's volume. Doing the integral over the directions of $\boldsymbol{p}$ gives a factor of 4π. There is another factor of 2 because photons have two possible polarizations.[41] Dividing by the volume to obtain energy density yields an expression that agrees with Planck's empirical formula for the experimentally observed thermal energy spectrum, Equation 1.19.

1.3.3′d The role of frequency

The main text regarded "frequency" as an arbitrary label to distinguish various kinds of light (Section 1.3.1, page 27). This attitude may leave doubts about the falsifiable content of the Einstein relation (Equation 1.6, page 34). Is ν nothing more than an abbreviation for $E_{\text{photon}}/(2\pi\hbar)$? But Section 1.3.3 hinted (and Chapter 4 will show) that ν can be regarded as the frequency that, in the classical model of light, would reproduce the interference behavior actually observed with a given source. With this definition, both sides of the Einstein relation are independently defined, so it does makes falsifiable predictions.

Einstein reasoned from his partial understanding in 1905, see Stachel, 1998, Pais, 1982, and Stone, 2013. Section 13.5.1 (page 403) gives a framework in which indistinguishability arises automatically.
[40]This statement is part of the Light Hypothesis part 1, Idea 1.11 (page 37).
[41]Section 13.5.1 (page 403) will discuss polarization in detail.

Moreover, in some situations, photon frequency can be measured more directly. For example, the radio band of the light spectrum involves frequencies much lower than that of visible light, so a lab instrument can simply count wave crests in a particular time window to get the frequency. The energy per photon is correspondingly small, but nevertheless certain nuclear spin transitions have energy splittings in this range, which indeed are found to be induced only by radiation with the frequency predicted by the Einstein relation. This observation ("nuclear magnetic resonance" or NMR) forms the basis for magnetic resonance imaging, which has become both an indispensable medical tool and a method to determine molecular structure.

T_2 **Track 2**

1.4′ Corrections to Poisson emission and detection

The main text pointed out that, when a light source of uniform intensity illuminates a detector, individual photon detection events follow a random process. The data shown were consistent with a Poisson process, and indeed this is usually an excellent approximation. For example, the distribution of photon counts in a fixed time interval follows Poisson statistics exactly for continuous-wave laser light. An incandescent source, such as the Sun, has a characteristic time scale called its "coherence time." When we count detector blips over a time window that is much longer than the source's coherence time, then again the counts are Poisson distributed.

Moreover, photon counts in different time windows separated by an interval τ are mutually uncorrelated for laser light, and also for sunlight when τ exceeds the coherence time. (For a discussion see Loudon, 2000, chapt. 6.) Matters are different, however, for more exotic light sources, for example, emission by a single fluorophore or cathode-ray tube.

The distribution of photon detection events depends on the detector as well as on the source. Many detectors have a significant "dead time" after registering a blip, during which they are insensitive to other photons, cutting off the distribution of observed waiting times. Some detector types can give artifacts such as random multiplicative factors, afterpulsing, and so on.

T_2 **Track 2**

1.5.1′a Gamma rays

Even before the invention of ultrasensitive electronic detectors for visible light, it was well known that some radioactive substances give off rays that behave like light ("gamma rays"), and that generate discrete blips in radiation detectors. Like the blips from a photomultiplier tube or avalanche photodiode, gamma-ray blips were also found to arrive in a Poisson process; they're easier to detect individually than visible photons because each carries much more energy.

In positron emission tomography (PET) scanning, a radioactive element is bound into a sugar molecule (typically fluoro-deoxy-glucose, made with the radioactive isotope fluorine-18), which accumulates in tissue wherever metabolism is intense. This particular kind of radioactivity involves the nucleus emitting a "positron" (the anti-matter counterpart of an electron). Before traveling very far, that positron finds an ordinary electron and they annihilate each other, giving off two photons in the gamma part of the spectrum.[42] The two gamma photons emerge back-to-back, by conservation of momentum. By detecting them both, the PET scan can therefore determine a line along which the original fluorine-18 nucleus was located. After enough such photon pairs have been detected, it becomes possible to infer a three-dimensional map of metabolic activity in the organ being studied, for example, a living patient's brain.

1.5.1'b More about the Light Hypothesis

1. The Light Hypothesis in the main text only attributes one property to a photon: its energy (or equivalently its frequency or vacuum wavelength). Section 1.3.3'b (page 51) also mentioned momentum, but said that it was determined once we state the photon's energy and direction of motion. In fact, however, one additional distinction can be made between photons of a given energy and direction: They can have either of two polarizations.[43] We will neglect this aspect of light until Chapter 13, because our eyes do not seem to detect polarization.[44] Light from special sources, like lasers, can have additional characteristics, for example, the degree of coherence *among* the photons, which lie beyond the scope of this book.

2. The Light Hypothesis states that photons interact with electrons. Photons also interact in the same way with any charged particle, for example, a proton or the positron mentioned in (a) above. For many purposes, however, we can neglect interactions with protons because their effects are suppressed by the proton's high mass.

3. The Light Hypothesis refers to photon absorption as involving a single electron. In fact, a free (isolated) electron cannot absorb a photon without reemission, because there is no way for that process to conserve energy and momentum. However, an *atom* or molecule can absorb photons. The atom or molecule recoils to conserve momentum, but is so much more massive than the electron that its recoil absorbs very little energy; thus, it is still a good approximation to say the photon "transfers all its energy" to the electron.

4. Even a free electron can absorb a photon, exist briefly in a state with energy and momentum values that are not normally allowed, then reemit a new photon with a different frequency from the initial one ("Compton scattering"). The difference in incoming and outgoing photon energies shows up in net kinetic energy lost or gained by the electron.

$\boxed{T_2}$ **Track 2**

1.5.3' Mechanism of DNA photodamage

Ultraviolet photons damage the DNA molecules of living organisms in various ways.

---

[42]Section 13.6.4 (page 409) will return to the creation and destruction of electrons and positrons.

[43]By a linear change of basis, we can equivalently speak of two "spin angular momentum" states for a photon. Electrons, too, are indistinguishable, apart from having two polarization states ("electron spin").

[44]Some other animals can detect polarization; see Section 13.7.2.

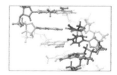

Fig. 1.7b (page 40)

In one common photodamage event (caused by shorter-wavelength UV light), two adjacent thymine bases bond covalently with each other, instead of forming hydrogen bonds to their partners on the other chain of the double helix (a "thymine dimer," Figure 1.7). But UV light below the threshold for such direct DNA damage can nevertheless induce "indirect damage," by exciting another molecule that in turn generates an excited state of an oxygen molecule (a free radical); the oxygen in turn can damage DNA.

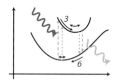

Fig. 1.9 (page 42)

T_2 **Track 2**

1.6.1′ Dense media

1. In dense media, such as water solution, we must extend the definition of the reaction coordinate y to include also the state of the surrounding water molecules. For example, those molecules can align, setting up a local electric field, which in turn can interact with the molecule of interest via the molecule's dipole moment.

2. The total energy functions $U(y)$ in diagrams like Figure 1.9 should also be interpreted as *free* energy functions, to acknowledge the effect of entropy changes in the surroundings that accompany rearrangement in the one molecule under study.[45]

3. Finally, the main text simplified by neglecting the kinetic energy of nuclear motion. In dense media, this sometimes makes sense (friction from surrounding molecules dominates inertial effects). However, for an isolated molecule in vacuum, and sometimes even in solution, the "vibrational spectrum" is significant and must be taken into account; see Section 1.6.3′b.

T_2 **Track 2**

1.6.2′a More about atoms and light

Section 1.6.2 claimed that, for an atom to revert spontaneously to its ground state, it must lose a sharply defined amount of energy, the difference between the two discrete energy levels. Two finer points must be mentioned.

1. The difference in energy levels may be slightly different from the energy of the emitted photon, because of the uncertainty relation for energy: Extremely short-lived states will have significant uncertainty, and therefore an intrinsic width to their emission spectrum. But for the long-lived states relevant for fluorescence, this source of spectral width is negligible compared to the broadening mechanism discussed in the main text.

2. In principle, the energy difference between levels could be shared between two or more photons of longer wavelength. Such processes are usually much less probable than single-photon emission, but nevertheless they lie at the heart of two-photon microscopy (Section 2.7, page 79) and spontaneous parametric down-conversion (Section 1.3.3′a, page 50).

---

[45]Free energy is discussed, for example, in Dill & Bromberg, 2010, and Nelson, 2014, chapt. 6.

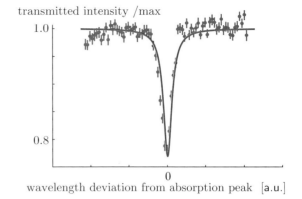

transmitted intensity /max

Figure 1.13: [Experimental data with fit.] **Gamma-ray absorption spectrum of iron-57.** Error bars reflect the standard deviation of count data (Section 0.5.3, page 16). See also Problem 1.5. The curve shows a constant minus a Cauchy distribution. [Data from Ruby & Bolef, 1960 (Dataset 3).]

wavelength deviation from absorption peak [a.u.]

1.6.2′b A Cauchy distribution in physics

Section 1.6.2 pointed out that the probability for an atom to absorb a photon depends on the photon's wavelength, and exhibits a *peak* at some optimal value. Similarly, the photons emitted by an atom as it transitions from an excited to the ground state will have a range of wavelengths, with a probability density function called the emission spectrum. Several real-world complications alter the apparent shape of atomic emission and absorption lines, but analogous spectra can be accurately obtained for some *nuclear* transitions. Figure 1.13 shows the intensity of gamma-ray light passing through (not absorbed by) a sample of iron-57. A monochromatic emitter was set in motion relative to the sample; the resulting Doppler shift effectively swept it through a narrow range of wavelengths. The solid curve shows a fit to a Cauchy distribution (Equation 0.40, page 13).

T_2 **Track 2**

1.6.3′a Born-Oppenheimer approximation

The main text subdivided a molecule into two sets of variables, describing electrons and nuclei respectively, and proposed that

- The nuclei set up an electrostatic potential energy landscape for the electrons, whose energy levels are then quantized as if the nuclei were fixed in space.
- The electronic energy levels, in turn, create an effective potential energy landscape for the slower motion of the nuclei.

The validity of this "Born–Oppenheimer approximation" (subdividing the system and solving one sector at a time) rests on the fact that an electron has much lower mass than any nucleus, and hence moves much more rapidly. The formal justification can be found in books on molecular quantum mechanics (for example, Atkins & Friedman, 2011; Atkins & de Paula, 2011). In the context of photophysics, this approximation amounts to saying that absorption of a photon instantaneously promotes the electrons to an excited state, and that during this event we may neglect nuclear motions (the "Franck-Condon principle"). Thus, the transitions in Figures 1.9 and 1.12 are shown as vertical lines; horizontal displacement in the figure corresponds to nuclear motion.

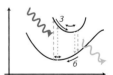

Fig. 1.9 (page 42)

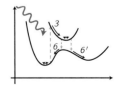

Fig. 1.12 (page 46)

1.6.3′b Classical approximation for nuclear motion

The main text described electrons in atoms and molecules as being subject to the Electron State Hypothesis (Idea 1.12, page 40), yet seemed to treat nuclei as classical objects, for example, "sliding with friction" on an effective potential energy landscape (arrows labeled *3*, *6*, and *6′* in Figures 1.9 and 1.12).

In fact, nuclei are subject to the same quantum physics as electrons; for example, their motions give rise to quantized energy levels in an isolated molecule, which can be measured by methods such as Raman spectroscopy. However, in a crowded environment such as water solution, a fluorophore is *not* isolated. Its constant interaction with neighboring molecules washes out fine structure in its energy levels and destroys its quantum-mechanical coherence on a characteristic time scale shorter than the one for fluorescence.[46] In such situations, we may treat the fluorophore's nuclear motion as effectively classical, which helps clarify key features of fluorescence such as the Stokes shift. This approximation breaks down for some ultrafast processes, apparently including those involved in photosynthesis and the photoisomerization of the visual pigment retinal.[47] We will continue, however, to treat it as a useful approximate guide.

1.6.3′c Debye relaxation

The main text mentioned one complication connected with fluorescence in solution: a "friction" effect that drains away energy from a molecule if its configuration is not close to the one minimizing the effective potential energy function. The origin of this friction was imagined as literal collisions with surrounding water molecules. A more realistic picture also acknowledges the electrostatic coupling between the dipole moment of the excited molecule and those of the highly polar water molecules in its neighborhood (Section 1.6.1′). The loss of energy from a molecule's internal degrees of freedom into the surroundings by this interaction is called "Debye relaxation."

$\boxed{T_2}$ **Track 2**

1.6.4′ Fast conformational changes

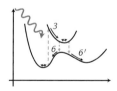

Fig. 1.12 (page 46)

The mechanism envisioned in Figure 1.12 for photoisomerization is in some cases an oversimplification. Particularly for molecules with very fast conformation changes, the excited- and ground-state energy curves can approach very close to each other near $\bar{y}_\star$. The corresponding electron energy states can then mix quantum-mechanically, giving a very rapid, nonradiative route to the electronic ground state (Bialek, 2012, chapt. 2).

---

[46]Chapter 14 will discuss the role of decoherence in FRET.
[47]Section 10.4.1 will introduce the photoisomerization of retinal. For its ultrafast dynamics, see Section 1.6.4′ below and Bialek, 2012.

PROBLEMS

1.1 *Thump*

Newton imagined light as a stream of tiny material particles obeying the same sort of laws as ordinary matter. Benjamin Franklin objected to this model; in 1752 he wrote in a letter "Must not the smallest particle conceivable, have with such a motion, a force exceeding that of a [cannonball]?" Suppose that a tiny particle, weighing just a picogram, could be brought up to the speed of light. Evaluate the Newtonian kinetic energy formula, $\frac{1}{2}mv^2$, for this particle, and comment on Franklin's assertion. Then compare your result to Equation 1.6, with frequency appropriate to visible light.

1.2 *Count distribution*

Prove Equation 1.9.

1.3 *Photon shot noise*

The audio file Media 1 represents a time series, the output of a light detector at very low illumination level, recorded for five seconds. Dataset 1 gives the same information numerically; it contains the arrival times of each individual photon blip.

a. Obtain Dataset 1. (See page xx for how to obtain the datasets.) The first column gives the arrival times of 290 blips, in units of 50 ns. Convert the times to seconds.

b. Convert to a list giving waiting times (time intervals between successive events).

c. Make a convenient choice of time bins and make a histogram of waiting times. Then draw another graph showing the probabilities for each of your bins. Make a third graph showing an estimate of the probability density function. Does your answer resemble one of the families of explicit PDFs in this book's Prologue? If so, add a second curve to your graph showing a member of that family that appears to match well.

d. Find the sample mean of the waiting times. Convert your list to a string of zeros and ones, where $s_i = 0$ if waiting time #i is smaller than the sample mean, and $s_i = 1$ if it's larger.

e. Find the probability distribution $\mathcal{P}(s_i)$, which is just two numbers because s_i takes only two values. Find the joint probability distribution $\mathcal{P}_{\mathrm{joint}}(s_i, s_{i+1})$ of neighboring pairs, which consists of four numbers. Comment on whether the successive waiting times appear to be statistically independent.

f. What else might you want to check to see if successive wait times are independent?

g. Go back to the original list of arrival times. Divide the 5-second total recording time into bins of width 0.1 s. Find how many events happened in each bin. Do a simple calculation to see if this time series might have come from a Poisson process, and comment.

h. Repeat (g) with bins of width 0.05 s and comment on how the sample mean and variance change.

i. The dataset has 290 blips in five seconds, or a mean rate of $58\,\mathrm{s}^{-1}$. Re-express your answers to (g, h) as follows: From the counts in one bin, you can make an estimate of the average rate of events in that bin. Which bin size, 0.1 or 0.05 s, gives a smaller spread of values around the "true" value, and why?

1.4 *Extreme sensitivity*

If you haven't done Problem 0.7 (touch sensitivity of skin), do it before this problem.

a. A source creates monochromatic light with vacuum wavelength 550 nm. How much energy does *one* photon of this light transfer to an electron when it is absorbed?

b. Chapter 9 will present evidence that your eyes can detect a flash of the light described in (a) consisting of only 100 photons. Assuming that this claim is true, how much more sensitive must your eyes be than your skin?

1.5 $\boxed{T_2}$ *Spectral line shape*

Obtain Dataset 3. Find a Cauchy distribution that gives a good fit to these data (see Section 1.6.2′b, page 57).

CHAPTER 2

Photons and Life

> Nor is it of much Importance to us to know the Manner in which Nature executes her Laws 'tis enough, if we know the Laws themselves. 'Tis of real Use to know, that China left in the Air unsupported, will fall and break; but how it comes to fall, and why it breaks, are Matters of Speculation. 'Tis a Pleasure indeed to know them, but we can preserve our China without it.
>
> — *Benjamin Franklin to Peter Collinson*

2.1 SIGNPOST: *SEEING AND TOUCHING*

Chapter 1 introduced a number of phenomena, for example, fluorescence, that show the lumpy aspects of light and of molecular energy levels. We found that many disparate phenomena can be regarded as consequences of just two hypotheses, which we called the Light Hypothesis and Electron State Hypothesis. We are now ready to understand many biological examples of the Light Hypothesis at work, including some new phenomena that, once again, fit better with a particulate than a wavelike physical model of light. These phenomena have led to indispensable lab techniques that allow us to reach into living cells to view, and even control, their internal processes. Such light-based techniques have revolutionized life science by offering precisely targeted ways to *see and touch.*

The Focus Question is

Biological question: How can you see tiny intramolecular motions using ordinary light?

Physical idea: Fluorescence resonance energy transfer lets us monitor the distance between two points on a macromolecule, in real time.

2.2 LIGHT-INDUCED DNA DAMAGE

Section 1.6.1 pointed out that each molecular species has a characteristic light absorption spectrum. This realization led F. Gates to a remarkably prescient experiment. As early as 1868, J. Miescher had obtained a molecule now called DNA from cell-nucleus extracts, but for a long time the role of this substance was not clear. Certainly a cell cannot live indefinitely without its nucleus, but the nucleus contains many other molecules as well, notably proteins. The importance of DNA was far from clear.[1]

Gates reasoned that the ability of ultraviolet light to kill bacteria might arise from photodamage to some molecule critical for life. The fact that visible light was not effective was therefore a clue to the identity of the most critical molecule for cellular

---

[1] The discovery of the key role of DNA is often attributed to an experiment by Avery, McLeod, and McCarty in 1944. But Gates had already made his own indirect determination, and interpreted it, 16 years earlier.

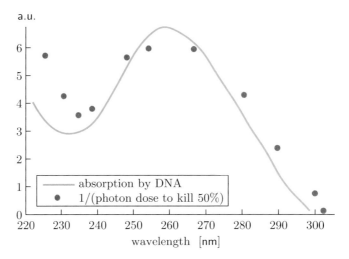

Figure 2.1: [Experimental data.] **Gates's demonstration that DNA is an essential molecule for cell survival.** *Circles:* The reciprocal of the number of ultraviolet photons per unit area required for the destruction of 50% of a colony of *Staphylococcus aureus*, as a function of wavelength of the light. This function is also called the "action spectrum" for cell death. [Data from Gates, 1930.] *Curve:* Light absorption spectrum (probability that a molecule will absorb an incoming photon) of bacterial DNA in the region studied by Gates. [Data from van Holde et al., 2006.]

function. To get more details, Gates decided to measure the dose of monochromatic light at each wavelength that sufficed to kill 50% of the bacteria in a sample. This function's reciprocal is called the **action spectrum** for that outcome (Figure 2.1). If the crucial molecular species absorbed light strongly at a particular wavelength, he continued, then fewer photons would be required at that wavelength to kill cells by photodamaging that molecule. Thus, the action spectrum should resemble the absorption spectrum of the key molecule.

Your Turn 2A

Why?

Gates's results were clear (Figure 2.1). As he put it, "The close correspondence between the curve of bactericidal action and the curves of absorption of UV energy by [DNA bases]... point[s] to these substances as essential elements in growth and reproduction." Particularly important, the *peak* of the action spectrum matched that of DNA absorption more closely than the absorption spectra of most proteins, pointing to DNA as the critical molecule whose damage can kill a cell.

2.3 FLUORESCENCE AS A WINDOW INTO CELLS

2.3.1 Fluorescence can discriminate healthy from diseased tissue during surgery

Autofluorescence

Lung cancer is the most common cancer in the western world. Generally it develops as lesions in the airways (bronchi), specifically in the boundary layer of cells (epithelium) separating the airways from interior tissue. The transformation of healthy epithelium

Figure 2.2: [Experimental data.] **Autofluorescence spectra of human bronchial tissue.** The curves show fluorescence spectra from healthy tissue (*dashed curve*) and early-cancer lesions (dysplasia and carcinoma in situ, *solid curves*). The different tissue types have different concentrations of naturally occurring fluorophores, giving rise to the observed spectral difference. Spectral photon arrival rates were measured using 405 nm wavelength (blue-violet) excitation light; all are scaled in the same way to allow comparison. Tissue types were independently determined by traditional pathology tests. [Data courtesy Georges Wagnières; see also Zellweger et al., 2001, and Media 4.]

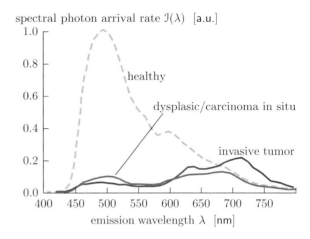

to cancer takes place in several steps, starting with a precancerous lesion (dysplasia) followed by an early stage (carcinoma in situ) before becoming invasive and disrupting surrounding tissue. Discovering and treating cancer in its early stages can greatly improve a patient's prognosis, but it is difficult: Even the advent of fiber-optic probes[2] (endoscopes) that let us see deep inside the lung was not helpful, because dysplasias and carcinomas in situ are visually indistinguishable from healthy tissue.

However, there are systematic differences between healthy and precancerous tissue when viewed in fluorescence mode. In some cases, the fluorescence is intrinsic to the natural molecules present in the tissue, so it is called **autofluorescence** ("self-fluorescence"). Figure 2.2 shows such a signature in the spectra of light emitted from various tissue types when they are illuminated with light of wavelength 405 nm. Noncancerous changes to the tissue, such as mechanical injury, were found to have little effect. Precancerous lesions, however, had major spectral differences that could even be assessed by eye.[3] Because the same endoscope can also perform surgery, a single instrument can be used to find and remove such lesions before they have a chance to develop further. It can also be used for followup screening after surgical removal of a more obvious tumor.

Induced fluorescence

Not all cancers have a useful autofluorescence signature. For example, bladder cancer is also reachable by endoscope, but its early stages look the same as healthy tissue, both under ordinary white-light illumination (Figure 2.3, left panel) and under fluorescence excitation. Nevertheless, precancerous lesions are metabolically distinct from their healthy neighbors.

One such difference lies in the cells' abilities to convert a precursor molecule called[4] 5-ALA into a fluorophore from a class called photoactive porphyrins. When the interior of the bladder is filled with a solution containing 5-ALA, some of that molecule is taken up into the cells, is metabolized, and temporarily results in an increase of fluorescence, especially in precancerous tissue. The endoscope illuminates

---

[2]Fiber optics will be discussed in Section 5.3.4.

[3] $\boxed{T_2}$ See also Section 3.8.4′a (page 135).

[4] $\boxed{T_2}$ The abbreviation stands for 5-aminolevulinic acid. Other derivatives of this molecule are also used.

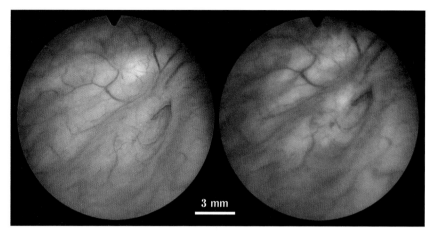

Figure 2.3: [Endoscopic images.] **Fluorescence detection of cancer.** *Left:* Human bladder wall observed after administration of a molecule related to 5-ALA, under white-light illumination. We see mainly scattered light, which swamps any fluorescence that may be present. Precancerous lesions are present but invisible. *Right:* The same site illuminated with 410 nm light. A filter has removed most of the scattered blue light, highlighting any red fluorescence that may be emitted. Two small early-stage cancers (carcinoma in situ) are clearly glowing red, due to fluorescence from a metabolic product of 5-ALA. [Courtesy Patrice Jichlinski, CHUV University Hospital; see also Wagnières et al., 2014, and Media 4.]

the interior of the bladder with blue light. A filter removes most of the blue in the resulting image, so that any red fluorescence is noticeable (Figure 2.3, right panel). When a patient undergoes surgery for a known tumor, this diagnostic tool can be used to examine the boundaries of the excised region, confirming that all the cancer cells are gone. It can also be used to find and eliminate any additional satellite colonies of precancerous cells that may also be present, improving long-term patient outcomes.

2.3.2 Fluorescence microscopy can reduce background and specifically show only objects of interest

A big part of the art of microscopy is making the objects of interest to us in a sample visible, while *not* showing the things that don't interest us. Historically, one approach to this problem involved the discovery of various staining techniques, which we can think of as attaching colored molecules (**chromophores**) to the objects of interest. But many other objects also scatter light along with the stained structures, and some of that light will coincide with the color of the stain.

Using fluorophores in place of chromophores can give a big improvement in the situation, by virtue of the Stokes shift. Section 1.2 mentioned filters that absorb photons of some wavelengths while transmitting others. We can imagine taking an ordinary (transmission) light microscope and installing two such filters:

- One filter, placed in front of the light source, transmits light in a narrow range of wavelengths lying within the excitation band of the fluorescent molecule of interest.
- Another filter, placed somewhere before the eyepiece, then blocks all light except for a *different* narrow band, corresponding to the peak of the emission spectrum. Because emission is Stokes-shifted, the image seen through the eyepiece will mostly consist of light emitted by fluorophores.

Figure 2.4: [Schematic.] **Epifluorescence (wide-field) microscopy.** White light from a source (*1*) passes through an "excitation filter" that blocks most long-wavelength photons. The remaining short-wavelength light (*2*) reflects off a dichroic mirror and enters the objective lens of a microscope, arriving at a specimen (*3*). Fluorescence emission from the specimen emerges in all directions (*4*), including back toward the microscope. The same objective lens captures some of this light, which passes freely through the dichroic mirror, eventually forming an image in a camera or observer's eye (*5*). Even if a small amount of long wavelength light gets through the excitation filter, it passes through the mirror and does not corrupt the image (*6*). Similarly, some of the excitation light gets scattered back to the microscope (*7*), but most of it bounces off the mirror and returns to the source. A final "barrier filter" removes residual excitation light that gets past the mirror. The two filters and the dichroic mirror are typically packaged into a single module, a "filter cube," specifically designed for a particular fluorophore. This schematic does not attempt to depict the focusing of light by the lenses; see Figure 6.13a (page 228).

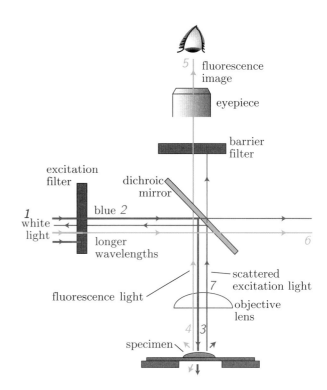

This straightforward scheme, however, turns out to be impractical. No filter is perfectly opaque at some wavelengths and perfectly transparent at others, so some of the exciting light always gets through the second filter to the observer. In order to see the light from a handful of fluorophores, we need to detect signals that are about 10^{-5} times as bright as the excitation light. Even if a little of the latter gets through, it can easily swamp the signal of interest.

J. Ploem pioneered an improved technique, now called **epifluorescence microscopy**.[5] This method involves a different kind of filter from the ones just imagined, called a **dichroic mirror**, which doesn't absorb incoming photons at all. Instead, photons are either transmitted or reflected, depending on their wavelength. Figure 2.4 shows a popular setup exploiting this technology. The epifluorescence method uses three successive means to reduce contamination of the image by stray excitation light: (*i*) Most excitation light is transmitted right through the sample, emerging out the bottom of the figure.[6] (*ii*) Excitation light scattered by the sample also emerges mainly in the downward direction, unlike fluorescence, which "forgets" the direction of the exciting light. (*iii*) The dichroic mirror and barrier filter also divert any back-scattered excitation light away from the microscope's eyepiece.

Fluorescence microscopy can also address the problem of labeling specificity mentioned above:

- Small fluorophores can be chemically attached to a particular biomolecule of interest in various ways, a procedure called fluorescent "tagging."

---

[5] Often this term is abbreviated to just "fluorescence microscopy."
[6] For thick samples, some of the excitation light is also absorbed by the sample.

Figure 2.5: [Micrograph.] **Multicolor epifluorescence image.** *Center:* An S2 cell from the fruitfly *Drosophila*, during cell division (mitosis). Different targeted fluorophores have been used to stain actin filaments, microtubules, and DNA. The light from these three classes of structures was spectrally separated by filters, leading to three images, which were then false-colored *red, green,* and *blue,* respectively, and reassembled into a single image. Other cells seen in the image are not undergoing division, and have a completely different internal organization. [Image courtesy Nico Stuurman.]

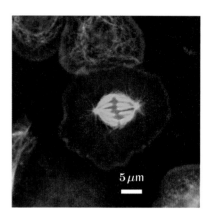

Fig. 1.11b (page 45)

- Perhaps most elegantly, some proteins are themselves fluorescent. A famous example is **green fluorescent protein**, or **GFP**, initially derived from a naturally occurring protein in the bioluminescent jellyfish *Aequorea victoria* (Figure 1.11b).[7] The genetic code for GFP can be appended to that of any other protein of interest by genetic engineering methods, creating a **fusion protein**. Every time the cell (or its offspring) manufactures a copy of that protein, it then automatically contains the fluorescent domain.

- More elaborate versions of the two-filter scheme allow separate visualization of two, or even more, different fluorescent tags, each targeted to a different class of object (Figure 2.5).

Section 2.3 has outlined some ways in which fluorescence helps us to see otherwise invisible things. Before we turn to "touching," we first review some ideas about how cells generate and use electric potentials.

$\boxed{T_2}$ *Section 3.8.4'b (page 135) says more about multicolor fluorescence.*

2.4 **BACKGROUND:** MEMBRANE POTENTIAL

Sections 2.5–2.6 will introduce light-based schemes for controlling and interrogating individual nerve cells in an intact, functioning brain. Before that discussion, this section will review some background about neural signaling.

2.4.1 Electric currents involve ion motion

You may be familiar with electric currents in wires: Some electrons enter one end and others exit the other end, while the nuclei of the metal atoms undergo no net motion. Electric currents are also possible inside cells, and in the spaces surrounding them, but they involve a more varied cast of characters. Whole atoms and molecules can migrate through such fluid solutions, but frequently they do so in a form that has been stripped of one or more electrons (such as the sodium, potassium, and calcium

---

[7]Several similar proteins have since been found with different fluorescent colors; they have names like "YFP, CFP, RFP" for yellow, cyan, and red emission bands, respectively. The palette of fluorescent proteins now spans a range of emission peaks from 424 to 637 nm.

ions, Na^+, K^+, and Ca^{++}), or that has acquired one or more extra electrons (such as the chloride ion, Cl^-). Either a net transfer of positive ions into a cell, or a net outward transfer of negative ions, is said to create an inward electric current.

2.4.2 An ion imbalance across the cell membrane can create a membrane potential

Any cell membrane constitutes a barrier to the free flow of ions. Thus, any living cell is a *capacitor,* analogous to the ones studied in first-year physics: An electrical conductor (the cell's interior fluid, or **cytosol**) is separated from another conductor (the surrounding fluid) by an insulator (the membrane). The resting electric potential is uniform throughout each conductor, but its values on either side of the insulating layer may be quite different. The difference is called the **membrane potential**.[8] A negative value for the membrane potential means that the cell interior is at lower potential than the exterior.

If a species of positive ions, say sodium, is present at greater concentration on one side of a membrane than the other, then that imbalance will generate a contribution to the electric potential that is higher on the first side. If the combined effects of all ion species do not cancel, then the resulting membrane potential tends to drive positive ions toward the side with lower potential.[9] At the same time, the ions also experience a thermodynamic force, just as molecules of the compressed gas in a balloon "want" to escape. Each ion species will leak (diffuse) slowly through the membrane in the direction determined by the net effect of the electric and thermodynamic forces, analogously to the slow loss of helium from a rubber balloon. We say that the membrane's ionic "conductance" is small but not zero, so an "electrochemical gradient" can drive an "ion flux."

2.4.3 Ion pumps maintain a resting electric potential drop across the cell membrane

If the preceding section were the whole story, then we might expect that eventually all ion concentrations would match on either side, and there would never be any jump in electric potential across a resting cell's membrane.[10] On the contrary, however, living cells maintain a nonzero membrane potential despite ion leakage, by constantly pushing ions across their membranes via **ion pumps**, molecular machines located in the cell's outer (plasma) membrane. The pumps consume molecules of adenosine triphosphate (ATP) and use the chemical bond energy stored in those molecules to drive specific ions across the membrane in specific directions. Their action overcomes the ions' tendency to leak in the opposite direction and ultimately leads to the observed steady resting membrane potential.[11]

---

[8] Another name for this quantity is "transmembrane potential," which emphasizes that it is not intrinsic to the membrane itself. Some authors call it "voltage drop" (because "voltage" is an informal name for electric potential), or "membrane polarization."

[9] As in our discussion of the photoelectric effect (Section 1.3.2), the potential energy of a sodium ion is $U = e\Phi$. Because the charge e of the ion is positive, a region of high electric potential Φ is a region with high potential energy for sodium ions, resulting in a force pushing them toward lower Φ.

[10] In the language of first-year physics, the small but nonzero conductance of the membrane means a large but finite resistance. Thus, although the relaxation time constant RC for the potential drop is large, it is not infinite.

[11] Another class of ion transporters, called exchangers, will be described in Chapter 10.

For example, T. Tomita, A. Kaneko, and coauthors found in the mid-1960s that the resting electric potential is about $40\,\text{mV}$ more negative inside a photoreceptor cell than outside. They performed this measurement by inserting a microelectrode directly into photoreceptor cells from fish eyes. For other classes of animal cells,[12] a more typical value is about $-70\,\text{mV}$.

$\boxed{T_2}$ Section 2.4.3′ (page 100) discusses additional points about equilibrium potentials.

2.4.4 Ion channels modulate the membrane potential to implement neural signaling

In addition to pumps, cell membranes also contain ion **channels**. Channels do not consume any energy; they passively allow ion flow. Many allow only ions of a specific species to pass, or a class such as all small, positively charged ions.

Some channels conduct only when they are in an "open" state. We will see in a moment that the opening of ion channels can depend on internal and external conditions. Other channels are always open and give rise to the "leak" conductances mentioned earlier.

2.4.5 Action potentials can transmit information over long distances

Section 2.4.3 described how cells arrive at a "resting" state, in which the effects of ion pumping and passage via leak conductance balance, leading to a steady membrane potential. Nerve cells, however, have found a way to exploit *changes* in membrane potential as a means of fast communication, in some cases over very long distances.

A generic nerve cell (**neuron**) has one long, thin, cylindrical projection called its **axon**. A well-understood example is the "giant" axon of the squid *Loligo forbesi*, so named because it can be up to a millimeter in diameter. The axon's membrane is studded with ion channels. One representative class of ion channels admit mainly sodium ions when they are open. They also have the property that their opening is controlled by the local membrane potential: They are closed in the neuron's resting state, but can open if the membrane potential ever rises sufficiently, that is, if the interior becomes less negative than usual compared to the exterior. We say they are **voltage gated** (Figure 2.6).

In the resting state, the neuron maintains its usual (resting) membrane potential by ion pumping, and the voltage-gated sodium channels remain closed. However, if for some reason the membrane potential at one end of the axon rises, then nearby voltage-gated channels open. Sodium ions then rush inward, because both their concentration gradient and the resting potential push them in that direction. Their influx is too rapid for the relatively slow ion pumps to remove them, so their flow increases the membrane potential still further. An increase in membrane potential toward zero (from its negative resting value) is called **depolarization**. It now spreads to nearby parts of the membrane, opening more channels and thus giving rise to a traveling *wave* of depolarization that continues all the way to the far end of the axon. A later recovery mechanism re-closes the ion channels, allowing the axon to restore its resting state, ready for another action potential. The entire sequence is called a "nerve impulse," an **action potential**, or a neural **spike**. Other animals (besides squid) use similar mechanisms, sometimes over enormous lengths; for example, individual axons stretch from a giraffe's spinal cord to its feet.

$\boxed{T_2}$ Section 2.4.5′ (page 100) mentions other uses for the resting potential.

---

[12]Many plant cells and bacteria also actively maintain membrane potentials.

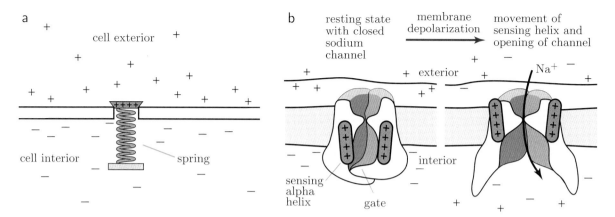

Figure 2.6: **Voltage-gated ion channel.** (a) [Conceptual model.] In the resting (polarized) state shown, external positive ions and internal negative ions create an electric field pointing downward. That field pushes on the positively charged valve, keeping it shut. If positive ions are allowed to enter the cell, however, that force is reduced and can be overcome by a spring that opens the valve. (b) [Sketch based on structural data.] A sodium channel. *Left:* In the resting state, positive charges in the channel protein's four "sensing" alpha helices are pulled downward by excess negative charges inside the cell. The sensing helices in turn pull the channel into its closed conformation. *Right:* Upon depolarization, the channel changes to its relaxed conformation. The sensing helices are linked to the rest of the protein, so when they move upward, the whole complex changes configuration, opening a channel for the passage of ions. See also Media 5. (*Not shown:* Sodium channels have an additional "channel inactivating segment" that later shuts down ion flow even though the membrane is still depolarized.) [(b) Adapted from Armstrong & Hille, 1998.]

2.4.6 Creation and utilization of action potentials

Input

The action potential is an elegant mechanism for fast communication over long distances, but to be useful, a neuron needs some means to initiate it at one end of the axon in response to an appropriate signal, and some other means to utilize the signal that arrives at the other end. Most neurons accomplish the first via another projection from the neuron called its **dendrite**. The dendrite generally splits into many branches, each studded with a class of ion channels different from the ones described in the preceding section. These **ligand-gated ion channels** respond to their chemical environment: When the right **neurotransmitter** molecule comes to them, it binds to them and induces a conformational change, opening the channel.[13] Thus, a sudden increase in neurotransmitter concentration depolarizes the dendrite's membrane. This change in membrane potential spreads throughout the dendrite and across the cell body. If enough neurotransmitters arrive closely enough in time, the resulting depolarization can trigger an action potential.

Output

The axon delivers information about neural excitation to its distant end (or multiple ends, if the axon branches). Each depolarization event at an axon terminal triggers release of neurotransmitter, the "output" of the neuron. Each axon terminal is usually

---

[13]A **ligand** is a generic term for a small molecule that sticks to (forms a complex with) another molecule. A neurotransmitter is a particular kind of ligand, associated to a particular channel type.

situated very close to another neuron's dendrite, with a gap between them (the neural **synapse**) that is only 10–20 nm wide. Neurotransmitter released from the axon terminal can quickly diffuse across the synapse to activate ligand-gated ion channels on the opposing dendrite.

Networking

Thus, most neurons respond to chemical signals (bursts of neurotransmitter molecules at the dendrite), and create similar chemical signals at their axon termini. This structure opens the way for construction of chains of neurons, each excited by its predecessor and in turn exciting its successor. Moreover, many axons can all connect to a single dendrite,[14] allowing one neuron to integrate many inputs, each with its own weighting. And one axon can branch many times, delivering its output signal to multiple other neurons' dendrites. By broadcasting each axon's signals to multiple target cells' dendrites, and integrating many inputs at each of those dendrites with specific weights, a network of neurons can perform complex computational tasks.

Alternative inputs and outputs

Even the most sophisticated computer would do us no good if it only connected to itself! So some specialized neurons have alternatives to the input scheme just mentioned: They modulate their membrane potential in response to touch, heat, light, or other sensory stimuli. Later chapters of this book will discuss light activation (**photoreception**) in detail.

Also, some neurons terminate not on other neurons, but on muscle cells; action potential-like depolarization of the muscle cell membrane triggers contraction. Branches from a single **motor neuron**'s axon can cause many muscle cells to contract. Other neurons trigger release of hormones into the blood, and so on.

2.4.7 More about synaptic transmission

Each neuron constantly synthesizes new neurotransmitter molecules, and may also scavenge used molecules from the synapse, left over from previous action potentials. The neuron packages the neurotransmitter molecules into small membrane-bounded bags (**vesicles**), which are then "docked" just inside the axon terminal, ready for release (Figure 2.7, top).

When an action potential arrives at the axon terminal, its depolarization opens another class of voltage-gated ion channels present there, which specifically allow calcium ions to pass. Intracellular calcium is normally maintained at a low concentration by ion pumps, so the action potential creates a sudden burst of intracellular calcium. That burst in turn triggers the fusion of the neurotransmitter vesicles' membranes with the outer membrane of the cell (Figure 2.7, center). Prior to this fusion, the neurotransmitters were inside a vesicle that was inside the cell; after fusion, the neurotransmitters find themselves liberated, outside the cell, and they proceed to diffuse across the synapse to meet their corresponding ligand-gated channels.[15]

In short, neurons communicate over long distances via discrete events (action potentials), which in turn trigger other discrete events (release of neurotransmitter packets).

---

[14]Purkinje cells in our brains have dendrites that integrate over 100 000 axonal inputs.
[15]Figure 10.13 (page 338) depicts another kind of neural synapse.

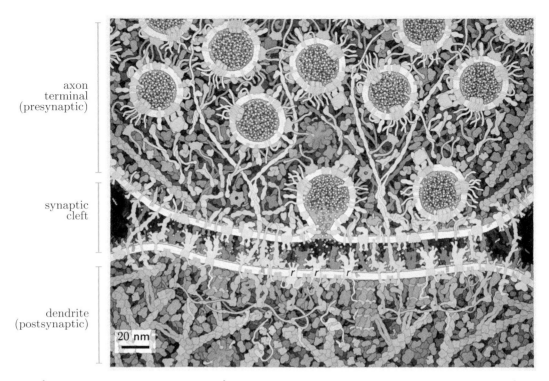

axon
terminal
(presynaptic)

synaptic
cleft

dendrite
(postsynaptic)

20 nm

Figure 2.7: [Drawing based on structural data.] **Cross section of a neural synapse.** The end of an axon (its presynaptic terminal) is shown at the top, with several synaptic vesicles full of neurotransmitter molecules (*yellow dots*) inside. One vesicle is caught in the process of fusing with the axon's outer membrane, delivering its contents into the synaptic cleft. The receiving (postsynaptic) dendrite is shown at the bottom. Neurotransmitter molecules diffusing across the cleft bind to receptor proteins (r) embedded in the dendrite's membrane. Typically these receptors are ligand-gated ion channels. Other proteins shown in blues, greens, and violet are involved in maintaining the spatial organization needed for efficient, repeatable signaling. [Art by David S Goodsell.]

Discrete release of vesicles

The discrete (integer) character of neurotransmitter release can be observed directly. Although the binding of a single neurotransmitter molecule to its target ion channel has a tiny effect on the receiving dendrite's membrane potential, nevertheless the release of an entire vesicle filled with thousands of neurotransmitters does have a measurable effect. Figure 2.8 shows that these events do come in roughly integer multiples of a basic magnitude ("quantal release"), supporting the picture described in the preceding section. That basic value is the effect of a single vesicle release.[16]

This background section has reviewed the role of electric potentials in neural signaling. With that foundation, we can now explore ways of manipulating those signals with light.

For further details on membrane potential, see the references listed at the end of this chapter.

---

[16] You'll analyze these data in Problem 2.1.

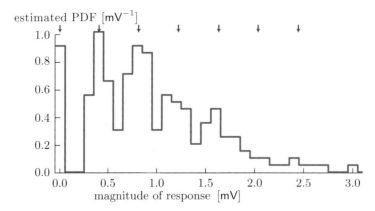

Figure 2.8: [Experimental data.] **Evidence for discrete vesicle release at a synapse.** This experiment monitored a muscle cell's membrane potential response upon receiving a single action potential (nerve impulse) from its motor neuron. The horizontal axis refers to the peak change of membrane potential, and the vertical axis gives the estimated probability density function based on 198 identical trials. In order to show more clearly the discreteness of the response, the mean number of neurotransmitter vesicles released in response to an action potential was reduced by raising the concentration of magnesium in the solution surrounding the cells. The muscle cell was itself prevented from firing any action potential; thus, its response was a proxy for the amount of neurotransmitter actually released from the axon. Bumps in the distribution of amplitudes occur at integer multiples of a basic value (*arrows*), illustrating the discrete character of the release. There is some spread in each bump, due partly to the distribution in the number of neurotransmitter molecules packaged into each vesicle. The peak at 0 mV indicates occasional failures to respond at all. See also Problem 2.1. [Data from Boyd & Martin, 1956, available in Dataset 4.]

2.5 OPTOGENETICS

2.5.1 Brains are hard to study

Section 2.4 outlined how neurons, properly connected into networks, can perform complex computational tasks. Most animals use such networks not only to interact with their environment, but even to keep different parts of their own bodies working harmoniously. It would be nice to understand how such networks work, but the obstacles are daunting. For example, a human brain contains about 10^{11} neurons, and they are not laid out in a planar arrangement like a printed electronic circuit. We can try to study the brain by observing the response (perhaps verbal) to various stimuli (perhaps visual), but there are a lot of levels between those two, and indeed after centuries of such study, our understanding remains primitive.

One way to begin might be to study a far simpler animal, perhaps a worm, but many psychiatric and other medical concerns only show up in organisms with complex brains. Another way could be to try stimulating the brain at various intermediate levels between the two endpoints mentioned above. For a long time, however, the main available method for this approach was to poke electrodes into the brain, and attempt either to pierce one neuron from a particular desired class, or to stimulate many neighboring neurons. The first method (intracellular stimulation) could be used to change the neuron's interior electric potential, either depolarizing it (raising the interior potential, and hence triggering action potentials) or hyperpolarizing it (lowering the potential, and hence suppressing action potentials that might otherwise be elicited from the cell's usual inputs). Prior to the 21st century, however, it was hard to know what class the target neuron came from, and usually impossible to choose that class in advance. The second method (extracellular stimulation) was even less specific. Moreover, the animal under study generally had to be immobilized, to avoid disturbing the electrode. Similar considerations applied to the problem of recording the activity of single neurons: Most are deeply packed among others, and hence difficult to reach with electrodes.

A powerful experimental technique was developed in the mid-2000s for controlling neural activity, based not on electrode implantation but instead on photoisomeriza-tion.[17] The story of this discovery, generically called **optogenetics**, is a case study in the value of fundamental research, because the early stages had no obvious connection with practical treatment of any disease, nor indeed anything connected with humans.

2.5.2 Channelrhodopsin can depolarize selected neurons in response to light

Many microorganisms have a primitive visual sense, which allows them to migrate to regions where light is more abundant (**phototaxis**). One example is the unicellular green alga *Chlamydomonas reinhardtii*, which expresses a protein, called **channelrhodopsin-2**, that embeds itself in a small patch of the cell's outer (plasma) membrane. Just below that patch, the cell contains an array with two or more layers of pigmented granules, which blocks light coming from behind it. The granules partly shield the light-sensitive channelrhodopsin, so that it responds to light from only certain directions. The whole complex is called the "eyespot."

Channelrhodopsin-2 consists mainly of seven helical segments that each span the cell membrane, forming a barrel structure with a channel through its center. The protein also contains an embedded small molecule (a "cofactor") called **retinal**. Blue light photoisomerizes the retinal. The resulting conformational change then pushes on the surrounding protein, distorting its shape and opening up its central channel to the passage of ions. That is, *channelrhodopsin-2 is a light-activated ion channel*. A few milliseconds later, the retinal reverts to its initial conformation and the channel closes. An entire family of related channelrhodopsin proteins is now known.

Like other cells, *Chlamydomonas* maintains a membrane potential by ion pumping. Channel opening triggered by light depolarizes the membrane everywhere, affecting other voltage-sensitive devices also embedded in the membrane and ultimately chang-ing the cell's swimming behavior (the phototactic response).[18]

Many researchers realized that, because human neurons also modulate their mem-brane potential to implement signaling, the addition of channelrhodopsin to a neuron would open the possibility of stimulating it externally, by using light. Remarkably, it proved possible to create genetically modified organisms in which the channelrhodopsin gene, although derived from a unicellular alga, was expressed in mammalian cells with-out toxic side effects. Moreover, the neurons transported it to their outer membranes, where it embedded and functioned normally. The channelrhodopsin gene can be deliv-ered to only a specific class of neurons, by using a virus that targets that class. More commonly, the gene can be added to the entire animal's genome, with appropriate control sequences that allow it to be expressed only in a desired cell type. In each case, the effect of light in the proper wavelength band is to open the channels, depolarizing the neuron just as in the natural context (*Chlamydomonas*). That depolarization can in turn trigger an action potential. Figure 2.9 shows that the response can track light flashes precisely, allowing researchers to send artificial signals ("spike trains") into the middle of a neural network with any desired time course.

---

[17]Section 1.6.4 (page 45) introduced photoisomerization.
[18]Section 10.3.3 (page 323) discusses phototaxis in simpler organisms.

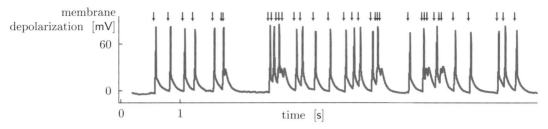

Figure 2.9: [Experimental data]. **Action potentials triggered by light.** Neurons from rat brain, not normally sensitive to light, acquired such sensitivity when a gene coding for channelrhodopsin-2 was added to their genome via a virus. In this figure, the vertical axis denotes cell depolarization, that is, the electric potential inside the cell minus that outside, compared to the resting value. The neuron was stimulated by a series of light flashes at times shown by the *arrows*. The pulses were delivered at random times drawn from a Poisson process, to simulate normal conditions inside the brain. [Data courtesy Edward S Boyden; see also Boyden et al., 2005.]

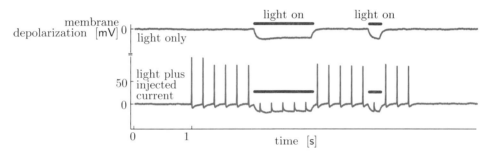

Figure 2.10: [Experimental data] **Silencing of a neuron by light.** Neurons from rat brain, not normally sensitive to light, acquired such sensitivity when a gene coding for halorhodopsin was added to their genome via a virus. In these figures, the vertical axis again denotes cell depolarization. *Top:* When the cell was resting, illumination with yellow light activated the halorhodopsin to pump chloride ions into the cell, inducing hyperpolarization. *Bottom:* The cell was electrically stimulated to elicit periodic action potentials. Illumination with yellow light eliminated the spiking response: The cell depolarized in response to stimuli, but did not reach the threshold needed to initiate a spike. [Data from Han & Boyden, 2007.]

Even greater specificity can be obtained by adding fine control of the illuminated zone, for example, by sending light to one particular brain region via a fiber-optic guide.[19] Many more examples of light-dependent ion channels are now known, and more are being engineered every day for specific properties such as high conductance, desirable action spectrum, faster opening and closing, and so on.

The effects of targeted stimulation can be dramatic and large scale. One demonstration showed that delivering light to a specific brain region of a freely moving mouse could promptly induce the animal to turn in one direction; delivering light to other regions of mouse brain promptly induced aggression or fear responses.[20]

2.5.3 Halorhodopsin can hyperpolarize selected neurons in response to light

The preceding section focused on light-induced depolarization, which can stimulate a cell to fire action potentials or (with weaker illumination) merely predispose it to fire more easily in response to its normal inputs. Artificially **hyperpolarizing** a

---

[19]Fiber optics will be discussed in Section 5.3.4.
[20]See Media 6.

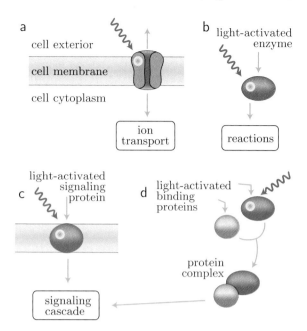

Figure 2.11: [Schematic.] **Four schemes for optogenetic control.** (a) Channelrhodopsins open in response to light, allowing ions to flow across an otherwise impermeable cell membrane and altering its membrane potential. Other microbial opsin proteins, for example, halorhodopsin, actively transport protons or other ions in response to light. (b) An enzyme within a cell can be rendered light-activated by engineering a fusion protein linking it with a light-sensitive domain. (c) A light-activated signaling protein can be expressed and coupled to an existing signaling pathway. (d) Light-driven binding of two proteins can bring them into close proximity, triggering activity. [See also Chow & Boyden, 2013.]

cell (making its interior potential even more negative than usual) should have the opposite effect. Again remarkably, another class of membrane-embedded proteins were found in archaea that serve as *light-powered ion pumps*. One such pump, known as **halorhodopsin**, normally imports chloride ions into cells of the salt-loving *Natronomas pharaonis* (initially found in the Dead Sea), helping it to maintain osmotic balance with the high-chloride exterior solution. This activity also hyperpolarizes the cell, because chloride is a negative ion. Indeed, mammalian neurons engineered to express halorhodopsin were found to be light sensitive: Exposure to light in the green-yellow range silenced such neurons, overriding the effect of injected depolarizing currents that otherwise would reliably induce spikes (Figure 2.10).

Another key feature of halorhodopsin is that its action spectrum, which peaks for wavelength in the range 525–650 nm (yellow light), is well separated from that of channelrhodopsin (blue light). Thus, light of two distinct colors can be independently used to stimulate and silence the same class of neurons.

In short,

> *Optogenetic control allows a specified class of target neurons to be activated, or silenced, by exposure to light.*

Optogenetic control enables systems neuroscientists to test causal ("A causes B") rather than correlative hypotheses ("B follows or accompanies A") about the functioning of brain cells, by up- or down-regulating specific points in a neural circuit and observing the response. The method combines the high speed of direct electrical stimulation with the precise targeting characteristic of some drug actions.

2.5.4 Other methods

Several other optogenetic control techniques have been developed (Figure 2.11).

2.6 FLUORESCENT REPORTERS CAN GIVE REAL-TIME READOUT OF CELLULAR CONDITIONS

2.6.1 Voltage-sensitive fluorescent reporters

Section 2.5 outlined light-based methods to create or repress neural signals, by introducing ion pumps and channels from other organisms. A few years after these discoveries, related progress was made in optical *readout* of neural activity via genetically encoded reporters.[21]

Again, the key step was to look for a promising membrane protein from the rhodopsin family, already naturally existing in microbes. In this case, J. Kralj and coauthors started with green-absorbing proteorhodopsin, a light-driven proton pump found in bacteria in the ocean. The researchers noted that this protein was fluorescent only when it bound a proton. Thus, they reasoned, any environmental influence that increases the concentration of protons on the side of the membrane with the proton-sensitive part of the protein (the inner, cytoplasmic side) would increase its probability to be in the fluorescent state.

Normally, we think of proton availability in chemical terms, quantified by a solution's pH value. But if the membrane is polarized, for example, by pumping positive ions out of the cell, then they will tend to "loiter" near the membrane on the outside, attracted to the negative charges left behind on the inside. Moreover, this cloud of excess exterior positive charges *repels* protons from the *inner* face of the membrane, leading to a local *decrease* in proton availability there. When the membrane depolarizes, the exterior charge is released, erasing the corresponding layer of depletion near the inner side. If the inner side of the membrane contains proteorhodopsin, then the researchers reasoned that its fluorescence should respond to this change, and indeed, they did find that depolarization enhanced its fluorescence. That result showed that proteorhodopsin could be used as a fast voltage reporter.[22]

In later work, Kralj and coauthors documented similar activity from mutants of another microbial ion-pumping protein named archaerhodopsin 3. Unlike proteorhodopsin, however, archaerhodopsin 3 and its mutants can be used in eukaryotic cells. The fact that it is not naturally present in mammalian cells allowed the researchers to target specific cell types by genetic means similar to those used for optogenetic control. Figure 2.12 shows that the resulting traces of cell activity, obtained without piercing or even touching the cell, track those of traditional methods closely.[23]

The next logical step was to *combine* optical stimulation and monitoring. D. Hochbaum and coauthors found a pair of channelrhodopsin and archaerhodopsin mutants with the property that the former's action spectrum peak, and the latter's excitation and emission bands, were well separated, allowing independent stimulation and measurement of activity in genetically targeted classes of neurons (Figure 2.13).

---

[21] $\boxed{T_2}$ Readout of membrane potential via externally introduced, voltage-sensitive dyes preceded the work described here; see Further Reading at the end of the chapter.

[22] $\boxed{T_2}$ Kralj and coauthors found that they needed to use a mutant proteorhodopsin in order for the transition between bright and dark states to occur at a membrane potential in the range of interest for their cells. As a bonus, the mutant also lacked the normal light-induced proton pumping activity, making it purely a voltage reporter.

[23] $\boxed{T_2}$ Another method involves the use of a fusion protein that combines a fluorescent protein with the voltage sensing domains of an ion channel.

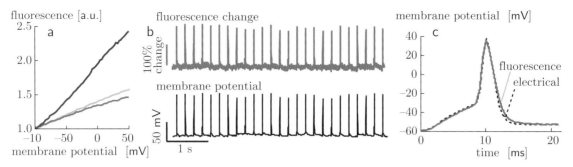

Figure 2.12: [Experimental data.] **Genetically encoded membrane potential indicators.** (a) Fluorescence versus membrane potential for three varieties of archaerhodopsin-3 expressed in human embryonic kidney cells. Each indicator gives a highly linear readout of externally imposed changes in membrane potential. (b) *Top:* Changes in observed fluorescence in a rat neuron as it fires a series of action potentials in response to electrical stimulation. *Bottom:* Simultaneous recording from the same cell made by the traditional electrode method. (c) Detail of a single action potential. *Black:* Electrical measurement. *Blue:* Fluorescence measurement, scaled to match the peak amplitude. [Data courtesy Adam E Cohen; see also Media 7a. Reprinted by permission from Macmillan Publishers Ltd: Hochbaum et al. All-optical electrophysiology in mammalian neurons using engineered microbial rhodopsins. *Nat. Methods* (2014) vol. 11 (8) pp. 825–833, ©2014.]

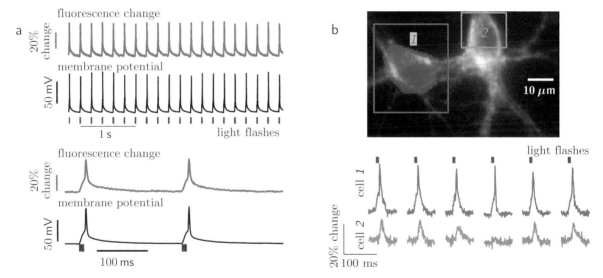

Figure 2.13: [Micrograph; experimental data.] **Optical initiation and monitoring of action potentials** in cultured rat neurons. (a) *Upper two traces: Red markers* show the times of illumination. *Blue traces* show the evoked changes in fluorescence. *Black traces* show electrical recording for comparison. *Lower two traces:* Blowup showing finer temporal structure. (b) *Top:* Two neurons joined by a synapse. The body of one neuron, highlighted in *blue,* was optically stimulated. *Bottom:* Optical recordings in both the stimulated cell (*1*) and neighboring cell (*2*) bodies. *Red bars* again indicate times of optical stimulation. Transmission failed on one of the instances shown, a normal feature of synapses. Addition of a chemical that blocks synaptic transmission abolished the response from (*2*) without affecting that of (*1*) (*not shown*). [(a) Data and (b) image courtesy Adam E Cohen; see also Media 7b. Reprinted by permission from Macmillan Publishers Ltd: Hochbaum et al. All-optical electrophysiology in mammalian neurons using engineered microbial rhodopsins. *Nat. Methods* (2014) vol. 11 (8) pp. 825–833, ©2014.]

2.6.2 Split fluorescent proteins and genetically encoded calcium indicators

Fluorescence depends on the energy levels of an entire group of atoms within a molecule (the fluorophore), not just an individual atom contained within it. Even domains of the molecule that are not directly part of the fluorophore can influence its energy levels. Thus, it is possible to modify or even abolish fluorescence by rearranging the groups within a molecule.

Split fluorescent proteins

An extreme example of the preceding idea involves designing two separate proteins, each containing just half of a complete fluorescent protein. When the proteins are in close proximity, they can find each other and bind noncovalently, temporarily creating a functional fluorescent protein. For example, the first one could be a fusion of something we wish to study, *A*, and half of a yellow fluorescent protein (YFP). The second could then be a fusion of something else, *B*, and the other half of a YFP. When *A* binds to *B*, their association brings the two YFP fragments close enough to combine, and the complex becomes fluorescent. Such **bimolecular fluorescence complementation** probes can report on where in a cell two proteins associate (and when), documenting not only the fact of association but also its correlation with externally imposed changes, internally programmed changes in cell state, and so on. That information, in turn, can lead to better models of signaling networks.

For example, K. Cole and coauthors used bimolecular fluorescence complementation to visualize the binding of Cdc42, a signaling molecule that controls cell cycle progression and other functions, to its regulator Rdi1p. They documented that in yeast, the complex of the two proteins was preferentially found at sites of polarized cell growth, for example, at incipient budding sites.

Bimolecular fluorescence complementation assays have advantages similar to those for the genetically encoded voltage indicators mentioned in Section 2.6.1: They are produced by the cells themselves, and their expression can be limited to only a class of cells that is of interest. However, they also suffer from slow response time (tens of seconds), as the two halves of the fluorescent protein do not find each other and begin to fluoresce immediately once their carriers associate.

Genetically encoded calcium indicators

Figure 2.14 shows another variation of this idea. It is not even necessary to split the fluorescent protein into two separate parts: In some cases, all that is needed to reduce or eliminate fluorescence is to *distort* it slightly. One way to do this is to rearrange the sequence of amino acids in the fluorescent protein, so that two residues that are normally covalently bound in the middle of the amino-acid chain are instead at its start and end. If those two ends are free to flop about, then they won't spend much time together, and fluorescence is lost. If, however, some other agency holds them together, then the protein adopts a form similar to its native conformation and fluoresces normally.

In Figure 2.14, such a "circularly permuted" green fluorescent protein, or cpGFP, has been fused with calmodulin, a calcium-sensing protein, attached to one of its ends. Intracellular calcium ions serve various signaling roles in cells, so it is useful to visualize their concentration in an individual cell. For example, when a neuron generates an action potential, there is usually a burst of intracellular calcium. Because the change occurs throughout the cell, and not just on its thin membrane, the resulting signals are potentially stronger than those available with membrane voltage indicators.

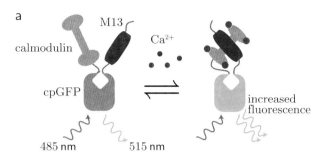

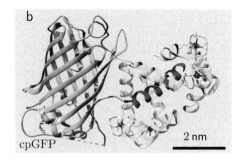

Figure 2.14: **The GCaMP sensing mechanism.** (a) [Cartoon.] A protein is expressed in a desired class of cells. It consists of three main domains. One of those domains is a circularly permuted, green fluorescent protein (cpGFP). Upon calcium binding, the calmodulin domain forms a complex with the third domain (M13), altering the local environment of the cpGFP and enhancing its fluorescence. (b) [Protein structure diagram.] Structure of calcium-bound GCaMP, showing the barrel-shaped cpGFP (*green*) sandwiched between calmodulin (*blue*) and M13 (*magenta*) domains and binding calcium ions (*red*). [(a) See also Broussard et al., 2014. (b) Courtesy Lin Tian.]

When calmodulin binds calcium ions, it undergoes a conformational change that enables it to bind to a short protein fragment,[24] which is covalently attached to the other end of the fluorescent protein. The association of calmodulin and its partner modifies the cpGFP's environment and conformation, enhancing its fluorescence whenever calcium is present in the cell. This idea forms the basis for GFP-calmodulin fusion protein (GCaMP), a genetically encoded calcium indicator. After extensive engineering, a variant called GCaMP6f was found to respond rapidly when calcium rises (150 ms), though not as fast when it falls (650 ms).

Section 2.6 has given some examples of light-based interrogation of a single cell's state. We now return to fluorescence microscopy, and describe another useful phenomenon that cannot be explained by the classical wave model of light.

2.7 TWO-PHOTON EXCITATION PERMITS IMAGING DEEP WITHIN LIVING TISSUE

2.7.1 The problem of imaging thick samples

Ordinary optical microscopy involves flooding an entire sample with light, then adjusting the microscope objective (lens) until the light scattered toward it by one particular plane of interest is in focus.[25] Light scattered from any other depth degrades the image, by adding unfocused background. One traditional way to address this difficulty is to cut the sample into extremely thin slices, but that procedure destroys normal biological function.

A similar problem applies to fluorescence microscopy: If we attempt to illuminate only one region of interest, instead we find an entire swath is illuminated, analogous to the hourglass-shaped glowing region in the sample shown in Figure 2.15. We saw previously that the fluorescence can be separated from scattered incoming light by

---

[24]Earlier GCaMPs used a calmodulin-binding domain called M13; others use alternative choices such as RS20.

[25]Chapter 6 will discuss the focusing of light; Chapters 4 and 6 will describe other strategies to eliminate background light from microscope images.

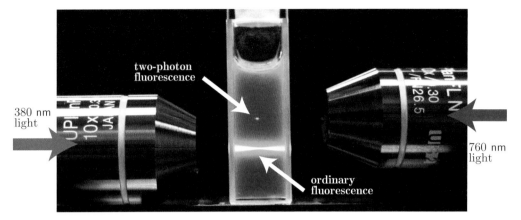

Figure 2.15: [Photograph.] **The difference between one- and two-photon excitation.** *Center:* A vial containing a solution of fluorescein dye is illuminated by two light sources. Short pulses of light with wavelength 380 nm emerge from the microscope objective on the *lower left.* The light is focused down nearly to a point at the center of the sample. But on either side of that point it spreads, forming the hourglass-shaped region of fluorescence shown. Similar pulses of light with wavelength 760 nm emerge from an identical objective on the *right.* Although this light also traverses an hourglass-shaped region, it only excites fluorescence in one tiny spot (*upper arrow*). To make this illustrative image, the beams were expanded to make spots large enough to see. Much narrower beams are used in microscopy. [Image courtesy Kevin D Belfield, Zhenli Huang, and Ciceron Yanez.]

using the Stokes shift, but still, fluorophores outside of the focal plane get excited and give off light that ends up as background. Clearly the situation would be improved if we could illuminate only a single point—but how could light magically appear at a point deep inside a sample, without also illuminating its entire incoming and outgoing paths?

2.7.2 Two-photon excitation depends sensitively on light intensity

There is a remarkable way out of the preceding dilemma, with roots going back to the early days of quantum physics. The Light Hypothesis emphasizes that a photon can interact with just one electron.[26] But this still leaves open the possibility that one electron could absorb energy from more than one *photon.* In her 1931 dissertation, M. Göppert-Mayer predicted theoretically that an atom or molecule could achieve the energy kick it needs to make an electronic transition in this way, even if the molecule in question has no "stepping-stone" state halfway between the starting and ending states. Later experiments confirmed her theory (Figure 2.16): Simultaneous absorption of two photons, each with only half the energy required for excitation, can excite a fluorophore.

Crucially for microscopy, the probability of this **two-photon excitation** depends on the *square* of the illumination intensity. To understand why, first imagine that you are standing still as a large crowd of people flows past you. One out of every hundred people in the crowd has a ticket; if you bump into such a person you can take the ticket and get some reward. There is probability per unit time β for this to happen,[27] so your expected waiting time is $1/\beta$. If you were to repeat the game in a similar crowd, but where only one out of every *two* hundred people had a ticket,

---

[26]The Light Hypothesis was stated in Idea 1.11 (page 37).
[27]Section 1.4.3 (page 36) connected waiting times to mean rates.

Figure 2.16: [Experimental data.] **One-photon versus two-photon excitation spectra.** The vertical axis gives the mean rate of fluorescence excitation in a sample exposed to exciting lights of various wavelengths, each delivering photons at the same rate. The data are for a fluorophore named mAmetrine. The main peak of two-photon excitation occurs at twice the wavelength (half the energy) of the one-photon excitation curve. (In other fluorophores this relation is not exactly obeyed, due to complications involving molecular vibrations.) The parts of the spectrum corresponding to one- and two-photon excitation have been normalized differently. [Data from Drobizhev et al., 2011.]

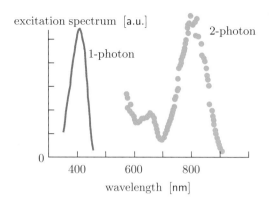

your expected waiting time would be twice as long. In the analogy to single-photon excitation, "you" represent the fluorophore, and the "people holding tickets" that you encounter are photons in your excitation band.

Now imagine another game. This time, to achieve your goal you must collect *two* tickets; moreover, each ticket expires a short time Δt after you collect it. Reducing the density of ticket-holders in the crowd by half now increases your waiting time by *much more* than twofold, because this time, two must be found in quick succession. In fact, your probability per unit time to get the reward is reduced by an additional Bernoulli trial, with probability $(\Delta t)\beta$. That is, the mean rate is proportional to $(\Delta t)\beta^2$. Thinning the crowd has a similar effect, because now a smaller fraction of people in it come close enough to bump into you.

Similarly, when a fluorophore needs two nearly simultaneously absorbed photons to fluoresce,[28] the waiting time will be about four times as long when the photon density is halved (two factors of 2). This analogy suggests that the two-photon excitation rate should vary as the *square* of the intensity of the exciting light, in contrast to the linear dependence expected for ordinary fluorescence.

$\boxed{T_2}$ *Section 2.7.2′ (page 100) gives more details about the β-squared rule.*

2.7.3 Multiphoton microscopy can excite a specific volume element of a specimen

Figure 2.15 illustrates these ideas. The lower microscope objective focuses a beam of light at a wavelength matched to the fluorophore's excitation peak. The apex of the glowing hourglass shape has the greatest density of photons, and hence a fluorophore located there has the greatest probability per unit time to fluoresce. As we move to the left or right of this point, the mean rate at which photons are presented to any particular fluorophore decreases gradually. Thus, the fluorescence also decreases gradually; as mentioned earlier, there is significant fluorescence emitted from regions outside the focus.

The upper microscope objective shown in the figure focuses a beam with twice the wavelength of the lower one, and hence half the energy per photon. In this case, fluorescence is possible only via two-photon excitation. As we move away from the

---

[28]Typically "nearly simultaneous" means that the absorptions must coincide to within 0.1 fs.

Figure 2.17: [Fluorescence micrograph.] **Two-photon imaging of an intact, living brain.** Besides neurons, mammalian brains contain other cell types, notably microglia that respond to injury. This image shows two kinds of cells in mouse brain cortex, just under the skull (*top*): Neurons expressed enhanced yellow fluorescent protein (*red, open arrowheads*), and microglia expressed GFP (*green, solid arrowheads*). Sections were scanned by the method described in the text, resulting in a stack of images. The side view shown was then created mathematically from those slices. Although brain tissue scatters visible light, the infrared light used for two-photon excitation allowed imaging deep within the tissue. This technique was used to show that microglia are highly dynamic; they constantly extend and retract thin branches to check nearby neurons for injury. [From Nimmerjahn, 2011; see also Media 8.]

50 μm

focal plane, the illumination intensity again falls off, but the *square* of that intensity falls off faster, and hence, the fluorescence decreases more sharply than in the case of ordinary (one-photon) excitation. Thus, only the part of the beam with the very highest light intensity gives appreciable fluorescence, as seen from the tiny dot of light alongside the upper lens in Figure 2.15.

Pioneering work by W. Denk and coauthors used this phenomenon as the basis for a breakthrough in fluorescence microscopy. To achieve the very high light intensity needed for two-photon excitation, the researchers used a laser that bunched the incoming photons into ultrashort bursts (each burst had duration less than a picosecond). The preceding discussion emphasized that, thanks to the fast falloff of excitation outside the focused spot, the method reduces stray light that would otherwise degrade the resulting image.[29] The method has several other benefits as well. For example, the long-wavelength (infrared) photons typically used for two-photon excitation fall into a band that penetrates tissue more readily, with less scattering and photodamage, than the visible light used for one-photon excitation. Hence, complex tissues such as lymphatic organs, kidney, heart, skin and brain can now be examined in detail at depths of up to 1 mm, while leaving the tissue intact. One millimeter may not seem like a great distance, but it is deep enough to image throughout the cortex of a small animal's brain (Figure 2.17).

Two-photon imaging can even record neural activity of an awake, behaving animal in real time, simultaneously in many neurons, without implanting electrodes. For example, C. Harvey and coauthors used an apparatus of this sort to record activity in a mouse's navigation region while it negotiated a virtual-reality maze.[30]

Restricting fluorescence excitation to the focal region brings another benefit:

---

[29] T_2 Similarly, if we wish to use light to trigger some other kind of reaction, such as uncaging a reagent, we again would like precise, three-dimensional control of illumination. A related idea has been proposed as the basis for high-density, 3D optical data storage.

[30] See also Media 9. Neural activity was monitored by using a reporter molecule from the GCaMP family described in Section 2.8.6.

Figure 2.18: [Experimental data.] **Spectral overlap.** *Curves on left:* Excitation and emission spectra of fluorescein, a fluorophore sometimes used as a FRET donor (and in some highlighter pens). *Curves on right:* Corresponding spectra of Texas red, a fluorophore sometimes used as an acceptor for fluorescein. When a solution containing both molecules is illuminated with light of wavelength shorter than 500 nm (*blue bar*), fluorescein molecules will be directly excited, but not those of Texas red. Nevertheless, excitation can be passed from donor to acceptor, resulting in acceptor fluorescence. How often this occurs is determined by the overlap between the donor's emission spectrum and the acceptor's excitation spectrum (*shaded*), and by their spatial separation. [Data from Johnson et al., 1993.]

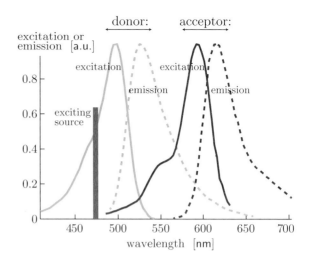

Most fluorophores photobleach after repeated excitation.[31] Suppose that we wish to build up a three-dimensional image. To do this, we could scan the focus through a stack of parallel planes in the sample and image each one. In ordinary fluorescence microscopy, the unwanted illumination outside the focal plane subjects fluorophores to photobleaching before their "turn" arrives, degrading image intensity in the later planes. Two-photon excitation is free from this limitation. Photodamage to the sample itself is also reduced.

Finally, transitions requiring more than two photons have been utilized to push the excitation wavelength even further into the infrared.

Section 2.7 has introduced two-photon excitation, an elaboration on the idea of fluorescence. We next turn to another variation, which again will prove to be qualitatively understandable based on the Light Hypothesis and Electron State Hypothesis. $\boxed{T_2}$ *Section 2.7.3′ (page 101) describes another feature of two-photon microscopy.*

2.8 FLUORESCENCE RESONANCE ENERGY TRANSFER

2.8.1 How to tell when two molecules are close to each other

Often when studying molecular mechanisms, we are frustrated by our inability to see events directly: Even the largest biomolecules are too small to image with visible light.[32] Chapter 7 will discuss one way around this limitation, but that method (localization microscopy) has the drawback that we must collect lots of photons; thus, the object to be imaged must be stationary, or at least slowly moving (to allow sufficient photon collection before the image changes). Many interesting molecular association events, however, are short-lived. In addition, gathering many fluorescence photons per image implies an even greater number of excitation photons, which cause photodamage. And yet, in many applications we wish to monitor a sample over a long time period. Some other method is needed for such measurements.

For example, we may wish to know when two molecules are bound together,

---

[31]Section 1.6.4 introduced photobleaching.
[32]Chapter 6 will discuss the limit of resolution imposed by diffraction.

Figure 2.19: [Cartoons; not to scale.] **Fluorescence resonance energy transfer.** The fluorophores mentioned in Figure 2.18. *Left:* When a donor molecule is excited, but is far from any acceptor molecule, then it just fluoresces as usual, emitting a Stokes-shifted photon in a randomly chosen direction. Even if an acceptor is present, there is little chance that the emitted photon will excite it, because most emitted photons are traveling in the wrong direction to intercept it. *Right:* If an acceptor is nearby, however, FRET can occur instead of donor fluorescence. The excited acceptor in turn then emits a photon, which can be distinguished from donor fluorescence by its longer emission wavelength.

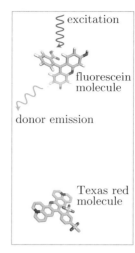
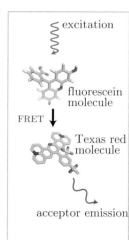

because we suspect that one of them regulates the other's action. It would be very useful if we could find a method that reports to us when the two specific molecules are in close proximity, without hindering their normal functions.[33] Similarly, in other cases we may wish to know when one part of a single large molecule moves relative to another part. Here, too, it would be useful to have a measurable signal that reports when the two parts are close to each other in space, and even quantitatively *how* close, with nanometer accuracy. Ideally, we'd like these reports with high time resolution, so that we can analyze the statistics of waiting times, or even correlate one kind of event with another.

Remarkably, a fluorescence-based mechanism allows us to obtain just this sort of information, in situations where we do not need a detailed image. **Fluorescence resonance energy transfer**, or FRET, involves two distinct fluorophore molecules, called the **donor** and **acceptor**. To observe relative motion of two molecules (or two parts of a single molecule), we arrange to attach the donor to one and the acceptor to the other, then illuminate with light that can excite the donor, but not the acceptor (Figure 2.18). When the two molecules are separated by more than a few nanometers, then we see the characteristic fluorescence of the donor and nothing from the acceptor, as expected. However, when the two molecules approach each other, the donor fluorescence drops sharply (Figures 2.19 and 2.20a), and is replaced by fluorescence with the characteristic spectrum of the acceptor. We thus get a real-time report of the proximity of the two molecules by sending the output light through two filters, designed to pass light only in the emission bands of the donor and acceptor, respectively. We can then collect the light from those two bands separately, and compare their intensities.

To define a quantitative measure of energy transfer, the **FRET efficiency** is defined as the fraction of fluorescence photons attributable to the acceptor:

$$\mathcal{E}_{\text{FRET}} = \frac{(\text{acceptor fluorescence photons})}{(\text{donor fluorescence photons}) + (\text{acceptor fluorescence photons})}. \quad (2.1)$$

With appropriate choices for the fluorophore pair, nearly complete conversion of donor

---

[33]Section 2.6.2 (page 78) described another method with this goal, but that one responds more slowly than the method discussed in this section.

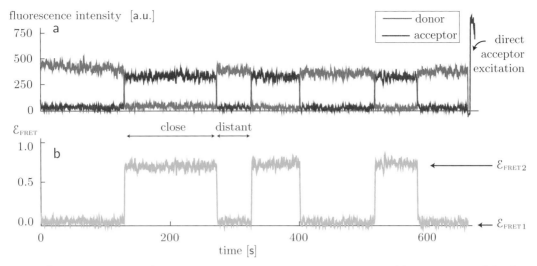

Figure 2.20: [Experimental data.] **Example of a single-molecule FRET dataset.** (a) Time series of the fluorescence intensities in the donor and acceptor emission bands, measured from a single DNA molecule. The DNA was labeled with a donor (Cy3) at one end and an acceptor (Cy5) at the other, and illuminated near the donor's excitation peak. There are distinct episodes indicating that the donor and acceptor are either close to, or far from, each other. The *arrow* shows a brief interval of illumination at the *acceptor's* excitation peak, to confirm that the acceptor had not yet photobleached. (b) Corresponding FRET efficiency defined in Equation 2.1. More details about this experiment appear in Section 2.8.5 (page 89). [Data courtesy Taekjip J Ha. From Fig. 3A, p. 1099 from Vafabakhsh and Ha. Extreme bendability of DNA less than 100 base pairs long revealed by single-molecule cyclization. *Science* (2012) vol. 337 (6098) pp. 1097–1101. Reprinted with permission from AAAS.]

to acceptor fluorescence is possible. For example, Figure 2.20b shows $\mathcal{E}_{\mathrm{FRET}} \approx 75\%$.
$\boxed{T_2}$ *Section 2.8.1′ (page 101) discusses a correction to Equation 2.1.*

2.8.2 A physical model for FRET

To summarize, we can find pairs of molecular species, called donor/acceptor pairs, with the property that physical proximity abolishes fluorescence from the donor. When such a pair are close, the acceptor nearly always pulls the excitation energy off the donor, before the donor has a chance to fluoresce. The acceptor may either emit a photon, or lose its excitation without fluorescence ("nonradiative" energy loss).

We can begin to understand FRET by using ideas introduced in this chapter. The left side of Figure 2.21 just reproduces Figure 1.9, our physical model for fluorescence from the donor molecule. From state *4*, the donor can emit a photon of energy ΔU_{D}, as usual. However, in the presence of the acceptor, the initially excited donor gets a second option for how to revert to its electronic ground state. If the acceptor molecule is chosen such that the energy ΔU_{A} needed to excite it matches ΔU_{D}, then the donor can transfer its excitation energy directly to the acceptor, instead of itself fluorescing. Later, the acceptor will fluoresce, but its Stokes shift implies that the resulting photon will have lower energy than ΔU_{D} (step *8* of Figure 2.21).

The story just outlined is a good start at understanding FRET in terms of ideas developed earlier. Here are two imagined students developing the idea further:

George asks: How can an excitation "hop" intact between two molecules? That is, what physical process is represented by the horizontal dashed arrows in Figure 2.21?

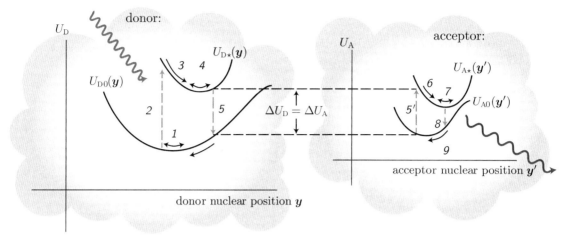

Figure 2.21: [Schematic energy diagram.] **Physical model of FRET,** an elaboration of Figure 1.9 (page 42). *1–4*: Initially the donor enters its excited state as usual. *5,5′*: The donor's momentary configuration y determines an energy release $\Delta U_D = U_{D\star}(y) - U_{D0}(y)$ for return to the ground state. When the acceptor is distant, then the donor can release this energy by emitting a photon, as usual. If the acceptor is nearby, however, and its momentary configuration y' determines a ΔU_A that is equal to ΔU_D, then the excitation can pass from donor to acceptor via direct electrostatic interaction, without any intermediate photon. *6*: The excited acceptor delivers some excess energy to its surroundings as heat, preventing return of its excitation back to the donor. *7–9*: Eventually, the acceptor emits a photon, with the same emission spectrum as if it had been directly excited.

Martha: The incoming photon gave the donor's electrons a kick, analogously to a guitar pick plucking a string and setting it vibrating. The vibrating electrons in turn create an oscillating electric field, which persists after the incoming photon is gone and which can be "felt" by other nearby electrons as long as the donor remains in its excited state. That is, the two molecules are connected by their surrounding electric fields, not by the emission and later reabsorption of any intermediary photon.

George: But why does the excitation hop only to a specific acceptor molecule, ignoring the many water molecules (and other species) that lie between donor and acceptor? Some of those molecules will be located closer to the donor than the acceptor, which may be several nanometers away. It's a bit like whispering in a crowded, noisy room and only being heard by *one* person on the far side.

Martha: Continuing the musical analogy, if we place a second guitar alongside the first one, then only one of its strings will respond strongly to the sound generated by the first. That's because only the string corresponding to the pitch that was initially struck is **resonant** with it. Analogously, in a FRET experiment, the crowd of other molecules present will generally not have any allowed electron states that give excitation energies in the donor's emission band. Each such molecule therefore has a much lower probability of accepting the donor's excitation than does the acceptor, which was chosen to have an excitation energy resonant with the donor.

George: Stop! The musical metaphor involves matching sound frequencies, but the physical application involves matching *energies*.

Martha: The only way for the transfer step to satisfy energy conservation is for the energies to agree. The frequency of light that *would have been* emitted by the donor, had FRET not taken place, must therefore match the frequency that *would* *have been required* to excite the acceptor by light, had FRET not taken place.

George: There's another discrepancy: A vibrating string gradually depletes its energy, by continuously radiating it into the room in the form of sound. But the Electron State Hypothesis says that an excited donor fluorophore can only lose energy in *discrete* transitions; it remains fully excited for a random waiting time, then discontinuously sheds all its excess energy, making a random choice (a Bernoulli trial) between emitting a photon versus exciting a nearby acceptor.

Martha: Yes, we must be careful not to push the analogy to an ordinary guitar string too far. And I admit that there are still some puzzling aspects. For example, how can the presence of a single acceptor molecule eliminate *nearly all* the donor's probability to fluoresce? Can't an excitation hop to the acceptor, then hop *back* to the donor, which could then fluoresce as usual?

George: Apparently excitation energy makes a *one-way* trip to the acceptor. That makes sense: Once the acceptor is excited, some of its energy gets irreversibly dissipated heating the surroundings as the acceptor configuration y' "slides down the hill" of potential energy (Figure 2.21, step *6*), arriving near the minimum of $U_{A\star}$. Although the acceptor's electrons are then still in an excited state, nevertheless that state is no longer in resonance with the donor (nor with any other nearby molecule). Thus, the excitation is stuck until it is finally released via acceptor fluorescence.[34]

$\boxed{T_2}$ *Chapter 14 gives another reason for the one-way character of the energy transfer.*

2.8.3 Some forms of bioluminescence also involve FRET

Many animals, protists, and fungi use some of their metabolic energy to emit light (bioluminescence). Some do this by the action of an enzyme (**luciferase**) that attaches an oxygen to a substrate called **luciferin**, creating the molecule oxyluciferin in an excited state. After its synthesis, the excited oxyluciferin falls to its electronic ground state, emitting a photon.

The reaction just described produces blue light, with spectrum peaking at 460 nm. However, coastal waters transmit green light farther than blue. Perhaps for this reason, the jellyfish *Aequorea* has an accessory protein, GFP, that acts as a FRET acceptor, Stokes-shifting blue light from the luciferin to about 509 nm (Figure 1.11, page 45).

2.8.4 FRET can be used to create a spectroscopic "ruler"

Figure 2.22 shows that, although the donor and acceptor need not touch each other, nevertheless energy transfer can only occur over a limited distance; indeed, we can use FRET to determine quantitatively the distance between two fluorophores. Let's try to understand the relation between distance and energy transfer.

To predict the FRET efficiency $\mathcal{E}_{\mathrm{FRET}}$, we must consider the interaction represented by the dashed horizontal lines in Figure 2.21. The electric force far from an object with net charge decreases with distance as a constant times r^{-2}, for distances r larger than the size of the object. However, the excited donor molecule is electrically *neutral* (it neither gains nor loses an electron). A neutral object can also have an exterior electric field, but that field's strength must decrease faster than r^{-2}; thus,

---

[34] $\boxed{T_2}$ As with any kind of fluorescence, there are also other deexcitation pathways besides photon emission. For lab work, we choose a fluorophore that doesn't follow those unwanted pathways very often.

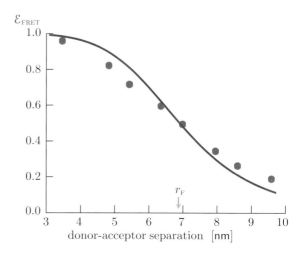

Figure 2.22: [Experimental data.] **FRET efficiency as a function of the separation between donor and acceptor.** $\mathcal{E}_{\text{FRET}} = 1$ corresponds to 100% probability that an excitation from a donor will be transferred to an acceptor (Equation 2.1). Here, the experimenters prepared a series of short DNA molecules each with a donor fluorophore at one end, but with its acceptor at various distances down the chain. *Circles* show single-molecule measurements of $\mathcal{E}_{\text{FRET}}$. The *curve* shows the value of $\mathcal{E}_{\text{FRET}}$ given by Equation 2.3 for each separation. The value of the Förster radius r_{F} in that formula can be obtained by fitting the data (see Problem 2.2). [Data from Lee et al., 2005, available in Dataset 5.]

generically the force that shakes the acceptor's electrons decreases[35] as a constant times r^{-3}.

You'll show in Problem 2.4 that in classical mechanics, a shaken oscillator gains energy at a rate proportional to the amplitude of the shaking force *squared*, times an overall factor. (In FRET, that factor includes the spectral overlap; see Figure 2.18, page 83.) We just argued that the force depends on distance as r^{-3}, so squaring it gives r^{-6}.

Ex. Use this analogy to get a formula for the dependence of the FRET efficiency on separation, assuming that every donor excitation leads to either a donor or an acceptor photon.

Solution: Suppose that at time zero, a donor molecule is in its excited state. If it survives to time t in this state, then during the interval from t to $t + \mathrm{d}t$ it may either

• Emit a fluorescence photon, with probability $\beta_{\text{D}}\mathrm{d}t$;
• Transfer energy to an acceptor, with probability $\beta_{\text{FRET}}\mathrm{d}t$; or
• Remain unchanged.

(The question told us to neglect the possibility of other forms of excitation loss.)

Thus, the conditional probability is[36]

$$\mathcal{P}(\text{FRET} \mid \text{deexcitation in } \mathrm{d}t) = \beta_{\text{FRET}}/(\beta_{\text{FRET}} + \beta_{\text{D}}) = (1 + \beta_{\text{D}}/\beta_{\text{FRET}})^{-1}. \quad (2.2)$$

Because we are assuming that every donor excitation ends up as a fluorescence photon, this quantity is also the FRET efficiency (the proportion of all fluorescent emissions that are from the acceptor).

The rate of donor fluorescence does not depend on the separation r, but we argued earlier that β_{FRET} is a constant times $(r^{-3})^2$. Hence, Equation 2.2 can be written as

$$\mathcal{E}_{\text{FRET}}(r) = \left[1 + (r/r_{\text{F}})^6\right]^{-1}, \quad (2.3)$$

where $r_{\text{F}} = r(\beta_{\text{FRET}}/\beta_{\text{D}})^{1/6}$ is a constant (independent of r), called the **Förster radius**.

---

[35] Students of electrodynamics may recall that an electric dipole creates an electric field that falls with distance in this way.
[36] Equation 0.6 (page 4) defined conditional probability.

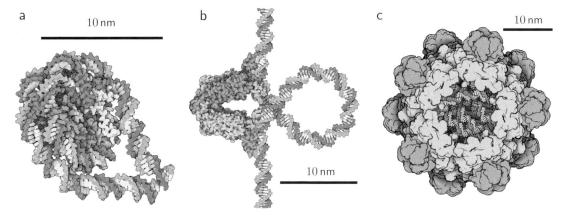

Figure 2.23: [Artist's reconstruction based on structural data.] **Biological examples of tightly bent DNA.** (a) DNA winds around a protein core (*blue*) to form the nucleosome. (b) The *lac* repressor protein (*left, blue*) forces DNA into a tight loop by binding to two specific sites. (c) A bacteriophage (virus) called ϕX174 packs 5386 base pairs of DNA (*shown in cutaway*), coding for 11 genes, into a small capsid (protein shell, *blue*). [Art by David S Goodsell.]

You can try elaborating this argument to account for a nonzero probability for the donor to lose its excitation by some means other than FRET.

The result in Equation 2.3 embodies several idealizations. For example, we have assumed that r is large enough to use the approximation of electric dipole interaction (at least several times as large as the fluorophores), and that other excitation loss processes are negligible.

Each donor/acceptor pair has its own characteristic value of r_F, which we can often look up in published tables. For example, fluorescein (Figures 2.15 and 2.18) has an emission peak at around 520 nm. It makes a good donor for rhodamine, whose excitation peak is around 550 nm; this FRET pair has $r_F \approx 4.5$ nm in water.

In short, the above discussion leads to two predictions:

> *A pair of fluorophores will engage in FRET if they are close enough in space, and the acceptor's excitation spectrum overlaps significantly with the donor's emission spectrum. The FRET efficiency falls with donor-acceptor separation roughly proportionally to r^{-6}, when r exceeds the pair's Förster radius r_F.*   (2.4)

Idea 2.4 implies that FRET can actually give a *quantitative* readout of the distance between two molecules—a "spectroscopic ruler." These ideas have been confirmed in experiments, for example, the one in Figure 2.22. They also guide our choices of fluorophores to use in experiments. In particular, some fluorescent proteins make good FRET pairs, as do many other fluorophores that can be attached to antibodies, covalently linked to particular points on proteins, and so on.

$\boxed{T_2}$ *Section 2.8.4′ (page 102) mentions other experimental tests. Chapter 14 gives a quantum-mechanical derivation of FRET.*

2.8.5 Application of FRET to DNA bending flexibility

DNA is normally represented as a perfect double helix: two sugar-phosphate chains winding around a straight central axis. In real cells, however, the extremely long DNA molecule will always be distorted from its ideal shape by thermal collisions with the surrounding molecules, and by forces exerted by DNA-binding proteins.

For example, when a bacterium's DNA is bound to the *lac* repressor protein LacI, it often forms a tight loop (Figure 2.23b). Loop formation is the result of LacI actually having *two* DNA-binding sites, each of which recognizes and sticks to a specific DNA sequence. In the genome of *E. coli*, those binding sequences are separated by 92 base pairs. Bringing them into close proximity, as needed for both to bind simultaneously to a single *lac* repressor, thus requires the DNA to adopt the looped conformation shown in Figure 2.23b. The figure also shows two other examples of tightly bent DNA arising in biology. All these examples were puzzling because the DNA molecule was once thought to be very resistant to bending; how, then, can it form such structures?

R. Vafabakhsh and T. Ha approached this problem by first showing that tight DNA loops do in fact form spontaneously. To do this, they used single-molecule FRET to monitor the overall conformations of short strands of DNA. Thus, they attached donor and acceptor fluorophores to the ends of short double-stranded DNA molecules (Figure 2.24a). At each end of their constructs, however, one strand extended farther than the other. These single-stranded protrusions were complementary to each other ("sticky ends"), so that if they came close together, they had some chance of binding, and hence of creating a tight circle. Figure 2.24b shows the fluorescence from many such constructs, in both the donor and acceptor emission bands. Under the conditions shown in panel (b1), none persistently exhibit FRET, that is, none are looped. However, after a change in solution conditions that favors sticking of the ends, nearly all of the constructs do eventually close into circles [panel (b2)]. Contrary to earlier expectations, the experiment showed that *short DNA does spontaneously form tight loops,* even without any proteins that might apply forces to it. The researchers then went on to quantify the dependence of loop formation on the length and sequence of the DNA construct, and to dissect its kinetics into separate contributions from the stickiness of the ends and from the intrinsic ability of the DNA to bend.

$\boxed{T_2}$ *Section 2.8.5′ (page 102) discusses a fine point about FRET efficiency.*

2.8.6 FRET-based indicators

Section 2.6 described fluorescent molecules that can serve as indicators, reading out moment by moment some internal variable describing cell state. Each such indicator has its merits—one choice may be more sensitive than another, less toxic, faster to respond, and so on. Most, however, require a difficult step for their interpretation: determination of normalization. For example, if an indicator lights up in response to a change of some condition, then the total fluorescent signal we see will be the product of a response curve (for example, Figure 2.12a) times other factors:

- The fluorescence excitation probability at the wavelength used;
- The arrival rate of exciting photons;
- The number of functional indicators present in the region under study.

To determine where we are on the response curve, we need to divide the observed fluorescence by these three factors, but the last two are often hard to estimate or control (and they may change over time).

One way out of this conundrum is to use FRET. Figure 2.25 shows two illustrative examples of calcium indicators. In each case, the cartoon is similar to Figure 2.14a, but now the conformational change in question serves to bring a donor and an acceptor into closer proximity. Such a construct will have the same excitation spectrum regardless

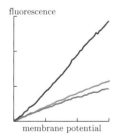

Fig. 2.14a (page 79)

Fig. 2.12a (page 77)

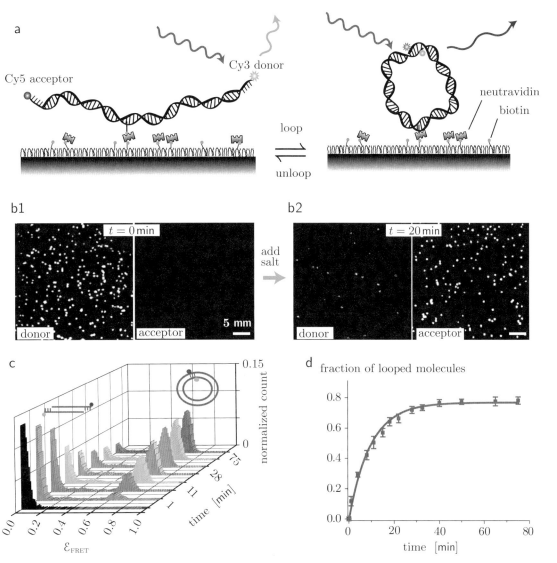

Figure 2.24: [Experimental data; micrographs.] **Single-molecule FRET applied to DNA looping.** (a) DNA molecules of length 91 base pairs, labeled at the ends with FRET pairs, were anchored to a surface via a specifically binding pair of molecules called biotin and neutravidin. Each end of the DNA was free. (b) *Left panels:* Fluorescence images of single DNA molecules, made with wavelength filters corresponding to the donor and acceptor bands, before adding high salt buffer. The low salt condition suppresses base pairing of the two ends: That is, even if they come together momentarily, they immediately separate again. In high salt concentration, however, such momentary appositions lead to long-lived looped states. *Right panels:* 20 min after adding high salt, nearly all DNA molecules have formed closed loops, as seen from the high FRET. (c) Histograms of FRET efficiency as a function of time show the time courses of circular (high FRET, *right*) and open (low FRET, *left*) subpopulations. (d) Fraction of circular, or "looped" DNA as a function of time. Fitting an exponential function to this curve gives the mean rate of conversion to the circular form. [From Fig. 1, p. 1098 of Vafabakhsh and Ha. Extreme bendability of DNA less than 100 base pairs long revealed by single-molecule cyclization. *Science* (2012) vol. 337 (6098) pp. 1097–1101. Reprinted with permission from AAAS.]

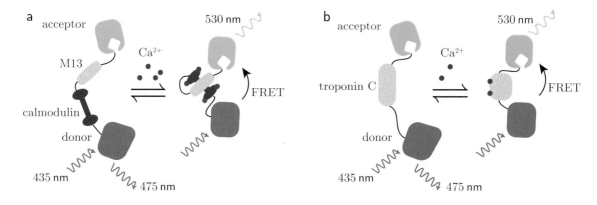

Figure 2.25: [Cartoons.] **FRET-based, genetically encoded calcium indicators.** (a) The cameleon family of calcium indicators. As in the GCaMP sensors (Figure 2.14, page 79), calcium levels affect the binding of calmodulin to its partner domain M13. Unlike those sensors, however, two attached fluorescent proteins form a FRET pair; calcium binding brings the pair into closer proximity, increasing energy transfer. (b) A similar strategy, using a conformational change in a single calcium-binding domain called troponin C. [Courtesy Lin Tian; see also Broussard et al., 2014.]

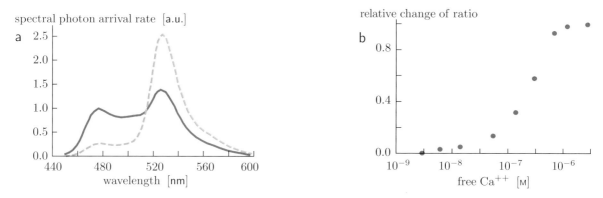

Figure 2.26: [Experimental data.] **Performance of FRET-based calcium indicators.** (a) Emission spectra of an indicator named "yellow cameleon 3.60" at zero (*solid*) and high (*dashed*) calcium levels. The excitation light had wavelength 435 nm. [Data from Miyawaki et al., 2011.] (b) Change in the ratio of photon arrival rates in the donor and acceptor wavelength bands when calcium is added, rescaled to the range from 0 to 1, for another indicator in this family. [Data from Truong et al., 2007.]

of state, but it emits light in one of two well-separated bands. To find the fraction of time that it spends in each state, we need only divide the photon arrival rates in the two bands. The unknown factors listed above all *cancel* from that quotient, leading to a "ratiometric" determination of the average state of the indicators. Figure 2.26 shows some data on performance of such indicators.

Other FRET-based indicators read out membrane potentials or the levels of specific ions such as chloride or molecules such as cyclic AMP (an internal signaling molecule used by many cells).

Section 2.8 has introduced FRET, another tool for looking inside cells that makes sense in the light of the Light Hypothesis and Electron State Hypothesis.

2.9 A GLIMPSE OF PHOTOSYNTHESIS

Earlier sections of this chapter have described how fluorescence lets us see what we want to see in cells, and not see the rest. But some living organisms don't want to "see" at all; instead, they need to capture and harness energy. To get started understanding that process, we'll now see that we must get accustomed to thinking of molecular excitations as *things* that can be created, tossed around, and destroyed by various means. Section 2.8 already began that shift of viewpoint in the context of FRET.

2.9.1 It's big

Photosynthetic organisms convert around 10^{14} kg of carbon from carbon dioxide into biomass each year. In addition to generating the food that we enjoy eating, photosynthetic organisms emit a waste product, free oxygen, that we enjoy breathing. They also stabilize Earth's climate by removing atmospheric CO_2.

Photosynthesis is complex; this book can only discuss some of its workings, mostly in the context of cyanobacteria. As their name implies, cyanobacteria are prokaryotes with a bluish-green color. They account for roughly a quarter of Earth's current photosynthetic productivity. Cyanobacteria are also thought to have been primarily responsible for transforming Earth's atmosphere from a "normal" planetary form, without free oxygen (like Venus and Mars), to the high-oxygen form observed today. Cyanobacteria live freely in both fresh and salt water. Their ability to harvest sunlight also makes them prized as symbionts, helping meet the energy needs of lichens, corals and other larger organisms. In fact, the light-harvesting organelles (**chloroplasts**) in plants, algae, and other photosynthetic eukaryotes are thought to have had their origin as symbiotic cyanobacteria.

2.9.2 Two quantitative puzzles advanced our understanding of photosynthesis

The idea of photosynthesis goes back at least to 1796, when J. Ingenhousz presented his experiments on the role of sunlight in oxygen production by plants, in the context of Lavoisier's new ideas about chemistry. We will pick up the story in the mid-20th century, describing two extraordinarily fruitful observations. By this time, it was clear that light stimulated the production of oxygen and carbohydrates via some mechanism involving **chlorophyll**, a chromophore (pigment) contained in the leaves of green plants.

The puzzle of low oxygen yield per chlorophyll

W. Arnold was an undergraduate student interested in a career in astronomy. In 1930, he was finding it difficult to schedule all the required courses he needed for graduation. His advisor proposed that, in place of Elementary Biology, he could substitute a course on Plant Physiology organized by R. Emerson. Arnold enjoyed the class, though he still preferred astronomy. But unable to find a place to continue his studies in that field after graduation, he accepted an offer from Emerson to stay on as his assistant.

Emerson and Arnold went on to do a series of seminal measurements. One of these involved giving the photosynthetic alga *Chlorella pyrenoidosa* a very short ($10\,\mu s$) flash of bright light. Although the metabolically important output of photosynthesis is the energy molecule ATP, the researchers found it easier to measure the production of free oxygen. Not surprisingly, for weak flashes the oxygen production was proportional to the intensity of the light: More photons can drive more photochemical reactions per light pulse. Also unsurprisingly, at higher illumination the oxygen production leveled off. Like many reactions, photosynthesis is carried out by an apparatus of molecules, each of which has a maximum processing speed. If photons arrive faster than this limiting rate, then the excess cannot be used: We say the reaction **saturates**.

Emerson and Arnold took a further, crucial step. They estimated the total number of chlorophyll molecules in their sample, and the total number of oxygen molecules produced per saturating light pulse, and here they did find a surprise. One might imagine that the ratio of these numbers would be some small integer, because after a saturating pulse every chlorophyll is activated, and it seems reasonable to guess that the reaction requires one or a few such activated chlorophylls for each product molecule Instead, however, Emerson and Arnold found just *one O_2 molecule per 2500 chlorophylls* after a saturating flash of light. Other plants were later found to have a similarly large ratio. *What were all the other chlorophylls doing?* Before we interpret this mysterious result, we turn to a related quantitative puzzle that appeared soon after.

The puzzle of the action spectrum of photosynthesis

Another clue to the mechanism of photosynthesis appeared as scientists began to measure and appreciate its spectral character. Chlorophyll absorbs light strongly in two bands, located in the blue and red regions of the visible spectrum (Figure 2.27a). It does not absorb light between these bands; most plants reflect that light, and therefore appear green. Conversely, green light is not needed for plants to thrive (see Figure 2.28). But it had been known for decades that some photosynthetic organisms do productively absorb light in other parts of the spectrum. How could that happen? Isn't chlorophyll the light-absorbing molecule in photosynthesis?

Emerson and C. Lewis decided to investigate this question by using the cyanobacterium *Chroococcus* and light intensities well below saturation. They defined the **quantum yield** of photosynthesis as the number of O_2 molecules produced per pulse of light, divided by the total number of photons absorbed in that pulse.[37] Figure 2.27b shows this quantity, for light of various wavelengths. The curve dips in the blue region, but this was not surprising: Figure 2.27a shows that chlorophyll does not absorb 480 nm light. Other **accessory pigments**, such as **carotenoids**, do absorb at this wavelength (Figure 2.27a), but they don't perform photosynthesis.

What *is* surprising about the curve is that the quantum yield is *not* low in the region 560–640 nm, even though Figure 2.27a shows that in this region nearly all absorption is done by another accessory pigment called **phycocyanin**—not chlorophyll.[38] Emerson and Lewis concluded that "the energy absorbed by phycocyanin must be available for photosynthesis."[39] Soon after this work, other researchers ruled out the possibility that phycocyanin might have its own separate photosynthetic system: Chlorophyll really is needed. Phycocyanin seemed to play the role of an "adaptor plug," converting light from a form not directly usable by chlorophyll into—what?

---

[37] The phrase "quantum yield" emphasizes that photons ("quanta") are considered as participants in a chemical reaction, with stoichiometry like any other chemical species. The quantum yield is smaller than 1 because it takes multiple photons to create one oxygen molecule, and also because some absorbed photons are lost to other processes, ultimately producing heat.

[38] More precisely, phycocyanin is a protein containing the relevant chromophore as a cofactor.

[39] You'll make a more quantitative estimate in Problem 2.3.

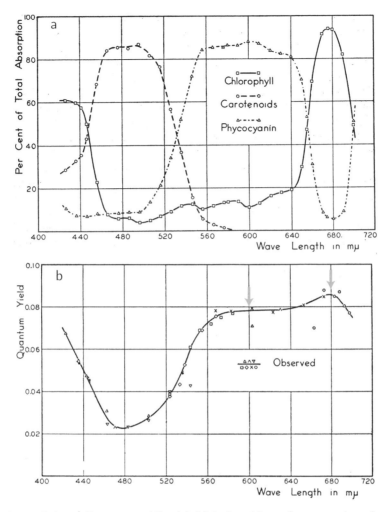

Figure 2.27: [Experimental data.] **Emerson and Lewis's historic evidence for energy transfer among photosynthetic pigments.** (a) Relative absorption of light by three of the chromophores present in cyanobacteria, at various visible wavelengths. The unit "mμ" is an obsolete notation for nm. (b) The overall quantum yield of photosynthesis in *Chroococcus* as a function of illumination wavelength. *Arrows* indicate the absorption maxima for phycocyanin and for chlorophyll. [Adapted from Emerson & Lewis, 1942. Data available in Dataset 6; see Problem 2.3.]

Figure 2.28: [Photograph.] A "pinkhouse" that grows 2.2 million plants under light from blue and red light-emitting diode lamps. [Courtesy iBio, Inc.]

2.9.3 Resonance energy transfer resolves both puzzles

The first step to resolving the two puzzles just outlined is to note that *chlorophyll is fluorescent.* You can readily confirm this by soaking plant leaves in a solvent like acetone for a day or two. The bright green color in the resulting solution under ordinary light arises from the absorption of blue and red by chlorophyll. When you illuminate the solution with ultraviolet light, however, you instead see chlorophyll's fluorescence, a dramatic, blood-red color. Other plant chromophores, such as phycocyanin, were also found to be fluorescent in solution. In contrast, intact green plants and cyanobacteria are *not* very fluorescent in their natural state.

Other pigments can transfer excitation to chlorophyll, resolving the spectral puzzle

Could phycocyanin absorb light energy and somehow *transfer* it to the chlorophyll system? One simple idea would be that phycocyanin, whose emission spectrum over-laps one of chlorophyll's excitation peaks, could give off appropriate light that then gets intercepted by chlorophyll. But most of the photons given off in this way would be emitted in the wrong directions to hit a chlorophyll. Instead, they would be lost, giving a net quantum yield much lower than the value observed when the plants are illuminated at chlorophyll's absorption peak. The experiments belied this pre-diction, however: They showed the two yields to be nearly equal (see the arrows in Figure 2.27b).

Arnold eventually left Emerson's lab to study elsewhere, but they stayed in contact. Emerson told him about the results with Lewis, and suggested that he think about the energy-transfer problem. Arnold had once audited a course on quantum physics, so he visited the professor for that course to pose the puzzle. The professor was J. R. Oppenheimer, and he did have an idea. Oppenheimer realized that a similar energy transfer process was known in nuclear physics; from this he created a complete theory of fluorescence resonance energy transfer.[40] Oppenheimer and Arnold also made quantitative estimates indicating that phycocyanin and chlorophyll could play the roles of donor and acceptor, and that this mechanism could give the high transfer efficiency needed to explain the data in Figure 2.27b.

Oxygen yield and the light-harvesting complex

The FRET-like mechanism for energy transfer also cleared up the other puzzle men-tioned earlier: that thousands of chlorophyll molecules, each driven to saturation by a short, intense flash of light, only create one oxygen molecule on average (Section 2.9.2).

In order to capture enough light for its needs, a photosynthetic organism must deploy a very large number of individual chromophores. Each chromophore, however, is idle most of the time:

Your Turn 2B

Solar energy arrives at Earth's surface carrying about $1.4\,\mathrm{kW/m^2}$ (at the Equa-tor, at noon, no clouds). Convert that figure to photons, with the simplifying assumption that all the photons have wavelength about $500\,\mathrm{nm}$. Your answer may seem like a big number, but express it in terms of a typical molecular size scale $(\mathrm{nm^{-2}s^{-1}})$. If a chromophore could process photons in, say, a microsecond, would it be very busy in such a light beam?

---

[40]FRET was introduced in Section 2.8 (page 83). This work came several years before T. Förster's independent discovery.

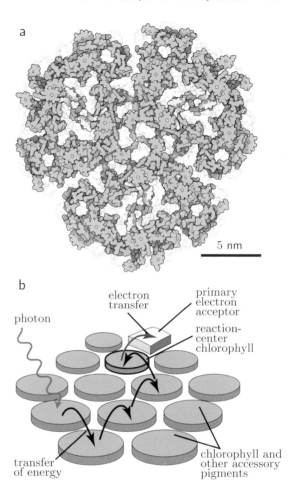

Figure 2.29: **Photosystem I** from the cyanobacterium *Synechococcus elongatus*. (a) [Artist's reconstruction based on structural data.] This illustration shows the chlorophylls (*green*) and other chromophores such as carotenoids (*orange*); in the real system, these are surrounded by other proteins, a few of which are shown in *gray.* The actual photochemistry begins when an excitation arrives at one of the three reaction centers (*yellow*). [Art by David S Goodsell.] (b) [Schematic.] When a chromophore absorbs a photon, the resulting excitation hops among many chromophores in the photosystem by a process similar to fluorescence resonance energy transfer, eventually reaching a reaction center. At the reaction center, photochemistry begins with the transfer of an electron from a chlorophyll molecule to a neighboring electron acceptor, initiating a chain of reactions that ultimately converts the original photon's energy to chemical form.

If the organism had to replicate the entire photosynthetic apparatus for each chromophore, then that apparatus, too, would be mostly idle—a very inefficient approach. Instead, plants and cyanobacteria create **light harvesting complexes** ("LHCs"), which in turn are assembled into elaborate "supercomplexes," each typically containing hundreds of chromophores. Photon absorption by a chromophore creates an excitation that gets transferred by a FRET-like mechanism, eventually landing on a **reaction center** in the complex that can initiate electron transfer and ultimately drive the photosynthesis reaction. When we disassemble the complex, for example, by dissolving out the chlorophyll in solvent, then the transfer chain is broken and the excited chlorophyll molecules must instead return to their ground state by ordinary fluorescence, as is observed.

When many photons arrive in a short, saturating burst, then the entire photosystem must wait for its reaction center to process the first excitation before it can make use of any of the others. During this time, the hundreds of chromophores in the complex are all unable to contribute their excitation energy to photosynthesis; eventually that energy gets dissipated by other, nonproductive processes. This observation, along with the fact that about eight transferred electrons are needed to create an O_2 molecule, roughly accounts for Emerson and Arnold's discovery that a short, saturating flash of light creates only one O_2 per 2500 chlorophylls.

Figure 2.29a shows an example (**Photosystem I**) found in cyanobacteria. Most of the chlorophyll molecules shown are unable to perform electron transfer; they simply capture energy and hand their excitation over to others, until it lands on one of the three reaction centers (Figure 2.29b).[41] Engineers are now starting to create artificial light-harvesting systems, also based on FRET, that can confer similar advantages for photovoltaic applications.

Section 2.9 has looked at just one aspect of photosynthesis, the initial stage. We found that a FRET-like mechanism was part of a highly effective light harvesting scheme.

$\boxed{T_2}$ *Section 2.9.3′ (page 102) gives more details about the photosynthesis apparatus. Section 14.3.3 outlines some recent developments concerning photosynthetic excitons.*

THE BIG PICTURE

This chapter introduced some techniques to *see and touch* the internal workings of a living cell, including devices that can be artificially introduced into living cells, in an intact organism, to create functions not given to those cells by evolution. Those functions include membrane potential sensing, light-activated ion channel opening, and even light-driven ion pumping. Chapters 10–11 will discuss the much more highly evolved phototransduction schemes in our eyes.

KEY FORMULAS

- *Action spectrum:* The action spectrum for some light-induced outcome, such as cell death, is a constant times the reciprocal of the number of photons needed to get that outcome in a specified fraction of instances, as a function of light wavelength.
- *FRET:* The FRET efficiency is defined as the acceptor fluorescence divided by the total (donor+acceptor) fluorescence. In an approximate treatment discussed in the text, it depends on distance via the formula $\mathcal{E}_{\mathrm{FRET}}(r) = (1 + (r/r_{\mathrm{F}})^6)^{-1}$, where the "F̲örster radius" r_{F} is a constant depending on the details of donor, acceptor, and intervening solvent molecules.
- *Quantum yield:* The quantum yield of a light-induced process is the ratio of the number of molecules in a sample that undergo the process to the number of photons absorbed by that sample. Thus, high quantum yield implies that most of the absorbed photons are productively absorbed.

FURTHER READING

Semipopular:
Fluorescence and bioluminescence in Nature: Johnsen, 2012.
Neural imaging: Schoonover, 2010.
Photosynthesis and its evolution: Lane, 2009; Morton, 2008.

---

[41] An even larger complex of chromophores feeds energy to Photosystem I.

Intermediate:

Fluorescence microscopy and FRET: Jameson, 2014.

Membrane potentials: Nelson, 2014, chapt. 11–12; Nicholls et al., 2012; Purves et al., 2012.

Two-photon microscopy: Jameson, 2014; Ustione & Piston, 2011; Mertz, 2010.

Photosynthesis: Steven et al., 2016, chap. 15; Phillips et al., 2012, chap. 18; Atkins & de Paula, 2011, chapts. 5 and 12; Nordlund, 2011, chapt. 10.

Technical:

Fluorescence endoscopy for cancer detection: Wagnières et al., 2014; Wagnières et al., 2003. Other forms of fluorescence-guided surgery: Lee et al., 2016; Yun & Kwok, 2017.

Fluorescence microscopy: Fritzky & Lagunoff, 2013.

Resting potential of photoreceptors: Tomita et al., 1967.

Optogenetics, historic: Genetically encoded reporters: Siegel & Isacoff, 1997. Discovery of algal sensory photoreceptors: Hegemann, 2008. Neurons driven by light: Zemelman et al., 2002. Channelrhodopsin-2 expressed in human cells: Nagel et al., 2003. Advances in 2005: Boyden et al., 2005; Ishizuka et al., 2006; Li et al., 2005; Nagel et al., 2005.

Optogenetic control, general: Klapoetke et al., 2014; Chow & Boyden, 2013. Control of freely behaving animals: Fang-Yen et al., 2015; Liu et al., 2012; Zhang et al., 2011; Lin et al., 2011; Gradinaru et al., 2007. Control of protein interactions and gene expression by light: Zhou et al., 2015.

Readout of membrane potential, historic: Salzberg et al., 1973. Recent: Hochbaum et al., 2014; Kralj et al., 2012; Kralj et al., 2011. Fusions of a voltage sensing domain with a circularly permuted fluorescent protein have also been used for this purpose: Yang et al., 2016; St-Pierre et al., 2014.

Calcium reporters and others: Newman et al., 2011.

Fast GCaMPs: Badura et al., 2014; Broussard et al., 2014. Sensitive GCaMPs: Chen et al., 2013. Other genetically encoded calcium indicators: Tian et al., 2011.

Application of a split fluorescent protein assay: Cole et al., 2007.

Two-photon excitation and imaging: Franke & Rhode, 2012; Harvey et al., 2012; Dombeck & Tank, 2011; Fisher et al., 2008; Denk et al., 1990. Two-photon endogenous fluorescence used to diagnose age-related macular degeneration: Palczewska et al., 2014.

Oppenheimer's treatment of FRET: Arnold & Oppenheimer, 1950; Oppenheimer, 1941.

Application of FRET to DNA looping: Vafabakhsh & Ha, 2012. Application of FRET to photodynamic cancer therapy: Kim et al., 2015.

Artificial analogs of light-harvesting complexes: Buhbut et al., 2010.

T_2 **Track 2**

2.4.3′ More about membrane potentials

Section 2.4.2 mentioned that every ion species contributes to a single overall electric potential, which then acts on all ions according to their charges. Ions of each species are also subject to a private thermodynamic driving force, related to only their own species' concentration. The two forces may have opposite directions; in that case, transport will be determined by their net effect. For example, potassium ions move *out* of a neuron when their channels open, even though their charge is the same as that of sodium ions, due to an interior "pressure" (concentration gradient) that overcomes electric attraction to the interior. Ultimately, the cell arrives at a steady state, in which the membrane potential and ion concentrations are such that each ion species' net transport, including contributions from pumps, exchangers, open channels, and leakage, equals zero. For more details, see Nelson, 2014, chapt. 11.

Section 2.4.3 implied that a cell will come to an equilibrium with membrane potential equal to zero if ion pumping stops. More accurately, the leak conductances for *small* ions like Na^+ are nonzero, but for macromolecules such as DNA and proteins, which are generally negatively charged, the leak conductance is essentially zero. The presence of those trapped negative macromolecules implies that even in equilibrium, there will be some potential drop across the cell membrane (the "Donnan potential"). However, it is still correct to say that active ion pumping maintains nonequilibrium ion concentrations and potential drop across the cell membrane, and hence that the opening of a previously closed channel will alter the flow of charge. In neurons, the resting state is such that opening sodium channels depolarizes the cell. Again, Nelson, 2014, chapt. 11 gives details.

T_2 **Track 2**

2.4.5′ Other uses for the resting potential

The original role of ion pumping may simply have been to maintain osmotic balance (Nelson, 2014, §11.2.3.). Bacteria, and some organelles within eukaryotic cells, such as mitochondria and chloroplasts, also utilize their membrane potential as an energy-distribution scheme to power their ATP synthases and flagellar motors, and some secondary active transporters in eukaryotes are also powered by electrochemical gradients. Finally, Section 2.5.2 mentions that some single-celled eukaryotes use changes in membrane potential for intracellular signaling.

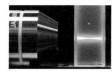

Fig. 2.15 (page 80)

T_2 **Track 2**

2.7.2′ The β-squared rule

Although most of the photons in Figure 2.15 pass through the sample without being absorbed, nevertheless their density varies as the beam constricts to the focus, then reexpands. Thus, the mean rate β at which photons are presented to any particular fluorophore varies along the beam.

The main text gave a qualitative argument for why the probability for two photons to arrive within a short time interval T should be proportional to β^2. To be more precise, suppose that a blip in a Poisson process arrives at some time. The probability that *no* blip arrives within an interval T of the first one is then the limit of $(1 - \beta \Delta t)^{T/\Delta t}$ as $\Delta t \to 0$, that is, $e^{-\beta T}$. Our model of two-photon excitation says that in such instances, the original blip gives no fluorescence (the "ticket expires" before it can be used). That is, a fraction $e^{-\beta T}$ from the original Poisson process get rejected. The non-rejected blips themselves form a Poisson process, with mean rate reduced by the factor $1 - e^{-\beta T}$. For $T \ll \beta^{-1}$, the new rate is $\approx \beta^2 T$, which indeed is proportional to the initial mean rate squared.

Why should the "expiration time" T be small? When a single photon arrives with the wrong energy to promote a molecule to any excited state, normally no excitation occurs and the photon is either unaffected or at most scattered. But a molecule's electrons can accept energy without "realizing" it, and during that time another photon can arrive. How big a delay is acceptable is controlled by a quantum mechanical uncertainty relation; for the fluorophores used in two-photon microscopy, it's indeed much shorter than the mean waiting time for photon arrivals.

$\boxed{T_2}$ **Track 2**

2.7.3′ More about two-photon imaging

The main text mentioned that the excitation light used in two-photon microscopy can be arranged to use a wavelength band that does not scatter much in tissue. However, the resulting fluorescence is shorter in wavelength, and does scatter. It may seem as though nothing has been gained: We still get a blurred image from the scatter of the outgoing light!

In fact, all that is needed for image formation is that *either* the incoming *or* the outgoing light travel on straight lines. Ordinary microscopy illuminates an entire sample with diffuse light, but registers the origin of each outgoing photon by focusing them (see Chapter 6). Thick-sample, two-photon microscopy uses the opposite approach: The excitation light is focused to excite fluorophores in only one tiny spot of the sample. We then collect *all* fluorescence light, regardless of what direction it appears to have come from, because we *know* that it must actually have come from the illuminated spot. By sweeping the focused spot through a plane, we build up an image of everything in that plane. By sweeping the spot through a solid volume, we can even build up a 3D reconstruction of what is in that volume.

$\boxed{T_2}$ **Track 2**

2.8.1′ About FRET and its efficiency

1. The name "fluorescence resonance energy transfer" can be confusing, because energy, not "fluorescence resonance energy," is being transferred. For this reason, some authors abbreviate the term still further, to just "RET." Others interpret FRET as "Förster resonance energy transfer" or "Förster radiationless energy transfer," to emphasize the contributions of T. Förster. Other modified versions of FRET have names like

bioluminescence resonance energy transfer (BRET) and lanthanide based luminescence resonance energy transfer (LRET).

2. Equation 2.1 neglects other channels for excitation loss, for example, conversion to heat energy. Also, in practice FRET measurements must correct for the fact that photons from the acceptor have some probability of being mis-classified as coming from donor fluorescence, and vice versa ("crosstalk").

$\boxed{T_2}$ **Track 2**

2.8.4′ Other experimental tests of FRET

1. Our discussion of FRET included the prediction that it depends on the overlap between the donor emission and acceptor absorption spectra.[42] More precisely, the prediction is that the effect should be proportional to the spectral overlap integral. This prediction has been tested experimentally (van der Meer et al., 1994, §2.6).

2. The predicted distance dependence was also tested (Sindbert et al., 2011).

3. The interaction of two electric dipoles also depends on their relative *orientation,* and hence so does r_F. This prediction was tested as well (Iqbal et al., 2008).

$\boxed{T_2}$ **Track 2**

2.8.5′ Why reported FRET efficiencies sometimes exceed 100%

Reported values of FRET efficiency are not always numbers between 0 and 1. To understand this phenomenon, we need to make our definition a bit more precise.

   Measured FRET signals, like any kind of fluorescence, are partly corrupted by stray "background" light not coming from the fluorophores in question. Thus, the true donor emission must be estimated as the actual number of photons observed in a time window minus that observed in a similar sample region containing no donor fluorophore, and similarly for the acceptor. All of these numbers, however, are subject to Poisson noise; moreover, the difference of two noisy quantities has even greater relative standard deviation than either by itself (Section 0.5.4, page 16). Thus, the apparent FRET efficiency has a random component, and can in some situations appear to be greater than 100% (or indeed smaller than zero).

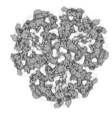

Fig. 2.29a (page 97)

$\boxed{T_2}$ **Track 2**

2.9.3′ More details about the photosynthesis apparatus in plants

Cyanobacteria, algae, and plants actually have a two-stage photosynthetic reaction, supplied with energy by *two* kinds of photosystems; Figure 2.29a shows the one that was discovered first. The other one, "Photosystem II," is also fed excitations from an even larger complex.

---

[42]That statement appears in Idea 2.4 (page 89). You'll work out some details in Problem 2.4.

PROBLEMS

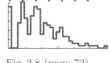

Fig. 2.8 (page 72)

2.1 *Discrete neurotransmitter release*

The goal of this problem is to test a physical model by making predictions and comparing to experimental data (Figure 2.8). First obtain Dataset 4, which contains binned data on how often various peak membrane potential changes, $\Delta\Phi$, were observed in a muscle cell stimulated by a motor neuron. In a separate measurement, the authors also studied spontaneous events (no stimulus), and found the sample mean of the peak potential change to be $\mu_{\Phi,1} = 0.40\,\text{mV}$ and its estimated variance to be $\sigma^2 = 0.00825\,\text{mV}^2$.

The physical model discussed in the text states that each response involves the release of an integer number, ℓ, of vesicles. But the measured quantity (peak membrane potential change) is continuous. To connect these two random variables, the model assumes that each released vesicle makes an additive contribution to the membrane potential change with the same distribution as that from the spontaneous events. (That is, each spontaneous event is assumed to involve release of exactly one vesicle.) Thus, for given ℓ the response is assumed to be the sum of ℓ independent, identically distributed continuous random variables. Assume that each of these constituents follows a Gaussian distribution with expectation and variance given by the values found for the spontaneous events.

a. Find the predicted distribution of the responses for an event that releases exactly 2 (or 3, ...) vesicles. You can neglect any source of variation other than number of neurotransmitter molecules per vesicle. [*Hint:* First work Problem 0.8 (page 20).]

b. The model also assumes that ℓ is itself a Poisson random variable. Form the sum of the distributions you found in (a), weighted by a Poisson distribution in ℓ, and explain why the result is a correctly normalized probability density function. (For $\ell = 0$, use some PDF that is very narrowly peaked about zero.)

c. Show that the expectation $\langle \Delta\Phi \rangle$ in the PDF that you just found equals $\mu_{\Phi,1}\mu_\ell$, and thus that μ_ℓ can be obtained by computing the sample mean of all the responses in the dataset, and dividing by the mean response from a single vesicle.

d. Plot the PDF you found in (b, c).

e. Your prediction for the estimated PDF contains *no* free fitting parameters. Superimpose its graph on a graph of the estimated PDF from the experimental data.

f. The researchers found that, in 18 out of 198 trials, there was no response at all. Compute $\mathcal{P}_{\text{pois}}(0; \mu_\ell)$ and comment.

2.2 *FRET versus distance*

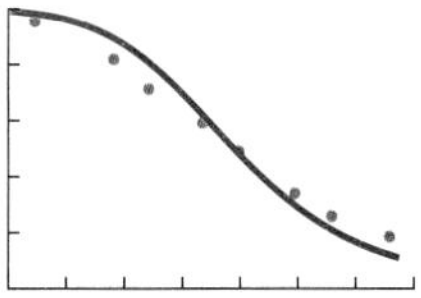

Obtain Dataset 5, which was used to generate Figure 2.22. Before you can compare these data with the prediction of Equation 2.3 (page 88), you must convert the separation in basepairs N, given in the dataset, into actual spatial separation r. To do this, use the formula given by the experimenters:

$$r = \sqrt{((0.34\,\text{nm})(N-1) + L)^2 + a^2},$$

Fig. 2.22 (page 88)

where $L = 0.4\,\text{nm}$ and $a = 2.5\,\text{nm}$. Then adjust the value of the Förster radius r_F to get a reasonable fit, and comment.

2.3 *Quantum yields in photosynthesis*

The main text proposed a FRET-type energy transfer mechanism in photosynthe-

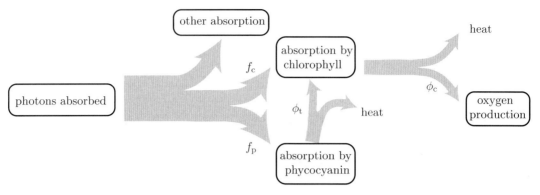

Figure 2.30: [Schematic.] **Overall quantum yield for two different pathways.** See Problem 2.3. The branching fractions f_p and f_c depend on the wavelength of the incoming light, but the quantum yields ϕ_c and ϕ_t do not.

sis, and gave some qualitative arguments for it. In this problem, you'll do a more quantitative exploration.

Emerson and Lewis measured the overall quantum yield ϕ_{tot} for photosynthesis in their organism, that is, the number of photons that must be absorbed in order to create one oxygen molecule (Section 2.9.2). However, they could not separately measure the quantum yield ϕ_c for chlorophyll alone, that is, the average number of oxygen molecules formed per excited chlorophyll molecule. Nor could they measure the probability ϕ_t that an excited phycocyanin will transfer its excitation to a chlorophyll.

Assume that once a chromophore has been excited, it retains no other memory of the exciting photon, as in fluorescence. In particular, the separate quantum yields ϕ_c and ϕ_t do not depend on the exciting photon's wavelength (Figure 2.30). Thus, the overall quantum yield at any wavelength is the average of ϕ_c and $\phi_c\phi_t$, *weighted* by the corresponding abilities of the two chromophores to capture photons of that wavelength.

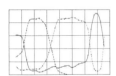

Fig. 2.27a (page 95)

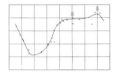

Fig. 2.27b (page 95)

a. Use Figure 2.27a to estimate $f_{1,c}$, the fraction of light at wavelength $\lambda_1 = 680$ nm that is absorbed by chlorophyll. Similarly, estimate $f_{2,c}$, the same quantity for wavelength $\lambda_2 = 600$ nm, and the two corresponding quantities $f_{1,p}$ and $f_{2,p}$ for phycocyanin.

b. Use Figure 2.27b to estimate the overall quantum yields at these two wavelengths.

c. Explain why the overall quantum yield reflects a weighted average of the two unknown quantities ϕ_c and $\phi_c\phi_t$, and in this way, obtain two equations in two unknowns.

d. Solve your equations to find ϕ_c and $\phi_c\phi_t$, and compare them.

e. Obtain Dataset 6, which contains the data in the figures. Continue to assume that ϕ_c and ϕ_t are independent of wavelength, and have the values you found in (d). Also assume that the quantum yield of the carotenoids is *zero* at every wavelength. Predict the overall quantum yield throughout the wavelength interval shown in Figure 2.27b, and plot your prediction along with the experimentally measured yield from the dataset. [*Hint:* In the dataset, the measurements on the different pigments were not all made at exactly the same values of wavelength. Find out how to get your computer to interpolate the given data, so that each curve may be evaluated at a common set of values.]

f. How could you improve your fit to the data?

2.4 $\boxed{T_2}$ Classical model of FRET

We can get insight into fluorescence resonance energy transfer by using ideas from classical mechanics.[43] Imagine an oscillator representing the donor molecule's electrons. These electrons can jiggle relative to the nuclei, and thus give rise to an electrostatic force on a second, nearby, oscillator, which represents the acceptor fluorophore. Suppose that this force $f(t)$ has angular frequency ω_D (determined by the donor's excited state), and strength J (determined by the donor's state and the distance to the acceptor):

$$f(t) = J \cos(\omega_D t). \tag{2.5}$$

We model the acceptor's electron cloud as an object with mass m. It's attached to the molecule's nuclei by a spring, with spring constant k. Moreover, the acceptor continually dissipates energy to "friction," a classical representation of energy loss from the acceptor, for example, by its fluorescence. Calling the friction constant ζ, Newton's law $f_{\text{tot}} = ma$ states that the donor's position $x(t)$ obeys

$$m\frac{\mathrm{d}^2 x}{\mathrm{d}t^2} = -kx - \zeta\frac{\mathrm{d}x}{\mathrm{d}t} + f. \tag{2.6}$$

To simplify this equation, define new symbols $\omega_A = \sqrt{k/m}$, $\eta = \zeta/m$, and $K = J/m$, and eliminate k, ζ, and J by writing them in terms of the new symbols.

a. Regardless of its initial condition, eventually the solution $x(t)$ will oscillate at the driving frequency ω_D. So consider trial solutions of the form $x(t) = A\cos(\omega_D t) + B\sin(\omega_D t)$. Find the constants A and B in terms of K, η, ω_D, and ω_A.[44]

b. In the steady state that we are studying, the rate at which the acceptor gets energy from the donor must equal the rate at which it loses energy to dissipation, which is $P_{\text{fret}} = \zeta(\mathrm{d}x/\mathrm{d}t)^2$. Evaluate this for your solution.

c. The quantity you found in (b) is always positive, but it oscillates. We only need its time average, however, which is given by a simpler expression than your answer to (b). Find that.

d. Section 1.6.3 emphasized that actually the donor and acceptor are not in precisely known states:[45] Rather, each moves within a *distribution* of possible states, with varying values of ω_D and ω_A. The average rate of energy transfer is then the average of the quantity you found in (c), weighted by the corresponding distributions $\wp_D(\omega_D)$ and $\wp_A(\omega_A)$:

$$\int \mathrm{d}\omega_D\,\wp_D(\omega_D) \int \mathrm{d}\omega_A\,\wp_A(\omega_A) \left[P_{\text{fret}}(\omega_D, \omega_A)\right]_{\text{time avg}}.$$

To simplify this expression, suppose that the damping η is very small. Then your result from (c) is very sharply peaked near $\omega_D = \omega_A$. Exploit this fact by letting

$$\omega_D = \bar{\omega} - \tfrac{1}{2}\Delta\omega; \quad \omega_A = \bar{\omega} + \tfrac{1}{2}\Delta\omega,$$

and change integration variables from ω_D, ω_A to $\bar{\omega}$, $\Delta\omega$. Then approximate your answer to (c) by replacing $\Delta\omega$ by 0 everywhere, except for the one term in the

---

[43] $\boxed{T_2}$ Chapter 14 gives a quantum-mechanical derivation.

[44] Section 4.4 will introduce a mathematical trick, involving complex numbers, that simplifies these calculations. If you already know that trick, you can use it instead of the method suggested here.

[45] Figure 2.21 (page 86) indicates this fact with double-headed arrows.

denominator responsible for the sharp peak. With this approximation, you can readily do the integral over $\Delta\omega$.

e. Section 2.8.4 argued that the force K is proportional to the inverse cube of the distance between the donor and acceptor. From this observation and your calculations, comment on the assertions in Idea 2.4 (page 89).

f. The donor has other pathways to lose its excitation energy, for example, by the direct emission of light (donor fluorescence).[46] Without calculating these energy-loss rates, we can say that they do not depend on r; such losses would happen even if the acceptor were not present at all. Use your result from (e) to get a general formula for the r-dependence of the FRET efficiency, that is, the ratio in Equation 2.1 (page 84). Compare the form of your answer with Equation 2.3 (page 88).

---

[46] You may ignore other possible deexcitation pathways, such as releasing energy as heat.

CHAPTER 3

Color Vision

> As it is almost impossible to conceive each sensitive point of the retina to contain an infinite number of particles, each capable of vibrating in perfect unison with every possible undulation, it becomes necessary to suppose the number limited, for instance to the three principal colours.
>
> — *Thomas Young, 1802*

3.1 SIGNPOST: *A FIFTH DIMENSION*

Living organisms are inference machines. They (we) constantly attempt to predict the future based on measurements made in the present and past.

Vision is one powerful mechanism for gathering information about the world. Our eyes intercept a stream of photons and deliver an estimate of the probability density for those photons to arrive from various directions, moment by moment in time. We also get information about the distances to objects from our stereo vision, so all together we can construct a mental model of the world in terms of the familiar four dimensions of first-year physics. Later chapters will discuss the details of how this sensory transduction works.

First, however, this chapter will discuss our ability to access a different aspect of the photon stream, its "color." Color amounts to a highly informative *fifth dimension* that we can use to interpret our world. As with other stories in this book, the main outlines were correctly grasped by imaginative scientists long before the technical tools needed to get direct confirmation became available.

Color is subtle. Even the ancient Greeks knew that color is not a purely objective quantity like mass. Similarly, Galileo wrote in 1630 that color "resides only in consciousness." But surely color has *something* to do with the physical stimuli entering our eyes! Let's see how much of color perception we can understand, starting from ideas developed in the preceding chapters.

Our Focus Question is

Biological question: How can a mixture of red and green light appear yellow?

Physical idea: Our eyes project light spectra down to a low-dimensional "color space," making a trade-off between color discrimination and spatial resolution.

3.2 COLOR VISION CONFERS A FITNESS PAYOFF

Color vision has various advantages for an animal. For example:

1. Our brains must take a two-dimensional projection of the world and somehow organize it into perceived *objects.* Color information can help us to distinguish two adjacent objects, even if they have similar brightness and texture.

2. Color also helps us discriminate between objects, even if they are not adjacent. An animal that can distinguish ripe from unripe fruit, without taking time to taste each one, can forage faster. Similarly, it's useful for an animal to be able to recognize that a potential prey is poisonous, and avoid it.

3. Color helps many animals to distinguish individuals of their own species (for example, possible mates) from others.

4. Some animals use color-based cues to broadcast their emotional state to others (for example, aggression), and to read others' states.

Clearly, color perception can deliver an evolutionary (fitness) payoff.

Color perception is not all-or-nothing. Most mammals (dogs, horses) have worse color discrimination than do humans. Some nocturnal animals have only single-color vision (raccoons, kinkajous), as do many deep-sea fish, octopus, and whales. Other species (some birds and crustaceans) have *better* discrimination than we do. So beyond asking about how color vision is possible, we must also ask: What physically determines an animal's ability to discriminate colors?

3.3 NEWTON'S EXPERIMENTS ON COLOR

Let's first summarize some of Isaac Newton's discoveries on light and color.

- Newton found that he could separate sunlight into a spectrum by passing it through a glass prism. That is, sunlight is a *mixture;* in his own words, "Blew rayes suffer a greater refraction than red ones.... [White light] consists of rayes differently refrangible.... To the same degree of refrangibility ever belongs the same colour, & to the same colour ever belongs the same degree of refrangibility."

- Newton also showed that he could arrange a second prism to take the spectrum of sunlight and *reassemble* it into white light, in all aspects identical to the light that entered the first one. Thus, the colors were not being created by the first prism: They were all present in the "white" sunlight from the start.

- In an experiment Newton called his "Experimentum Crucis," he singled out just one color of the spectrum, by blocking all but one band of light coming from a prism. When Newton passed such monochromatic light through a second prism, he found he could not split it any more, and moreover, the second prism bent the light path by the same amount as the first.

If we think of a beam of light as a stream of photons, each of which carries a particular amount of energy, then we can interpret Newton's results by saying that a prism somehow *sorts* the stream, directing each photon type into a slightly different spatial direction.[1] The spectral photon arrival rate,[2] $\mathcal{I}(\lambda)$, then gives the mean rates at which photons arrive at different locations on a projection screen that intercepts the spread-out beam. Figure 3.1 illustrates these ideas in two examples. Newton also studied what happened when ordinary (nonluminous) objects were illuminated by various kinds of light:

- When monochromatic light shines on an object, what comes back to our eyes

---

[1] Section 5.3.5 will return to the question of how a prism accomplishes this.
[2] Section 1.5.2 (page 38) defined spectral photon arrival rate.

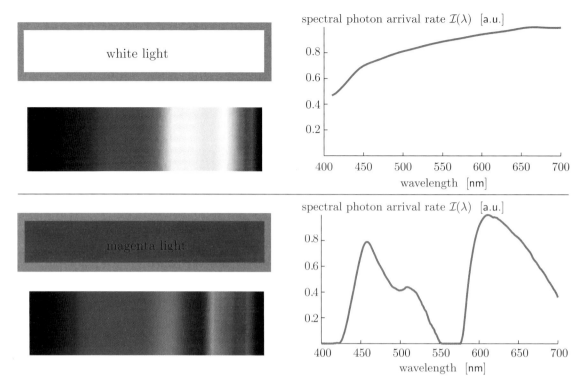

Figure 3.1: [Color patches, spectra obtained with a prism, and experimental data.] **Light spectra.** *Top left:* Sunlight, or the light from a white-hot object, consists of a broad spread of photon energies. When passed through a prism, it separates into a continuous smear of colors. *Top right:* We can represent the spectrum by a spectral photon arrival rate function that is nearly constant in the range of visible light. *Lower left:* Magenta colored light, in this case from a computer projector, also consists of a broad spread of energies, but with less green relative to red and blue than is found in sunlight. *Lower right:* This time, the spectrum has *two* peaks, in the blue and red, with a dip between them corresponding to the absent green. [Data courtesy William Berner.]

is again monochromatic, with the same wavelength,[3] regardless of how the illuminated object looks in daylight (Figure 3.2): "I have reflected [such light] with bodies which in day light were of other colours; I have... transmitted it through coloured mediums & through mediums irradiated with other sort of rayes, & diversly terminated it, & yet could not produce any new colour out of it."

In short, Newton concluded that color is a property of the *light* that enters our eyes. As he put it, the world appears colorful because it consists of bodies "variously qualified to reflect one sort of light in greater plenty than another." That is, objects *selectively reflect* light of particular wavelengths from the streams that impinge on them: We observe the combined effect of the distribution of photon types arriving at the object, and the distribution of its reflection probabilities for each type of photon. This insight simplifies our task: If we can first understand our discrimination of colored *lights,* then that will also help us to understand colored *objects.*

---

[3]Actually, exceptions to this rule were later discovered, though they are rare: For example, a few substances display fluorescence, discussed in Chapter 1.

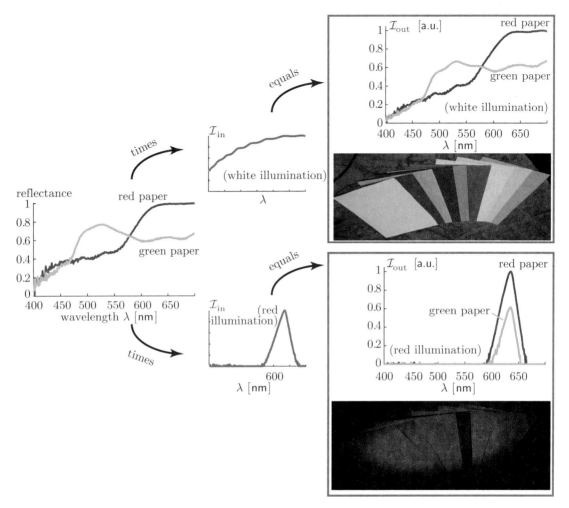

Figure 3.2: [Experimental data; photographs] **Color perception requires a range of incoming wavelengths.** *Left:* Reflectance spectra of two colored papers, that is, the relative probabilities for a photon to be reflected as functions of wavelength. *Center:* Two spectra of light used to illuminate the red and green colored papers. *Right:* A variety of colored objects as they appear in the two illuminations shown. The two objects characterized in the graphs at *left* are shown in the middle of each photo. Every one appears to be some shade of red when illuminated with nearly monochromatic red light. The graphs at *right* show the spectra of light reflected by red and green paper under each illumination. [Data, photos, and realization courtesy William Berner.]

3.4 **BACKGROUND:** MORE PROPERTIES OF POISSON PROCESSES

Section 1.4 (page 35) introduced the general notion of a random process, and one particularly simple example, called the Poisson process, that describes many steady, monochromatic light sources. Here we briefly state two additional properties of Poisson processes that will be needed later.

3.4.1 Thinning property

Suppose that we have a Poisson process with some mean rate β, for example, the arrivals of photons from a light source. Each draw from this random process is a series of "blip" times, and in any interval Δt much shorter than β^{-1}, the probability of a blip is $(\Delta t)\beta$. We now define a new random process by giving a procedure to obtain a draw from it:

- First draw a series of blip times from the original Poisson process;
- Then subject each blip to a Bernoulli trial that keeps a fraction ξ and discards the rest. These Bernoulli trials are all independent of one another.
- The sequence of surviving blip times constitutes a draw from the new random process.

You can readily confirm that the new process is again Poisson. The probability to get a blip in Δt equals $(\Delta t)\beta$ times ξ, so

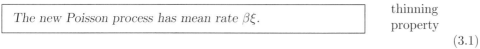

The new Poisson process has mean rate $\beta\xi$.	thinning property

(3.1)

We already used this property implicitly in our discussion of two-photon excitation (Section 2.7).

In contrast, had we started with a *periodic* (regularly repeating) series of blips with period β^{-1}, then randomly deleting a fraction of them would *not* have led to a new series of the same general form (that is, periodic).

3.4.2 Merging property

Next, suppose that we begin with *two* independent Poisson processes, with mean rates β_1 and β_2. We now create a new random process by again giving a procedure to obtain its draws:

- First draw a series of blip times from each of the two parent processes;
- Then interleave the lists to create a list of times when *either one* had a blip.

Again, the new process turns out to be Poisson. To see this, consider a Δt that is much smaller than either β_1^{-1} or β_2^{-1}. Then the probability to get *neither* kind of blip in Δt is the product

$$\mathcal{P}(\text{no blip of type 1})\mathcal{P}(\text{no blip of type 2}) = (1 - \beta_1\Delta t)(1 - \beta_2\Delta t) \approx 1 - (\beta_1 + \beta_2)\Delta t,$$

because we may drop the small Δt^2 term. So the probability to get a blip of either type is just the sum of those for the two parent processes, and we have shown that

The new Poisson process has mean rate $\beta_1 + \beta_2$.	merging property

(3.2)

In contrast, had we started with two *periodic* series of blips, then interleaving them would be very unlikely to lead to a new series that was of the same general form (that is, periodic).

The merging property fits with a result about convolution obtained earlier: Prior to merging, the numbers of blips counted in a fixed interval T are Poisson distributed with expectations $\beta_1 T$ and $\beta_2 T$. After merging, the total number of blips in the same time interval is the convolution of those two Poisson distributions, that is,[4] Poisson with expectation $(\beta_1 + \beta_2)T$. That's the same result we'd have found had we first merged the original Poisson processes according to Equation 3.2, then asked for the distribution of blip counts.

3.4.3 Significance for light

These mathematical properties of Poisson processes let us answer some potential objections to the Light Hypothesis. If we were to model light of constant intensity as a sequence of projectiles arriving uniformly in time, that is, with always the same time intervals between successive lumps, then randomly deleting a fraction of the lumps would destroy that property, changing the character of light. Even if we follow a strict rule ("delete blips whose sequence number is a multiple of three"), the resulting time series is not uniform. But light of constant intensity does *not* consist of lumps arriving with uniform intervals, and the thinning property does imply that it preserves its character when we delete some fraction of the lumps in the series. Indeed, when we put on sunglasses a constant-intensity light source appears to be less intense, but otherwise has the same character as before.

Similar remarks apply to, say, doubling a constant-intensity beam by turning on a second, identical source. If each light consisted of blips arriving at fixed time intervals, then the merged time series would not have that character. However, the merging property says that interleaving the blips from two Poisson processes does lead to another Poisson process. Indeed, when we combine two identical, constant-intensity light sources, the result is more intense, but otherwise has the same character as before.

This background section has developed the idea of Poisson processes by stating two of their key properties. This chapter, and later ones, will now translate those mathematical properties into physical statements about color vision.

For further details about merging and thinning, see the references listed at the end of this chapter.

3.5 COMBINING TWO BEAMS CORRESPONDS TO SUMMING THEIR SPECTRA

When we combine two independent beams, for example, by letting them land on the same spot of a viewing screen, the merging property says that the combined effect of $\mathcal{I}_1$ and $\mathcal{I}_2$ is the same as that of a single beam with spectrum $\mathcal{I}_{\text{tot}}$, where[5]

$$\mathcal{I}_{\text{tot}}(\lambda) = \mathcal{I}_1(\lambda) + \mathcal{I}_2(\lambda). \quad \text{spectrum of combined light} \qquad (3.3)$$

We will abbreviate formulas like this one, writing $\mathcal{I}_{\text{tot}} = \mathcal{I}_1 + \mathcal{I}_2$ to indicate a relation that holds at each value of wavelength.

---

[4]See Equation 0.49 (page 17).

[5]Two beams of light with a common source can behave in a more complicated way, displaying interference. Chapter 4 will discuss this situation.

Your Turn 3A

Interpret Equation 3.3 as a consequence of the merging property of Poisson processes (Section 3.4.2).

$\boxed{T_2}$ *Section 3.5′ (page 133) defines some other quantities used to describe spectra, and gives another viewpoint on the combination rule (Equation 3.3).*

3.6 PSYCHOPHYSICAL ASPECTS OF COLOR

The previous section discussed aspects of light that can be objectively measured by instruments, without any human making judgments. The rest of this chapter is concerned with understanding the relationship between the spectrum $\mathcal{I}(\lambda)$ and our subjective *perception* of color.

3.6.1 $R+G$ looks like Y

Like many phenomena that we observe every day, color has some surprises when examined carefully. Suppose that we project pure 500 nm and pure 650 nm lights onto separate areas of a screen, for example, by sending white light through colored filters. We observe patches of green and red light. When we swing one beam so that it overlaps on the screen with the other, however, many people are surprised that they observe a clear yellow—not some sort of "reddish green." Figure 3.3a gives a less vivid version of this experiment, in which the mixing of two colors of light is accomplished by using the limitations of our eyes' spatial resolution. How could our eyes be so bad that they can't even tell spectral yellow from red+green?

3.6.2 Color discrimination is many-to-one

Perhaps equally surprising, when we add a third color of light (blue) to the red and green, the result appears approximately *white* (Figure 3.4), a perception that many would refer to as "no color at all." In short,

> Our eyes **discard some information** about the spectrum of light.   (3.4)

However, any two light sources that are perceptually different, in a controlled color-matching experiment, really do have different spectra.

Idea 3.4 is a big clue to how color vision works. Before we can use it, however, we need to make it more precise and systematic.

3.6.3 Perceptual matching follows quantitative, reproducible, and context-independent rules

Our perception of color is complex and subjective. A professional trained to distinguish and name a hundred distinct colors will "see" differently from untrained subjects. Even a single individual will report different color perceptions of the same light, depending on its surroundings, overall intensity, and so on. Much of this complexity arises from neural processing that happens after the initial absorption of light in the retina, which we will not pursue in this chapter.

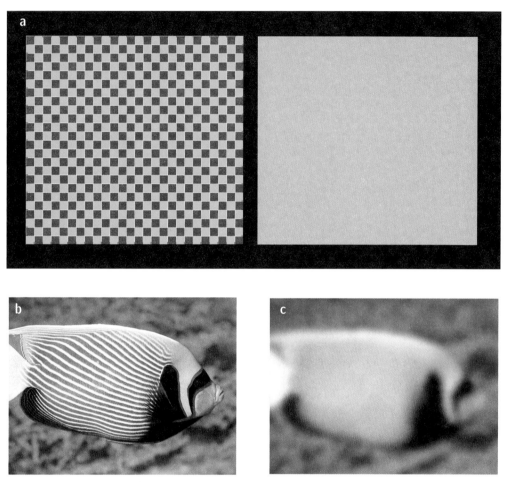

Figure 3.3: (a) [Visual illusion; photographs.] **Additive color.** (a) When viewed up close, the left box is seen to consist of small red and green squares. When viewed from a few meters away, however, each photoreceptor in your eye receives light from *both* red and green squares, whose spectra merge. (Try removing your eyeglasses, if you wear them.) The resulting percept is closer in color to the right box than it is to either pure red or green— a "pointillist" synthesis of yellow from two other colors, illustrating the claim that, perceptually, $R + G \sim Y$. (b) The angelfish *Pomacanthus imperator*. These and other reef fish exhibit narrow yellow and blue stripes that are conspicuous at close range. (c) The same photograph as (b), but blurred to approximate a small reef fish's visual acuity at $2\,\mathrm{m}$ distance. Note how the blue-yellow striped area has combined to form a dull gray-green, helping the fish blend with its surroundings. [(b,c) Photo by Steve Parish.]

Figure 3.4: [Photographs.] **"White" lights.** *Top:* Broad spectrum obtained from sunlight by using a transmission grating. *Bottom:* The spectrum from a white compact-fluorescent light obtained with the same grating consists of distinct bands of red, green, and blue; yet this light, too, is perceived as "white."

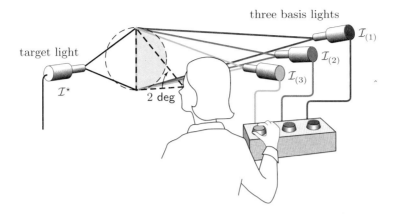

Figure 3.5: [Schematic.] **Color-matching experiment.** A "target" light is projected onto the left half of a screen. A subject attempts to obtain a perceptual match by adjusting the intensities of three otherwise fixed basis lights that converge on the right half of the screen.

The majority of observers *will*, however, agree on the yes/no question of whether two adjacent, uniformly lit patches of the visual field *match* when presented side-by-side (Figure 3.5). If the two light sources have identical spectra, then certainly all subjects will agree that they match. But we are interested in the fact that two *different* light spectra will sometimes also give a perceptual match, a phenomenon called **metamerism**. Two physically different spectra that give a perceptual match are called **metamers**; we will denote this situation by the equivalence symbol $\mathcal{I} \sim \mathcal{I}'$. We would like to know what mathematical property of two spectra determines whether they are metamers, and what the answer can tell us about the biophysical machinery that implements color discrimination.

Much experimentation in the 18th and 19th centuries revealed some regularities about metamerism, long before anything was known about the underlying mechanisms of vision:

1. *Observer independence:* All (normal-vision) observers agree on the question of whether two spectra give a perceptual match. Also, a single observer will make the same judgment about identical presentations repeated over time.

2. *Persistence of matches:* Lights that match under one set of viewing conditions continue to match when the conditions are changed. Thus, for example, we can change the border surrounding the two patches of light under study without destroying a match (or creating one where there was previously no match).

3. *Reflexivity, transitivity:* If we reverse the positions of two lights $\mathcal{I}_1$ and $\mathcal{I}_2$, this does not affect whether they match. Moreover, if $\mathcal{I}_1 \sim \mathcal{I}_2$ and $\mathcal{I}_2 \sim \mathcal{I}_3$, then $\mathcal{I}_1 \sim \mathcal{I}_3$.

4. *Grassmann's law:*

$$\text{If } \mathcal{I}_1 \sim \mathcal{I}_2, \text{ then } (\mathcal{I}_1 + \mathcal{I}_3) \sim (\mathcal{I}_2 + \mathcal{I}_3) \text{ for any other } \mathcal{I}_3. \qquad (3.5)$$

In the last formula, the plus signs denote the combination of two beams of light (Equation 3.3).

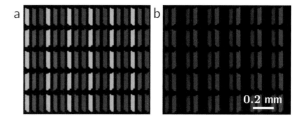

Figure 3.6: [Photographs.] **The pixels of a computer's display screen.** Each pixel is a small rectangle emitting a fixed spectrum of light with adjustable light intensity. (a) A "white" region of a screen image. (b) A "magenta" region. [Photos courtesy Mojca Čepič.]

Your Turn 3B

Use the preceding rules to show that, if $\mathcal{I} \sim \mathcal{I}'$, then $2\mathcal{I} \sim 2\mathcal{I}'$.

More generally, we can scale any two perceptually equivalent lights by any common factor and obtain two other equivalent lights.[6]

One particularly informative version of the matching experiment holds one light's spectrum $\mathcal{I}^\star$ fixed (the "target"), while the human subject adjusts the other one, seeking a perceptual match (Figure 3.5). The experimenter supplies three **basis lights** with fixed spectra; the subject can adjust only the intensities of these lights, which are then projected onto a viewing screen. Color scientists found three additional empirical rules in this situation:

5. *Trichromacy:* Just three basis lights are enough to give perceptual matches to a wide range of target lights. Using fewer than three basis lights results in a large region of color space that cannot be matched. We say that humans, and other old-world apes, are **trichromatic**.[7] Most other mammals are *di*chromatic; for them, two basis lights suffice to give perceptual matches. Most birds require *four* different basis lights.

6. *Uniqueness:* Once we make a choice of three fixed basis lights, then the match to a particular target light is unique: No other settings for the intensities of those basis lights will give a match.

7. *Basis freedom:* There are many different choices for the spectra of the three fixed basis lights, any of which will allow matching a wide range of colors.

Our goal for this chapter is explain these observations as mathematical consequences of a physical model, and moreover to make quantitative predictions for color matching that can be compared to experiments.

Item **7** above may surprise you, if you have been told that "the primary colors of light are red, green, and blue." In fact, that traditional set is just one convenient choice of **additive primary colors**, so called because "mixing" them means presenting the eye with the *sum* of their spectra (Equation 3.3, page 112). The traditional set is a popular choice because the range of colors that it can match (its **color gamut**) is slightly larger than some other choices. Thus, television and computer screens consist of tiny dots, or pixels, each with a spectrum peaked in red, blue, or green (Figure 3.6). By varying the relative pixel intensities, the screen can generate sensations corresponding to millions of distinct colors.

Section 3.6 has summarized some systematic features of color vision in animals.

---

[6]However, $2\mathcal{I} \not\sim \mathcal{I}$: Lights differing in overall intensity are not considered to be matches.
[7]Trichromacy in humans had been observed or at least suspected as early as 1708.

$\boxed{T_2}$ *A small subpopulation of observers do not agree with others about color matching;
they are either color-blind, or have other inherited anomalies in color vision. See
Section 3.6.3' (page 133).*

3.7 COLOR FROM SELECTIVE ABSORPTION

3.7.1 Reflectance and transmittance spectra

Newton's experiments led him to conclude that the spectrum of light coming to our
eyes from a nonluminous object depends on both the object and the spectrum of the
illuminating light $\mathcal{I}_{\mathrm{in}}$ (Section 3.3, page 108). To make this idea more precise, suppose
that the probability for each photon to be reflected depends on its wavelength λ.
Taken together, the probabilities constitute the object's "reflectance spectrum" $\mathcal{R}(\lambda)$;
what reaches our eyes are photons with the reduced spectral arrival rate $\mathcal{I}_{\mathrm{refl}}(\lambda) =
\mathcal{I}_{\mathrm{in}}(\lambda)\mathcal{R}(\lambda)$. Figure 3.2 illustrates this multiplication.

> **Your Turn 3C**
> Interpret the last statement as a consequence of the thinning property of Poisson
> processes.

Similar remarks apply to transparent substances that *transmit* light with probability that depends on wavelength, leading to a "transmittance spectrum."

3.7.2 Subtractive color scheme

Instead of red/green/blue, you may instead have been told that "the primary colors
are red, yellow, and blue." Such statements refer to *paint* or *ink*, not light itself. The
difference between the two color statements arises because colored objects (such as
ink on paper) absorb the components of white light selectively, giving rise to a new
system called the **subtractive color** scheme.

A more precise version of the subtractive scheme gives the primaries as cyan,
magenta, and yellow, and states, for example, that "mixing yellow and cyan yields
green." To understand this statement, first imagine sending a beam of white light
through a filter that transmits a band of wavelengths centered on spectral yellow
(Figure 3.7a). What emerges is a mix of spectral green, yellow, and red (that is, the
filter blocks blue and cyan). Similarly, sending white light through a "cyan" filter
leaves a mixture of blue, cyan, and green [panel (b)]: This time, the filter blocks yellow
and red. So if we send white through both filters in succession, they block everything
except green [panel (c)].

Similarly, when we spread a mixture of yellow and cyan pigments on paper, then
shine white light on it, the incoming light bounces around in the rough paper surface,
repeatedly encountering pigment molecules (Figure 3.7d,e,f). Bits of the spectrum get
absorbed by both yellow and cyan pigments, like the two filters discussed above, so
again yellow mixed with cyan ink reflects only green light. Similar logic applies to
other combinations of inks, allowing us to generate many color sensations, so we call
cyan, magenta, and yellow a set of **subtractive primary colors**.

Mixing all three primary colors of ink leads to a substance that absorbs light of *all*

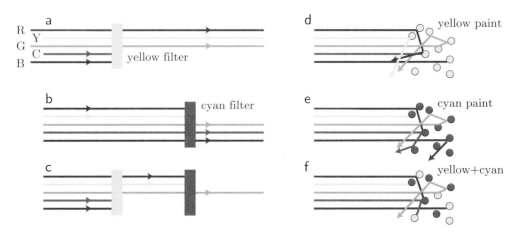

Figure 3.7: [Schematics.] **Subtractive color combination.** (a) A "yellow" filter transmits a band of wavelengths including spectral green, yellow, and red. It blocks other wavelengths, including those corresponding to blue and cyan. (b) Similarly, a "cyan" filter instead blocks red and yellow light. (c) Placing these two filters in succession blocks everything except green light. (d,e,f) Yellow and cyan chromophores are distributed within the fibers of a piece of paper. Multiple scattering of light combines their effects, similarly to (c).

wavelengths, and so approximates *black*.[8] To get white, we must either use a broadly reflecting paint, or simply apply *no* paint to white paper.

Your Turn 3D

Explain which of the primary paint colors cyan, magenta, and yellow should be mixed to obtain "red" and "blue."

$\boxed{T_2}$ *Section 3.7′ (page 134) mentions how higher visual processing compensates for changes in illumination spectrum.*

3.8 A PHYSICAL MODEL OF COLOR VISION

Section 3.6.3 mentioned that our color discrimination has systematic, repeatable, and sometimes surprising features. So we should attempt a quantitative physical model, confront it with data, and see if it can explain those features (Section 3.8.8).

3.8.1 The color-matching function challenge

We'd like to find a set of simple hypotheses, grounded in known physical facts, that predict all the empirical observations **1–7** (page 115). We will propose such a model, then ask, "Are the parameters directly measurable by physical measurements? If so, can we make successful quantitative predictions that reach from the basement level of physics, through chemistry, through neuroscience, all the way up to psychology (verbal reports from human subjects)?" It sounds like an ambitious goal.

---

[8]Consumer color printers actually have a fourth ink cartridge with black ink (abbreviated K), in part to reduce the expense of consuming all three colors of ink for every black and white document. Thus, they are said to use the **CMYK color system**. Other printers have even more inks, in order to fill out a larger color gamut.

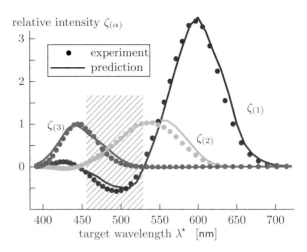

Figure 3.8: [Experimental data, with theoretical prediction.] **Color-matching functions.** *Dots*: Results of a color-matching experiment similar to the one sketched in Figure 3.5. In each trial of this experiment, the target light was monochromatic, with various values of the target wavelength $\lambda^\star$ (see Figure 3.5). *Solid curves:* Results predicted from the measured spectral sensitivities of cone cells (see Section 3.8.5). No matches were possible for monochromatic target lights in the *hatched* region; see text for the meaning of the dots in this region. [Data from Stiles & Burch, 1959, prediction from Problem 3.5.]

We can state the last part of this challenge more precisely as follows: Suppose that we have chosen three basis lights with spectra $\mathcal{I}_{(\alpha)}(\lambda)$, labeled by the index $\alpha = 1, 2,$ or 3. For any given "target" spectrum $\mathcal{I}^\star(\lambda)$, we wish to predict three numbers with the property that an experimental subject will report that the corresponding combination of basis lights matches the target. In a formula, then, we want to solve

$$\mathcal{I}^\star \sim \zeta_{(1)}\mathcal{I}_{(1)} + \zeta_{(2)}\mathcal{I}_{(2)} + \zeta_{(3)}\mathcal{I}_{(3)} \tag{3.6}$$

for the three relative intensities $\zeta_{(\alpha)}$. We must do this for every target light used in the experiment (Figure 3.5).

Figure 3.8 shows some empirical psychophysical data on color matching. In this experiment, the target lights were all monochromatic (sharply peaked at just one wavelength $\lambda^\star$). The basis lights were also monochromatic, with fixed wavelengths $\lambda_{(\alpha)} = 645, 526,$ and $444\,\text{nm}$ for $\alpha = 1$, 2, and 3, but relative intensities $\zeta_{(\alpha)}$ adjustable by the subject. That is, basis light α had mean photon arrival rate equal to a fixed overall value Φ_p, times an adjustable scaling factor $\zeta_{(\alpha)}$; the three sets of dots show the subject's choices of the three ζ's as functions of the target wavelength. The figure shows a range of target wavelengths that could not be matched with the basis lights chosen. Nevertheless, experimental data were obtained in this band by adding the red basis light to the *target* side, adjusting all three basis light intensities to find a match, and reporting red's intensity as a negative quantity. That is, in this range the matching condition Equation 3.6 was replaced by

$$\mathcal{I}^\star + |\zeta_{(1)}|\mathcal{I}_{(1)} \sim \zeta_{(2)}\mathcal{I}_{(2)} + \zeta_{(3)}\mathcal{I}_{(3)}, \tag{3.7}$$

where $\zeta_{(1)}$ is a negative quantity.

Fig. 3.5 (page 115)

3.8.2 Available wetware in the eye

To formulate a physical model, we first need some appreciation of the organization of the vertebrate retina.

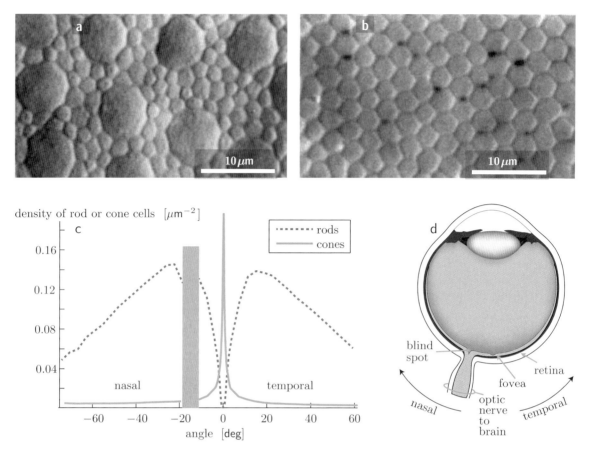

Figure 3.9: **Cones versus rods.** (a) [Optical micrographs.] Packing of cone (*large*) and rod cells (*small*), viewed end-on, in the periphery of the human retina. (b) Packing of cone cells in the human fovea. The cone cells are small, and rods are absent, yielding color vision with the greatest possible acuity (spatial detail) but poorer sensitivity to dim light than in the periphery. (c) [Experimental data.] Area densities of rod and cone cells in the human eye. The horizontal axis gives angular position in the field of view, centered on the fovea and moving outward to left and right. The *gray region* denotes the optic disk ("blind spot"), which has no rods or cones. Figure 6.6 (page 219) shows retinal anatomy in greater detail. [Images and data from Curcio et al., 1990.] (d) [Anatomical sketch.] Diagram explaining the horizontal axis of (c). This sketch depicts the right eye in cross section, viewed from above.

Anatomical

The back of the eye is lined with a mosaic of light-sensitive cells called **photoreceptors**, analogous to the pixels in a digital camera. Figure 3.9a shows two visibly different types of photoreceptors. The figure shows that one class, the **rod cells**, outnumber the other **cone cells** in the periphery of our visual field. However, a retinal region corresponding to the center of the visual field (the **fovea**) is tightly packed with small cone cells only (see Figure 3.9b,c).

Nerve axons from each part of the retina converge to form a bundle called the **optic nerve**, which extends to the brain. The retinal region where this bundle exits the eye is so crowded with axons that there is no room for photoreceptors; this region forms a **blind spot** in the visual field.

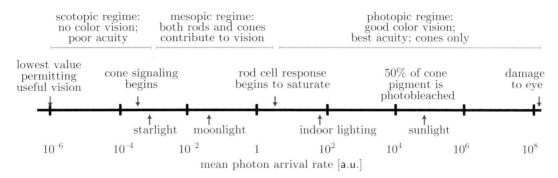

Figure 3.10: **Our eyes give useful information in environments spanning over 10 orders of magnitude in illumination level.** (Problem 11.1 will give details about the units used in this scale.)

Functional

A dim star in the sky seems to disappear when you look straight at it: The fovea is less sensitive to dim light than the surrounding regions. However, the fovea has better color discrimination than other regions. Combined with the information in Figure 3.9c, these facts suggest that

- Rod cells are very sensitive to faint light, but do not help the brain to discriminate colors.
- Cone cells are less sensitive than rods, but do discriminate color.

Further support for these claims comes from the fact that we are all effectively color-blind in dim light (**scotopic vision**, Figure 3.10), because only rod cells can function in that regime. Although our rod cells' sensitivity is wavelength-dependent (it peaks in the blue-green region), we cannot use this fact to discriminate colors, because all rod cells have *the same* spectral sensitivity. Suppose that the eye reports a stronger rod signal to the brain from one part of the visual field than from another. That information is ambiguous: It may be that both regions are equally bright but the second one is redder, or alternatively both regions are the same color but the first is brighter. To discriminate color, we must *compare* the relative amounts of two or more different wavelengths, which requires separate reports to the brain from subpopulations of photoreceptor cells with different spectral responses.

In daylight (**photopic vision**), the rod cells' responses "saturate"; that is, they all give their maximum response, which is not informative. In this situation, the less sensitive cone cells are responsible for our vision. The fact that the photopic regime is also where we have color vision thus suggests that *there must be more than one kind* of cone cell.

3.8.3 The trichromatic model

By around 1800, scientists interested in color had split into two camps. One group held to Newton's idea that light carries a continuously variable quality, which we now describe as wavelength (or frequency). The others cited the trichromacy phenomenon;[9] from this they concluded that there were just three kinds of light.

---

[9]Point **5** on page 116 introduced trichromacy.

The scientists in the second group were making a "category error": They supposed that the trichromacy of color mixing was a property of light physics, when in fact it is a property of a light-sensing *organ* (the retina). Had they studied vision in chickens, they might have concluded that there were *four* kinds of light! In a remarkable lecture in 1802, Thomas Young placed the emphasis properly, proposing that our retinas have three kinds of light detectors (now called cone cells) for daylight vision, each *tuned* to resonate with one of three wavelength bands. Each receptor type reports how much light of its preferred wavelength is present. The brain then attempts to interpret this incomplete information, resulting in a color sensation.

Translated into photon language, Young's proposal is that

a. "Color" is determined by the relative arrival rates at the eye of photons with various wavelengths.

b. The eye contains a mosaic of photoreceptor cells, each sensitive to a particular band of wavelengths.

c. The photoreceptors relevant to human color vision (cone cells) come in just three distinct classes. Each cell has exactly the same color preference as all the others in its class.

d. We can express a cell's preference by its **spectral sensitivity** function $\mathcal{S}(\lambda)$: Each photoreceptor "catches" or "misses" each photon in a Bernoulli trial, with probability to catch given by the value of its spectral sensitivity function at that photon's wavelength.[10]

e. The neural signal elicited in a cone cell by a flash of light is determined by the mean rate of "catches."

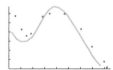

Fig. 2.1 (page 62)

The cone cell classes are now called L, M, and S type. The terminology reflects the fact that the first class's sensitivity peaks at the l̲ongest wavelength of the three, the second at m̲iddle wavelengths, and the last at the s̲hortest wavelengths. The corresponding sensitivity functions $\mathcal{S}_L$, $\mathcal{S}_M$, and $\mathcal{S}_S$ have broad maxima (see Figure 3.11), like the absorption bands of molecules.[11]

Why is the number of receptor classes so small? Young reasoned that the retina must contain an array of many detectors, in order to capture fine spatial information.[12] If there were just one type of photoreceptor, then we would have monochromatic vision. It may seem advantageous to have 20 or even 100 different sorts of detectors, in order to record the full spectrum of light at each point on the image. But each pixel type takes up space. Too many types would give each one a sparse array of pixels, degrading image resolution in each color channel. Moreover, the brain would have to process more complex information in that case. Evolution has selected for a limited ability to discriminate spectra, in exchange for finer spatial discrimination.[13]

The next sections will derive predictions from Young's hypothesis, and compare them with the color-matching data (Figure 3.8).

[T₂] *Section 3.8.3′a (page 134) describes the experimental measurement of the sensitivity functions. Section 3.8.3′b (page 135) contrasts the situation with audition (hearing).*

---

[10] Equivalently, $\mathcal{S}(\lambda)$ can be regarded as a kind of action spectrum for eliciting a certain level of cone cell response, analogous to the one we studied for DNA photodamage (Figure 2.1).

[11] Section 1.6.3 (page 42) discussed why molecular absorption bands are broad.

[12] See this chapter's epigraph (page 107).

[13] [T₂] The mantis shrimp, however, appears to have many more receptor spectral classes than any mammal, for unknown reasons; see Section 10.4.1′ (page 344).

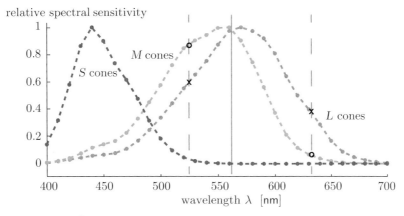

Figure 3.11: [Experimental data.] **Relative spectral sensitivities of cone cells.** The *solid dots* show spectral sensitivity $\mathcal{S}(\lambda)$ in the macaque monkey, relative to its maximum value. Human cone cells are similar. The sensitivities were determined from the electrical response of single photoreceptor cells (see Chapter 9), corrected to account for absorption in the eye lens and elsewhere, and arbitrarily rescaled to make each curve's peak value equal to 1. All three curves are nearly zero outside the "visible" range 400–700 nm. The *dashed vertical lines* represent two spectrally pure lights that appear green and red, respectively. When presented together in the proper ratio, they can generate a perception indistinguishable from spectrally pure yellow (*solid line*). This can happen when the weighted sum of the values shown as *open circles* equals that of the values shown as *crosses*. [Data from Baylor et al., 1987, available in Dataset 7.]

3.8.4 The trichromatic model explains why $R+G\sim Y$

Young's physical model for color discrimination was essentially correct, despite the fact that he conceived it in the framework of the classical wave theory. This leap is all the more impressive when we recall that photoreceptor cells themselves had not yet been seen in 1802. When images such as Figure 3.9a,b became available, they gave no hint that cone cells come in three types; even today, two of the types are anatomically indistinguishable. It took over *160 years* to confirm Young's hypothesis directly!

The first such confirmation involved measuring the light absorption of individual cone cells by directing tiny beams of light at them (microspectrophotometry). A newer method measures the spectral sensitivity directly.[14] Either technique shows that human cone cells do fall into three distinct classes.

Later sections will develop a quantitative test of Young's hypothesis, based on the curves shown in Figure 3.11. First, however, let's see how Young and his successors explained qualitatively the phenomenon of ambiguous color discrimination (the paradox raised in Section 3.6.1).

The three cone cell sensitivity curves are shown in Figure 3.11; in particular, note that the L and M curves are broad and overlap. These curves intersect for light with wavelength ≈ 560 nm. If instead we present a light at 525 nm, then the M cells will respond more strongly than the L cells, and vice versa for $\lambda = 630$ nm. But if we *mix* those two lights appropriately, we can arrange for the L and M cells to respond as they do for 560 nm light. Because the outputs of the photoreceptor cells are the only color information available to the brain, it cannot distinguish the spectral yellow light from such a mixture. A similar argument explains why the two spectra in Figure 3.4 can both appear "white."

Fig. 3.4 (page 114)

$\boxed{T_2}$ Section 3.8.4'a (page 135) discusses a medical application of these ideas. Sec-

---

[14]Chapter 9 will explain how this method works. $\boxed{T_2}$ See also Section 10.4.1' (page 344).

tion 3.8.4′b (page 135) discusses an artificial system with better color discrimination than the human eye, and its usefulness as an analytical tool.

3.8.5 Our eyes project light spectra to a 3D vector space

Interpretation of spectral sensitivity functions

Young proposed that there are three classes of photoreceptor cells responsible for color vision; within each class, all cells have the same sensitivity function $S_i(\lambda)$, where $i = L$, M, or S. We can interpret his hypothesis part **c** (page 122) in molecular terms by supposing that

c′. Each photoreceptor class expresses only one of three types of light-sensitive protein (retinal pigment).

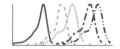

Fig. 1.10a (page 43)

This claim explains the observed spectral sensitivity functions (Figure 3.11) in terms of the excitation bands for light-induced reactions in the three pigments (compare Figure 1.10a). That is, the spectral sensitivity function can be interpreted as the probability $S(\lambda)$ that a photon of wavelength λ, incident on a photoreceptor cell, will be productively absorbed by one of its pigment molecules, that is, absorbed in a process that actually generates a corresponding signal.

The "one cone/one pigment type" picture also suggests a molecular interpretation of parts **d–e** of Young's hypothesis:

d′. Our response to light begins when one of those pigment molecules absorbs a photon and initiates a chain of events that culminates with a signal sent to the brain. The mean rate of capturing photons in this way is determined by the probability for one pigment molecule, the number of such molecules in the photoreceptor cell, and the mean rate of photon arrivals at the cell.

e′. Each photoreceptor somehow computes the rate of productive absorptions, and reports that to the brain.

These hypotheses imply that,[15]

> *The wavelength of a photon controls the* **probability** *that a particular retinal pigment will absorb it, but that is all. The color information forwarded to the brain consists only of the rates of productive photon absorption in each of the three classes of photoreceptor cells.*

univariance principle

(3.8)

We now use the ideas just described in a chain of reasoning (Figure 3.12):

- A stream of photons of various wavelengths enters each photoreceptor cell.
- In any small range $\Delta\lambda$, photons arrive in a Poisson process with some mean rate $\mathcal{I}(\lambda)\Delta\lambda$ (Section 1.5.2, page 38).
- Each of these photons may or may not be productively absorbed, depending on the outcome of a Bernoulli trial. The probability of success is the spectral sensitivity $S_i(\lambda)$, where the subscript $i = L$, M, or S.
- If we block all light except those photons in a range $\Delta\lambda$ about some λ_0, then by the thinning property, productive absorptions will also occur in a Poisson process, but with reduced mean rate[16] $S_i(\lambda_0)\mathcal{I}(\lambda_0)\Delta\lambda$.

---

[15]Section 9.4.4 (page 307) will describe evidence for univariance at the single-photoreceptor level.
[16]Section 3.4.1 (page 111) introduced the thinning property.

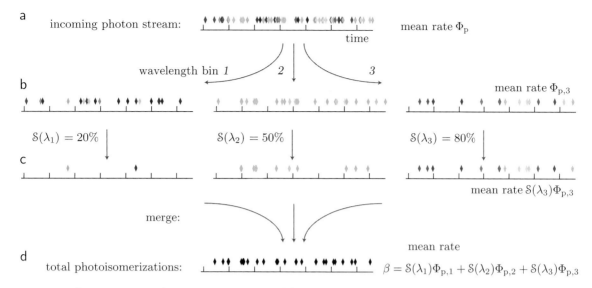

Figure 3.12: [Simulated data.] **Selective absorption.** (a) An incoming stream contains photons with various wavelengths. (b) We imagine separating the incoming stream into substreams consisting of those photons in each of several ranges of wavelength. The mean rate $\Phi_{p,1}$ of the first substream is given by $\mathcal{I}(\lambda_1)\Delta\lambda$ and so on. (c) A particular retinal pigment has a sensitivity function $\mathcal{S}(\lambda)$, which determines what fraction of photons of each type it will absorb. The figure illustrates the case of a pigment that preferentially absorbs long-wavelength light. (d) The net effect of all photons is to give photoisomerizations in a Poisson process with mean rate β. Each class of cone photoreceptors has its own sensitivity function, and hence will respond differently to the same incoming light spectrum. Taking the limit of many narrow bins yields Equation 3.9.

- If we *don't* block any colors of light, then by the merging property, productive absorptions will occur in a random process whose mean rate β_i is the sum of those for each frequency slice, that is,[17]

$$\beta_i = \int \mathrm{d}\lambda\, \mathcal{S}_i(\lambda)\mathcal{I}(\lambda). \tag{3.9}$$

- Once we know the rates of productive photon absorption in each class of cone cell, for a particular region in the visual field, then we know everything the brain can know about the color of light in that region. That is, the brain must conclude that two colors match ($\mathcal{I} \sim \mathcal{I}'$) if the three corresponding rates β_L, β_M, and β_S all agree (Idea 3.8).

Equation 3.9 is a consequence of the independence of the random photon arrivals, and that of the Bernoulli trials that determine which photons give rise to productive absorptions. It says that the mean rate of productive photon absorption depends *linearly* on the spectrum of the incoming light.

T2 *Section 3.8.5'a (page 138) gives more details about the notion of "projection." Section 3.8.5'b (page 138) mentions the role of nonproductive absorption. Section 3.8.8'a (page 139) discusses the linearity of Equation 3.9 in greater detail.*

---

[17]Section 3.4.2 introduced the merging property.

3.8.6 A mechanical analogy for color matching

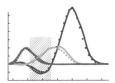

Fig. 3.8 (page 119)

Young's hypothesis appears promising: It explains the $R + G \sim Y$ puzzle, and it can be interpreted in the framework of other things we know about light (Section 3.8.5). But to confirm that it describes our vision, we must show that it correctly predicts the systematics (Section 3.6.3, page 113), as well as the quantitative data obtained from color-matching experiments (Figure 3.8). We can see how to proceed by first making an analogy to a purely mechanical system.

Imagine an air-hockey game. The puck is a disk floating on a thin layer of air. It can move without friction in two directions x and y; in the third direction, however, it is confined to the plane $z = 0$. We can apply a force $\boldsymbol{F}$ to the puck and watch it respond. When we do this, we see that it obeys Newton's law of motion:

$$\boldsymbol{a}_\perp = \boldsymbol{F}_\perp / m,$$

where $\boldsymbol{a}$ is the acceleration vector and the subscript $\perp$ means projection to the xy plane. Let Π denote the projection operation:

$$\boldsymbol{F}_\perp = \Pi(\boldsymbol{F}) = \begin{bmatrix} F_x \\ F_y \end{bmatrix}. \tag{3.10}$$

The component F_z is irrelevant to the puck's motion.[18] To indicate this, we introduce the equivalence symbol $\sim$; the formula $\boldsymbol{F} \sim \boldsymbol{F}'$ means that two forces generate the same puck motion, whether or not they are equal. The preceding discussion implies that this statement is the same as saying that $\boldsymbol{F}_\perp = \boldsymbol{F}'_\perp$:

Two forces are equivalent ($\boldsymbol{F} \sim \boldsymbol{F}'$) if they have the same projection, that is, if $\Pi(\boldsymbol{F}) = \Pi(\boldsymbol{F}')$. (3.11)

The relation $\sim$ has some mathematical properties reminiscent of the ones in Section 3.6.3:

3′. It is reflexive and transitive.

4′. It obeys a Grassmann-type law (page 115).

5′–6′. We can choose two fixed "basis" forces $\boldsymbol{F}_{(1)}$ and $\boldsymbol{F}_{(2)}$. Then any other "target" force $F^\star$ can be uniquely expressed as equivalent to a linear combination of those two:

$$\boldsymbol{F}^\star \sim \zeta_{(1)} \boldsymbol{F}_{(1)} + \zeta_{(2)} \boldsymbol{F}_{(2)}. \tag{3.12}$$

7′. There are many allowed choices for the two basis forces. For example, they need not be chosen perpendicular to each other, nor even lie in the xy plane. They must not, however, be parallel to each other, nor may either one be purely vertical.

The origin of all these properties is that the $\sim$ relation amounts to equality *of the projections,* and projection is a linear function:

$$\Pi(a\boldsymbol{F} + b\boldsymbol{F}') = a\Pi(\boldsymbol{F}) + b\Pi(\boldsymbol{F}'). \tag{3.13}$$

---

[18]This statements assumes that the force is not too large. Too large a downward force would create friction with the table; too large an upward force would lift the puck off the table.

Your Turn 3E

Confirm that property **4′** follows from the fact that Π is linear. That is, show that if $\boldsymbol{F}_1 \sim \boldsymbol{F}_2$, then $\boldsymbol{F}_1 + \boldsymbol{F}_3 \sim \boldsymbol{F}_2 + \boldsymbol{F}_3$.

Ex. Derive properties **5′–7′**.

Solution: We wish to show that Equation 3.12 has a unique solution. First, rephrase it as an ordinary vector equation by applying Idea 3.11 and linearity, Equation 3.13:

$$\begin{bmatrix} F_x^\star \\ F_y^\star \end{bmatrix} = \begin{bmatrix} \zeta_{(1)}F_{(1)x} + \zeta_{(2)}F_{(2)x} \\ \zeta_{(1)}F_{(1)y} + \zeta_{(2)}F_{(2)y} \end{bmatrix}. \tag{3.14}$$

These are two linear equations in the two unknowns $\zeta_{(1)}$ and $\zeta_{(2)}$; every other symbol denotes a known quantity. Solving them in the usual way establishes **5′**, and is possible as long as the two basis forces have nonzero projections $\boldsymbol{F}_{(1)\perp}$ and $\boldsymbol{F}_{(2)\perp}$ that are not parallel to each other. The other properties can be obtained by extending this argument.

To restate the Example in a more general way, we can reduce our problem to a set of two ordinary linear equations once we know $\Pi(\boldsymbol{F}^\star)$ (the left side of Equation 3.14) and the two vectors $\Pi(\boldsymbol{F}_{(\alpha)})$ (coefficients of $\zeta_{(\alpha)}$ on the right side, for $\alpha = 1, 2$). If you are familiar with matrix notation, you'll recognize Equation 3.14 as the equation

$$\begin{bmatrix} F_x^\star \\ F_y^\star \end{bmatrix} = \begin{bmatrix} F_{(1)x} & F_{(2)x} \\ F_{(1)y} & F_{(2)y} \end{bmatrix} \begin{bmatrix} \zeta_{(1)} \\ \zeta_{(2)} \end{bmatrix}. \tag{3.15}$$

The condition for a solution to exist is just the statement that the 2×2 matrix in Equation 3.15 should be invertible; moreover, that condition guarantees a unique solution, obtained by multiplying both sides by the inverse matrix.

3.8.7 Connection between the mechanical analogy and color vision

How does the air-puck analogy help us to understand color?

Section 3.8.5 argued that the rate of productive photon absorptions is a linear function of the spectrum (Equation 3.9), analogous to the projection Π in the mechanical analogy (Equation 3.10). In the analogy, Π projects three-dimensional forces to the 2D space of (F_x, F_y) values. In human vision, Equation 3.9 projects the *infinite*-dimensional space of spectra to the *three*-dimensional space of productive absorption rates: the $(\beta_L, \beta_M, \beta_S)$ values. Despite these differences, we can again obtain the color-matching facts in Section 3.6.3 as consequences of the property that the projection operation is linear in $\mathcal{I}$.

Your Turn 3F

Show that the empirical color-matching rules **3–4** (page 115) follow from our model.

3.8.8 Quantitative comparison to experimentally
observed color-matching functions

The quantitative challenge announced in Section 3.8.1 was: Given a target light spectrum $\mathcal{I}^\star$ and three basis lights with spectra $\mathcal{I}_{(\alpha)}$, find the amount of each basis light, $\zeta_{(\alpha)}$, needed to match a specified target (or state when no such match is possible).

Before proceeding, we first summarize notation introduced earlier, and extend it somewhat:

$\Phi_p^\star$	mean photon arrival rate of a target light (units s^{-1})
$\Phi_{p(\alpha)}$	mean photon arrival rate of basis light α, where $\alpha = 1, 2$, or 3
$\mathcal{I}^\star$	spectral photon arrival rate ("spectrum") of a target light to be matched (units $s^{-1}nm^{-1}$)
$\lambda^\star$	wavelength in the special case of a target light that is monochromatic (units nm)
$\mathcal{I}_{(\alpha)}$	spectrum of basis light α (units $nm^{-1}s^{-1}$)
$\lambda_{(\alpha)}$	wavelength in the special case that basis light α is monochromatic
$\zeta_{(\alpha)}$	mixing coefficient for basis light α (dimensionless)
β_i	mean rate of productive photon absorptions by cones in class $i = L, M$, or S (units s^{-1})
$\mathcal{S}_i(\lambda)$	spectral sensitivity function for cone class i (dimensionless)
$\beta_i^\star$	projection of the target light spectrum onto cone sensitivity i (units s^{-1})
$B_{(\alpha)i}$	projection of basis light α's spectrum onto cone sensitivity i

To address the challenge, we now implement the condition given in Equation 3.6:

$$\mathcal{I}^\star \sim \zeta_{(1)}\mathcal{I}_{(1)} + \zeta_{(2)}\mathcal{I}_{(2)} + \zeta_{(3)}\mathcal{I}_{(3)}. \qquad \text{[3.6, page 119]}$$

As in the mechanical analogy, first apply the projection (Equation 3.9, page 125) to $\mathcal{I}^\star$, obtaining the three numbers

$$\beta_i^\star = \int d\lambda\, \mathcal{S}_i(\lambda)\mathcal{I}^\star(\lambda), \quad i = L, M, \text{ or } S.$$

Similarly, define a 3×3 array of numbers $B_{(\alpha)i}$ by applying the same projection to each of the $\mathcal{I}_{(\alpha)}$. Analogously to Equation 3.14, Equation 3.6 then yields

$$\begin{bmatrix} \beta_L^\star \\ \beta_M^\star \\ \beta_S^\star \end{bmatrix} = \begin{bmatrix} \zeta_{(1)}B_{(1)L} + \zeta_{(2)}B_{(2)L} + \zeta_{(3)}B_{(3)L} \\ \zeta_{(1)}B_{(1)M} + \zeta_{(2)}B_{(2)M} + \zeta_{(3)}B_{(3)M} \\ \zeta_{(1)}B_{(1)S} + \zeta_{(2)}B_{(2)S} + \zeta_{(3)}B_{(3)S} \end{bmatrix}. \qquad (3.16)$$

To find the required values of $\zeta_{(\alpha)}$, we solve this set of three linear equations in the three unknowns $\zeta_{(\alpha)}$. All other symbols in Equation 3.16 represent known quantities.

Your Turn 3G

Derive Equation 3.16 from the preceding formulas.

Mathematically, Equation 3.16 always has a unique solution, as long as the three basis lights are suitably generic.[19] However, unlike forces on an air puck, there is no such thing as negative light. Thus, physically all three coefficients $\zeta_{(\alpha)}$ must be *nonnegative* numbers. If a particular target light leads to a solution of Equation 3.16 with any negative values of $\zeta_{(\alpha)}$, then that target cannot be matched with the chosen basis lights.[20] For this reason, the color gamut of a computer or television monitor is incomplete; the spectral colors that you see from a prism are more vivid than anything you perceive when looking at a computer screen or printed page.

The experiment of interest to us actually involves a special case of Equation 3.16, which is simpler than the general case: The target lights used to obtain Figure 3.8

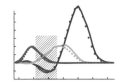

Fig. 3.8 (page 119)

---

[19] If there are fewer than three basis lights, or any basis light is a multiple of another, or one is a combination of the other two, then many target lights cannot be matched. (Mathematically, we say the basis lights must be chosen such that the matrix of $B_{(\alpha)i}$ values is invertible in order to avoid this problem.)

[20] We already encountered this failure in the experimental data (Figure 3.8, page 119).

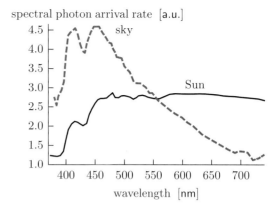

Figure 3.13: [Experimental data.] **Sun and sky.** *Solid curve:* Spectral photon arrival rate of sunlight arriving at Earth's atmosphere. *Dashed curve:* Spectral photon arrival rate of the blue sky. [Data from Smith, 2005].

were monochromatic. Denote the mean total photon arrival rate of a target light by the symbol $\Phi_p^\star$. Then we do not need to integrate over λ when evaluating the left side of Equation 3.16:

$$\beta_i^\star = \mathcal{S}_i(\lambda^\star)\Phi_p^\star \quad \text{for } i = L,\ M,\ \text{or } S. \tag{3.17}$$

Your Turn 3H

In the experiment, the basis lights were also monochromatic. Obtain a formula analogous to Equation 3.17 for the $B_{(\alpha)i}$ in this case.

You'll finish deriving a prediction for the color-matching experiment in Problem 3.5; that is, by using the empirical sensitivity functions shown in Figure 3.11, together with Equations 3.16–3.17, you'll predict the values of $\zeta_{(\alpha)}$ as functions of $\lambda^\star$, and compare them to the experimental data. The result, shown as the solid curves in Figure 3.8, was the main goal of this chapter.

Section 3.8 has shown that the detailed empirical facts about color matching emerge naturally from the physical model of light based on the Light Hypothesis and Electron State Hypothesis, plus the one additional hypothesis that humans determine color by using three classes of photoreceptor cells.

$\boxed{T_2}$ *Section 3.8.8′a (page 139) explains why we may use relative sensitivity functions in this derivation in place of the absolute ones. Section 3.8.8′b (page 140) discusses the limited color gamut in more detail.*

3.9 WHY THE SKY IS NOT VIOLET

It is often said that a cloudless sky appears blue because Earth's atmosphere scatters blue light more than it does longer wavelengths. Thus (unlike on the Moon), we see light in directions other than the line of sight to the Sun, but that light has a different spectrum from direct sunlight. Figure 3.13 shows this difference: Light from the sky has more photons at the shortest visible wavelengths than at any longer wavelength.

Fig. 3.11 (page 123)

Faced with that observation, however, we may wonder: If we mostly get violet photons, then why doesn't the sky appear *violet?* To answer that question, return

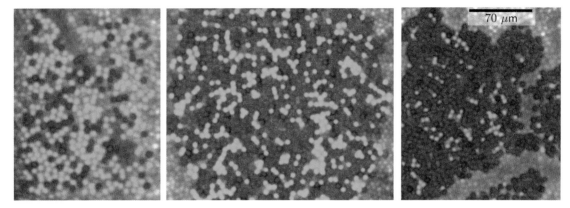

Figure 3.14: [Adaptive optics micrographs.] **Individual variation in cone densities.** False color images showing the arrangement of L (*red*), M (*green*), and S (*blue*) cones in the retinas of different human subjects, all of whom had normal color vision. The relative densities of L and M cones vary widely. (The retinal regions shown contained few rod cells.) [Courtesy Heidi Hofer; see also Hofer et al., 2005, and Hofer & Williams, 2014.]

to the cone cell spectral sensitivity functions (Figure 3.11). Monochromatic violet light, say at 440 nm, gives negligible response from M and L cone cells. Thus, to get a perception of "violet" requires activity in the S cells *and not* in the L or M cells, that is, a light spectrum without long wavelengths. The sky spectrum in Figure 3.13 does not have this property, so it does not appear violet. Indeed, it has substantial strength in long wavelengths, giving the sensation of unsaturated (pale) blue.

3.10 DIRECT IMAGING OF THE CONE MOSAIC

This chapter has presented strong, but indirect, evidence for the trichromatic hypothesis. It would be valuable to see the placement of each cone type directly in an intact, living eye. For example, the arrangement and densities of each cone type on the retina must affect our vision. But L and M cone cells are morphologically identical, and so are indistinguishable even in the electron microscope. In principle, it should be possible to distinguish them by the spectral differences in their absorption of light. But in practice, the differences are subtle, and aberrations (defects in the eye's optics) distort ordinary micrographs, making them hard to interpret.[21]

A. Roorda and D. R. Williams overcame this limitation by using adaptive optics methods borrowed from astronomy, which faces a similar distortion problem from Earth's atmosphere. In addition to removing aberration, the researchers enhanced the subtle differences in absorption spectrum between cone types by comparing images taken before and after selectively photobleaching some of the visual pigments with bright light. For example, exposure to bright light with wavelength 550 nm photobleached both the L and M cone cells, with little effect on the S cones. Subtracting the intensities at each image pixel before and after the photobleach gave a signal proportional to the density of either L or M pigment at that pixel, discriminating those types from S cones. Using other photobleaching wavelengths gave other image pairs that further discriminated between L and M cones.

Figure 3.14 illustrates the fact that all human retinas were found to have a very sparse array of S cones (blue dots). Remarkably, however, different human subjects,

---

[21]Chapter 6 will discuss aberration.

all with normal color vision, were found to vary widely in the density of their L and M cone cells (red and green dots). Moreover, all subjects were found to have large patches in which either M or L cones are missing. In such patches, each of us is partially colorblind; we reconstruct a colorful world by constantly shifting our gaze as we look at a visual scene.

THE BIG PICTURE

This chapter described wavelength as an informative *fifth dimension* potentially available to our eyes. Although we argued that animals make incomplete use of this information, nevertheless it is extremely useful in constructing our mental model of the outside world.

We interpreted the observed phenomena of color matching in the photon framework introduced in Chapter 1: We imagined that a stream of photons arrives at each region of our eye's retina, that photoreceptor cells located there signal the brain to indicate the mean rates at which they productively absorb photons, and that the cells come in three classes, each with a characteristic spectral sensitivity function. We imagined a mechanical model, involving forces acting on an air-hockey puck, and used it to guide a calculation of predicted color-matching functions that agreed with experiments.

Although our model seems consistent with known aspects of light, however, it also raises many questions. Chapters 4–6 will discuss how our eyes form images on the retina in the first place. Then Chapters 9–11 will discuss the transformation of light into the neural signals that eventually arrive at the brain.

Moreover, color *matching* is only a small aspect of the much bigger field of color *perception*. Books cited at the end of this chapter pick up the story, discussing color illusions and other subtleties arising in this fascinating field.

KEY FORMULAS

- *Thinning property:* Suppose that we make a draw from a Poisson process. From the resulting time series, we discard some of the blips by subjecting each to a Bernoulli trial. If β denotes the original mean rate, and ξ the fraction of blips not discarded, then the resulting time series is itself a draw from another, "thinned," Poisson process with $\beta' = \xi\beta$.
- *Merging property:* Suppose that we make a draw from each of two Poisson processes, with mean rates β_1 and β_2. We make a new time series by interleaving the blips from the two that we drew. Then the resulting time series is itself a draw from another, "merged," Poisson process with $\beta_{\text{tot}} = \beta_1 + \beta_2$.
- *Matching:* Color matching obeys some empirical rules, including Grassmann's law: If $\mathcal{I}_1 \sim \mathcal{I}_2$, then $\mathcal{I}_1 + \mathcal{I}_3 \sim \mathcal{I}_2 + \mathcal{I}_3$ for any other $\mathcal{I}_3$.
- *Cone photoreceptors:* Cone cells of type i are characterized by a linear projection that converts the spectrum $\mathcal{I}$ of incident light to the mean rate β_i of productive absorptions. It can be expressed in terms of a spectral sensitivity function $\mathcal{S}_i$ as $\beta_i = \int d\lambda\, \mathcal{S}_i(\lambda)\mathcal{I}(\lambda)$. The index i runs over L, M, and S, which stand for the long-, medium-, and short-wavelength sensitive cone cell populations. The brain must conclude that two colors match if the three corresponding rates β_L, β_M, and β_S all agree (Idea 3.8).

- *Matching, quantitative:* A typical color-matching experiment presents three monochromatic "basis" lights, with wavelengths $\lambda_{(\alpha)}$, $\alpha = 1, 2, 3$, with which the subject tries to match a "target" light $\lambda^{\star}$. Define $\beta_i^{\star}$ to be the productive absorption rates corresponding to the target light, and a 3×3 array $B_{(\alpha)i}$ of productive absorption rates for each of the three basis lights. Then the relative amounts $\zeta_{(\alpha)}$ of each basis light needed for the match are obtained by solving

$$
\begin{bmatrix} \beta_L^{\star} \\ \beta_M^{\star} \\ \beta_S^{\star} \end{bmatrix} = \begin{bmatrix} \zeta_{(1)}B_{(1)L} + \zeta_{(2)}B_{(2)L} + \zeta_{(3)}B_{(3)L} \\ \zeta_{(1)}B_{(1)M} + \zeta_{(2)}B_{(2)M} + \zeta_{(3)}B_{(3)M} \\ \zeta_{(1)}B_{(1)S} + \zeta_{(2)}B_{(2)S} + \zeta_{(3)}B_{(3)S} \end{bmatrix} . \qquad \text{[3.16, page 128]}
$$

$\boxed{T_2}$ This equation can be compactly expressed as the matrix formula $\boldsymbol{\beta}^{\star} = \mathsf{B}^{\mathrm{t}}\boldsymbol{\zeta}$, where B is the matrix with $B_{(\alpha)i}$ in row α and column i.

FURTHER READING

Semipopular:
Mahon, 2003; Livingstone, 2002; Hubel, 1995.
Newton's experiments: Johnson, 2008.
Evolution of color vision: Carroll, 2006.

Intermediate:
General: Snowden et al., 2012; Packer & Williams, 2003; Rodieck, 1998.
Merging and thinning: Nelson, 2015; Blitzstein & Hwang, 2015.
Matrices and systems of linear equations: Felder & Felder, 2016; Shankar, 1995.
Animals that use color: Cronin et al., 2014.

Technical:
Brainard & Stockman, 2010.
Why we have separate rod and cone systems: Lamb, 2016.
Cone sensitivity curves and other current data: `http://www.cvrl.org/` .
Cone mosaic: Hofer & Williams, 2014; Roorda & Williams, 1999.
$\boxed{T_2}$ Spectral karyotyping: Garini et al., 2006; Fauth & Speicher, 2001; Schröck et al., 1996.
$\boxed{T_2}$ Section 3.8.8′b gave a greatly simplified account of color space. For details about the standard chromaticity diagram, see Shevell, 2003.

$\boxed{T_2}$ **Track 2**

3.5′a Flux, irradiance, and spectral flux irradiance

Vision scientists sometimes refer to the total mean photon arrival rate, Φ_p, by the term "photon flux." This usage can cause confusion, because physicists generally reserve the word "flux" for a rate *per unit area,* in this case Φ_p/A, where A is the cross sectional area of a beam of light. We won't give that quantity any particular name at all; vision scientists generally call it the **photon flux irradiance**.

Both quantities can be further broken down into their spectra. Just as the main text introduced the spectral photon arrival rate $\mathcal{I}(\lambda)$, vision scientists define the **spectral photon flux irradiance**: $E_{p\lambda}(\lambda) = \mathcal{I}(\lambda)/A$. Thus, typical units for $E_{p\lambda}$ are $\mathsf{s}^{-1}\,\mathsf{nm}^{-1}\,\mu\mathsf{m}^{-2}$. In this chapter, we are imagining a beam focused to land on just one photoreceptor, and we want the total rate at which photons land anywhere on it. Thus, $E_{p\lambda}$ always appears multiplied by A; $\mathcal{I}$ is a useful abbreviation for that product.[22]

Physical measures of illumination like the ones just described are called "actinometric quantities." An alternative group, the "radiometric quantities," incorporate the energy per photon, introducing "radiant flux" (power, in W), "irradiance" ($\mathsf{W\,m}^{-2}$), and "spectral irradiance" ($\mathsf{W\,m}^{-2}\,\mathsf{nm}^{-1}$).

Vision scientists have also created an entirely separate set of quantities describing *apparent* illumination intensity. This system is called "photometry," with corresponding "photometric quantities" and units. Problem 11.1 (page 375) introduces some of these; see also Peatross & Ware, 2015, §2.A, Bohren & Clothiaux, 2006, chapt. 4, or Packer & Williams, 2003, §2.8.

3.5′b Combining spectra is a linear operation

The linearity of Equation 3.3 (page 112) can be thought of as a consequence of the Light Hypothesis, which implies that photons don't interact with one another in a beam passing through vacuum. They only interact with *electrons* (and to a lesser extent, protons) in the matter they strike.

$\boxed{T_2}$ **Track 2**

3.6.3′a Variation of color matching

Figure 3.11 presented some spectral sensitivity functions, but these must be regarded as approximate, because individual subjects differ. There are dozens of "normal" variants of the L and M cone pigment genes, which generate small variations in color-matching functions. Individuals also differ in the absorption spectrum of their eye lenses, for example, with aging. Also, a layer of "macular pigment" covers the central part of the retina, with its own wavelength-dependent absorption; this, too, can differ among individuals.

Fig. 3.11 (page 123)

There are other subtle effects not mentioned in Section 3.6.3. For example, color-matching functions depend on the intensity of light, due to the different effects of cone pigment photobleaching at different wavelengths. See Packer & Williams, 2003, §2.5.3, for a discussion of this and other aspects of the problem.

---

[22]Some vision scientists instead use the symbol $\Phi_{p\lambda}$ and call it the "spectral photon flux."

3.6.3′b Colorblindness

A larger effect arises when an individual completely lacks a functioning gene for one of the cone cell pigments. In that case, the brain receives a projection of physical spectra to a *two*-dimensional space, and many color pairs become metamers (indistinguishable), even though they appear distinct to normal individuals.

The most common form of this phenomenon involves the M cone pigment. Because its gene lies on the X chromosome, females have two opportunities to receive a functioning copy from their two parents, and the condition is uncommon. Among males, however, some 4% of the population is affected. The term "colorblindness" is a misnomer: These **dichromats** have normal L and S pigments, so their color discrimination, although reduced, is still partly present. Other more severe deficits are much more rare.

3.6.3′c Tetrachromacy

The discussion in (a,b) above raises an intriguing possibility: What happens if a female human receives two functioning, but slightly *different*, genes for M cone pigments on her two X chromosomes? Each cone cell silences one of its two copies of the X chromosome, chosen at random, so such an individual would have *four* distinct classes of cone cells, and hence would project light spectra to a four-dimensional space of cone signals. In principle, that would allow her to discriminate colors that appear to others to be metamers—but can the human brain actually make use of this additional information?

G. Jordan and coauthors studied a number of female subjects whose male children were known to have slightly different M cone sensitivity functions: One son inherited a normal cone pigment gene; the other an anomalous, but functioning, one. Thus, these women were known to carry both versions of the gene on their two X chromosomes, and hence they were candidates to be "tetrachromats." Not all of them could discriminate a four-dimensional space of colors, but at least one did (Jordan et al., 2010). Similarly, a genetically modified mouse with one extra cone type was found to have enhanced color discrimination, compared to the wild-type (Jacobs et al., 2007).

$\boxed{T_2}$ **Track 2**

3.7′ Perceptual color

Section 3.7.1 (page 117) spoke of the color of an illuminated object as a sort of product between the spectrum of the illumination and an intrinsic reflectance function of the object. Remarkably, however, what we *perceive* is largely independent of the illuminating spectrum. Neural circuits in our higher visual regions combine many cues to compensate for the spectral differences between predawn, morning, and midday sunlight, the many kinds of artificial light that we use, and so on. Only in extreme cases, for example, purely monochromatic illumination, do we lose our color discrimination (Figure 3.2, page 110).

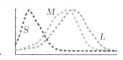

$\boxed{T_2}$ **Track 2**

3.8.3′a Determination of sensitivity functions

To obtain the sensitivity curves shown in Figure 3.11, D. Baylor, B. Nunn, and J. Schnapf monitored the responses of single cone cells to flashes of monochromatic light. The general technique will be described later, in Chapter 9; it involved observing the extracellular electric current I just outside the cell, and its peak change ΔI_{peak} after a flash.

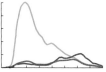

Fig. 3.11 (page 123)

    Rather than attempting to observe single-photon responses, the experimenters noted empirically that ΔI_{peak} could be fit to a standard function of flash strength, $I_0\big(1-\exp(-\mathcal{S}\mu)\big)$, where μ is the mean number of photons delivered in the flash and $\mathcal{S}$ is a wavelength-dependent fitting parameter, a measure of the sensitivity. Because each individual cell differed in its overall sensitivity, the functions thus obtained were first normalized, then averaged over several representative cells in each of the three cone classes.

3.8.3′b Contrast to audition

Our *hearing* does not have spatial resolution, so Young's constraint does not apply to it. In fact, our ears do analyze the full spectrum of sound using an array of many receptors, each tuned to a narrow range of frequencies. Thus, when we play the musical notes C and E together we hear a chord, not a note that lies midway between their frequencies.

$\boxed{T_2}$ **Track 2**

3.8.4′a Enhancement of color contrast in autofluorescence endoscopy

Section 2.3.1 (page 62) described the empirical observation that precancerous cells in the bronchi are distinguishable by their fluorescence spectrum (Figure 2.2). However, the difference mainly appears as a *loss* of fluorescence emission around 500 nm. Thus, diseased tissue will appear to the eye as simply dark, which could have other explanations, for example, poor illumination in one region (see Figure 3.15a,b). To detect the diseased state, we could try to normalize the light signal to account for such nonuniformities, by using some sort of mathematical image processing.

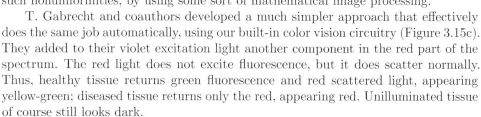

Fig. 2.2 (page 63)

    T. Gabrecht and coauthors developed a much simpler approach that effectively does the same job automatically, using our built-in color vision circuitry (Figure 3.15c). They added to their violet excitation light another component in the red part of the spectrum. The red light does not excite fluorescence, but it does scatter normally. Thus, healthy tissue returns green fluorescence and red scattered light, appearing yellow-green; diseased tissue returns only the red, appearing red. Unilluminated tissue of course still looks dark.

3.8.4′b Spectral analysis can discriminate many fluorophores and their combinations

One reason to study color vision could be to learn how to invent an artificial color vision system better than our own.

    Certain inherited abnormalities in humans manifest themselves as gross aberra-

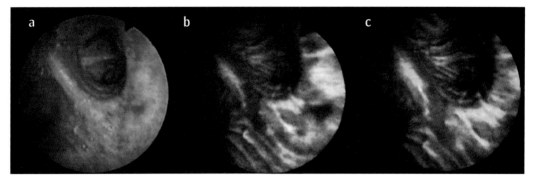

Figure 3.15: [Endoscopic images.] **Enhanced color contrast in diagnostic imaging.** Each image shows the same tissue (human bronchi in vivo), viewed through the same fiber-optic instrument. The diameter of the tube on the upper right is about 12 mm. *Left:* White-light illumination. *Center:* Pure violet light illumination; the excitation light has been filtered out, showing only fluorescence. *Right:* As in the center panel, but with added red light illumination. A dysplasic lesion is now clearly visible. [From Gabrecht et al., 2007.]

tions in single chromosomes. For example, a germ cell may contain an entire extra copy of one chromosome, leading to Down syndrome. More subtly, a large domain belonging on one chromosome may be exchanged with one normally belonging to a different chromosome. In each case, a copy of every essential gene is present, but delicate regulatory relationships may be disrupted by having the wrong copy number of the affected genes, or by having them be physically misplaced in the genome.

A traditional method to visualize defects of this sort involves staining the chromosomes and viewing them under a microscope (**karyotyping**). One needs to examine all the chromosomes, identify which is which, and then decide which ones do not have the right overall shape, size, and banding pattern. Although the method is highly developed, sometimes it fails to disclose abnormalities.

We can imagine a complementary method, in which each part of the genome is labeled according to which chromosome normally contains it. Then each normal chromosome will be uniformly labeled, whereas defective ones will have elements that clearly belong elsewhere. One way to implement this idea involves creating a set of fluorescent labeling molecules (tags) that bind only to DNA sequences known to belong to chromosome #1;[23] observing that they really do localize to a single chromosome would imply that it was the normal form of #1. Such a measurement would need to be performed 23 times, however, in order to check all 23 pairs of human chromosomes, a time-consuming task.

A faster procedure would involve 23 *distinct* sets of fluorescent labels, one for each of the distinct normal chromosomes. We could then examine them all in parallel, just by viewing a color image of the labeled cell. But this approach faces a practical problem. There is a limited supply of candidate fluorophores, and each has a rather broad emission spectrum. Moreover, human color vision discards quite a lot of information about the spectrum of light in the range for which convenient fluorophores are available,[24] making it difficult to discriminate more than two or three kinds of label (Figure 3.16a). Standard cameras are designed to mimic our eye in this regard, because there is little point in gathering spectral information that our eye will later discard.

---

[23] One version of this technique is called fluorescence in situ hybridization, or "FISH."

[24] Section 3.6.2 (page 113) introduced the idea of discarding information.

Figure 3.16: [Experimental data.] **Spectral karyotyping technique.** (a) Full color fluorescence micrograph of three pairs of human chromosomes (numbers 4, 7, and 13), each uniquely labeled by fluorescence in situ hybridization (FISH). Both copies of chromosome #7 are labeled by the fluorophore Cy3; #13 are labeled by the fluorophore Texas red; #4 are labeled with both Cy3 and TR. When viewed in an ordinary microscope, all six chromosomes appear similar, because they are morphologically similar and each one's label mainly excites only our L receptors. (b) However, each of Cy3, TR, and their combination has a distinct spectral fingerprint. (The spectra differ from the ones in Figure 1.10, because a filter was used here that removed photons with wavelengths in bands that were needed for excitation.) (c) Each pixel in panel (a) was decomposed into its spectrum. Each such spectrum was compared with the three in (b), and the best fit was found, labeling the pixel as belonging to chromosome #4, #7, or #13. Each of the three options was assigned a "false color," and a new image was created by giving each pixel its corresponding false color. The method consistently assigns an identity to every part of every normal chromosome. (d) The same procedure was applied to chromosomes prepared from blood cells of a father of a child with a serious genetic error. Two pairs of chromosomes are shown. Although these chromosomes looked normal in ordinary karyotyping, the spectral technique shows that a fragment of DNA normally found on chromosome #1 (*yellow*) is exchanged with a fragment normally found on #11 (*blue*). [(a–c) Images and data courtesy Yuval Garini. (d) From Fig. 3a, p. 496 from Schröck et al. Multicolor spectral karyotyping of human chromosomes. *Science* (1996) vol. 273 (5274) pp. 494–497. Reprinted with permission from AAAS.]

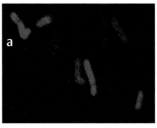

spectral photon arrival rate [a.u.]

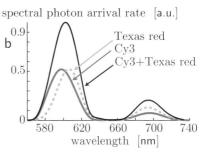

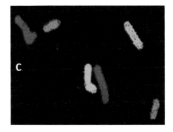

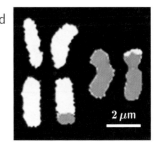

E. Schröck and coauthors overcame this limitation, inventing a technique called **spectral karyotyping** (Schröck et al., 1996; Lindner et al., 2016). They chose a set of five fluorophores with distinct emission spectra, then created a "binary code": Each chromosome's tag had each fluorophore either present or absent, for a total of $(2^5 - 1)$ possibilities, more than enough to assign each chromosome a unique label set. Rather than taking an ordinary color micrograph of the labeled chromosomes, however, the researchers passed their microscope image through a spectrometer, which yielded the *full spectrum* of light present in each image pixel. They then compared the spectrum in the image with those expected for each of the 23 labels, and chose which was the best fit for each pixel. To present the resulting data, they then created a new set of 23 "false colors" chosen to be readily distinguishable by humans. Replacing each pixel by its assigned false color then gave an informative image.

Figure 3.16 shows a dramatic illustration of a result obtained by this method. Panel (a) shows an ordinary fluorescence microscope image of three pairs of labeled chromosomes. Although they are partly distinguishable by their gross anatomy (length), all three appear similar in color. However, panel (b) shows that the three different labels used had distinct spectra. Panels (c–d) show the power of this method to assign identities to both normal and abnormal chromosomes, for example, disclosing a defect in a case where the chromosome's overall shape and size appeared normal.

$\boxed{T_2}$ **Track 2**

3.8.5′a Photoisomerization rate regarded as an inner product

Equation 3.9 (page 125) gives the mean rate of cone cell signaling as an integral of the product of two spectra. Although this formula may look a bit like a convolution,[25] there is an important difference: The convolution of two probability density functions involves evaluating them at *different* values, whereas Equation 3.9 involves the product of absorption spectrum and spectral photon arrival rate at the *same* wavelength. Also, the convolution of two probability density functions is another PDF, whereas the result of Equation 3.9 is *one* number.

A better analogy for this formula is as a generalization of the dot product of two vectors (sometimes called the "inner product"). In ordinary three-dimensional space, the dot product of v and w is the single number $v_1w_1 + v_2w_2 + v_3w_3$. In higher-dimensional spaces, we have more terms, but always we multiply corresponding elements of v and w before summing. Equation 3.9 can be regarded as a limiting case, with a continuous index λ. Thus, we may think of each cone cell's spectral sensitivity as defining a *projection* from the infinite-dimensional space of light spectra to a single number, just as $v \cdot w$ can be regarded as the component of w that lies along v (scaled by the length of v). The three classes of cones project any spectrum down to a three-dimensional space, as discussed in Section 3.8.5 (page 124).

3.8.5′b Correction to predicted color matching due to absorption

Other, "nonproductive," absorptions and light scattering take place in the eye, with probabilities that may depend on photon wavelength. For example, although visual pigments can respond to UV light, nevertheless, such photons mostly fail to arrive at the retina, due to absorption in the cornea and lens. In fact, people who have had an eye lens removed surgically have enhanced UV sensitivity.

Baylor and coauthors accounted for these effects as reductions of the effective sensitivity function of the entire cone cell; Figure 3.11 gives these corrected functions.

Fig. 3.11 (page 123)

---

[25]Section 0.5.5 (page 16) defined convolution.

$\boxed{T_2}$ **Track 2**

3.8.8′a Relative versus absolute sensitivity

Equation 3.17 (page 129), and the corresponding formula for $B_{(\alpha)i}$, purport to give the mean rates of productive photon absorptions in terms of the spectrum of a light source. One may object that a cone cell sends a signal to the brain that is related to, but not the same as, the rate of productive absorptions—it may even be a nonlinear function of that rate.[26] Worse, the main text did not even present the actual sensitivity functions: Figure 3.11 shows only the spectral sensitivities each multiplied by a different overall constant to enforce the arbitrary convention that it peaks at the value 1. That is, the curves are *relative* spectral sensitivities, accurately showing only the ratios of the true sensitivities at different wavelengths. The true, or "absolute," sensitivities are not even fixed! For example, the population of the light-sensitive molecules in a photoreceptor cell may fluctuate depending on the observer's recent history, differences between observers, and so on.

Fortunately, however, we do not need to understand any of the preceding fine points in order to analyze color-matching experiments. Equation 3.9 (page 125) gives the rates of productive absorption β_i as linear functions of the spectrum of light $\mathcal{I}(\lambda)$, determined by the absolute sensitivities.[27] Let's write a similar formula involving the relative sensitivities, distinguished from absolute ones by an overbar:

$$\bar{\beta}_i = \int d\lambda\, \bar{\mathcal{S}}_i(\lambda)\mathcal{I}(\lambda). \qquad (3.18)$$

Because each $\bar{\mathcal{S}}_i(\lambda)$ is a constant times $\mathcal{S}_i(\lambda)$, the same is true for the $\bar{\beta}$s. So there must exist three functions of $\bar{\beta}_i$ that give the signals sent to the brain:

$$\text{signal } i = f_i(\bar{\beta}_i) \quad \text{for } i = L,\ M,\ \text{and } S. \qquad (3.19)$$

The key point is that, although each f_i may be nonlinear and may vary across observers and over time, it does *not* depend on wavelength (the univariance principle, Idea 3.8, page 124). Each cone cell's wavelength dependence is set by the absorption spectrum of its chromophores, and each class of cone cells expresses only one type. Nor do the f_i vary from one cone cell to the next within each of the classes.

A color match, $\mathcal{I} \sim \mathcal{I}'$, occurs when all three of the signals agree across two visual regions. According to Equation 3.19, however, this can occur only when all three of the $\bar{\beta}_i$ are also equal to their counterparts $\bar{\beta}'_i$. Thus, we may assess color matching by substituting the relative sensitivity in the formulas of Section 3.8.8;[28] we do not need to know the constants connecting them to absolute sensitivity, or anything else about the functions f_i.

Your Turn 3I

Give a simplified version of the preceding argument that explains why it was permissible to use relative instead of absolute sensitivities when we explained $R+G \sim Y$ in Section 3.8.4 (page 123).

---

[26] Section 3.8.3′a (page 134) gave an example of such a nonlinear function.

[27] See the table on page 128 for notation.

[28] For brevity, the main text and Problems 3.5 and 3.6 make this substitution implicitly, that is, we drop the overbars from the notation.

3.8.8′b Simplified color space

This section will outline the idea of a "color space," glossing over many fine points; for more details, see Shevell, 2003, or Peatross & Ware, 2015, §2.A.

The main text argued that perceived "color" involves three signals that are sent to the brain to represent each region of the visual field. This is an oversimplification; in fact, some spatial processing occurs already in the retina, prior to transmission.[29] But regardless of what kind of processing occurs, it remains true that the raw data from which we make color-matching assessments consists of sets of three numbers. [Section (a) above gave an example of why such "downstream" modifications don't affect matching.]

Thus, color perception involves a three-dimensional vector space: A point β in that space represent one possible signal from which a percept can be obtained. But not all points in this space actually correspond to colors that we see. For example, the overlaps between the sensitivity curves (Figure 3.11) imply that *no* kind of light will excite only the M cone cells; thus, vectors of the form $\beta = (0, x, 0)$ represent signals that never get generated in normal circumstances. So we are interested only in the part of β space that corresponds to physiologically possible signals, those actually generated by monochromatic lights or their combinations.

A three-dimensional color space is redundant. If we take a spectrum and multiply it at every wavelength by, say, 1.1, then we get a slightly more intense light, but not a new "color."[30] So to study chromaticity, we should examine a 2D surface in β space consisting of lights of approximately equal overall intensity ("luminance").

One useful outcome from constructing such a 2D "color space" is that we can use it to characterize the color signals made by mixtures of three basis lights, and investigate how well they fill the complete space of possible signals. You'll do this in Problem 3.6; a typical result is shown in Figure 3.17. In this figure, the large triangular region represents all β vectors whose components sum to 1, a rough proxy for the set of signals of equal strength. Within this triangle, the solid black curve represents the responses to monochromatic lights. All realizable color signals correspond to linear combinations of the points on this curve. For example, any combination of two monochromatic lights will be represented by a point somewhere along the line joining the corresponding two points on the solid black curve. Such points fill out a rounded triangular region, shown in white on the figure.

The smaller, colored region in the figure represents three basis lights as its vertices. Points inside this triangle represent the responses to all possible combinations of the basis lights; that is, it represents the color gamut of that choice (Section 3.6.3). Although the basis lights chosen for this illustration do not correspond exactly to those used in computer monitors, still the figure makes the point that there are a lot of realizable color signals (and hence color sensations) outside the gamut. Monochromatic lights are the hardest to match by three-color synthesis, because they lie at the boundary of the realizable region. That's why artificial reproductions, contrived from basis lights (or inks), appear less vivid than the spectrum cast by a prism, or the deep blue of a real butterfly wing.

Another feature of Figure 3.17 is that the colors appear more washed-out as we move toward the "white point" near the center. This observation suggests that we

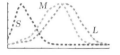

Fig. 3.11 (page 123)

---

[29]Section 11.4.1′b (page 370) gives some details about this processing.
[30]Actually, there are perceptual differences depending on intensity; for example, we perceive "brown" and "orange" as distinct categories, even though the difference is one of intensity. So this reduction to a two-dimensional surface in color space is an approximation.

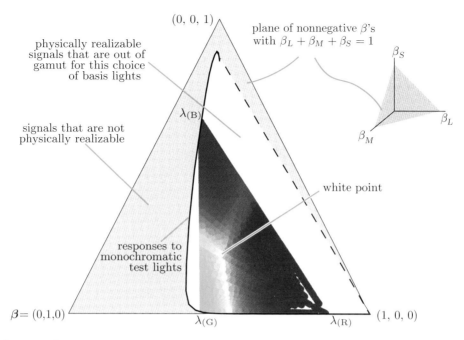

Figure 3.17: [Drawing.] **Simplified chromaticity diagram.** The interior of the outer equilateral triangle represents sets of three numbers $(\beta_L, \beta_M, \beta_S)$ whose sum equals 1 (see *inset*). Thus, the center corresponds to the point $(1/3, 1/3, 1/3)$, and the corners have the coordinates shown. Such triples represent the responses of cone cells to lights of roughly the same intensity. The *black curve* represents the signals generated by various monochromatic lights, with the longest wavelengths at the far right and shortest at the top. Thus, the region of the triangle corresponding to realizable color stimuli lies in the area bounded by this curve and the *dashed line* (the "convex hull" of the black curve). The responses to three monochromatic basis lights are shown as the corners of the colored region, labeled $\lambda_{(B)}$, $\lambda_{(G)}$, and $\lambda_{(R)}$. Other colored spots represent mixtures; they are painted with RGB values indicating the relative intensities $(\zeta_1, \zeta_2, \zeta_3)$ of the basis lights. Thus, the "white point" near the center represents the response to the mixture with $\zeta_1 = \zeta_2 = \zeta_3$. See Problem 3.6.

introduce polar coordinates about that point. The radial distance from the white point is then roughly analogous to a color specification called **saturation**; monochromatic lights have 100% saturation. The angular position is roughly analogous to a color coordinate called **hue**. A third coordinate, called **value**, is roughly analogous to the overall intensity that we suppressed earlier. Together, hue, saturation, and value form the basis of the "HSV color system," an alternative to specifying the amounts of red, green, and blue from some set of basis lights.

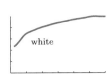

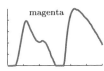

Fig. 3.1 (page 109)

PROBLEMS

3.1 *Four-color monitor*

A manufacturer of televisions introduced a high-end model containing pixels that were red, green, blue, and yellow. But "yellow" can be synthesized from red+green. So what might be the advantage of this scheme?

3.2 *Spectral conversion*

Figure 3.1 shows graphs of two spectral photon arrival rates. Section 1.5.2 (page 38) defines this quantity as a constant times the probability density function $\wp(\lambda)$ for an arriving photon to have a particular wavelength. Some scientists, however, prefer to describe the same light by the probability density function expressed in terms of frequency. Find the relation between these two descriptions. [*Hint:* Use the result in Section 0.5.1 (page 15).]

3.3 *Thinning property*

Section 1.4.3 stated that the waiting times in a Poisson process are draws from an Exponential distribution. One way to simulate a Poisson process is to turn this result around: Each draw from a Poisson process is a sequence of times that can be obtained as successive cumulative sums of Exponentially distributed numbers.

a. Implement this idea on your computer. First, generate some Uniformly distributed random numbers x between zero and one. Then, confirm that the numbers $y = -\beta \ln x$ are in fact Exponentially distributed with expectation β. Choose a value for β and compute the cumulative sums to get simulated blip arrival times.

b. Next, explore the thinning property (Section 3.4) by deleting a randomly chosen subset of your blip times, for example, half of them. Make a histogram of the new resulting waiting times and comment.

3.4 *Extended Grassmann law*

Use the color-matching rules (page 115) to prove an extended form of Grassmann's law:

$$\text{If } \mathcal{I}_1 \sim \mathcal{I}_2 \text{ and } \mathcal{I}_3 \sim \mathcal{I}_4, \text{ then } (\mathcal{I}_1 + \mathcal{I}_3) \sim (\mathcal{I}_2 + \mathcal{I}_4).$$

3.5 *Color-matching functions*

a. In this problem, you will use numerical software to solve systems of linear equations. To warm up, use a computer to solve the following three equations:

$$\begin{bmatrix} x+y \\ x-2y \\ 2z \end{bmatrix} = \begin{bmatrix} 1 \\ 3.14 \\ 2.71 \end{bmatrix}. \tag{3.20}$$

Then use your own brain to solve the equations, and iterate until you and your computer agree.

Obtain Dataset 7 and load it into your computer. It contains an array `lambdas` (a list of wavelengths in `nm`), and another array `sensitivity` containing the sensitivity functions $\mathcal{S}_i$ for $i = L, M,$ and S cone cells to monochromatic light at those wavelengths (see Figure 3.11).[31]

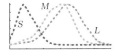

Fig. 3.11 (page 123)

---

[31] $\boxed{T_2}$ The dataset, and Figure 3.11, have been corrected for the wavelength dependence of light absorption in the eye lens and in other elements prior to the retina. The dataset gives relative sensitivity curves; see Section 3.8.8′a (page 139).

Consider a color-matching experiment with three basis lights that are monochromatic, with wavelengths $\lambda_{(\alpha)} = 645$, 526, and 444 nm. Each basis light has the same mean photon arrival rate Φ_p as the others, and every target light used also has that same Φ_p.

b. Estimate the sensitivities at the basis wavelengths by interpolating from nearby entries in the sensitivity dataset. Set up three linear equations whose solution tells us how much of each basis light is needed to match any given monochromatic light (see Equations 3.16 and 3.17, page 129). Get a numerical answer for matching monochromatic light of wavelength $\lambda^{\star} = 560$ nm.

c. Solve your equations many times, for monochromatic target lights with wavelengths throughout the range 400 nm $< \lambda^{\star} <$ 650 nm, and make a graph showing the color-matching functions, similar to Figure 3.8.

d. There is a range of spectral colors that *cannot* be matched, even approximately, by this set of three basis lights. What is it? [*Hint:* Specifically, consider a light to be "approximately matchable" if every computed ζ is bigger than -0.05.]

e. Suppose that we change the matching problem somewhat: Instead of taking the target colors to be monochromatic, we "dilute" them by adding a fixed amount of white light to each one. Here "white" light is taken to be light with a uniform spectrum throughout the visible region. Thus, instead of red we have slightly pink red, and so on; we say the target colors are less **saturated** than the ones in (b). Explain qualitatively why, with enough added white, we can match every target color in the spectrum.

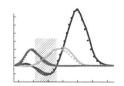

Fig. 3.8 (page 119)

3.6 $\boxed{T_2}$ *Simplified chromaticity diagram*

Obtain Dataset 7 to get the relative sensitivity curves for each of the three types of cone photoreceptor. (Problem 3.5 describes the dataset.) A discrete version of Equation 3.9 (page 125) then gives a point $\boldsymbol{\beta} = (\beta_L, \beta_M, \beta_S)$ in a three-dimensional space corresponding to any desired light spectrum.

Although $\boldsymbol{\beta}$ space is three-dimensional, one dimension simply corresponds to the overall level of illumination. The remaining two dimensions correspond to what we normally think of as the "color" of the light (chromaticity), and so the range of possible colors can be displayed in a two-dimensional **chromaticity diagram** (Figure 3.17). In this problem, you will construct a simplified version of the standard diagram.

Fig. 3.17 (page 141)

a. Define three basis lights, which are monochromatic with wavelengths $\lambda_{(1)} = 630$ nm, $\lambda_{(2)} = 540$ nm, and $\lambda_{(3)} = 470$ nm, and which all have the same mean photon arrival rate. Calculate $\boldsymbol{\beta}$ for each of these basis lights.

b. Consider a series of monochromatic target lights, of various wavelengths, but each with the same mean photon arrival rate as the basis lights. For each such light, use the sensitivity functions to construct a corresponding $\boldsymbol{\beta}$. Make a 3D plot of these points as a curve in $\boldsymbol{\beta}$ space. Highlight the three points on this curve corresponding to the choice of three basis lights given above. Rotate your plot in space (change its 3D viewpoint) to best visualize the objects you placed on it.

c. Any combination of basis lights can be specified by three numbers $\zeta_{(1)}, \zeta_{(2)}, \zeta_{(3)}$ via Equation 3.6 (page 119). Construct a three-dimensional grid of such triples, with each $\zeta_{(\alpha)}$ in the range from 0 to 1, find the corresponding points in $\boldsymbol{\beta}$ space, and add them to your plot.

d. Section 3.8.8′ (page 139) argued that rescaling all three components of $\boldsymbol{\beta}$ corresponds roughly to a change in the overall brightness (value) of a light, without changing its chromaticity. Take each of the points in (c) and rescale it to enforce

$\beta_{\text{tot}} = 1$, where $\beta_{\text{tot}} = \beta_1 + \beta_2 + \beta_3$. Do the same for the curve of monochromatic lights, and make a second 3D plot. Choose a nice viewpoint that shows the space of constant β_{tot} without distortion.

e. Make a third 3D plot by coloring the one in (d). That is, assign RGB values to each dot in your grid according to the relative amounts $\zeta_{(\alpha)}$ of each of the basis lights.

f. Add an asterisk to your chromaticity diagram corresponding to the spectrum of white light (Uniform distribution of photon probability over the visible wavelengths).

g. Is there a region of $\boldsymbol{\beta}$ space that does not correspond to any physically realizable light? Is there a region that is physically realizable, but that doesn't correspond to any color perception that can be created by combination of the three chosen basis lights?

CHAPTER 4

How Photons Know Where to Go

> I admired Dick [Feynman] tremendously, but I did not believe he could beat Einstein at his own game. Dick fought back against my skepticism, arguing that Einstein had failed because he stopped thinking in concrete physical images and became a manipulator of equations. I had to admit that was true.
>
> — *Freeman Dyson*

4.1 SIGNPOST: *PROBABILITY AMPLITUDES*

Chapter 1 proposed that light should be considered as a stream of tiny lumps—photons. Extraordinary claims like that require extraordinarily convincing proof. We began by sharpening the photon idea into the Light Hypothesis (Section 1.5.1), and discussed some biophysical phenomena that seemed to fit it (Chapter 2). Then we saw how some experiments on human color vision can also be understood in detail by using the photon model (Chapter 3).

But we know that light also displays *wavelike* phenomena (Figure 4.1). Even within your own eyes, light enters, bends, and focuses to form an image. Your visual acuity is limited in part by diffraction, which seems to be inherently wavelike.[1] We cannot have different, contradictory physical models of light for the lens in the front of your eye and the photoreceptors in the back!

Turning from natural systems to artificial ones, a single fluorophore molecule appears smeared out when viewed in an ordinary light microscope. But if light consists of lumps, each of which interacts with a single electron, then why can't we eliminate this blurring? It won't do to say simply, "You cannot resolve things that are closer than the wavelength of light," because no notion of "wavelength" appears explicitly in the Light Hypothesis.

Must light literally resemble *either* a water wave or a stream of sand grains? It's pleasant and convenient when a physical phenomenon has an analogy to everyday experience. For example, millions of years of throwing rocks have made projectile motion feel reasonable to our brains. Millions of years of dealing with water have made classical wave behavior feel reasonable. But Nature does not really care about our sense of what is "reasonable." Any time we find a reduced representation—a physical model—that is simple and fits all the observations, then we've done some physics, even if the model involves some new conceptual categories not hardwired into our brains by long experience. It's remarkable that humans could ever transcend the conceptual limitations of our daily experience by careful observation—but sometimes it works.

---

[1] Chapter 6 will discuss the role of diffraction in vision and microscopy.

Figure 4.1: **Metaphors for light.** This chapter will develop the second scientist's idea. [Cartoon by Larry Gonick.]

This chapter will develop a physical model for light that is neither a stream of material particles nor a classical wave. It may not be the last word on the subject. In particular, our model requires us to introduce a new description of reality, involving a quantity (the *probability amplitude*) with no precedent in pre-20th century physics. Nevertheless, quantum physics is conceptually minimal and makes testable, quantitative predictions—including many that are relevant to biology. Later chapters will explore some biological systems using these insights.

If you have studied physical chemistry, you probably approached quantum physics via the Schrödinger equation, which at first appears to be quite different from this book's formulation. Each approach has its strengths for particular kinds of problems.[2] For now, keep an open mind and see how the viewpoint advocated here can help us to understand some phenomena that are important for understanding life processes—and the experimental techniques used to study them.

The Focus Question is

Biological question: If light really consists of particles, then how can our eye lens focus it?

Physical idea: The stationary-phase principle tells photons where to go.

4.2 SUMMARY OF KEY PHENOMENA

Chapter 1 noted that

1. "Light" is a set of phenomena that involve transmitting energy through vacuum or through certain substances. The overall process of emitting and absorbing light consists of discrete events, for example, the blips that come from a sensitive light detector.

2. Those events are random, no matter how hard we try to create a stable source.

------

[2] $\boxed{T_2}$ Chapter 12 will outline why they are in fact equivalent.

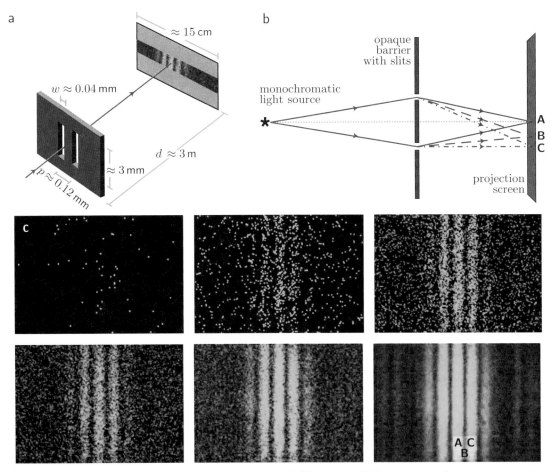

Figure 4.2: **Experiments to display diffraction patterns.** (a) [Schematic.] The passage of photons is partly obstructed by an opaque barrier, which has two narrow openings ("slits"). Not drawn to scale. (b) [Path diagram.] The same experiment, represented by a cross sectional diagram [top view of (a)]. Each opening is narrow in the plane of the page, but long in the direction out of the page. Three points on the projection screen are labeled; of these, **A** lies on the line that is perpendicular to the central screen and midway between the slits. A light detector placed at any of these points counts the arrivals of photons from the source at *left.* (c) [Experimental data.] Illumination pattern in an experiment equivalent to (a,b), recorded using an electronic camera sensitive to single photons. Laser light with wavelength 537 nm can take two routes to the camera's detector plane. Successive panels show data collected over increasing exposure times: The last panel's exposure is 500 000 times as long as the first one's. The illumination is so faint that there is never more than one photon in the apparatus at a time; nevertheless, a pattern of blips eventually emerges. Three screen points corresponding to those labeled in (b) are shown in the last frame. See also Media 10. [(c) From Dimitrova & Weis, 2008.]

The intensity of the light is actually the mean rate of a random process describing photon emission.

3. An "image" is a spatial nonuniformity of that mean rate. If the image of a point source of light is blurry, that implies a spread in distribution of the mean rates, not any enlargement of the sizes of individual photons themselves.

These points suggest that, when seeking to understand light, we should not try to predict where any individual photon will go—an impossible task, apparently. Instead, we should try to predict the *probability* of photon arrivals at various locations on a

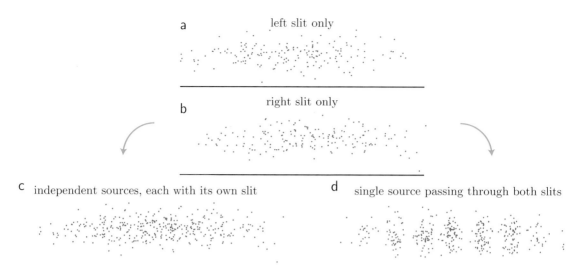

Figure 4.3: [Simulated data.] **Two-slit interference.** (a) "Cloud" representation of the probability density function for photon arrivals when a spot of light from a distant source passes through a narrow, horizontal slit that itself is far from a projection screen. Diffraction makes the distribution much wider than the slit. (b) Similar representation, for a situation in which the slit has been moved a short distance to the right. (c) Distribution obtained by summing the probability density functions in (a,b). This distribution really is observed when each of two distant, independent light sources shines through its own slit, and the lights are combined on a screen. (d) Distribution observed when a *single* monochromatic light source illuminates two narrow, closely spaced slits, as in Figure 4.2. This situation differs from (a) by the introduction of an additional pathway from source to screen; nevertheless, at some points the probability of photon arrival *decreases* to zero.

projection screen or detector array (such as our eye's retina). This chapter will argue that that task is *not* impossible, and will find a physical law that accomplishes it.

In addition

4. In everyday experience, light appears to travel along straight lines. For example, when you hold your hand in direct sunlight, you see a sharp shadow of your hand on the ground.

5. But actually light can **diffract** (bend its path). Although we are usually unaware of this effect, it's not hard to find situations where it's visible. For example, we can pass sunlight through a slit and let it land on a distant screen. As we narrow the slit, its image at first narrows, but then *widens* and becomes fuzzy.

Water waves do diffract (bend) as they pass by an obstruction, but it's not easy to see how a stream of independently moving particles ("sand grains") could do this.

To put our task into sharpest relief, here is a diffraction phenomenon that's particularly hard to understand with a "sand grains" metaphor. Imagine taking light from a distant source, selecting a single wavelength, then passing it through a narrow slit as in point **5** above. When the light lands on a distant screen, we see a pattern of illumination that has a maximum in the center, with intensity that falls off as we move to either side. If we block the slit and open another, identical one, for example 0.5 mm to the left of it, we get a similar pattern on the screen, imperceptibly shifted to the left. But when we open *both* slits, we get a surprise: Instead of two bars, or a single merged blur, experiment gives a pattern of *light and dark bands*—a **two-slit**

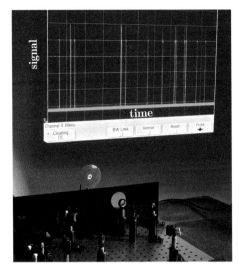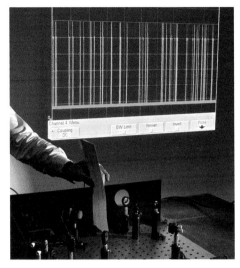

Figure 4.4: [Photos.] **Nonlocal character of light.** Two light paths create an interference pattern. A light detector capable of registering individual photons is placed near one of the minima of that pattern. *Left:* Signals from the detector are shown as discrete spikes in a time series projected on the screen, when both alternative pathways are open. The observed blips form a draw from a Poisson process, as in Figure 1.2 (page 27). *Right:* When one of the alternative paths is blocked, without moving the detector, the mean rate of blips (photon arrivals) *increases*. [Courtesy Antoine Weis; see also Dimitrova & Weis, 2008.]

interference pattern. Figures 4.2c and 4.3d show such patterns.[3] If we switch to a light source that is closer to the blue end of the spectrum, then the same slits create a new pattern, similar to but smaller than the original.

Somehow, in the spaces between the bright bands, *closing* one slit *increases* the probability that a photon will arrive there (Figure 4.4)! Closer examination shows what happened to the "missing" light (Figure 4.3c,d): With two slits, the bright bands are more than twice as bright as the corresponding locations in the single-slit case, as though light that "ought" to have gone to the dark bands has instead been "shifted," and added to the bright bands. In the wave model of light (Section 1.2.2), this behavior is not surprising. Something analogous happens when two sets of concentric ripples collide on the surface of water: Wherever a wave crest (maximum water height) from one set coincides with a trough (minimum water height) from the other one, the resulting wave amplitude is zero. A floating object at such a point will not bob up and down; it will not be able to absorb any energy from the wave, analogously to the absence of energy in the dark bands of a two-slit interference pattern. Similarly, when two crests coincide, the result is a greater water displacement than with one ripple source.

The problem with the preceding interpretation is that we have argued that light consists of photons, and its intensity must be interpreted as their mean rate of arrival. Two-slit interference implies that those photons must be making some very odd, and highly structured, "decisions" about where to go. One might try to argue that the interference pattern arises from photons "bumping into" one another. But, just a few years after Einstein's original photon proposal, G. Taylor showed that two-slit interference patterns can be generated even with light so faint that *at most one photon is ever inside the apparatus at any time!* Indeed, the images in Figure 4.2c were made

---

[3]See also Media 10.

under such conditions.

Section 4.2 has set out the challenge ahead of us: Apparently, light cannot be literally regarded as *either* a material particle or a wave. The rest of this chapter will propose a third physical model for light, in a formulation developed by Richard Feynman in the mid-20th century. Our proposal will involve extending the Light Hypothesis, in a way that tells photons where to go.

$\boxed{T_2}$ *Section 4.2′ (page 172) says a few words about unintuitive theories.*

4.3 THE PROBABILITY AMPLITUDE

Faced with the problem of explaining planetary orbits, Newton found a way forward by investigating something simpler and deeper: a *force* acting between any two bodies. So it may seem inevitable to ask, "When photons don't travel on straight lines, what force pushes them around?" But this line of inquiry has not proven to be fruitful. So instead of postulating some new layer of physical mechanism, our attitude will be: Don't try to explain why; just describe *what*. That is, instead of saying, "There's a new random force whose origin is...," we will just seek a framework that can predict the photon arrival probabilities that we observe. Once we have this language, we'll extend the Light Hypothesis by adding a rule to calculate the desired probabilities. Then we can explore how that rule reconciles wavelike and particle-like phenomena, even if we can't obtain it as a consequence of something deeper and more intuitive.

4.3.1 Reconciling the particle and wave aspects of light requires the introduction of a new kind of physical quantity

We wish to reinterpret the concept of "intensity" in terms of a probability distribution for photon arrival. When we try to formulate a rule for those probabilities, however, we hit an immediate obstacle: Probabilities must be *nonnegative numbers*.[4] There is no way to add two positive probabilities and obtain zero, as seems to be needed to explain the dark bands of a two-slit interference pattern (compare Figure 4.3c to Figure 4.2c). Instead, the architects of quantum physics were forced to suppose the existence of a new sort of physical quantity, now called the **probability amplitude** and denoted by the letter Ψ.[5] The probability amplitude need not be positive, so two contributions can cancel, just as $(1) + (-1) = 0$. Max Born proposed in 1926 that the probability amplitude be interpreted as the quantity whose *square* at any location tells us the probability that a photon will be observed there. With this rule, the probabilities we compute are never negative.

It is not enough just to admit negative values of Ψ, however. We wish to explain the apparently wavelike behavior of light. Suppose that we try to describe light emerging from a point source by the probability amplitude function $\Psi = \sin\varphi$, where φ is related to the distance r from the source. It is true that adding two such functions can result in cancellation, an "interference" effect. But even for a *single* source, the function $(\sin\varphi(r))^2$ periodically goes to *zero*, which makes no sense as a photon probability: We don't expect a point source of light to appear dark at certain distances! On the

---

[4]Section 0.2 (page 2) made this point.

[5]It's important to distinguish the probability amplitude from the ordinary amplitude of a classical wave, which is a positive real quantity that can in principle be measured directly, for example, an electric field or air overpressure.

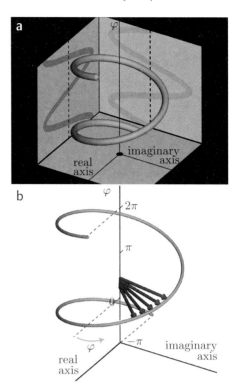

Figure 4.5: [Mathematical functions.] **Euler formula.** (a) The *vertical axis* represents various values of a variable φ, which increases from $-\pi$ to 2π. For each value of φ, one point is marked in the horizontal plane at that height. That point can be interpreted as the tip of a vector that starts at the vertical axis and lies in this plane. Alternatively, the point can be interpreted as the value of the complex function $Z(\varphi) = \cos\varphi + \mathrm{i}\sin\varphi$. Either way, the marked points sweep out the helical curve shown. *Shadows* on the walls show that the real and imaginary parts of $Z(\varphi)$ really are $\cos\varphi$ and $\sin\varphi$, respectively, but the overall function encircles the origin (*vertical axis*) without ever touching it. (b) In this panel, the *arrows* represent the same function as (a), at a series of specific φ values. The lengths of these arrows represent the modulus of the function (see Appendix D), and illustrate geometrically that it is always equal to 1. The angle that each arrow makes with the real axis is the phase φ of $Z(\varphi)$.

contrary, as we move a few hundred nanometers toward or away from a distant light source, the probability of catching photons hardly changes at all.[6]

We seem to have hit an impasse: We are looking for a probability amplitude function with the property that, for a single source, its square is nearly constant. But that function must periodically go from negative to positive values, in order to generate the interference patterns observed with a second source. How can it do that without passing through zero along the way?

The way out of this dilemma requires that we interpret Ψ as a function whose values are more general than ordinary real numbers. One generalization of ordinary numbers is the space of two-dimensional *vectors*. Like ordinary real numbers, it's possible for two nonzero vectors to sum to zero. Unlike real numbers, however, a vector may "point" in many directions (not just left and right on a number line). Thus, as we consider various locations on a projection screen, the value of Ψ may *encircle* the origin, periodically reversing its direction without ever changing its *length* (see the arrows in Figure 4.5b).

Combining ideas from the preceding paragraphs, we can say that the following proposal is free from some of the more egregious defects found earlier:

> *The value of the probability amplitude at any point is a two-dimensional vector. If light can take multiple pathways to get from one point to another, as in the two-slit experiment, the corresponding vectors are summed. The probability of arrival is the squared length of the resultant vector.* (4.1)

---

[6]If you stand one meter from such a point source, then move away by an additional half a wavelength of light, the decrease in intensity due to the $1/r^2$ law is negligible.

Two-dimensional vectors have some other simple properties that make them particularly useful for describing waves. Section 4.4 and Appendix D review those properties.

Section 4.5 will sharpen Idea 4.1 into a more precise extension of the Light Hypothesis. Already, however, we can see how it differs from the naive picture of photons somehow "interfering with one another," which was invalidated by Taylor's observation of single-photon interference.[7] Instead, it says that

*Alternative photon **paths** are what give rise to interference.*   (4.2)

Vectors are often used for quantities that point in some direction in space—north, south, east, west, up, down, and combinations of these. But the vectors used to describe probability amplitudes have *no such interpretation*. They "point" in an abstract plane, with no ordinary spatial meaning.[8]

The probability amplitude does not correspond to anything tangible in everyday experience. Whenever we admit this kind of abstraction into a physical model, we must be especially careful to test the model's predictions experimentally. We must also look hard for an alternative description of Nature that eliminates the need for the abstract element. No such alternative to the probability amplitude has yet been found and widely accepted, however, despite immense effort. The introduction of a quantity that cannot be directly measured appears to be the price we must pay in order to accommodate both the wavelike and particle-like aspects of light.

Section 4.3 has made a case for introducing a physical quantity with no precedent in classical physics. We now review some math that will make it easier to manipulate quantities of this sort.

4.4 **BACKGROUND:** COMPLEX NUMBERS SIMPLIFY MANY CALCULATIONS

A two-dimensional vector can be represented by two real numbers, for example, its components along two axes. But our formulas would become very long and awkward if represented in this way. Fortunately, there is an algebraic system that treats a two-component vector as a single **complex number**, simplifying our formulas considerably. Appendix D reviews the definition and properties of complex numbers; this section presents a key result specifically useful for wavelike phenomena.

The horizontal slices of Figure 4.5a,b show the complex number $Z(\varphi)$ with real part $\cos\varphi$ and imaginary part $\sin\varphi$, for various values of φ. Panel (b) of the figure illustrates the key feature that every time we add a constant increment to φ, the function turns through the same *angle*. Because rotating a complex number about the origin corresponds to multiplication by another complex number,[9] we see that the function in the figure has the property that adding an increment to any initial value φ just multiplies the function by a constant (independent of φ):

$$Z(\varphi + \Delta\varphi) = CZ(\varphi) \qquad \text{for all } \varphi.$$ (4.3)

This is the defining property of an exponential function, so $Z(\varphi) = \mathrm{e}^{A\varphi}$ for some constant A. Now consider φ close to zero. Comparing the behavior of $\cos\varphi + \mathrm{i}\sin\varphi$

---

[7] Taylor's experiment was mentioned in Section 4.2.

[8] $\boxed{T2}$ In particular, the direction of Ψ in its internal space is not related to the polarization of a photon. Even the Higgs boson, which has no polarization, has a probability amplitude of this type.

[9] You'll establish this property in Your Turn DA (page 450).

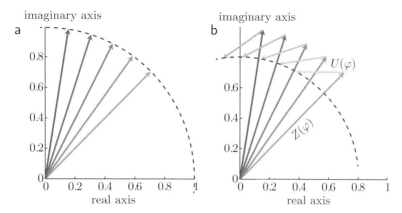

Figure 4.6: [Mathematical functions.] **Partial interference.** (a) Endpoints of *blue arrows:* The same function $Z(\varphi) = \cos\varphi + i\sin\varphi$ shown in Figure 4.5, evaluated at the same illustrative values of φ (in the range $0.25\pi \le \varphi \le 0.45\pi$). The function sweeps out a unit circle (*dashed*) in the complex number plane. (b) This panel represents the sum of the function in (a), plus the related function (*shorter arrows*) $U(\varphi) = (1/3)(\cos(\varphi + \varphi_0) + i\sin(\varphi + \varphi_0))$. Here $\varphi_0 = 3\pi/4$ is a constant. Each point on the *dashed curve* is thus the sum of the two vectors shown. The endpoints again sweep out a circular path around the origin, but with a reduced radius compared to (a).

to that of $e^{A\varphi}$ shows that $A = i$, or[10]

$$\boxed{\cos\varphi + i\sin\varphi = e^{i\varphi}. \quad \text{Euler formula}} \tag{4.4}$$

The dimensionless quantity φ appearing in this formula is called the **phase** of the complex number; it is also the angle between the real axis (with $\varphi = 0$) and the vector representing that number (see Figure 4.5).

For our purposes, the Euler formula is useful because it gives a compact formulation of the sort of periodic function that could represent photon behavior. It also shows us how to evaluate partial interference. For example, Figure 4.6b represents the sum of two functions $Z(\varphi)$ and $U(\varphi)$, which we'll soon interpret as a total probability amplitude. In the situation shown, the resultant vector always has length smaller than 1: That is, there is some destructive interference. Computing the resulting function of φ by using trigonometric identities would be a tedious exercise. Instead, we can use the Euler formula:

$$Z(\varphi) + U(\varphi) = e^{i\varphi} + \tfrac{1}{3}e^{i(\varphi + 3\pi/4)} = e^{i\varphi}\left(1 + \tfrac{1}{3}e^{i3\pi/4}\right).$$

That is, the sum is again the periodic function $e^{i\varphi}$, but multiplied by the constant factor $(1 + (1/3)e^{i3\pi/4})$.

Ex. Evaluate the modulus of that factor. Explain geometrically why it's smaller than 1. *Solution:* First use Equations D.3, D.9, and D.7 (pages 450–451) to get the modulus squared:

$$|Z+U|^2 = |e^{i\varphi}|^2|1+\tfrac{1}{3}e^{i3\pi/4}|^2 = (1+\tfrac{1}{3}e^{i3\pi/4})(1+\tfrac{1}{3}e^{-i3\pi/4}) = 1+\tfrac{2}{3}\cos\left(\tfrac{3\pi}{4}\right)+\tfrac{1}{9} \approx 0.64.$$

---

[10]The constant C in Equation 4.3 is therefore $\exp(i\Delta\varphi)$. Appendix D gives a more detailed discussion.

The modulus is the square root of this quantity, or 0.80. Geometrically, because $3\pi/4$ is bigger than a right angle, the sum of the arrows always partly "doubles back on itself," making the resultant vector shorter than its first term (see Figure 4.6b).

For further details, see Appendix D.

4.5 LIGHT HYPOTHESIS, PART 2

The first part of the Light Hypothesis said that monochromatic light consists of lumps each carrying the same energy E_{photon} (Section 1.5.1, page 37). We now extend our statement of the Hypothesis, in a formulation due to Richard Feynman, by specifying the recipe that will let us calculate probabilities of photon arrival.

> **Light Hypothesis, part 2a:**
> - *The probability to observe a photon is the square of the modulus of the probability amplitude Ψ; that is, $\mathcal{P} = |\Psi|^2$.*
> - *If there are multiple paths (or processes) by which a photon could make the trip from source to detector, and we cannot observe which one was taken, then each path makes an additive contribution to the total probability amplitude.*
> - *For monochromatic light traveling in vacuum, the **phase** of each contribution to Ψ equals $E_{\mathrm{photon}}/\hbar$ times the transit time for the path, where $\hbar$ is the reduced Planck constant.*
> - *For a straight-line path, the **modulus** of the path's contribution is an overall constant divided by the path length. For a path consisting of straight-line segments, the modulus is the product of one such contribution for each segment.*

(4.5)

In the first point, we usually want the probability as a function of position, for example on a viewing screen, and so Ψ, too, will depend on position.[11]

We will see in the next section how the second point in Idea 4.5 allows for the possibility of two-slit interference. This point will require some careful interpretation in Section 4.7.2. Then it will prove to be the key to understanding reflection, refraction, diffraction, and other optical phenomena.

The third point of Idea 4.5 gives the phase of a path as the product of two quantities. The significance of the first factor, $E_{\mathrm{photon}}/\hbar$, is that it is related to the kind of light being studied, and it has the dimensions $\mathbb{T}^{-1}$ needed to give a phase when multiplied by transit time. In vacuum, the second factor is (transit time) = (path length)$/c$, because light travels at the fixed speed c in vacuum.

The inverse-length factor mentioned in the last point of the Light Hypothesis is important, in that it makes the intensity of light fall off with distance from a point source.[12] But to simplify our math, this chapter and the next will consider situations

---

[11] For this reason, Ψ is sometimes called the "wavefunction," a term that we will not use. Strictly speaking, we usually wish to find a PDF, though often we will use the generic symbol $\mathcal{P}$.

[12] $\boxed{T2}$ Chapter 12 derives this $1/r$ factor from a more general starting point.

in which these factors are nearly equal for every path under consideration. In such situations, the $1/r$ factors effectively become an overall constant, which we may lump with other constants as a normalization factor.

Ex. Make that claim more precise, in the context of the two-slit experiment.
Solution: We are studying expressions of the form

$$\frac{1}{L} e^{i(E_{\text{photon}}/\hbar)(L/c)}.$$

It's convenient to lump together factors appearing in the exponential, by introducing the abbreviation $\lambda = 2\pi\hbar c/E_{\text{photon}}$. Then the expression just given reduces to $L^{-1}e^{2\pi i L/\lambda}$. Thus, λ is the distance over which the phase changes by 2π.

The claim is that the L dependence of the $1/L$ factor is less important than that of the exponential. Let's evaluate the derivative with respect to L:

$$\left(-\frac{1}{L} + \frac{2\pi i}{\lambda} \right) \frac{1}{L} e^{2\pi i L/\lambda}.$$

The first term in parentheses comes from the variation of $1/L$; the second comes from the variation of the exponential. In Figure 4.2 (page 147), $L \approx 3\,\text{m}$ and for visible light $\lambda \approx 500\,\text{nm}$. So indeed, the second term dominates the first.

Idea 4.5 applies in vacuum, which is not a very biological situation. For most purposes, however, we may also use it for light moving through air. Moreover, in a uniform, transparent medium (such as glass or water), it's often a good approximation to continue to use Idea 4.5, with the modification that the speed of light is reduced to c/n, where n is a constant called the medium's index of refraction.[13] Sometimes we also need to consider mixed situations, in which light passes from one medium to another, for example, from air to the fluid inside our eyes. For such cases, we augment Idea 4.5 with some additional points:

Light Hypothesis, part 2b:
- *When light passes from one uniform, transparent medium to another, the energy per photon does not change.*
- *Idea 4.5 remains valid in such a medium, but the transit time must account for the reduced speed c/n.*
- *When light arrives at a boundary between transparent media, its trajectory can either proceed through the boundary or "bounce," returning into the medium from which it came.*   (4.6)
- *If a trajectory traveling perpendicular to such a boundary "bounces," and if the photon was initially heading from a medium with smaller n toward one with larger n (for example, from air to water), then its contribution to the total probability amplitude gets multiplied by an extra factor of -1.*

In the last point, instead of multiplying by -1 we may equivalently add π to the phase associated to the path in question. There is no such extra phase, however, for a photon path that bounces when heading from larger toward smaller n.

---

[13]Section 1.7 (page 47) introduced the slowing of light by a medium. $\boxed{T_2}$ Chapter 12 will show that it is not really an independent hypothesis, but rather can be obtained from a more basic starting point.

Like the first part of the Light Hypothesis, the addenda proposed in this chapter make no explicit mention of any wave character of light; for example, neither mentions any "wavelength" or "frequency." Our goal in the following sections will be to see how the full Light Hypothesis implies that, in some ways, light can act *as though it were* a wave with frequency $\nu = E_{\mathrm{photon}}/(2\pi\hbar)$ (and hence with wavelength $\lambda = c/\nu$). That is, we will derive the apparent wavelike behavior, and the Einstein relation, as mathematical consequences of the Light Hypothesis. We'll also find out *when* light behaves in this way, and when it crosses over to more particle-like behavior.

As with the first part of the Light Hypothesis, we will not attempt to justify Ideas 4.5–4.6. Instead, we regard them as a small number of principles from which many experimentally testable conclusions can be drawn. The following sections will show that they look promising as a way to reconcile some wavelike phenomena with the photon concept.

Section 4.5 has formulated the Light Hypothesis without much motivation. We must now see how this principle accommodates apparently wavelike phenomena such as interference and diffraction. Later sections will systematically explore additional phenomena useful for any living organism trying to see—and for scientists trying to see better.

$\boxed{T_2}$ *Section 4.5′ (page 172) will describe some finer points about Ideas 4.5–4.6. Chapter 12 shows how they can be regarded as consequences of a more general principle, not limited to the special case of monochromatic light. Also, the Light Hypothesis as stated in this section neglects polarization effects, and so the diffraction theory we will derive is an approximation ("scalar diffraction theory"). Chapter 13 outlines a more complete approach.*

4.6 BASIC INTERFERENCE PHENOMENA

4.6.1 Two-slit interference explained via the Light Hypothesis

Fig. 4.2b (page 147)

Let's go back to the two-slit experiment (Figure 4.2b), and try to understand it by using our extension of the Light Hypothesis. Throughout this chapter and the next, we will neglect the $1/(\text{length})$ factor appearing in the last part of Idea 4.5, because it will be roughly constant; instead, we focus attention on the rapidly varying phase factors.

Idea 4.5 applies to the overall process of emission/transmission/detection. In the context of a two-slit experiment, it says that there are two contributions to the probability amplitude for photons to arrive at a given point on the projection screen. Consider these examples:

- Detection at point **A** in the figure involves two contributing paths of exactly the same length. So

$$\Psi(\mathbf{A}) = C(\mathrm{e}^{\mathrm{i}\varphi} + \mathrm{e}^{\mathrm{i}\varphi}), \tag{4.7}$$

  for some constants C and φ. The modulus squared of this function is $4|C|^2$.

- At point **B**, the upper path is slightly longer, and the lower path slightly shorter, than at **A**. Suppose that their transit times differ by one half of the quantity $2\pi\hbar/E_{\mathrm{photon}}$. That difference is insignificant for the slowly varying modulus C, but the phase difference *is* important:

$$\Psi(\mathbf{B}) \approx C\left(\mathrm{e}^{\mathrm{i}\varphi'} + \mathrm{e}^{\mathrm{i}\left(\varphi' + (E_{\mathrm{photon}}/\hbar)(\pi\hbar/E_{\mathrm{photon}})\right)}\right), \tag{4.8}$$

for some constant φ'. This quantity equals $Ce^{i\varphi'}(1+e^{i\pi})$, which equals zero; no photons will arrive at **B**, as is observed.

- Even farther off axis, at **C**, the transit times will differ by a full $2\pi\hbar/E_{\text{photon}}$. Then

$$\Psi(\mathbf{C}) \approx C\left(e^{i\varphi''} + e^{i\left(\varphi'' + (E_{\text{photon}}/\hbar)(2\pi\hbar/E_{\text{photon}})\right)}\right), \qquad (4.9)$$

for some constant φ''. This time, the probability amplitude equals $Ce^{i\varphi''}(1+e^{i2\pi})$. Its modulus squared is then $4|C|^2$, the same as at **A**.

Similar remarks apply if we move our light detector in the other direction on the projection screen (upward in the diagram). In short, the Light Hypothesis has predicted the pattern of light and dark bands observed on the screen when the second slit is opened (Figure 4.2c):

> The two-slit diffraction pattern arises by interference between the contributions to the probability amplitude associated with alternate paths that each photon may take.

Ex. Find the complete pattern of illumination on the projection screen.
Solution: First, note that the transit time difference can be written in terms of Δ (the path-length difference) as Δ/c. In terms of Δ, then, the same reasoning as above gives the probability amplitude to arrive at a point on the viewing screen as an overall constant times $1 + e^{i(E_{\text{photon}}/\hbar)(\Delta/c)}$. Again introducing the abbreviation $\lambda = 2\pi\hbar c/E_{\text{photon}}$, the probability density for photons to arrive at a particular location equals a constant times

$$\left|1 + e^{i2\pi\Delta/\lambda}\right|^2 = \left|(e^{i\pi\Delta/\lambda})(e^{-i\pi\Delta/\lambda} + e^{i\pi\Delta/\lambda})\right|^2.$$

Equation D.7 (page 450) helps us to simplify this expression:

$$= 1 \times \left|2\cos\frac{\pi\Delta}{\lambda}\right|^2 = 4\cos^2\frac{\pi\Delta}{\lambda}. \qquad (4.10)$$

Because Δ depends on position on the screen, this expression shows how the probability of photon arrival rises and falls smoothly as we move a detector across the screen. Our previous results (Equations 4.7–4.9) are the special cases where Δ is zero, $\lambda/2$, or λ, respectively.

The wave model of light predicts the *same formula* Equation 4.10 for the intensity pattern of light, though with a very different interpretation.[14] Thus, the Light Hypothesis subsumes the Einstein relation: It predicts that monochromatic light with a given E_{photon} will give diffraction (interference) patterns similar to those expected for a classical wave with wavelength $\lambda = 2\pi\hbar c/E_{\text{photon}}$.

Equation 4.10 embodies some approximations, so it cannot be taken literally outside the domain where those approximations are valid. For example, the real diffraction pattern in Figure 4.2c falls off in intensity after the first few fringes. The formula in Equation 4.10 does not have this property, and indeed cannot be normalized. The

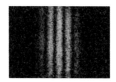

Fig. 4.2c (page 147)

---

[14]You'll find a more detailed form of the illumination pattern in Problem 4.3.

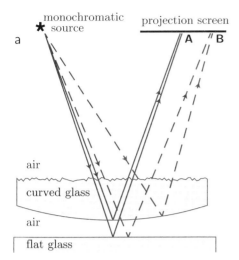
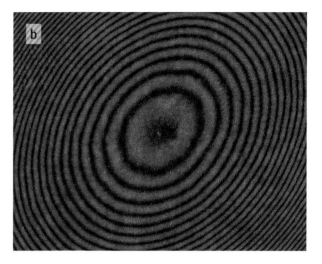

Figure 4.7: [Path diagram; photograph.] **Newton's rings.** (a) Side view of the setup. The bottom surface of the upper piece of glass is shaped like part of a sphere. Two different paths contribute to light observed at **A** (*solid lines*); two other paths contribute to light observed at **B** (*dashed lines*). For simplicity, the figure does not show the refraction (bending of light) at the air-glass interfaces. Reflections from the rough top surface of the glass go in random directions (*not shown*), and so may be neglected. The figure is not to scale; the curvature of the top piece has been exaggerated. Additional paths exist that bounce multiple times between the glass surfaces (*not shown*). Chapter 5 will show how to account for their contributions. (b) Top view of the interference pattern when monochromatic light shines on the setup in (a). The approximate axial symmetry in the apparatus sketched in (a) gives rise to a roughly circular arrangement of bands in (b). [Photo courtesy Robert D Anderson.]

reason for this discrepancy is that so far, we have taken each slit to be infinitesimally narrow, unlike those in real experiments.[15] For now, we just note that the provisional formula correctly accounts for the periodic structure in the center of the diffraction pattern.

$\boxed{T_2}$ *Section 4.6.1′ (page 173) says more about two-slit interference.*

4.6.2 Newton's rings illustrate interference in a three-dimensional setting

When we place two nearly flat, polished plates of glass together, the light reflected from them often shows interesting and colorful patterns. Similar patterns are also seen in the light reflected by a thin layer of oil or detergent on water, or in light reflected from a soap bubble.

Robert Hooke studied such phenomena in a simple geometry:[16] On top of a flat, polished glass plate we place another plate, polished but slightly curved on its lower face (Figure 4.7a). We then illuminate the plates with a distant, monochromatic light source, and observe the light reflected onto a screen.

The experimental geometry represented by Figure 4.7a is more complicated than the slit geometries that we have considered so far. In the slit case, the object under study has structure in one direction but is uniform in the other: The slits are straight and much longer than their spacing. We will call such situations "two dimensional,"

---

[15]Section 4.7.3 and Problem 8.2 discuss finite-width slits.
[16]Isaac Newton later described the same phenomenon in his book *Opticks*, and it has come to be called "Newton's rings."

because we can ignore one dimension in their analysis. In contrast, the Newton-rings apparatus involves a surface that is curved in both directions; there is no uniform direction. Such problems are truly three dimensional; they give rise to illumination patterns like Figure 4.7b, that is, more complex than the bars that appear in a 2D problem (Figure 4.2c).[17]

Fig. 4.2c (page 147)

Figure 4.7a shows the situation in cross section. Photons can arrive at **A** either by bouncing off the lower, flat surface, or by bouncing off the curved glass→air interface. As with two-slit interference, the probability of photon arrival at **A** is controlled by the difference in length $\Delta(\textbf{A})$ between these two paths.[18] At a different point **B**, $\Delta(\textbf{B})$ will have a different value, leading to alternating light and dark bands on the screen.

For any point on the projection screen, Idea 4.6 (page 155) states that one of the two paths will make a contribution with an additional minus sign, because it corresponds to a process in which photons in air are bouncing off a glass surface. In Figure 4.7a, the path that bounces from the flat plate is of this type. Thus, Equation 4.10 changes slightly, to

$$\left|1 - e^{i2\pi\Delta/\lambda}\right|^2 = \left|(e^{i\pi\Delta/\lambda})(e^{-i\pi\Delta/\lambda} - e^{i\pi\Delta/\lambda})\right|^2$$
$$= 1 \times \left|2i\sin\frac{-\pi\Delta}{\lambda}\right|^2 = 4\sin^2\frac{\pi\Delta}{\lambda}. \tag{4.11}$$

Equation 4.11 superficially resembles the two-slit formula (Equation 4.10), but this time, the path length difference Δ depends on both the x and y coordinates on the projection screen, not on x alone. In most demonstrations of the effect, the curved surface actually touches the flat one. Thus, close to the point of contact, the two interfering paths have the *same* length, so $\Delta = 0$. Equation 4.11 then predicts that the center of the ring pattern will be *dark*, as seen in Figure 4.7b. Moving in any direction away from screen point **A** increases Δ, so there are concentric rings on the screen along which Δ/λ is an integer; these are the dark bands in Figure 4.7b.

A layer of oil on water also generates interference patterns, again because its thickness will in general be different at different locations. Generally, we view such layers under white light, and the reflections appear colored. This happens because the photons interfere individually, not with each other,[19] and for white light each photon has its own wavelength. The distinction between light and dark bands depends on both the layer thickness, and also the wavelength of each photon (see Equation 4.11). Thus, each subset of photons corresponding to each wavelength has a different pattern of dark bands. The bands missed by red light from the source will appear cyan, and so on.

$\boxed{T_2}$ *Section 4.6.2′ (page 174) says more about the minus-sign rule used above and other fine points.*

4.6.3 An objection to the Light Hypothesis

The preceding sections described two striking phenomena that may seem to support the Light Hypothesis. But on further consideration, uncomfortable questions emerge:

---

[17]When we study lenses in Chapter 6, and x-ray diffraction in Chapter 8, we will again begin the discussion with the 2D case, to keep the math simpler.

[18]The extra length for the longer path is almost entirely due to its segments in air, so the index of refraction does not enter this simplified calculation, other than to introduce the extra minus sign in one of the paths' contribution.

[19]Taylor's experiment (Section 4.2, page 146).

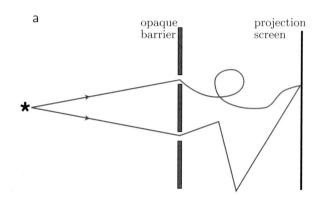

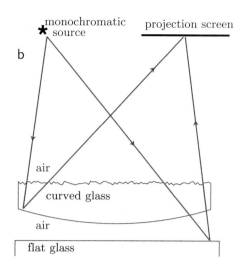

Figure 4.8: [Path diagrams.] **Crazy paths.** (a) Examples of photon paths that were not considered in Figure 4.2b. (b) Examples of photon trajectories that were not considered in Figure 4.7a. Unlike the trajectories considered there, these ones do not obey the "law of reflection" discussed in Section 5.3.1.

Fig. 4.2b (page 147)

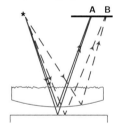

Fig. 4.7a (page 158)

- The Light Hypothesis (Idea 4.5, page 154) seems a bit vague about the "multiple routes" that should be considered. Why did we only draw paths consisting of *straight-line* segments in Figures 4.2b and 4.7a? What principle told us not to include paths like the curved one in Figure 4.8a? Is this some new, unstated aspect of the Light Hypothesis? If so, then what about the case of light passing through a medium of variable density? We know that in that case, it *should* bend!

- Even if we agree to limit attention to piecewise-straight paths, still there are many such paths from the source to point **A** other than the ones shown in Figure 4.7a. Why did we not consider paths like the jagged line in Figure 4.8a, nor the paths shown in panel (b)? Is this yet another new rule?

Feynman's resolution to this puzzle was elegant:

> *There is no extra rule that requires us to omit the crazy paths. But there **is** a mathematical principle that explains why, in macroscopic experiments, they make negligible contributions to our answers, and hence **may** be omitted, simplifying our calculations.*   (4.12)

The next section will develop the key mathematical idea alluded to in Idea 4.12: the stationary-phase principle. Section 4.7.2 will explain how this result justifies Feynman's remarkable insight. Then we will see how the same idea gives us quantitative results about the crossover between our everyday experience with light and the diffractive effects seen in microscopes.

Section 4.6 has begun our journey from the Light Hypothesis to image formation (in Chapter 6).

$\boxed{T_2}$ *Section 4.6.3′ (page 174) tackles two more objections to the Light Hypothesis. In fact, Feynman's idea instructs us to include even crazier paths than the ones discussed so far; Chapter 12 explains why these, too, may be neglected in macroscopic situations.*

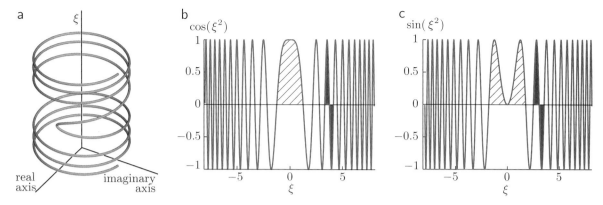

Figure 4.9: [Mathematical functions.] **Fresnel integral.** (a) The function $\exp(i\xi^2)$, in a representation similar to Figure 4.5. The value of ξ is represented by distance along the vertical axis; it covers the range $-4.5 < \xi < 4.5$ (b) The real part of this function. When we integrate it over a region away from $\xi = 0$, the positive contributions (for example, the *upper red* area) approximately cancel the negative ones (*lower red* area). (c) A similar cancellation occurs with the imaginary part of the function. In each case, however, the contributions from regions close to 0 do *not* cancel (*hatched* areas).

4.7 THE STATIONARY-PHASE PRINCIPLE

4.7.1 The Fresnel integral illustrates the stationary-phase principle

The Light Hypothesis seems to be telling us that to compute the probability amplitude, we must add a lot of contributions—one for "each path." Each contribution to Ψ is a complex number of approximately the same modulus, but each has a different phase. The preceding subsection suggested that instead of considering just *two* paths from source to detector, we must include a continuous *family* of paths. Instead of tackling this big problem head-on, we should first consider the simplest possible calculation of this type: evaluation of the integral $\int_{-\infty}^{\infty} d\xi \, e^{i\xi^2}$. A. Fresnel investigated this problem around 1819, long before the advent of quantum physics, so integrals of this type are called "Fresnel integrals." Let's understand this expression as a purely mathematical problem; later, we'll see its relevance to our physical model of light.

Evaluating the Fresnel integral may seem to be a nightmare. Normally we are taught that an integral to infinity is meaningless unless the integrand falls off rapidly with ξ; but this integrand is everywhere a complex number of modulus *one!* Worse, a look at the integrand shows that it is very rapidly oscillating for large $|\xi|$ (Figure 4.9). But take a closer look at the figure. In any region far away from $\xi = 0$, for example from $\xi = 5$ to 7, the real part of the integrand is positive about as often as it is negative (Figure 4.9b), so we may hope that the integral in such regions will largely *cancel*, leaving a finite and well-defined net value from the region near $\xi = 0$. Similar remarks apply to the imaginary part (Figure 4.9c). That is, we may hope that the integral will "oscillate to death" for large $|\xi|$, and so be well defined. Let's see if this hope can be justified.

Figure 4.10 gives a different graphical representation of an integral of this sort. We imagine taking an interval and dividing it up into small regions of width $\Delta\xi$. The integral of $\exp(i\xi^2)$ is then approximated as the sum of its values on each small region, times $\Delta\xi$. Each contribution to this sum is a complex number, $\Delta\xi e^{i\xi^2}$. The figure shows three arrows, with angle equal to ξ^2, for each of three values of ξ. The sum is

Figure 4.10: [Vector sum.] **Approximation of a complex integral by a finite sum.** For each value of ξ, the value of the integrand in the Fresnel integral has been represented by a *black arrow* in the complex plane. Each arrow has the same length $\Delta\xi$, but each has a different angle ξ^2 relative to the real axis. The sum of those arrows is computed by shifting each to the endpoint of its predecessor, and finding the resultant (*long arrow*). For clarity, this figure divides the range $0.66 < \xi < 1.2$ into just three terms; Figures 4.11–4.12 give the sums over some ranges of interest to us, with a fine enough division of the range to give accurate answers.

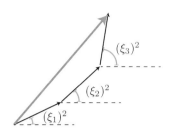

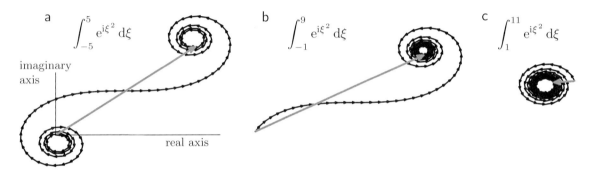

Figure 4.11: [Vector sums.] **Evaluation of the Fresnel integral over various regions in ξ.** Each panel shows the same sort of sum as in Figure 4.10, but with a larger integration region and more finely subdivided. Each *small arrow* is a contribution to the integral in the complex number plane; the *long arrow* shows the resultant, or sum, representing the value of the integral in this discrete approximation. The full integral from $-\infty$ to $+\infty$ resembles panel (a) and is called the **Cornu spiral**. Panels (a–c) depict integrals over wide ranges, all of width $\xi_{\max} - \xi_{\min} = 10$. They show that when the range of integration is large, the magnitude of the total depends strongly on whether the integration range includes the stationary-phase point $\xi = 0$ [panels (a,b)] or does not [panel (c)]. See Idea 4.13 and Problem 4.5.

obtained by laying successive arrows tail-to-head as usual. The colored arrow is the resultant vector.

Applying this procedure to the full Fresnel integral gives the picture shown in Figure 4.11a. The region near $\xi = 0$ appears at the center of this panel. Here the phase ξ^2 is changing slowly, so all the arrows point nearly in the same direction, and give a large total excursion across the page. But at large $|\xi|$, the curve curls up tightly. It doesn't much matter if we stop the integration at $\xi = 5$ (as shown), or 50, or 5000; the end point of the colored arrow will be little changed. Hence, there is a good limit as we send the range of integration to infinity: The full integral equals $(1 + \mathrm{i})\sqrt{\pi/2}$.

Ex. Connect the visual features of Figure 4.11a to the observation in Figure 4.9b–c that the red regions make nearly canceling contributions to the integral.
Solution: The successive endpoints of arrows in Figure 4.11a represent the integral up to various endpoints. Each curly end of the curve keeps returning to nearly the same point. Figures 4.9b–c show the same behavior for the area under the curve:

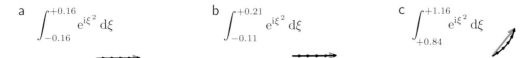

a $\displaystyle\int_{-0.16}^{+0.16} e^{i\xi^2}\,d\xi$ b $\displaystyle\int_{-0.11}^{+0.21} e^{i\xi^2}\,d\xi$ c $\displaystyle\int_{+0.84}^{+1.16} e^{i\xi^2}\,d\xi$

Figure 4.12: [Vector sums.] **Continuation of Figure 4.11.** The three panels all depict the same integrand as Figure 4.11, but integrated over narrow ranges, all of width $\xi_{max} - \xi_{min} = 0.32$. They show that when the integration range is small, the magnitude of the total vector (*orange*) does *not* depend very much on whether the integration range includes the stationary-phase point $\xi = 0$ [panels (a,b)] or does not [panel (c)]. See Idea 4.13 and Problem 4.5.

|| The positive and negative excursions nearly cancel in pairs, once we are far from $\xi = 0$.

Not all oscillatory integrals converge, but this one does, as we see from the inward spiral of the endpoints in the figure.

   The remaining panels of Figures 4.11–4.12 show some other important features of the Fresnel integral by evaluating it for various combinations of end points. You can see that

- *If we evaluate the Fresnel integral over any large range of ξ that includes the stationary-phase point $\xi = 0$, then we get a large resultant vector (Figure 4.11a,b). If we evaluate the integral over an equally large range that does not include 0, the resultant is much smaller [panel (c)].*   (4.13)
- *If we evaluate the Fresnel integral over any small range, then the magnitude of the result doesn't depend much on whether that range includes $\xi = 0$ (Figure 4.12).*

To understand these observations, note that the point $\xi = 0$ is special, because that's where the phase ξ^2 doesn't change rapidly with ξ. In fact, its rate of change, $d(\xi^2)/d\xi$, is *zero* at $\xi = 0$. Thus, the angular rotation of $e^{i\xi^2}$ *pauses* at this point: We call $\xi = 0$ a **stationary-phase point** of the integrand. Near a stationary-phase point, the arrows representing the integrand align, and can make a large resultant (vector sum). If our integral includes a significant range about this point (Figures 4.11a,b), then we'll get a much bigger answer than if it doesn't [panel (c)]. In short,

   In the limit where an oscillatory integral oscillates many times over its range of integration, its value is dominated by only those contributions that are close to stationary-phase point(s).   stationary-phase principle

(4.14)

In the contrary case (slow oscillation or narrow range of integration), then the stationary-phase points have no special significance.

   There is another useful way to think about stationary-phase points: If we look near a generic point, say $\xi_0 = 1$, then we find that on one side (ξ slightly greater than 1), the phase function ξ^2 is slightly greater than at ξ_0, whereas on the other side it is slightly smaller. But the stationary-phase point is different: The phase function is slightly greater on *either* side of $\xi = 0$ than its value right at 0. This is just the familiar result from calculus that a place where the derivative equals zero is generally also an **extremal point** (in this case, a local minimum).

Fig. 4.8a (page 160)

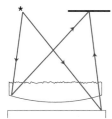

Fig. 4.8b (page 160)

4.7.2 The probability amplitude is computed as a sum over all possible photon paths

Armed with the stationary-phase principle, we can return to our urgent questions:

1. Should we include all the crazy paths in Figure 4.8?
2. Why does light (usually) (seem to) go on (pretty) straight lines? Why does it sometimes fail spectacularly to behave in that way?

We cannot digress to handle the mathematics of the first question in full. Nevertheless, the key idea is not hard to state. We arrived at the Light Hypothesis[20] by starting with the idea of summing over a *space of all paths* from source to detector. That prescription seemed paradoxical, but Feynman argued that it makes sense: Including the crazy paths does no harm (Idea 4.12, page 160). Now we can explain his reasoning:

- The integrand has the special property of being oscillatory.
- The preceding section showed that a much simpler integral of this type, the Fresnel integral in just one variable, can sometimes be approximated by keeping only contributions from its stationary-phase point.
- We can therefore hope that the full integral of interest to us, over all *paths,* will have similar behavior.
- That is, we will explore the idea that our sum over paths will be dominated by contributions from one or a few special *paths.* By analogy to the Fresnel integral, we'll call them **stationary-phase paths**.

Section 4.7.1 noted that the stationary-phase points in an integral of Fresnel type are generally also points where the phase is extremal. Because the Light Hypothesis gives the phase as a constant times the transit time, this result is often called the **principle of least time** (Fermat's principle).

With these insights, let's return to Figure 4.8a, and notice that *neither of the crazy paths shown is extremal* (that is, minimal) in length, and hence neither is a stationary-phase path. For example, we can stretch or shrink the looped path to generate similar paths of slightly greater or shorter length. As we traverse this family, the phase does not pause as we pass through the path shown. Similar remarks apply to the kinked path in Figure 4.8a.

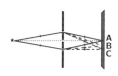

Fig. 4.2b (page 147)

In contrast, the paths we did consider in two-slit diffraction (Figure 4.2b) *are* extremal—they are the paths that a taut rubber band would follow if required to start at the source, end at a specified point on the projection screen, and pass through one of the slits. These are the paths that dominate the probability amplitude calculation. Later, we will see that the paths we considered in Newton's rings are also stationary phase, whereas the ones in Figure 4.8b are not.[21]

That is, *in the macroscopic world, including the crazy paths does not affect the full amplitude,* as claimed in Idea 4.12 (page 160). Indeed, when we try to look out of a room through a small window, much of the outside world is not visible: Light seems to travel on straight lines in a macroscopic apparatus. The next section will connect this observation to the Light Hypothesis by identifying those straight lines with stationary-phase paths, and will then see to what extent light can reach us via other paths. We'll see that, perhaps surprisingly, for the case of a tiny aperture the

[20]Idea 4.5 (page 154) states the Light Hypothesis, part 2a.
[21]See Section 5.3.1 (page 190).

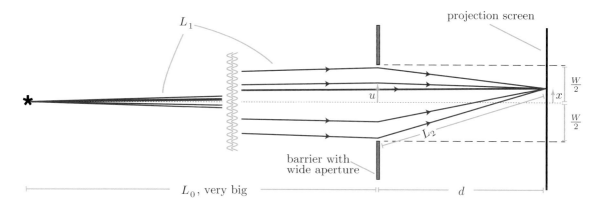

Figure 4.13: [Path diagram.] **Diffraction through a wide aperture.** Light from a distant, monochromatic source arrives at an aperture of width W, illuminating a screen at distance d from the barrier. The drawing is not to scale; actually W is much smaller than d. The *wavy lines* indicate that part of the sketch is missing; actually, the source is very far from the barrier, and the five red lines are all continuous. Positions on the screen are labeled by the distance x from the centerline (*dotted*); negative values of x describe the lower region in the figure. Five possible paths for photons are shown. Only one of these is a stationary-phase path (*heavy line*). The text describes how to calculate the screen illumination (probability of photon detection) at various observation positions x. We can identify a specific path by giving the height u above the centerline at which it passes through the aperture. In the limit of a very distant source, the stationary-phase path has $u = u_* \approx x$. (In a real setup, with finite source distance, $|u_*|$ is smaller than $|x'|$.)

stationary-phase principle does not apply, and light that originated from any direction can reach our eye.

$\boxed{T_2}$ *Section 4.7.2′ (page 175) discusses the problem of summing over all the many paths that are close to the stationary-phase path, and the situation of a medium with continuously varying index of refraction.*

4.7.3 Diffraction through a single, wide aperture

The ideas of the preceding section become clearer when we see them at work in a specific situation (Figure 4.13). The figure depicts light passing through a single **aperture** (opening in a barrier) of width W, that may or may not be very narrow. In this situation, even when we restrict attention to piecewise-straight paths, there are nevertheless many paths to consider.

In a typical classroom demonstration, the distance from aperture to screen might be $d = 4\,\mathrm{m}$. If the aperture's width is $W = 1\,\mathrm{cm}$, then everyday experience leads us to expect a sharply defined, uniform stripe of illumination on the screen, of width W. But as we decrease the slit width, for example below $0.5\,\mathrm{mm}$, we begin to see unusual phenomena. We would like to know if the Light Hypothesis can correctly predict both regimes of behavior, and the value of W near which the transition occurs.

To find the probability of photon arrival at position x, we must first compute $\Psi(x)$. We do this by adding contributions from all the paths under consideration. Each path shown in the figure can be described by the height u at which it passes through the aperture. The lengths L_1 and L_2 of the path segments depend on u. Hence, we integrate over u:[22]

$$\Psi(x) = C \int_{-W/2}^{W/2} du \, \frac{1}{L_1(u)} e^{2\pi i L_1(u)/\lambda} \frac{1}{L_2(u,x)} e^{2\pi i L_2(u,x)/\lambda}, \qquad (4.15)$$

where C is a normalization constant. As in the Example on page 155, we now make the approximation that the $1/L_i$ factors are also constant (independent of u and x). This claim is justified because those factors vary much more slowly than the exponential, due to the latter's factor of $1/\lambda$. In fact, even in the exponential, L_1 varies with u much more slowly than L_2, because $L_1 \gg L_2$. Hence, this part of the exponential may also be replaced by the constant $e^{2\pi i L_0/\lambda}$ and taken outside the integral.

Your Turn 4A

Evaluate the derivative with respect to u of the integrand in Equation 4.15 and justify the claims just made.

Writing $L_2(u,x)$ explicitly in the exponential then gives

$$\Psi(x) = C' e^{2\pi i L_0/\lambda} \int_{-W/2}^{W/2} du \, \exp\left[(2\pi i/\lambda)\sqrt{d^2 + (u-x)^2}\right], \qquad (4.16)$$

where $C' = C/(L_0 d)$.

We could now ask a computer to perform the integral numerically. But many situations of interest allow us to use a simplified, approximate form of the result. For example, in a classroom demonstration, we are generally interested in a pattern of illumination on the screen that is no wider than about $10\,\text{cm}$, so $|x| \ll d$ throughout the image. Moreover, typically the slit width is no larger than about $2\,\text{mm}$, so we also have $W \ll d$. In these conditions, the square root can be approximated by using its Taylor expansion (page 18). Let $R = \sqrt{d^2 + x^2}$ be the distance from the center of the aperture to the observation point; then

$$\sqrt{d^2 + (u-x)^2} = \sqrt{(d^2 + x^2) + (-2ux + u^2)} \approx R\left(1 + \frac{1}{2R^2}(-2ux + u^2) + \cdots\right). \qquad (4.17)$$

Dropping the higher-order terms[23] then gives

$$\Psi(x) = C' e^{2\pi i (L_0 + R)/\lambda} \int_{-W/2}^{W/2} du \, \exp\left[\frac{2\pi i}{2R\lambda}(-2ux + u^2)\right].$$

Complete the square inside the exponential to find

$$= C' e^{2\pi i (L_0 + R - x^2/(2R))/\lambda} \int_{-W/2}^{W/2} du \, \exp\left[\frac{\pi i}{R\lambda}(u - x)^2\right]. \qquad (4.18)$$

We can lump all the factors in front of the integral into a single symbol C''. This factor depends on x; however, its modulus does not.

The formula just obtained is starting to look familiar. To make it look exactly like the Fresnel integral, define a dimensionless variable $\xi = (u-x)/\sqrt{R\lambda/\pi}$, and yet

---

[22] We used the Light Hypothesis part 2a, Idea 4.5 (page 154).
[23] $\boxed{T_2}$ You'll explore this approximation in Problem 4.8.

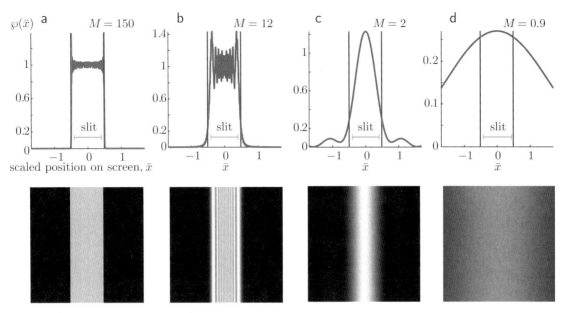

Figure 4.14: [Mathematical functions.] **Diffraction as aperture width decreases.** *Top:* Each panel shows the probability density for photon arrival (a constant times $|\Psi|^2$), as a function of $\bar{x} = x/W$, for various values of the parameter M defined in Equation 4.20. Rescaling x in this way means that in each case, the image of the slit in ray optics would correspond to the same range $-1/2 < \bar{x} < 1/2$ (*red lines*). See Problem 4.6 for details of the calculation. The separate values of the parameters d, λ, and W are irrelevant for these graphs; all that matters is the single composite quantity M. *Bottom:* Each panel shows the same information as in the one above it, but as a simulated diffraction pattern: The gray level corresponds to the probability density function $\wp(\bar{x})$, times an overall scaling factor.

another abbreviation $C''' = C''\sqrt{R\lambda/\pi}$, to obtain

$$\Psi(x) = C''' \int_{\xi_{\min}}^{\xi_{\max}} d\xi \, \exp(i\xi^2), \qquad (4.19)$$

where $\xi_{\min} = -(\frac{1}{2}W + x)/\sqrt{R\lambda/\pi}$ and $\xi_{\max} = (\frac{1}{2}W - x)/\sqrt{R\lambda/\pi}$. The probability of photon arrival is the squared modulus of this expression. The prefactor $|C'''|^2$ is nearly constant over the few-centimeter range of x, so we will drop it, because we are only interested in the relative intensity of light at various places on the screen.

The integrand in Equation 4.19 is now precisely of Fresnel form, so we know how the integral behaves from Section 4.7.1. Define the parameter M as the width of the integration region:

$$M = \xi_{\max} - \xi_{\min} = W/\sqrt{R\lambda/\pi} \approx W/\sqrt{d\lambda/\pi}. \qquad (4.20)$$

Because M is dimensionless, it makes sense to speak of it as "large" or "small"; the implied comparison is to the number 1.

We will refer to the pattern of illumination from a point source of light as the lens system's **point spread function**. Figure 4.14 shows our result (the modulus squared of Equation 4.19), for four illustrative values of M, and illustrates that

- If M is large, then our discussion of the Fresnel integral[24] implies that the prob-

---
[24]Idea 4.13 (page 163) states our conclusions about the Fresnel integral.

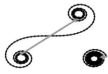

Fig. 4.11 (page 162)

Fig. 4.12 (page 163)

Fig. 4.1 (page 146)

ability amplitude is large when the stationary-phase point $\xi = 0$ lies within the integration range; otherwise, it is much smaller (Figure 4.11). That is, we get bright illumination at screen positions x that lie between $\pm W/2$. That's the image we would get if light traveled only on straight lines (Figure 4.14a).

- If M is small, then the probability amplitude is not concentrated within a band corresponding to the straight-lines expectation (Figure 4.14d and Figure 4.12); the resulting pattern is a broad smear of light.

In other words, the Light Hypothesis has explained both regimes of optics outlined in points **4** and **5** on page 148, and also the crossover between them. In the metaphor of Figure 4.1, the scientist expounding the Cornu spiral has the most general viewpoint, the one from which the others can be obtained as special cases.

Equation 4.20 sharpens several of our intuitive ideas about diffraction: For example, increasing the aperture width, W, eventually brings us to the large-M or **ray-optics regime**, where we get ordinary (nondiffractive) behavior. Because diffractive effects involve the finite wavelength of light, we expect that taking $\lambda \to 0$ holding other things fixed will also bring us to the ray-optics regime, and indeed M does become large in this limit. Finally, Equation 4.20 shows that increasing the distance d to the projection screen decreases M, moving the system toward the **diffractive regime**.

> **Your Turn 4B**
>
> For red visible light, and $d = 3\,\mathrm{m}$, estimate the slit width at which ray optics breaks down and diffractive effects begin to be prominent.

Other aperture shapes

A similar calculation can be done in three dimensions for other aperture shapes, for example, a rectangle. Figure 4.15 shows experimental results in this case. The aperture of an animal eye is called its **pupil**. Some animal eyes do have rectangular pupils (for example, goats), but our own are circular.[25]

4.7.4 Reconciliation of particle and wave aspects

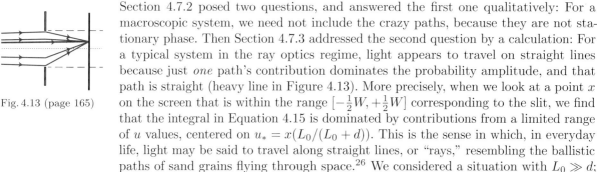

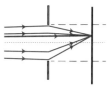

Fig. 4.13 (page 165)

Section 4.7.2 posed two questions, and answered the first one qualitatively: For a macroscopic system, we need not include the crazy paths, because they are not stationary phase. Then Section 4.7.3 addressed the second question by a calculation: For a typical system in the ray optics regime, light appears to travel on straight lines because just *one* path's contribution dominates the probability amplitude, and that path is straight (heavy line in Figure 4.13). More precisely, when we look at a point x on the screen that is within the range $[-\frac{1}{2}W, +\frac{1}{2}W]$ corresponding to the slit, we find that the integral in Equation 4.15 is dominated by contributions from a limited range of u values, centered on $u_* = x(L_0/(L_0 + d))$. This is the sense in which, in everyday life, light may be said to travel along straight lines, or "rays," resembling the ballistic paths of sand grains flying through space.[26] We considered a situation with $L_0 \gg d$; in that case, we got the simpler result $u_* \approx x$.

We have come to the Inner Chamber at the heart of physics, the reconciliation of wave and particle behavior. And like many archaeologists before, you may be

---

[25] $\boxed{T_2}$ You'll study rectangular apertures in Problem 4.6e and round apertures in Problem 4.7. Remarkably, in bright light, the cuttlefish *Sepia officinalis* narrows its pupils to the shape of the letter ω, perhaps for reasons related to its specific ecological niche.

[26] Section 6.5.1 (page 224) will further develop the idea of "rays" of light.

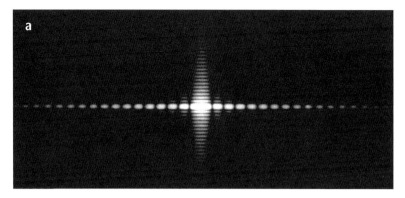
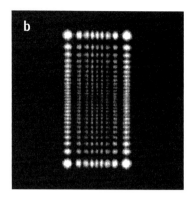

Figure 4.15: [Photographs.] **Rectangular aperture.** (a) Diffraction pattern cast on a distant screen (diffractive regime, small M) by light passing through a rectangular aperture, oriented with its long axis vertical. The smaller, horizontal constriction spreads the light more than the wider, vertical one. (b) Pattern cast by the same aperture, but on a near screen (almost the ray-optics regime). The image is now nearly the rectangle predicted by ray optics. [From Cagnet et al., 1962.]

disappointed at what we have found there: "There's nothing here but some dusty old mathematics!" But we should pause to appreciate how narrowly we have escaped contradiction, how predictive our answer is, and how universally applicable:

- *Escape contradiction:* Summing over paths prior to squaring gives us the diffraction and interference (wavelike) phenomena that we observe, without contradicting the other observed fact that light interacts locally, with only one electron at a time (particle-like).

- *Predictive:* The complete Light Hypothesis quantitatively explains phenomena at the two extremes of behavior:

  - Particle-like: In everyday life, sunlight casts sharp shadows and so on.
  - Wavelike: Light can display interference (Newton's rings) and diffraction (spreading after passing through a slit).

  The Light Hypothesis also lets us quantitatively predict what happens in the *crossover* between the two regimes (Figure 4.14b–c).

- *Universal:* Remarkably, electrons, quarks, and *all other known fundamental particles* also display analogous behaviors:

  - Particle-like: In a macroscopic apparatus, free electrons travel on straight lines, like light. If acted on by an external force, they obey the same laws of Newtonian mechanics as sand grains (or planets).
  - Wavelike: In microscopic contexts, however, electrons display interference behavior, also like light.[27] When spatially confined, they also show phenomena analogous to acoustic resonance, leading to the quantization of their energy levels (the Electron State Hypothesis, page 40).

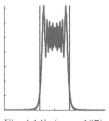

Fig. 4.14b (page 167)

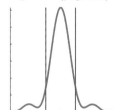

Fig. 4.14c (page 167)

$\boxed{T_2}$ *The Light Hypothesis needs little modification to cover electrons. Chapter 12 outlines this extension. Other versions of quantum mechanics, for example, the Schrödinger equation, then emerge as equivalent formulations, useful for some specialized classes of problems.*

---

[27]See Media 11 and Chapter 12.

THE BIG PICTURE

Chapter 1 emphasized one random aspect of light: the temporal randomness of the Poisson process of photon arrivals. The present chapter extended that idea to describe spatial randomness, and the diffractive phenomena that display it. The rule that we were led to adopt for light required us to introduce the *probability amplitude*. This physical quantity attracted much criticism when it was first proposed, and it still seems very weird. Fortunately, however, it is not singular—in essential respects, electrons, and all the other fundamental particles, obey the *same (weird) rule* as light. The rule introduces randomness in an intrinsic way, quite different from thermal molecular agitation, which drives phenomena like Brownian motion. Unlike thermal motion, quantum randomness cannot be adjusted by changing the temperature, or anything else.

Although we have seen how Feynman's rule can reconcile the lumpy character of light with some optical phenomena, this chapter has stopped short of explaining image formation—and image formation is central to our own vision. Chapters 6–8 will enter those waters, at the same time introducing technology for the superhuman vision obtained with microscopes and x-ray diffraction. First, however, we will explore some more physical and biophysical phenomena that are traditionally explained in terms of the classical wave model of light, but which turn out also to make sense using the more broadly applicable rules of quantum physics.

KEY FORMULAS

- *Complex numbers:* Probability amplitudes are complex numbers, which can be visualized as 2D vectors. A complex number can always be written as $Z = a + ib$ where $i^2 = -1$ and a, b are ordinary real numbers. Equivalently, we can write Z in terms of its length and angle, as $Z = re^{i\varphi}$. The angular quantity φ is called the phase of the complex number Z (mathematicians sometimes call it the "argument of Z"). The real number r is called the "modulus," "length," "magnitude," or "absolute value" of Z; it is also written $|Z|$.

  To convert between these representations, use $r = \sqrt{a^2 + b^2}$, $\varphi = \tan^{-1}(b/a)$ or the inverse relations $a = r\cos\varphi$, $b = r\sin\varphi$. A compact restatement is the Euler formula

  $$e^{i\varphi} = \cos\varphi + i\sin\varphi,$$

  or

  $$\cos\varphi = (e^{i\varphi} + e^{-i\varphi})/2, \quad \sin\varphi = (e^{i\varphi} - e^{-i\varphi})/(2i).$$

  To multiply complex numbers, use $(a + ib) \cdot (a' + ib') = (aa' - bb') + i(ab' + a'b)$; equivalently

  $$re^{i\varphi} \cdot r'e^{i\varphi'} = (rr')e^{i(\varphi+\varphi')}.$$

- *Light Hypothesis, part 2a:* The probability to observe a photon is the square of the modulus of the probability amplitude Ψ; that is, $\mathcal{P} = |\Psi|^2$. If there are multiple paths (or processes) by which a photon could make the trip from source to detector, and we cannot directly observe which one was taken, then each path makes an additive contribution to the total probability amplitude. For monochromatic light traveling in vacuum, the phase of each contribution to Ψ equals $E_{\mathrm{photon}}/\hbar$ times the transit time for the path, where $\hbar$ is the reduced Planck constant. For a straight-line path, the modulus of the path's contribution is an overall constant divided by the path length. For a path consisting of straight-line segments, the modulus is the product of one such contribution for each segment.

- *Light Hypothesis, part 2b:* When light passes from one uniform, transparent medium (or vacuum) to another, the energy per photon does not change, but the transit time must account for the reduced speed c/n. If a path "bounces," then its contribution to the total probability amplitude gets multiplied by an extra factor of -1 if it was initially heading from a medium with smaller n toward one with larger n (for example, from air to water); equivalently, the quantity π can instead be added to the phase associated to its path. There is no such extra factor, however, if the photon was heading from larger toward smaller n.

- *Two narrow slits:* The screen illumination intensity is a constant times $\cos^2(\pi\Delta/\lambda)$, where λ is the wavelength of the monochromatic source and Δ is the path-length difference.

- *Newton's rings:* The screen illumination intensity is a constant times $\sin^2(\pi\Delta/\lambda)$.

- *Fresnel:* The integral $\int_{-\infty}^{\infty} \mathrm{d}\xi \, \cos(\xi^2)$ is not infinite. It's approximately equal to the integral $\int_{-5}^{5} \mathrm{d}\xi \, \cos(\xi^2)$. That is, contributions to the integral from the rest of the integration region (far from the stationary-phase point $\xi = 0$) mostly cancel. Similar remarks apply to the integral with sine in place of cosine.

- *One wide slit:* Suppose that light of wavelength λ travels a long distance from a point source, hits an opaque barrier with a slit of width W, then proceeds a distance d to a screen. The image of the slit on the screen will be a sharply defined bar of width W if $M = W/\sqrt{\lambda d/\pi}$ is much larger than 1. That's the "ray-optics" regime. Otherwise, we'll get a broader smear, possibly with some additional structure (the "diffractive" regime).

FURTHER READING

Semipopular:

Feynman, 1985, Feynman, 1967, and Styer, 2000, all discuss quantum physics in general, and two-slit diffraction in particular. Our discussion follows Feynman, 1985.

Intermediate:

Feynman et al., 2010a; Lipson et al., 2011; Hecht, 2002. Aspden et al., 2016 and Pearson & Jackson, 2010, describe how to build a single-photon apparatus. For a roundup of modern experiments see Townsend, 2010, chapt. 1.

Complex numbers: Shankar, 1995.

Technical:

Feynman et al., 2010c.

Wavelike interference patterns have been observed with complete atoms, and even with giant molecules consisting of over 800 atoms (Eibenberger et al., 2013).

Interference phenomena involving many photons are traditionally taught as applications of classical electrodynamics. Many outstanding books are available; one that covers many real-world phenomena is Zangwill, 2013.

$\boxed{T_2}$ **Track 2**

4.2′ On philosophically repugnant theories

Einstein's hypothesis about the lumpy character of light met with stiff resistance; even his strongest supporters believed for a long time that it was a mistake.[28] Moreover, even Einstein never accepted that quantum randomness was an unavoidable implication of his idea. Indeed, the foundations of quantum theory are still under intense debate.

But it is worth remembering that many of today's scientific theories were rejected as philosophically repugnant, at first. For example, scientists in Europe deplored Newton's gravitation law—it seemed absurd for two bodies to pull each other without touching. As Richard Feynman put it, "An important point about [scientific] intelligence is that it should not be sure ahead of time what must be."

Although quantum randomness is repugnant to some, it offers the simplest known explanations of many observed phenomena, in quantitative detail. It also fulfills the scientist's preference that the most characteristic phenomena should have explanations that are hardwired into the theory, not the result of some complicated set of logical or mathematical steps applied to it. But—like any scientific theory—it is always subject to revision following some new experimental observation.

For our purposes, the facts that light is received in lumps, and that those lumps' arrivals are random in both space and time, are of great practical importance both to biophysical instrumentation, and to our own vision. The quantum theory of light incorporates those facts as essential ingredients.

$\boxed{T_2}$ **Track 2**

4.5′a More about the Light Hypothesis

The discussions of slit diffraction in the main text may seem a bit magical: Supposedly the barrier with the slits is creating the diffraction pattern, and yet we never explicitly introduced any interaction between the light and the barrier! Moreover, a little thought brings us to another paradox: We know that matter is mostly empty space, and the Light Hypothesis tells us to sum over "all paths," so why not include all the many paths that thread their way right through the electrons and nuclei that constitute any real barrier?

Starting with the second of those questions, imagine an opaque barrier with no slits. A lot of complicated interactions with the electrons in the constituent atoms end up with the simple result that the probability amplitude for photons to be detected on the other side is zero—that's the meaning of "opaque." Thus, the paths that pass through the barrier without any interactions are indeed present, but their contributions are cancelled by those of other paths.

Now remove one or more narrow strips of the material that constitutes the barrier, creating slit apertures. Between the slits, we have the same cancellation as before.[29] Within the slits, however, photon paths that previously involved interactions with

---

[28]See the epigraph to Chapter 1, page 23.
[29]We are neglecting the effects of paths that interacted with material both inside and outside of a strip, so this discussion is approximate. See the discussion in Feynman et al., 2010b, §31-6, p. 31-10.

the material no longer exist. Hence, we are left with only contributions from the paths that pass through the slits and do not interact with the barrier material. These contributions are the ones discussed in the main text.

The discussion of Newton's rings appears equally magical: Supposedly the two pieces of glass are creating the ring pattern, and yet again we never explicitly introduced any interaction between the light and the glass, other than the crucial minus sign! See the discussion in Section 4.6.2′ (page 174) below.

4.5′b More about uniform media

Idea 4.6 (page 155) invokes the concept of a "uniform" medium, but real media like glass and water, like all matter, consist of lumps—molecules. More precisely, we expect this conclusion to be a good guide for a medium that is effectively unform when *averaged* over length scales shorter than the wavelength of the light in question. Chapter 8 will investigate what we can learn by probing matter with much shorter wavelength light.

$\boxed{T_2}$ **Track 2**

4.6.1′ Which slit?

Our two imagined students are back, this time thinking about two-slit diffraction patterns.

George: If a photon interacts locally (with a single electron), then surely its passage through one slit or the other are mutually exclusive possibilities. Hence, the probability of arrival at a particular point on the screen must follow the addition rule (Equation 0.3, page 4), and in particular may never decrease when a second pathway is opened.

Martha: Experiment contradicts your reasoning! Section 4.2 (page 146) described how opening a second slit can *decrease* the probability of finding a photon at certain points on a detector screen. We must conclude that the two paths are *not really mutually exclusive alternatives*. Indeed, when the detector responds to a photon, we have no way to say which path it took.

George: But we can imagine altering the experimental arrangement to add that capability. Suppose that we use high-energy photons, and introduce some electrons into the space near each slit (for example, by adding some gas molecules). Then an incoming photon might collide with an electron during its passage, knocking it out of its normal place and into a detector, which would confirm that the photon definitely went through a particular slit.

Experiments equivalent to George's proposal have actually been done. What is observed is that, when we have enough scatterers to determine in principle which slit was taken by every photon, then *interference is abolished*, regardless of whether we actually make use of that information. When we have fewer scatterers, so that some photons proceed incognito, then we get partial interference; as we dial down the number of scatterers to zero, the full interference pattern gradually emerges (Buks et al., 1998).

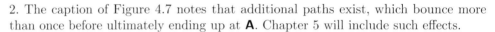

T_2 **Track 2**

4.6.2′ More about reflection

1. The Light Hypothesis includes a special rule about light reflecting from an interface,[30] which we needed in order to understand the Newton rings phenomenon.

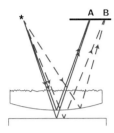

Fig. 4.7 (page 158)

One way to understand the need for this rule is to notice that a single sheet of glass will reflect light from both its front and back faces. If we imagine a series of thinner sheets of glass, eventually getting thinner than the wavelength of light, then we ought to find no reflection from the thinnest sheets, because in the limit there is no glass at all. The Light Hypothesis gives the overall probability amplitude for reflection as the sum of two terms, and in the limit the terms have equal path length, because the distance traveled by one path inside the glass goes to zero. In order to get these terms to sum to zero, then, we need the minus-sign rule. Chapter 12 shows how to obtain this rule, the slowdown of light by a medium, and other phenomena, all from a master principle (Idea 12.3, page 384).

2. The caption of Figure 4.7 notes that additional paths exist, which bounce more than once before ultimately ending up at **A**. Chapter 5 will include such effects.

3. Each path shown in Figure 4.7 is a sequence of events, for example: travel through space, enter glass, travel through glass, bounce, travel through glass, exit glass, travel through space to **A**. We have temporarily simplified by dropping numerical factors associated to some of these steps, for example the "reflection factors" associated to the bounces, because they are common to the pairs of paths whose interference interested us. For glass, the reflection factors are small, so actually most of the light from the source in Figure 4.7 is not reflected at all, but rather passes straight through the apparatus and out the bottom plate.[31] However, we were interested in the variation between reflections to different screen points, not the relative amounts of transmitted and reflected light, so we dropped this overall factor.

4. Actually, one of the factors we omitted is *not* common to both paths: One path must reenter the curved glass, incurring a "transmission factor," whereas the other one does not. However, for glass the transmission factor is nearly equal to 1. Again, see Chapter 5.

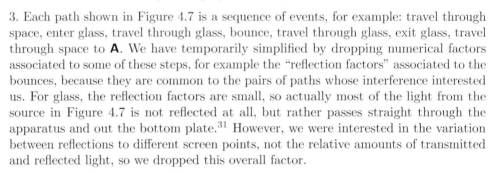

T_2 **Track 2**

4.6.3′ More objections

1. The idea of a single photon forming an interference pattern seems paradoxical. Consider two paths from source to detector with different lengths. Because light always travels at a universal speed in vacuum, it follows that the transit times are different, too. When a detector emits a blip, that happens at a single moment of time. So the two photon trajectories that supposedly interfered with each other must

---

[30]Idea 4.6 (page 155) gives the Light Hypothesis, part 2b.
[31]See Idea 5.1 (page 184). In the limit where we send the index of refraction of glass $n \to 1$, the reflection factor goes to zero.

have had *different* emission times at the source; how, then, could they be "the same" photon?

To reply, note first that a monochromatic source is an idealization for which the frequency of emitted light is perfectly constant, and hence the *time* of emission is completely uncertain. Paths of unequal length, interpreted in terms of trajectories with the same end time, indeed must have different start times, but this does not prevent interference as long as the source is close enough to being monochromatic.[32] Chapter 12 will explain more explicitly how, even for a single path, it is essential to integrate over all possible emission times.

2. Turning from physical objections to mathematical ones, the Light Hypothesis (part 2a, Idea 4.5, page 154) may at first seem self-contradictory. Consider the straight-line path that joins two points located a distance a apart. We are told to assign this path a weighting factor of $a^{-1}e^{i\omega a/c}$, times a constant, where $\omega = E_{\text{photon}}/\hbar$. But suppose that we subdivide the same path into two segments, of lengths b and $a - b$ respectively. Then we seem to be told to assign the total path, which is still a straight line of length a, the weighting factor $b^{-1}e^{i\omega b/c}(a - b)^{-1}e^{i\omega(a-b)/c}$. The new factor disagrees with the previous form, and not just by an overall constant.

What's missing in the reasoning just given is that having introduced an intermediate point, we must integrate over all of its possible positions. Choose coordinates in which the original straight-line path lies on the z axis, from the origin to the point $(0, 0, a)$. Next, consider intermediate points located throughout a plane that is perpendicular to the original path and that passes through the point $(0, 0, b)$. Any point on this plane can be specified by a two-component vector $\boldsymbol{u}$ via (u_1, u_2, b), so we can write the overall contribution from the two subsegments as

$$\int d^2\boldsymbol{u} \; \frac{1}{\sqrt{b^2 + \boldsymbol{u}^2}} \frac{1}{\sqrt{(a-b)^2 + \boldsymbol{u}^2}} \exp\left[i\omega(\sqrt{b^2 + \boldsymbol{u}^2} + \sqrt{(a-b)^2 + \boldsymbol{u}^2})/c\right].$$

Among the many paths considered, the only stationary-phase path is the one with $\boldsymbol{u} = 0$. Approximating the integral by the contributions from paths close to this one gives

$$b^{-1}(a-b)^{-1} \int d^2\boldsymbol{u} \; \exp\left[i\omega(b + \tfrac{1}{2}\boldsymbol{u}^2/b + \cdots)/c\right] \exp\left[i\omega(a - b + \tfrac{1}{2}\boldsymbol{u}^2/(a-b) + \cdots)/c\right].$$

We can change variables to $\bar{\boldsymbol{u}} = \boldsymbol{u}\sqrt{\omega(b^{-1} + (a-b)^{-1})/(2c)}$, which converts the integral into Fresnel form; its value is a constant. Dropping it, and other factors independent of a and b, gives

$$b^{-1}(a-b)^{-1}\frac{1}{b^{-1} + (a-b)^{-1}}e^{i\omega a/c} = \frac{1}{b + (a-b)}e^{i\omega a/c} = \frac{1}{a}e^{i\omega a/c},$$

which does agree with the first expression we considered.

---

[32]A more technical discussion would quantify "close enough" in terms of "coherence length," a measurable property of the source.

$\boxed{T_2}$ **Track 2**

4.7.2′a The neighborhood of the stationary-phase path

Our discussion of integrals over "all" paths has in fact been restricted to piecewise-straight paths with one kink—a one-parameter family of paths. But just as a small but finite range of u values contributes to the integral in Equation 4.15 (page 166), so also many paths that are slightly wiggly (not precisely made up of two straight segments) will make contributions to the full probability amplitude for light propagation. A full analysis of the needed infinite-dimensional, Fresnel-type integral would go beyond the scope of this book, but we can at least motivate an answer. Section 4.7.3 made some approximations and ultimately reduced Equation 4.15 to the form Equation 4.19 (page 167). In the ray-optics regime, the integral turned out to be either approximately zero (if there is no stationary-phase path), or else approximately a *universal constant,* depicted by the red arrow in Figure 4.11a (if there is one stationary-phase path). That constant is multiplied by a prefactor C''', which contains phase information based only on the stationary-phase path itself. Similarly, when we sum the contributions of *all* the nearly stationary-phase paths, we get an answer that is some other universal constant times $\mathrm{e}^{\mathrm{i}\varphi_*}$, where φ_* is again the phase associated with the stationary-phase path.

4.7.2′b Nonuniform media

In our examples, the stationary-phase paths have always been straight lines, or made up of straight-line segments. Actually, they may also be curved, if the medium's index of refraction is nonuniform. For example, the air near the ground can be hotter than the air higher up, if the Sun is heating the ground. Then light that arrives nearly parallel to the ground can have a curving stationary-phase path, giving rise to the phenomenon of mirages.[33]

---

[33]You'll study gradient-index materials in Problems 5.6 and 6.14.

PROBLEMS

4.1   *Euler formula*
Define $f(\varphi) = \cos\varphi + i\sin\varphi$ and $g(\varphi) = e^{i\varphi}$. Give an alternative proof of the Euler formula (Equation 4.4) $f = g$, as follows. Notice that $f(0) = g(0)$. Then express g as a Taylor series about the point $\varphi = 0$ (see page 18), rearrange by gathering together all terms with an even number of powers of i, plus all terms with an odd number, and identify the Taylor series for $\cos\varphi$ and $i\sin\varphi$, respectively.

4.2   *Complex numbers*
a. Derive Equation D.8 (page 451) as a consequence of Equation D.2, and express Equation D.8 in words. [*Hint:* You'll need to use two famous identities from trigonometry.]
b. Show that your result in Your Turn DA(c) (page 450) can also be regarded as a consequence of Equation D.8.

4.3   *Two-slit interference fringe spacing*
Make a quantitative prediction for the spacing between bright lines ("interference fringes") in the experiment described in Figure 4.2a. Use slit spacing $p = 0.3\,\text{mm}$ and screen distance $d = 3\,\text{m}$, and assume that the incoming light is monochromatic with wavelength 550 nm. Following the derivation of Equation 4.10 (page 157), assume that the slits themselves are very narrow. Thus, the probability amplitude $\Psi(x)$ will be the discrete sum of just two terms. Express its modulus squared in terms of x, d, and p, stating any approximations you made. Then substitute numbers to make a prediction for the spacing between fringes.

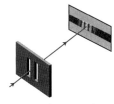

Fig. 4.2 (page 147)

4.4   *More slits*
Light from a distant, monochromatic source hits an opaque barrier with narrow, parallel slits. The slits are displaced by distances $\pm p/2$ from the centerline of the setup. You found the resulting illumination pattern for two slits in Problem 4.3. In this problem, assume slit spacing $p = 0.3\,\text{mm}$, distance $d = 9\,\text{m}$ to the screen, and visible light with $\lambda = 600\,\text{nm}$.

a. Get a computer to make a graph of light intensity versus screen position x for the two-slit problem.
b. Now consider a new situation: The same light, in the same overall geometry, hits an opaque barrier with *three* narrow, parallel slits. One slit is at the centerline of the setup; the others are distances $\pm p$ from the centerline. Find a formula for the resulting illumination pattern on a distant projection screen, get a computer to graph intensity versus x, and describe in words what it would look like. [*Hint:* If you wish to drop a term involving p^2, justify doing so.]
c. Consider a third situation: The same light hits an opaque barrier with *four* narrow, equally-spaced, parallel slits, displaced by distances $\pm p/2$ and $\pm 3p/2$ from the centerline. Make a third graph. If you see a trend emerging, describe it.

4.5   *Fresnel integral*
Figures 4.11–4.12 illustrate the stationary-phase idea by giving graphical representations of several Fresnel integrals. Use a computer to make such figures for yourself. Subdivide the integration ranges shown into enough slices to get a nice representation. Get the computer to draw arrows corresponding to the complex integrand's value in each slice. Arrange the arrows with each one's tail at its predecessor's head. Try the

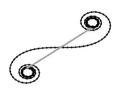

Fig. 4.11 (page 162)

Fig. 4.12 (page 163)

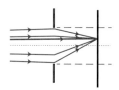

Fig. 4.13 (page 165)

integration ranges $-5 < \xi < +5$, $-0.5 < \xi < 9.5$, and $15 < \xi < 25$, each of which is of length 10.

4.6　Diffraction

Consider a point source at infinity that shines light with wavelength λ onto a barrier with a slit of width W (see Figure 4.13). A projection screen is located a further distance d away. We ask for the intensity of light as a function of position x on the screen. More precisely, assume as in the figure that the light source and the center of the slit lie on a line perpendicular to the screen. Let x be the distance of a point on the screen from that line.

Thus, in the ray-optics regime the shadow of the slit is the region where $|x| > W/2$. We are studying all paths that photons may take that consist of a straight line from the source to the plane of the barrier, followed by another straight line to the observation point. There is a family of such paths, each characterized by the fact that it passes through the slit at some vertical distance u from the centerline. To sum the paths, then, we must integrate u from $-W/2$ to $+W/2$.

Suppose that $d \gg x$ and $d \gg W$. Section 4.7.3 used this limit to simplify the probability amplitude (Equation 4.15), arriving at Equation 4.19 (page 167). The quantity M defined in Equation 4.20 controls how "wavy" the problem is. For example, the value of $|\Psi|^2$ depends on whether x lies in the ray-optics shadow or not: We argued qualitatively that the transition from light to shadow is abrupt if M is big, and gradual if M is not-big. In this problem, you will check this intuition on a range of M values.

a. Learn how to use a mathematical software package to perform numerical integration. First try it with an example where you know the answer: Use it to compute $\int_0^1 \xi^2 \, d\xi$.

b. Now do the required integral and find $|\Psi|^2$ for $M = 12$ and various nonnegative values of x in the range from 0 to W. (Why don't you need to consider negative values of x?) The Light Hypothesis doesn't tell us the overall normalization factor, so work out what it must be in order for your answer to be a probability density function. Graph your result as a function of x/W and comment.
   [*Hint:* Many numerical integration algorithms accept a parameter, sometimes called the "tolerance." Experiment with its value; as you make it smaller, the calculation will take longer but give a more accurate answer. Find a small enough value to get reasonably stable results.]

c. Repeat using $M = 2$ and $M = 0.5$, and comment.

d. Some typical numbers for a classroom demonstration are $d = 5\,\text{m}$, $W = 0.1\,\text{mm}$, $|x| < 20\,\text{cm}$, and $\lambda = 600\,\text{nm}$. What would you expect to see for this geometry?

e. $\boxed{T_2}$ The pupil of your eye is not an infinitely long slit; it is a circular aperture. The math gets a bit complicated in that case, but there's a related case that's easier. Consider light passing through a *square* hole of side W. For this problem, we must examine a *grid* of points on the screen labeled by x, y; similarly, each path is labeled by two quantities u, v. It may seem doubly difficult to do such a double integral, but actually you can recycle your effort from (b,c). Explain why, and then compute and display your predicted diffraction pattern for the same three values of M as before.

4.7　$\boxed{T_2}$　Circular aperture

Section 4.7.3 outlined a two-dimensional calculation, appropriate for diffraction from a slit. The calculation involved a single integral. If we wish to study diffraction in our eyes, however, we must acknowledge that our pupil is a small *circular* aperture. Set up the corresponding 3D calculation for this case (a double integral), to find out how much blurring we would expect when 600 nm light from a distant point source passes through such a hole and lands on a screen. Evaluate the pattern of light intensity (the point spread function) numerically for the case of a pinhole of diameter 0.5 mm and screen distance $d = 3$ m. Then describe your answer in words. [*Hint:* The illumination intensity pattern will be circularly symmetric, so you can replace Figure 4.14 by a graph of relative intensity versus distance from the center of the pattern.]

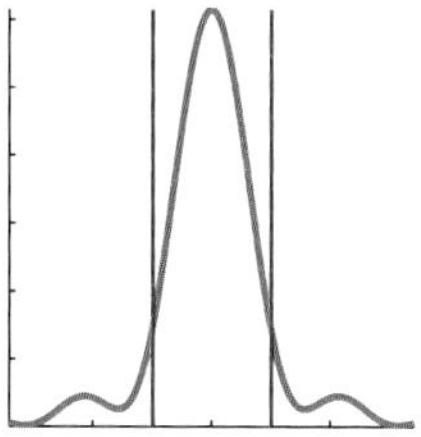

Fig. 4.14 (page 167)

4.8 $\boxed{T_2}$ *Justify an approximation*

Passing from Equation 4.16 to Equation 4.18 required the use of a Taylor series expansion, Equation 4.17. The leading term in this expansion was a constant (independent of the integration variable u).

a. The second term was retained, even though it is much smaller than the first one. On other occasions, we have simply dropped such subleading terms. For example, when passing from Equation 4.15 to Equation 4.16 we replaced $1/L_2(u)$ by $1/L_0$ and $1/L_2(u, x)$ by $1/d$ and lumped these factors into the overall constant. Why can't we do that here?

b. Third-order and higher *were* dropped in the derivation, however. Substitute some typical numbers $d = 5$ m, $W \leq 1$ cm, $x \leq 10$ cm, $\lambda \approx 600$ nm, and justify dropping these terms.

CHAPTER 5

Optical Phenomena and Life

> Color is the quintessential part of art, which holds the magical gift. Subject, form, and line speak primarily to the intellect; color has no meaning for the mind, but it is all-powerful over emotion.
>
> — *Eugène Delacroix, 1798–1863*

5.1 SIGNPOST: *SORTING AND DIRECTING*

Chapters 1–3 concentrated on the particle-like aspects of light. Chapter 4 explained how, and when, we can instead expect wavelike behavior. We are now ready to understand many biological applications of these ideas, as well as some other phenomena that were once regarded as evidence for the classical wave model. These involve *sorting* a stream of photons by wavelength, and *directing* it in sophisticated ways.

The examples in this chapter consider only the stationary-phase path contributions to probability amplitudes.[1] Later chapters, however, will return to the analysis of diffractive phenomena.

The Focus Question is

Biological question: Why do some butterfly wings lose their brilliant color when saturated with an appropriate liquid, then regain color when the liquid evaporates?

Physical idea: Many organisms generate their colors via the interference of photon paths.

5.2 STRUCTURAL COLOR IN INSECTS, BIRDS, AND MARINE ORGANISMS

Chapter 3 pointed out that some animals display vivid colors, for purposes such as identifying potential mates. Many of these colors involve pigment molecules that selectively absorb light. But some colors, for example, those on certain butterfly wings, beetle wing cases, and bird plumage, have a very different character. For one thing, most pigments bleach upon exposure to sunlight or chemical agents. But colors on some insects remain brilliant a century or more after the insect was killed and mounted, and they also resist attack with chemical bleaching agents.

5.2.1 Some animals create color by using nanostructures made from transparent materials

Here is another clue: Figure 5.1 shows two butterflies. In each specimen, the right wing is in its natural state. The left wing has been soaked in the solvent toluene. The

---

[1] Section 4.7 explains this approximation.

a

b

⊢— 5 cm —⊣

⊢— 2 cm —⊣

Figure 5.1: [Photographs.] **Structural color versus pigments.** (a) The dorsal (top) side of a male *Morpho rhetenor* butterfly. (b) The dorsal side of a male *Cymothoe sangaris* butterfly. In both photos, the wings on the right-hand side are in their natural state, but the wings on the left have been saturated with a liquid whose index of refraction matches that of the wing scale material. The pigment-based color of *Cymothoe* is barely affected, while the structural blue of *Morpho* is lost. The effect is reversible: When the liquid evaporates, the color returns. [Courtesy Glenn S Smith; see Smith, 2009.]

specimen on the right is hardly affected in appearance by this preparation, whereas the one on the left has completely lost its blue color. One might think that in this case the solvent had dissolved the colorant away, but remarkably, once the toluene evaporates, the wing returns to its original appearance.

In fact, the wings of *Morpho* butterflies are covered with scales made mainly of a *transparent* substance (called cuticle) whose index of refraction nearly matches that of toluene. This observation led scientists to suspect that the scales contained a complex structure with alternating layers of cuticle and air, and that somehow the contrast between the two indices was responsible for the color. Replacing the air layer with an index-matching fluid eliminates the contrast, and reversibly destroys the color. That is, the origin of these colors is *structural* in character, arising from the architecture of the wing scale and not from the absorption properties of any chromophore.

After the electron microscope was invented, it became possible to confirm the structural-color hypothesis directly. Figure 5.2 shows a series of images at increasing magnification. Each wing is covered by the thin scales seen in panels (a,b). The top surface of each scale contains parallel ridges, and each ridge has a "bookshelf" structure, shown in cross section in panel (c). Each shelf runs roughly parallel to the scale surface, and is roughly uniform in thickness.

Our discussion of Newton's rings gives a hint as to how a layered nanostructure could give rise to color. We supplemented the main Light Hypothesis by an addendum concerning how light behaves in transparent media:[2] A uniform, transparent medium slows light down, and reflection from an interface sometimes generates an extra contribution of π to the phase in a probability amplitude. Now consider a single planar layer of medium with index of refraction n_2 (for example, soapy water), surrounded

---

[2]Section 4.6.2 (page 158) studied Newton's rings. The Light Hypothesis part 2a is Idea 4.5 (page 154) and part 2b is Idea 4.6 (page 155).

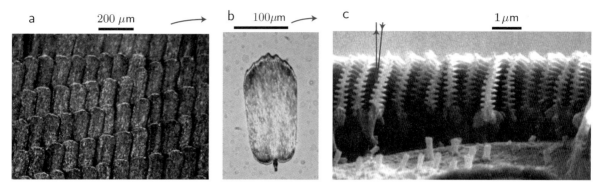

Figure 5.2: **Scales on the wing of the butterfly *Morpho rhetenor.*** Successively magnified images. (a) [Light micrograph.] Overall arrangement of scales. (b) [Light micrograph.] An individual scale observed via reflection microscopy, top view. (c) [Scanning electron micrograph.] Cross section of a scale (side view). Long ridges made of transparent cuticle project upward, out of the surface. Each ridge is ribbed, with "bookshelves" roughly periodically spaced in height. In normal viewing, light arrives from above in this image, as shown, and bounces off one or more of the shelves. [(a,b): From Kambe et al., 2011. (c) From Kinoshita et al., 2002.]

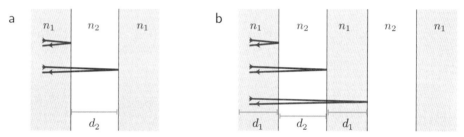

Figure 5.3: [Path diagrams.] **Reflection from a multilayer structure.** (a) A single layer of transparent medium with index of refraction n_2 is surrounded by another medium with $n_1 < n_2$. Each *line* is the stationary-phase representative of a class of paths. Not all classes of paths are shown. (Figures 5.4b and 5.5b will give a more complete description.) (b) Two layers with n_2 are separated, and surrounded, by another medium. The denser layers could represent the "shelves" of the structures shown in Figure 5.2c; the less-dense layers could represent the air between those shelves.

by another medium with smaller n_1 (for example, air).[3] Figure 5.3a shows two of the paths that an incoming photon could take to return to its origin. The top path bounces from the first interface; the corresponding contribution to the probability amplitude contains a phase π from the reflection. The second path shown bounces at the $2{\to}1$ interface, which incurs no phase change, but the path does have to travel through twice the layer thickness, introducing a phase $2\pi(2d_2)\nu/(c/n_2)$ relative to the first path. The corresponding two contributions to the probability amplitude will reinforce if their phases match (or differ by any multiple of 2π), so the reflection will depend on the frequency of the light.[4]

For a stack of many layers (sometimes called a **Bragg stack**), Figure 5.3b shows a third contribution to the probability amplitude. Like the first one, it has a phase π from the reflection. Like the second one, it has an additional phase $2\pi(2d_2)\nu/(c/n_2)$

---

[3]In this chapter, we will neglect the possibility that the index may itself depend on the frequency of the light. Section 6.5.4 will return to this "dispersion" phenomenon.
[4]Figure 5.4d (page 185) shows the resulting colors.

from double passage through medium *2*. And it has one more phase, $2\pi(2d_1)\nu/(c/n_1)$, from double passage through medium *1*. All three contributions will mutually reinforce if the two conditions

$$\pi = 4\pi d_2 n_2 \nu/c + \text{multiple of } 2\pi$$
$$= \pi + 4\pi d_2 n_2 \nu/c + 4\pi d_1 n_1 \nu/c + \text{multiple of } 2\pi$$

are satisfied. For example, we may have $d_1 = \lambda_1/4$ and $d_2 = \lambda_2/4$, where $\lambda_i = c/(n_i\nu)$. A reflector of this sort is called a **quarter-wave stack** for light with the given frequency, because each layer's thickness is 1/4 the wavelength of the light in that medium.

The semiquantitative discussion just given yields some intuition for how a layered structure, consisting of two transparent and nearly colorless materials (air and cuticle), can preferentially reflect some wavelengths of light and so create the colors observed on insect wings. But our analysis is still far from being complete. For one thing, there are many more paths that must be considered than the ones shown in Figure 5.3. And real structural colors found in Nature involve layer geometries that are not perfect quarter-wave stacks.

Your Turn 5A

Some typical numbers for butterfly wing scales are $n_1 \approx 1$ (air), $d_1 \approx 155\,\text{nm}$, $n_2 \approx 1.56$, and $d_2 \approx 65\,\text{nm}$. For what vacuum wavelengths will each layer be quarter-wave, and what are the corresponding colors?

Our preliminary discussion so far also points out another aspect of structural color: The conditions for strong reflection are geometrical (they depend on distances); they will change if the light enters and leaves at some angle other than perpendicular. The word **iridescence** is used to describe color that changes as the angle of view or the angle of illumination changes. Historically, the dependence of apparent color on the illumination and viewing angles was another clue that some animal coloration was structural in character.

5.2.2 An extension of the Light Hypothesis describes reflection and transmission at an interface

Chapter 4 gave some examples of how the interference between multiple alternative light paths can sort and direct lumps of light. We also know that living organisms are very good at forming spatially nanostructured materials, offering a rich variety of alternative paths, and Section 5.2.1 gave some preliminary calculations to show that colors can arise in this way. But a full listing of all those paths looks pretty complicated; accordingly, the next few sections will need to develop some math tools to handle it.

First, however, we must acknowledge that although the Light Hypothesis part 2b gave us the right answers for problems like Newton's rings, still it is incomplete.[5] When light enters glass from air, some *fraction* is reflected and the rest is transmitted, but the Light Hypothesis as it currently stands doesn't tell us that fraction. To handle

---

[5]Idea 4.6 (page 155) states the Hypothesis. T₂ Section 4.6.2′ (page 174) explained why we did not need the detailed form of the reflection factor when we studied Newton's rings.

such questions, we need to add a codicil to the addendum:

> **Light Hypothesis, part 2c:**
> *Consider two transparent media with indices of refraction n_1 and n_2, joining at a flat boundary surface.*
>
> - *When a photon path passes through the boundary between two transparent media, traveling perpendicular to it from 1 to 2, the path's contribution to the probability amplitude acquires an extra "transmission factor" $t_{1\to2} = 2\sqrt{n_1 n_2}/(n_1 + n_2)$.*
> - *When a photon path bounces off the boundary, staying in medium 1, its contribution to the probability amplitude acquires an extra "reflection factor" $r_{1\to2} = (n_1 - n_2)/(n_1 + n_2)$.*

(5.1)

Idea 5.1 fits with the preceding part of the Light Hypothesis, because when a path passing through air bounces off a water surface, for example, the factor $n_1 - n_2 \approx 1 - n_{\text{water}}$ is negative, and the phase of a negative real number is π. In the opposite situation, a path passing through water bouncing off a boundary with air on the other side, the factor is $\approx n_{\text{water}} - 1$, which is positive, and the phase of a positive real number is 0. These facts agree with Idea 4.6 (page 155), but they go beyond that statement to include also the moduli of the reflection and transmission factors. In particular, if $n_1 = n_2$, then $r_{1\to2} = 0$: As far as light is concerned, there is no boundary at all, and hence no reflection.

Because we are neglecting the possibility of absorption (we assumed transparent media), every photon must be either reflected or transmitted.[6]

Your Turn 5B

Check that indeed $|t_{1\to2}|^2 + |r_{1\to2}|^2$ always equals 1.

Your Turn 5C

Light arrives perpendicular to a flat glass surface. The glass has index of refraction 1.5. Find the fraction of incoming photons that are reflected, and the fraction that are transmitted.

T2 *Section 5.2.2′ (page 199) connects Idea 5.1 to a similar result from classical electromagnetism.*

5.2.3 A single thin, transparent layer reflects with weak wavelength dependence

Before we tackle reflection from a Bragg stack in detail, let's warm up with a simpler situation. Soap bubbles are nearly transparent, but sometimes we can see faint colors in the film, despite the fact that it consists of a colorless, transparent material (soapy water). Similarly, some insect wings consist of a single thin layer of transparent material; they, too, display interference colors (Figure 5.4d). The patterns of color are conserved as each species evolves, and seem to play a role in signaling.

Figure 5.4a represents a slab of material with index n_2 sandwiched between layers with n_1. To keep things simple, we will suppose that the light source and detector lie on a line that crosses the surface at right angles as shown. Material *1* may be air;

---

[6]Another example of an interface is a flat metal surface (a "mirror"), which idealize as having $|r|^2 = 1$.

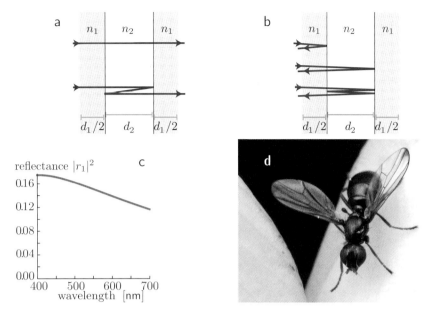

Figure 5.4: **Transmission and reflection from a single layer.** (a) [Path diagrams.] Transmission. The first two of an infinite set of light paths are shown. Reflection is possible at the two physical interfaces (*solid black lines*). For a single layer, medium *1* extends to infinity to the right and left. (b) Reflection. The first three of an infinite set of light paths are shown. (c) [Mathematical function.] The computed probability that a photon traveling perpendicular to a single layer will be reflected, as a function of its wavelength in air. One medium is air ($n_1 \approx 1$) and the other is wing cuticle ($n_2 = 1.56$) with thickness $d_2 = 65\,\text{nm}$. (d) [Photo.] A fly from the family Sepsidae. The fly's wings are transparent, so when viewed against a light background we mostly see transmitted light, and don't notice the fainter reflected light (*right*). When viewed against a dark background, however, subtle interference colors are apparent (*left*). [(d) Courtesy Alex Wild.]

material *2* may be soapy water or insect wing cuticle. Monochromatic light arrives from a source on the left, and we'd like to know how much will be observed on the right. The difficulty comes when we realize that there are many paths that a photon could take: It could pass straight through (upper path shown), or it could bounce any even number of times before finally emerging (lower path shown). How are we to sum all these contributions?

To get started, let's write the first two terms of the probability amplitude for transmission, corresponding to the paths shown in Figure 5.4a. Define the convenient abbreviation

$$k_i = 2\pi n_i \nu/c, \qquad i = 1, 2,$$

which is called the **wavenumber** of light in each medium.[7] Also, abbreviate $t_{1\to2}$ as t_0 and $r_{1\to2}$ as r_0. Then the first contribution is the product of factors corresponding to transit through *1*, transmission at the first interface, more transit through *2*, transmission at the second interface, and finally more transit through *1*, that is,

$$\left(e^{ik_1 d_1/2}\right)\left(t_0\right)\left(e^{ik_2 d_2}\right)\left(t_0\right)e^{ik_1 d_1/2}.$$

---

[7]The vector with this magnitude, pointing along the direction of the light's motion, is called its **wavevector**. Some authors define wavenumber without the factor of 2π; to avoid confusion, you can quote numerical values with the units rad/m or cycles/m respectively to indicate which definition you intend.

The second contribution involves the same steps, plus reflection at the second interface, transit back through medium *2*, reflection at the first interface, and transit yet again through *2*:

$$e^{ik_1 d_1/2} t_0 e^{ik_2 d_2} \left[\left(-r_0 \right) \left(e^{ik_2 d_2} \right) \left(-r_0 \right) \left(e^{ik_2 d_2} \right) \right] t_0 e^{ik_1 d_1/2}.$$

Note that in the preceding formula, both internal reflections contribute factors of $-r_0$:[8] In each case, the path is traveling in medium *2* and bounces from an interface with *1*, so each factor is $r_{2 \to 1}$, or $-r_{1 \to 2} = -r_0$.

A pattern is emerging: The third contribution will be like the second, but with another round of internal reflection. In fact, the full probability amplitude involves a geometric series:[9]

$$t_{1 \text{ layer}} = (t_0)^2 e^{i(k_2 d_2 + k_1 d_1)} \frac{1}{1 - r_0^2 e^{2ik_2 d_2}}. \tag{5.2}$$

Ex. ǁ Now that we have found the trick, work out the overall reflection factor for the layer. *Solution:* See Figure 5.4b. This time, the first term is not part of the geometric series:

$$r_{1 \text{ layer}} = r_0 e^{ik_1 d_1} \left[1 - \frac{t_0^2 e^{2ik_2 d_2}}{1 - r_0^2 e^{2ik_2 d_2}} \right]. \tag{5.3}$$

The reflection probability, $|r_{1 \text{ layer}}|^2$, is sometimes called the **reflectance**. Figure 5.4c shows it as a function of wavelength for some particular parameter values. We can see that, although a single thin layer doesn't reflect very efficiently, nevertheless its reflection is wavelength dependent, due to interference between the various allowed light paths. As promised in Section 5.2.1, there is stronger reflection at wavelength $\lambda_1 = 4d_2 n_2/n_1 \approx 410 \,\text{nm}$ than at other wavelengths.

Your Turn 5D

Chapter 4 mentioned in the context of Newton's rings that, as the layer thickness decreases to zero, the effect of the layer should disappear. Now that we have done a more complete calculation, verify that the transmission and reflection amplitudes behave properly in the limit $d_2 \to 0$.

5.2.4 A stack of many thin, transparent layers can generate an optical bandgap

Let's now upgrade our analysis to a repeating structure with many layers. It may seem too difficult to catalog all the many paths available when there are more than two interfaces, and find all the contributions to the probability amplitude that must be added. It's true that the list of tortuous zigzag paths is very complicated. But a bookkeeping trick reduces the calculation to only a small additional step after the one we just completed.

We imagine building up the many-layer structure by stacking n copies of the cell shown in Figure 5.4. Figure 5.5 represents the entire stack as a heavy, dashed box, and one additional copy as a heavy solid box. Panel (a) represents two large families of

---

[8] In the situation we are considering, r_0 is a negative quantity, so the factors $-r_0$ appearing in the formula are positive, consistent with the Light Hypothesis, Idea 4.6 (page 155).
[9] The geometric series formula appears on page 18.

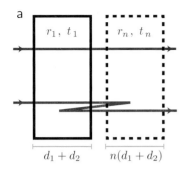
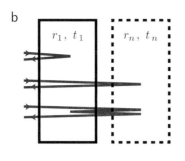
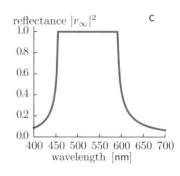

Figure 5.5: **Recursion relations for reflection and transmission.** (a) [Path diagrams.] Transmission. The *left box* now represents the entire compound object in Figure 5.4; the *right box* represents a stack of n such objects. Each *arrow segment* now represents an infinite class of photon paths. For example, segments in the left box represent classes sketched in Figure 5.4a,b. (b) Reflection. (c) [Mathematical function.] The corresponding probability for an infinite stack of such layers from the solution to Equation 5.6. For comparison with Figure 5.4c, again one medium is air ($n_1 \approx 1$) and the other is butterfly wing cuticle ($n_2 = 1.56$). Each cuticle layer has thickness $d_2 = 65\,\mathrm{nm}$, and the parallel layers are separated by $d_1 = 155\,\mathrm{nm}$ air gaps. (In real butterfly wing scales, the number of layers is not infinite; see Problem 5.1.) [The calculation follows Amir & Vukusic, 2013.]

allowed paths. A heavy line passing through the box on the left represents *all* the paths that contributed to transmission through a single layer (Figure 5.4a); a heavy line bouncing inside it represents all the paths that contributed to reflection (Figure 5.4b). We have just computed the sums that give the corresponding probability amplitudes. Let's abbreviate them, writing t_1 instead of $t_{1\,\mathrm{layer}}$ and so on.

The dashed box on the right of Figure 5.5a represents the sum over all the many complex paths available for a stack of n layers. We don't know its reflection and transmission amplitudes, so we just give them symbolic names r_n and t_n. The key to making progress comes when we realize that we can classify the paths through $n + 1$ layers according to how many times each path crosses the boundary between the solid and dashed boxes. In the first class (paths that cross once), we can then uniquely account for every path by selecting a path through the first layer and following it by a path through the remaining n layers; the top line in the figure represents this entire class. The next class crosses three times and is represented by the zigzag line in the figure, and so on. Thus, we can write the probability amplitude for transmission through $n + 1$ layers as

$$t_{n+1} = t_1 t_n + t_1 r_n r_1 t_n + \cdots = \frac{t_1 t_n}{1 - r_1 r_n}. \tag{5.4}$$

Similarly, Figure 5.5b represents the sum we must do to compute the reflection amplitude for a stack, and we find

$$r_{n+1} = r_1 + t_1 r_n t_1 + t_1 r_n r_1 r_n t_1 + \cdots = r_1 + \frac{t_1{}^2 r_n}{1 - r_1 r_n}. \tag{5.5}$$

Instead of tackling our problem head-on, we have found *recursion relations* for the desired quantities: We can start with one layer, then apply Equations 5.4–5.5 as many times as needed. Moreover, the answer becomes simpler, not harder, when the number of layers is very large! In that case, we may expect that after a while, adding one more layer won't change the reflection much, so that $r_n \approx r_{n+1}$. We can set both of those quantities equal to a common, unknown value r_∞; then Equation 5.5 becomes

$$r_\infty = r_1 + \frac{t_1{}^2 r_\infty}{1 - r_1 r_\infty}. \tag{5.6}$$

Rearranging that formula gives us a quadratic equation for r_∞. The modulus squared of the physical solution to this equation is shown in Figure 5.5c.

Something remarkable has happened: In the limit of many layers, a whole band of wavelengths has acquired *perfect reflectance.* We say that the multilayer structure has an **optical bandgap**. Photons with wavelengths outside that band are partially transmitted. In *Morpho,* any transmitted photons pass through the upper scales to a layer containing dark pigment (visible in the left wing of Figure 5.1a, page 181), where they are absorbed. The net effect is wavelength-selective reflection, leading to the observed color.

Ex. Can we apply the same logic to Equation 5.4?

Solution: We may be tempted to set $t_{n+1} = t_n = t_\infty$, then cancel it from both sides of the formula. What remains would be a new constraint on r_∞, not something we could solve for t_∞. But unlike the case with reflection, we have found that in the bandgap the transmission goes to *zero* in the limit of many layers. Dividing both sides of an equation by zero usually leads to incorrect conclusions, as it does in this case.

Indeed, any real material will have at least some absorption, which we have been neglecting so far. Thus, the transmission amplitude for an infinite stack must be zero even outside the bandgap. In contrast, reflection from an infinite stack is dominated by the first few layers, so it's a reasonable approximation to neglect absorption, as was done when deriving Equation 5.6. (You'll check this claim in Problem 5.1.)

Although the model discussed in this section is very crude, and we only considered light incident and reflected at right angles to the structure, still we found a sharply defined range of perfect reflectance in the blue part of the spectrum, dependent on the contrast of refractive index between two transparent media. Beyond explaining the coloration of some animals, this phenomenon is also used in the construction of the dichroic mirrors at the heart of fluorescence microscopes (Figure 2.4, page 65). $\boxed{T_2}$ *Section 5.2.4′ (page 200) outlines a generalization of this section's result to more general layered structures.*

5.2.5 Structural color in marine organisms

Some animals can prosper in nutrient-poor environments by harboring symbionts that can perform photosynthesis. A remarkable example are the "giant" clams in the family *Tridacna,* which maintain colonies of photosynthetic algae. We might expect those clams to present a drab green color, like the algae themselves, but instead they are brightly colored, and indeed iridescent (Figure 5.6a). The coloration comes from spherical cells called iridocytes in the outermost layer of the clam's soft tissue [panel (b) of the figure]. The iridocytes in turn are filled with parallel layers, alternating between transparent media with different indices of refraction [panel (c)], and hence act as Bragg stacks, producing the observed colors. But A. Holt and coauthors asked *why* the clams have iridocytes. These clams are immobile, so they are not using their coloration to find or attract potential mates. What biological role does it play?

We get a clue when we note that the clams deploy their algal symbionts in a fairly deep layer just below the iridocytes. Ordinarily, this would be an inefficient

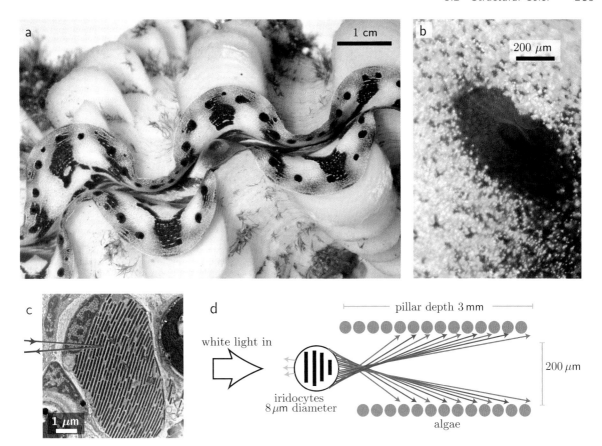

Figure 5.6: **Structural color in a marine organism.** (a) [Photograph.] The giant clam *Tridacna derasa*, showing vivid colors in its soft tissue. (Black spots along the rim are eyes.) (b) [Darkfield micrograph.] The clam's colors come from small individual particles called iridocytes (*colored dots*), each about 8 μm in diameter, seen here in the region surrounding an eye. (c) [Electron micrograph.] A single iridocyte (side view) contains parallel, proteinaceous platelets ≈ 100 nm thick. The platelets are normally transparent; here they have been stained for visualization. (d) [Cartoon.] The iridocytes reject unneeded green light and bend other wavelengths so that they enter the columns of photosynthetic algal cells all along their length. Not to scale. [Images courtesy Alison M Sweeney; see also Holt et al., 2014.]

arrangement: The uppermost layer of algae would be exposed to more light than they can handle, and so would suffer photodamage, whereas the lowest layers, shadowed by the upper layers, would receive little light. The researchers suspected that the actual structure overcomes this limitation. For one thing, they found that the stacked layers inside the iridocytes are always oriented perpendicular to the incoming light, and reflect back mainly the green and yellow wavelengths that are not used in the algal photosynthesis. Thus, some light that is not useful for photosynthesis, and that might otherwise have caused photodamage, is instead rejected.

A second clue came when the researchers noted that the algae are not uniformly distributed: They are arranged into deep pillars, oriented parallel to the incoming light by membranes generated by the clam. Some of the incoming light therefore passes deeply into the tissue through the spaces between pillars. The iridocytes at the top scatter the light they transmit slightly, so that it is no longer quite parallel to the pillars. Thus, it enters the pillars all along their length (Figure 5.6d). In this way, light

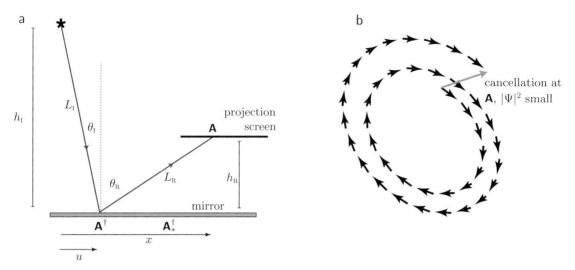

Figure 5.7: **Geometry of reflection in air or vacuum.** (a) [Path diagram.] A mirror reflects light from a point source (*top left*) to a projection screen, where we place a detector at point **A**. For a macroscopic mirror, the law of reflection says that the path shown will not contribute much to the light seen at **A**, because the angles θ_I and θ_R are not equal. Thus, $\mathbf{A}^\dagger$ lies on a part of the mirror that could be removed without affecting the light intensity at **A**. (b) [Vector sum.] The *black arrows* represent contributions to the probability amplitude from paths that strike the mirror close to $\mathbf{A}^\dagger$. Their sum (*colored arrow*) is small.

that would have been stopped in the first few micrometers instead gets distributed throughout a deeper slab of tissue. Spreading the light in this way helps assure an optimal photon arrival rate for algae throughout the slab, maximizing the nutrition that the clam can obtain by harvesting sunlight.

Section 5.2 has surveyed a few of the ways in which living organisms use photon interference to create vivid colors and otherwise alter the flow of light in adaptive ways. Now we turn to some other optical phenomena with implications for life science.

5.3 RAY-OPTICS PHENOMENA

5.3.1 The reflection law is a consequence of the stationary-phase principle

Figure 5.7a shows a light source, a mirror, and a projection screen. Light leaves the source at many angles and bounces off the mirror; the reflected light illuminates the entire screen. If we place a detector at position **A** on the screen, everyday experience tells us that only a small part of the mirror, close to $\mathbf{A}_*^\dagger$, is really necessary in order to reflect light to **A**. And yet, the Light Hypothesis says that we must include *all* paths, including for example the one shown in the figure.

To investigate, consider a family of paths. Each path intersects the mirror at a point $\mathbf{A}^\dagger$ that we can label by u, the horizontal distance from the light source to the point. Let x denote the horizontal distance between the source and observation point **A**. Then the total length of a path is $L_\mathrm{tot}(u) = \sqrt{h_\mathrm{I}{}^2 + u^2} + \sqrt{h_\mathrm{R}{}^2 + (x - u)^2}$. We now ask, at what value of u is L_tot stationary? To answer, compute the derivative of

L_{tot} with respect to u and require that it equal zero:

$$0 = \frac{2u_*}{2\sqrt{h_{\text{I}}^2 + (u_*)^2}} + \frac{2(x - u_*)(-1)}{2\sqrt{h_{\text{R}}^2 + (x - u_*)^2}}. \tag{5.7}$$

The stationary-phase path corresponds to the value u_* that solves this equation.

We can simplify Equation 5.7 by looking again at Figure 5.7a: The first term is $\sin\theta_{\text{I}}$, and the second is $-\sin\theta_{\text{R}}$. Thus, the condition that these two terms sum to zero is the familiar **law of reflection**:

> *The stationary-phase path is the one for which the **angle of incidence** θ_I equals the **angle of reflection** θ_R.* (5.8)

Although all paths contribute to $\Psi(\mathbf{A})$, we know that most of these contributions cancel. For example, paths close to the one drawn in Figure 5.7a make contributions that are indicated schematically in panel (b) of the figure.[10] The exceptions are the path that obeys the law of reflection and its immediate neighbors; these account for most of the light received at $\mathbf{A}$.

$\boxed{T_2}$ *Section 5.3.1′ (page 200) discusses an additional factor in the analysis of reflection.*

5.3.2 Transmission and reflection gratings generate non-ray-optics behavior by editing the set of allowed photon paths

We have obtained the law of reflection as a consequence of the stationary-phase principle (Idea 4.14, page 163). But the stationary-phase path only dominates when we integrate an oscillating function over many periods (recall Figure 4.14, page 167). To illustrate what can happen in other situations, Figure 5.8a shows a modified mirror, interrupted by evenly spaced, light-absorbing strips. The interruptions eliminate some of the contributions to the probability amplitude. The result of this selective removal of paths is to retain only those arrows contained in the green dotted line in panel (b) of the figure; these can sum to a nonnegligible total as sketched in panel (c), even when the law of reflection is not satisfied.[11]

Ex. Explain qualitatively why, for a given spacing of reflecting strips and a given monochromatic illumination, there can be *multiple* different values of the reflection angle for which light is seen.
Solution: The contributions from paths that land on one strip of mirror will have the same phase as those from paths landing on the next strip (Figure 5.8d) if the two classes of paths differ in length by a wavelength, or twice the wavelength, and so on.

The device in Figure 5.8 is called a **reflection grating**. The preceding Example implies that it can separate white light into a series of spectra, because the locations of each light band depend on wavelength. You can observe this effect yourself by reflecting a narrow beam of white light off a digital video disk, which has concentric circular tracks spaced $740\,\text{nm}$ apart.

---

[10]This derivation also justifies our omission of the paths shown in Figure 4.8b (page 160) when we studied Newton's rings. You'll explore just how much of the mirror we may remove, without significantly changing the illumination at $\mathbf{A}$, in Problem 5.3.

[11]You'll work out the details in Problem 5.4.

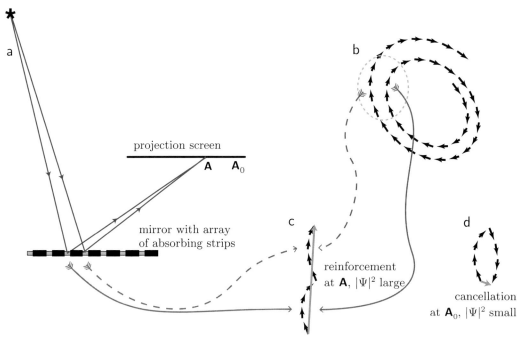

Figure 5.8: **Principle of the reflection grating.** (a) [Path diagram.] The geometry is similar to Figure 5.7, but with much of the mirror blocked by a periodic array of absorbing strips (*black bars*). A more detailed version of this figure appears in Figure 5.14 (page 203). (b) [Vector sums.] The periodic blocking eliminates most of the contributions shown in Figure 5.7b, leaving only those arrows enclosed by the *green dotted line.* (c) The sum of that subset of contributions can be nonnegligible, even if $\theta_\mathrm{I} \neq \theta_\mathrm{R}$. The lower four black arrows represent contributions from paths close to the one on the left in (a); the upper four arise from paths close to the one on the right in (a). In the situation shown, the paths in those two classes differ in length by about the wavelength of the light. (d) However, if we move the detector to a different location $\mathbf{A}_0$ on the projection screen, contributions analogous to those in (c) will *cancel:* The screen will be dark at $\mathbf{A}_0$.

Your Turn 5E

Think about an array of opaque lines drawn on the surface of a *transparent* medium (not a mirror), and invent a device that could be called a **transmission grating**.

5.3.3 Refraction arises from the stationary-phase principle applied to a piecewise-uniform medium

In a transparent material, light travels at a speed less than c. We defined the index of refraction n to be the ratio of c to the actual speed—a dimensionless quantity. For example, in w̲ater, $n_\mathrm{w} \approx 1.33$.[12] Various kinds of glass have n in the range 1.45–1.55.

Suppose that a light source is located in vacuum, with $n = 1$ (or air, which has almost the same value), and a detector is in water. The air/water interface is planar (Figure 5.9). The detector will see light from the source. We'd like to know "what path the light takes." As always, our answer is "*All* paths contribute to the probability amplitude." But we have already seen in other contexts that we can give a more

---

[12]Section 1.7 (page 47) introduced the index of refraction. Again, the initial letter of the word "water" is underlined in this sentence to alert you that the same letter in the symbol n_w refers to water.

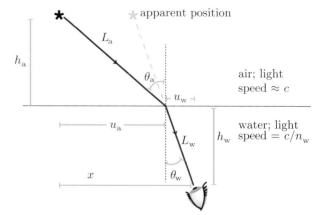

Figure 5.9: [Ray diagram.] **An experiment to demonstrate refraction.** Light from a source (*top*) crosses from air to water before being detected (*below*). The angles θ_a and θ_w are subject to the law of refraction, Equation 5.10. The *dashed line* shows the apparent direction of the source to an observer who assumes that light travels on straight lines.

informative answer, if the system is macroscopic. After all, if we hold an opaque card with a small hole at the water surface, we expect to find that the card blocks all light to the fixed observer unless the hole is in just the right position. We'd like to know what's the right position.[13]

Let's approach the problem by again asking, "What is the stationary-phase path?" To find it, notice that the Light Hypothesis gives the photon phase upon arrival at the detector as $2\pi\nu$ times the total transit time.[14] That time is the sum of L_a/c (the transit time of the part of the path in air) plus $L_w/(c/n_w)$ (from the part in water). The total horizontal separation x between source and detector is fixed (see Figure 5.9). It's the sum of a part u_a, the horizontal distance from the source to the point where the path enters the water, and $u_w = x - u_a$, the horizontal distance from the point of entry into the water to the observer. Thus, the phase is stationary when

$$0 = \frac{d}{du_a}\left[\sqrt{(h_a)^2 + (u_a)^2} + n_w\sqrt{(h_w)^2 + (x - u_a)^2}\right]. \tag{5.9}$$

Ex. The condition Equation 5.9 will be satisfied only for one path. Carry out the differentiation and rearrange your answer to rephrase the condition as $0 = (u_{a,*}/L_a) - n_w(u_{w,*}/L_w)$.
Solution: The chain rule from calculus gives the derivative as

$$\frac{1}{2}\left((h_a)^2 + (u_a)^2\right)^{-1/2}(2u_a) + n_w\frac{1}{2}\left((h_w)^2 + (x - u_a)^2\right)^{-1/2}\left(-2(x - u_a)\right).$$

Setting this expression equal to zero gives the requested relation.

The angles must therefore be related by $\sin\theta_a = n_w\sin\theta_w$, or more generally,[15]

$$\boxed{n_1\sin\theta_1 = n_2\sin\theta_2. \quad \text{law of refraction}} \tag{5.10}$$

---

[13] Alternatively, we may place a box around the detector, so that it can only "look" in one direction; then we can ask which way to point the detector in order to receive light from the source.
[14] Idea 4.5 (page 154) gives the Light Hypothesis part 2a.
[15] The law of refraction is often called "Snell's law," but Ibn Sahl had published this result in the year 964, long before W. Snell's work.

Figure 5.10: [Ray diagrams.] **An instrument to prevent total internal reflection.** (a) One way to diagnose and classify glaucoma involves measuring the angle between the eye's iris and cornea. However, their conjunction cannot be directly seen, because normally light originating in this part of the eye is totally internally reflected. (b) In order to view the area, a prism called a gonioscope is placed against the eye, replacing the cornea-air interface by a cornea-glass interface and hence suppressing total internal reflection there. (The light ray is also below the critical angle at the following glass-air interface.)

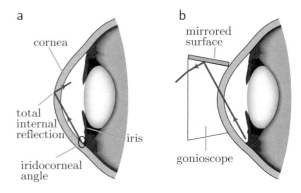

Thus, another macroscopic optical phenomenon follows from the Light Hypothesis. $\boxed{T_2}$ *Section 5.3.1′ (page 200) discusses an additional factor in the analysis of refraction.*

5.3.4 Total internal reflection provides another tool to enhance signal relative to noise in fluorescence microscopy

Refraction can give rise to surprising illusions. The next time you are in a swimming pool, submerge your head and look upward. Objects above you appear to be crowded toward the vertical direction (a "fisheye" distortion) because your brain expects light to travel on straight lines (see Figure 5.9). Humans may not understand refraction instinctively, but the archerfish does: It accurately aims a jet of water to bring down an insect in the air above it.[16]

Something even more surprising happens when you look at larger angles. Still underwater, direct your gaze farther from the vertical (that is, to larger angles θ_w). At a **critical angle**, $\sin^{-1}(1/n_w)$, the angle of incidence becomes 90 deg. For θ_w values beyond the critical angle, light from outside cannot reach you at all; here the water's surface resembles a mirror, only reflecting light from inside the pool to your eye.

Everything said about refraction in the preceding section also applies if we reverse the arrows in Figure 5.9, considering a light source in the water. Rays of light now bend *away* from the perpendicular as they emerge from denser water to less dense air. Again something interesting happens as θ_w, which this time plays the role of angle of incidence, increases past the critical value: There is no stationary-phase path for transmission, and hence very little light is transmitted.

More generally, when light passes through an interface moving from a transparent material with index n_{from} to one with a smaller index n_{to}, the critical angle is

$$\theta_{crit} = \sin^{-1}(n_{to}/n_{from}).\qquad(5.11)$$

At incidence angles beyond the critical value, the interface displays **total internal reflection**.

Medical applications

Figure 5.10 shows a medical application of these ideas, a simple device that eliminates total internal reflection in order to view an otherwise hidden region of the eye. Total

---

[16]See Media 12.

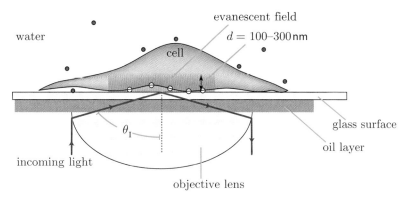

Figure 5.11: [Ray diagram.] **One form of total internal reflection fluorescence (TIRF) microscopy.** A light beam enters from lower left, bends upon entering the objective lens, and illuminates an interface between the glass coverslip and a specimen in water with angle of incidence that is greater than the critical angle. (The oil layer matches the index of the glass, so there is little reflection or refraction at the oil-glass interfaces.) The probability for photons to arrive at the specimen decreases exponentially with distance from the interface, with a decay length d that depends on the angle of illumination. The *white and gray spheres* represent fluorophores that, in TIRF, are illuminated or dark, respectively. Light from these point sources is collected and imaged by the same lens as is used to create the TIRF illumination. Not drawn to scale. [Adapted by permission from Macmillan Publishers Ltd: Sako and Yanagida. Single-molecule visualization in cell biology. *Nat. Rev. Mol. Cell Biol.* (2003) vol. 4(Suppl.) pp. SS1–SS5, ©2003.]

internal reflection is also (part of) the secret to how an optical fiber can conduct light many kilometers without appreciable loss. Light is initially directed along the axis of a thin fiber of high-index material. If the fiber is bent, the light may hit its surface. If the bend is gentle, however, then the angle of incidence will be large and the light will be totally internally reflected back into the fiber. Such reflections occur over and over, until ultimately the light emerges nearly undiminished at the far end. A bundle of such fibers can independently direct light along each one, so that the spatial arrangement of light entering one end (an image) is preserved at the other end, despite bends along the way. Endoscopes use this principle to allow minimally invasive visual inspection of internal organs,[17] or even direct intense light to a precise location for laser surgery.

Total internal reflection fluorescence microscopy

Chapter 2 observed that a major issue in microscopy is the problem of *not* seeing the many objects in a sample that are not of interest, and introduced the technique of fluorescently tagging interesting molecules. Often, however, there is still "background" (unwanted) fluorescence degrading the image. Section 2.7 discussed one way to reduce background (two-photon excitation), but total internal reflection offers another approach that, for some applications, is simpler.

    To understand <u>t</u>otal <u>i</u>nternal <u>r</u>eflection <u>f</u>luorescence (TIRF) microscopy, recall that our derivation of the law of refraction made use of the ray-optics regime: We only retained contributions from the stationary-phase path. A more detailed calculation shows that there is some probability for photons to penetrate a short distance into the region of lower index, even when the incident light is nearly parallel to the interface. Thus, in the situation shown in Figure 5.11, most of the specimen is unilluminated, *except* for a layer 100–300 nm thick near the glass surface. If we wish to see only the

[17]See Figure 2.3 (page 64).

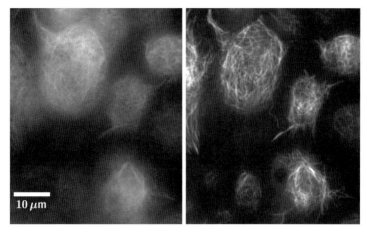

Figure 5.12: [Fluorescence micrographs.] **Illustration of the benefits of TIRF imaging.** *Left:* Living cells, express-ing a fusion of GFP with the structural protein alpha tubulin, viewed in conventional widefield epifluorescence microscopy. Even though these cells are fairly flat, contrast is degraded by background light from out-of-focus regions. *Right:* The same sample, viewed in TIRF microscopy (see Figure 5.11). [Images courtesy Nico Stuur-man.]

fluorescent molecules in this "evanescent field" region (for example, tagged receptor molecules in a cell's membrane), then TIRF microscopy is a good approach. Stray flu-orescent molecules not in this region will not be seen, because they are not illuminated.

Your Turn 5F

Won't the same argument imply that the fluorophores' light will be trapped, and hence not visible in the microscope?

Figure 5.12 illustrates the power of TIRF microscopy to reduce background fluores-cence.

5.3.5 Refraction is generally wavelength dependent

In vacuum, light of all wavelengths travels at the universal speed c. This is generally not true in a medium; that is, the index of refraction $n(\lambda)$ may depend on the wavelength. The law of refraction then predicts that different wavelengths of light will be bent by differing amounts when a mixture, such as white light, enters a medium. In this way, a prism can separate white light. If that's what we want, good—we can use this **dispersion** effect to build a spectroscope, or to get a beautiful rainbow.

If we *don't* want color separation, then we call the same phenomenon by a dep-recatory name: **chromatic aberration**. For example, this effect makes it hard to focus white light with a lens:[18] The resulting images suffer from colored fringes, unless we use monochromatic light or complex lenses designed to cancel the effect.

Section 5.3 has studied some familiar phenomena that can be understood by examining one stationary-phase path at a time, and some of their applications to imaging and other technology.

---

[18]Chapter 6 will discuss the focusing of light.

THE BIG PICTURE

This chapter has barely scratched the surface of plant and animal strategies to *sort and direct* light. Organisms excel at creating nanostructured materials with precisely controlled physical properties (for example, bone), so it should perhaps be no surprise to find those skills being put to use to create useful photonic structures as well. Some animal eyes contain internal reflectors to eject unwanted scattered light; others have sophisticated antireflection coatings for stealth, and so on. The rest of this book would hardly suffice to discuss all of these adaptations and their mechanisms! Instead, we now turn to the most important light-based adaptation of all: visual image formation.

KEY FORMULAS

- *Structural color:* A "quarter-wave stack" consists of alternating layers of two transparent media with indices of refraction and thicknesses related by $c/(n_1\nu) = 4d_1$ and $c/(n_2\nu) = 4d_2$. Such a stack will reflect light of frequency ν strongly.
- *Light Hypothesis, part 2c:* When a photon path passes through the boundary between two transparent media, traveling perpendicular to it from *1* to *2*, its contribution to the probability amplitude acquires an extra "transmission factor" $t_{1\to 2} = 2\sqrt{n_1 n_2}/(n_1 + n_2)$. When a photon path bounces off the boundary, staying in medium *1*, its contribution to the probability amplitude acquires an extra "reflection factor" $r_{1\to 2} = (n_1 - n_2)/(n_1 + n_2)$.
- *Law of reflection:* Consider a macroscopic situation where we may neglect diffraction, so that light may be thought of as traveling on piecewise-straight paths (the ray-optics approximation). When light bounces off a mirror (or partially bounces off any other flat interface), the incoming and reflected rays make equal angles with the line perpendicular to the interface.
- *Law of refraction:* Again restricting to the ray-optics approximation, when light passes from air or vacuum to another transparent medium with index n (such as water), it is partly reflected and partly transmitted into the medium. Define the angle of incidence, θ_a, as the angle between the direction of the incoming photons (in air) and the direction perpendicular to the interface. Similarly θ_w, the angle of refraction, involves the direction of the transmitted ray.
  If the light enters perpendicular to the interface, then $\theta_a = 0$; in this case, the angle does not change. More generally, $\sin\theta_a = n\sin\theta_w$.
  The index of refraction n may depend on the wavelength of the light, giving rise to dispersion effects such as chromatic aberration in lenses.
- *Total internal reflection:* When light passes through an interface, heading from a transparent material with index of refraction n_{from} toward one with a smaller index n_{to}, the critical angle is $\theta_{\text{crit}} = \sin^{-1}(n_{\text{to}}/n_{\text{from}})$.

FURTHER READING

Semipopular:
Johnsen, 2012.

Intermediate:
Index of refraction: Feynman, 1985, chapt. 3.

Total internal reflection in physiology and medicine: Yildiz & Vale, 2011; Franklin et al., 2010; Amador Kane, 2009; Ahlborn, 2004.

Structural color, natural and artificial: Kinoshita, 2008, Nassau, 2003. The trick of writing a recursion relation is explained in Amir & Vukusic, 2013.

Technical:

Feynman et al., 2010c.

Butterfly wing color: Smith, 2009. In transparent insect wings: Shevtsova et al., 2011.

$\boxed{T_2}$ **Track 2**

5.2.2′ Transmission and reflection in classical electromagnetism

The Light Hypothesis part 2c (Idea 5.1, page 184) gave formulas for the transmission and reflection at an interface between two transparent media. If you're familiar with the classical wave theory of light, you may have seen these formulas in a different-looking form (the "Fresnel equations" for light traveling perpendicular to the surface). Although this book does not cover the wave theory, here is the connection.

Suppose that a plane electromagnetic wave travels in the $\hat{z}$ direction through an infinite, transparent medium with uniform index of refraction n_1. Its electric and magnetic fields are then

$$\boldsymbol{E}(t,\boldsymbol{r}) = A\hat{\boldsymbol{x}}\mathrm{e}^{\mathrm{i}(kz-\omega t)} + \text{c.c.}, \qquad \boldsymbol{B}(t,\boldsymbol{r}) = A\frac{k}{\omega}\hat{\boldsymbol{y}}\mathrm{e}^{\mathrm{i}(kz-\omega t)} + \text{c.c.},$$

where "c.c." denotes the complex conjugate of the preceding term. These expressions solve Maxwell's equations in the medium, provided that $\omega/k = c/n_1$: The wave moves at that reduced speed. The overall prefactor A sets the intensity of the light.

If instead the wave encounters an interface, we must supplement the solution just given by superposing a similar solution in medium *1* moving in the opposite direction (reflected wave), and one in medium *2* moving in the same direction (transmitted wave). If there are no free charges nor currents at the interface, then the electric and magnetic fields must match there. Those two conditions allow us to solve for the reflected and transmitted waves, whose prefactors equal

$$A\frac{n_1 - n_2}{n_1 + n_2} \quad \text{and} \quad A\frac{2n_1}{n_1 + n_2},$$

respectively. The first of these formulas agrees with the one in Idea 5.1, but the second does not.

To resolve the discrepancy, we need to know that a wave carries energy proportional to $n|A|^2$. Thus, in our situation, the reflected wave's power per area is a constant times

$$|A|^2 n_1 \frac{(n_1 - n_2)^2}{(n_1 + n_2)^2}, \tag{5.12}$$

whereas the transmitted wave carries

$$|A|^2 n_2 \frac{4{n_1}^2}{(n_1 + n_2)^2}. \tag{5.13}$$

You can check that these quantities do add up to the incoming energy, which is proportional to $n_1|A|^2$.

Dividing everything by the incoming energy, $\propto n|A|^2$, gives the probabilities that any one photon will be reflected or transmitted; those results do agree with the moduli squared of the expressions appearing in Idea 5.1.

More complicated formulas are needed for light paths that do not arrive at right angles to the interface. For the geometries shown in Figures 5.4 and 5.5, however, such paths would not be extremal, so we are neglecting them in any case. (Also, in this situation we need not worry about refraction.)

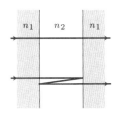

Fig. 5.4a (page 185)

$\boxed{T_2}$ **Track 2**

5.3.1' Fine points about reflection and refraction

1. The discussion of the reflection law in Section 5.3.1 (page 190) omitted a factor in the probability amplitude associated with the bouncing of the path from the surface. This omission is justified in macroscopic situations like the one considered, where this r factor is slowly varying over the range of paths needed to get cancellation; in that case, it is effectively just an overall multiplicative factor.

2. Similarly, the discussion of the refraction law in Section 5.3.3 (page 192) omitted a factor in the probability amplitude associated with the passage of the path through the surface. This omission is again justified in macroscopic situations like the one considered, where this t factor is slowly varying over the range of paths needed to get cancellation; in that case, it is again effectively just an overall multiplicative factor.

$\boxed{T_2}$ **Track 2**

5.2.4' More complicated layers

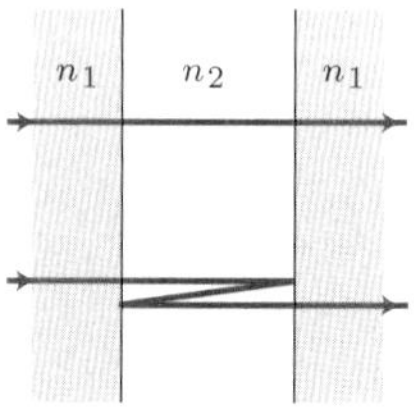

Fig. 5.4a (page 185)

The analysis of Section 5.2.4 was greatly simplified by the careful choice of unit cell in Figure 5.4a: By cutting through the middle of a layer of type 1, we got a unit cell that was reflection symmetric. Later, when we summarized the entire cell by overall reflection and transmission coefficients, this symmetry meant that the same factors controlled left-moving and right-moving path segments. For the case of a more elaborate unit cell (for example, a triple layer), a more elaborate form of the calculation is needed.

PROBLEMS

5.1   *Finite stack*

The main text showed how to calculate the transmission and reflection from an infinite stack of layers. Use a computer to recompute a result like the one in Figure 5.5c, with numerical values given in the figure's caption, but for a more realistic stack of just eight layers. As in the text, you may ignore the possibilities of dispersion (that is, take the index of refraction of cuticle to be independent of wavelength) and absorption (every photon is either reflected or transmitted).

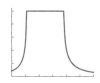

Fig. 5.5c (page 187)

5.2   *Total probability*

Review Equations 5.2 and 5.3 and confirm that $|r_{1\text{ layer}}|^2 + |t_{1\text{ layer}}|^2 = 1$.

5.3   *Law of reflection*

We should expect the law of reflection (Idea 5.8, page 191) to *fail* in some situations, due to diffraction. In this problem you'll upgrade the main text's derivation, which only considered the stationary-phase paths.

Figure 5.13 shows a cross section through an apparatus that is uniform in the y direction (out of the page). The surface of a mirror occupies a part of the x axis of width W; thus, u lies between $-W/2$ and $W/2$. A point source of light sits at $x = -a$, $z = h$. A detector sits at $x = +a$, $z = h$. To be concrete, let $a = h = 1\,\text{m}$. An opaque barrier shields the detector from seeing light directly from the source.

Everyday experience suggests that only a little bit of the mirror is needed to make the reflection. In fact, ray optics says that only an *infinitesimal* strip of the mirror near $u = 0$ is enough to bounce that light. But our study of diffraction suggests that if W is too small, then light will go everywhere, and much of it will miss the detector. That's the subject of this problem.

As usual, approximate the sum over all paths, replacing it by the sum over just those paths consisting of a straight line segment from the source to a point u on the mirror, followed by another straight line segment to the detector.

Suppose that the light is monochromatic, with wavelength 500 nm. We imagine dividing the mirror into little strips, each of width Δu, so that there are $W/\Delta u$ of these strips. Try taking Δu to be 30 times the wavelength of the light, and initially suppose that $W = 3\,\text{mm}$.

Each path contributes a term of the form $\exp(2\pi\mathrm{i}(L_{I,j} + L_{R,j})/\lambda)$ to the proba-

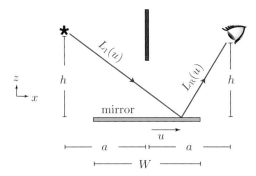

Figure 5.13: [Path diagram.] **See Problem 5.3.** An opaque barrier (*top*) prevents light from the source from directly reaching the detector.

bility amplitude. Here $L_{I,j}$ is the length of the first ("<u>i</u>ncident") part of path number j, and $L_{R,j}$ is the length of the second ("<u>r</u>eflected") part.

a. Obtain a formula for $L_{I,j} + L_{R,j}$, in terms of h, a, λ, and the location u_j at which path j strikes the mirror.

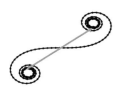

Fig. 4.11a (page 162)

b. Note that h and a are much larger than any value of $|u_j|$ that we are including. Use this fact to find a simplified approximate form for your answer in (a), retaining terms up to second order in u_j.

c. Write computer code to evaluate the discrete sum over paths numerically. Produce a plot analogous to Figure 4.11a showing the arrows contributed by each individual path, together with their sum. Find the length-squared of the total arrow. [*Hint:* The length-squared you compute here is equal to the light intensity measured at the detector times a constant normalization factor. You need not worry about this factor, because in this problem, we are only concerned with *relative* changes in intensity due to changes in the mirror width W.]

d. Now see what happens when you consider a narrower mirror: Include only the contribution from a limited range of the integration, say $W = 0.75\,\mathrm{mm}$, or less. At first, it doesn't matter much, but as W decreases, eventually your length-squared will become significantly smaller, say $1/4$ of its original value. For what values of W does that happen?

e. [T₂] Adapt the discussion that led to Equation 4.20 (page 167) to find a single dimensionless combination of parameters that controls how diffractive the result will be. Evaluate that parameter for the situations you studied earlier, that is, for $W = 3\,\mathrm{mm}$ and $0.75\,\mathrm{mm}$, and comment.

5.4   *Reflection grating*

Figure 5.14 shows a distant, monochromatic point source of light cut down to a beam by an aperture (an opening in a barrier). The aperture is wide enough that we may neglect diffraction effects when light emerges from it; however, it does restrict light to a spot of width W on the mirror, centered a distance d_I from the left. Reflected light is observed on a projection screen at point **A**.

We assume that W is much smaller than the distances L_I and L_R from the spot center to either the source or observation point. Thus, the angles of <u>i</u>ncidence are all approximately $\theta_I = \sin^{-1}(d_I/L_I)$, and the angles of <u>r</u>eflection are all $\approx \theta_R = \sin^{-1}\big((x - d_I)/L_R\big)$.

The light illuminates an array of narrow, uniformly spaced mirror strips (coming out of the page). The strips have periodic spacing p that is smaller than W, so that the incoming beam illuminates at least several of them. If this were an ordinary mirror, we wouldn't see any light at **A**, because the path that obeys the usual reflection law would have to bounce at **B**†, a point that's not illuminated. We are interested in a modification to the usual law of reflection (Idea 5.8, page 191) that comes about because of the interruptions in the mirror.

Consider only paths that each consist of two straight segments, like the ones shown, and suppose that the mirror strips are so narrow that all paths reflected from segment j have essentially the same length. Thus, you may approximate the integration over bounce locations by a discrete *sum* over j.

a. Label the middle path shown as $j = 0$, flanking paths $j = \pm 1$, and so on. Let $L_I(j)$ denote the <u>i</u>ncident path lengths, and $L_R(j)$ the <u>r</u>eflected path lengths. Write an exact expression for $L_{\mathrm{tot}}(j)$, the total path length in terms of h_I, h_R, d_I, x, and j.

There will be bright illumination at point **A** if the paths arriving there all have

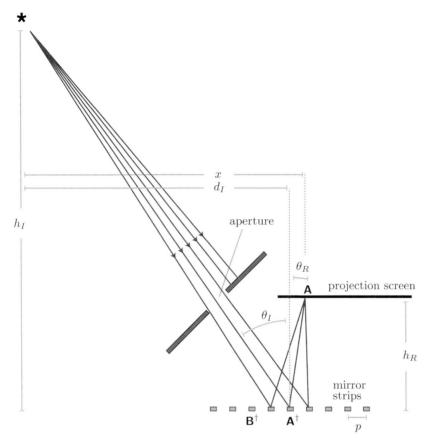

Figure 5.14: [Path diagram.] **Reflection grating.** See Problem 5.4. A series of mirror strips (*bottom, emerging from the page*) reflect light from a distant source (*top*). The drawing is not to scale: The distances h_I, h_R, x, and d_I are all macroscopic, whereas the spacing p between mirror segments is so narrow that the beam of light emerging from the aperture illuminates several of them.

approximately the same phase, or if they all have phases that differ by integer multiples of 2π. This condition in turn will be met if the total path lengths all obey $L_{\text{tot}}(j + 1) - L_{\text{tot}}(j) = m\lambda$, for some integer m.

b. Simplify the square roots in your answer to (a) by using a Taylor expansion about the middle path, finding L_{tot} to lowest nontrivial order in powers of W/d, where d is any of the macroscopic parameters.

c. What condition must be satisfied in order for neighboring mirror segments to make reinforcing contributions, and hence for point **A** to be illuminated? What location(s) x on the screen correspond to solutions to that condition? What is special about the case $m = 0$?

d. Make an Insightful Comment about the role of color in this problem.

5.5   *Total internal reflection*

The human eye is filled with a transparent medium with index of refraction $n = 1.34$. Assume that the cornea also has roughly this same index. Light rays inside the eye that hit the cornea in a certain range of angles are trapped and cannot exit; find that range. This effect can interfere with an ophthalmologist's inspection of the eye, unless some special measures are taken to overcome the problem (Figure 5.10).

Fig. 5.10 (page 194)

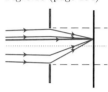

Fig. 4.13 (page 165)

Fig. 5.9 (page 193)

5.6 $\boxed{T_2}$ *Inhomogeneous medium*

The main text considered light traveling through a homogeneous medium, such as air or water. In such a medium, the stationary-phase path between two points is a straight line. When there are obstructions (opaque barriers), the stationary-phase path (if any exists) is still straight (heavy line in Figure 4.13). Section 5.3.3 (page 192) considered a more complex situation, involving blocks of two different media, each homogeneous but with a sharp interface between them (Figure 5.9). In that situation, we assumed that the stationary-phase path is piecewise straight. We then found that it has a kink at the interface such that the angles θ that each segment makes to the perpendicular direction have the property that $n \sin \theta$ does not change (Equation 5.10, page 193).

Something more interesting can happen when the index of refraction changes *continuously* through the medium. For example, Chapter 6 will mention animals whose eye lenses have continuously graded index. Suppose that $n(\boldsymbol{r})$ is some known function of position, and consider all possible curves that start at a "source" point **A** and end at another "observer" point **B**. We'll describe such a curve parametrically by a vector function $\boldsymbol{\ell}(\xi)$ with $\boldsymbol{\ell}(0) = \mathbf{A}$ and $\boldsymbol{\ell}(1) = \mathbf{B}$. The generic parameter ξ is not necessarily arclength s along the curve; instead, we have

$$\mathrm{d}s = \sqrt{\|\mathrm{d}\boldsymbol{\ell}/\mathrm{d}\xi\|^2}\,\mathrm{d}\xi. \tag{5.14}$$

We want to know which path between the specified points has stationary phase, so consider *two* paths that differ infinitesimally: $\boldsymbol{\ell}(\xi)$ and $\boldsymbol{\ell}(\xi) + \delta\boldsymbol{\ell}(\xi)$. The variation $\delta\boldsymbol{\ell}(\xi)$ must equal zero at the endpoints $\xi = 0$ and 1, because those points are fixed to be **A** and **B**.

a. The transit time Δt for a path is given by the integral of $n(\boldsymbol{\ell}(\xi))\mathrm{d}s/c$ from $\xi = 0$ to 1. Find an expression for the first variation of this quantity, that is, the difference $\delta(\Delta t)$ between its values for the two nearby curves, to first order in $\delta\boldsymbol{\ell}(\xi)$.

b. Show that the condition that $\boldsymbol{\ell}(\xi)$ must satisfy, in order for the transit time to be unchanged to first order in $\delta\boldsymbol{\ell}(\xi)$ by this variation, is

$$\boxed{\frac{\mathrm{d}}{\mathrm{d}s}\left(n\frac{\mathrm{d}\boldsymbol{\ell}}{\mathrm{d}s}\right) = \boldsymbol{\nabla}n \quad \text{ray equation}} \tag{5.15}$$

at every value of ξ.

Now consider the situation in which the index is a function that depends only on height z above the ground (Figure 5.15). This problem is mathematically simpler than the general case, in part because some solutions to Equation 5.15 stay strictly in the xz plane. Light is emitted by a source at height z_0 and detected somewhere else, also at height z_0 but a distance D away. We can describe a curve in the xz plane by its height function, $z = h(x)$, where $h(\pm D/2) = z_0$.

c. Take the dot product of Equation 5.15 with the unit vector $\hat{\boldsymbol{x}}$. Let $\theta(x)$ be the angle that the curve makes to the z axis, express your formula in terms of θ, and connect it to the ordinary law of refraction.

d. Show that the equation you found in (c) has two unsurprising solutions: One is a straight, horizontal line: $h(x) = z_0$, $\theta(x) = \pi/2$. The other is a straight, vertical line: $\theta(x) = 0$. But there can also be solutions that are *curved*. Suppose that the profile $n(z)$ is strictly increasing as z increases, and that θ starts out tilted downward as

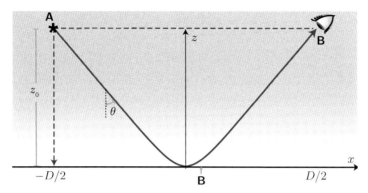

Figure 5.15: [Mathematical function.] **Curved stationary-phase path in an inhomogeneous medium.** See Problem 5.6. The x and z axes are not drawn to the same scale. *Dashed lines* show the simple solutions mentioned in part (d) of the problem. An observer who assumes straight-line propagation will interpret light from **A** as having come from **B**.

shown in the figure $(0 < \theta < \pi/2)$. Then θ can increase as h decreases, potentially even leveling off $(\theta \to \pi/2)$, as shown. At what height z_{bounce} will this happen? (Find a symbolic expression.)

The formation of mirages is a familiar example of this situation.[19] On a long, flat stretch of highway, solar heating creates a layer of air near $z = 0$ that is hotter than elsewhere. Hotter air is less dense, so we have the situation in (d). Thus, it can happen that, when we direct our gaze downward (toward the road) we'll see light originating from the sky that has traveled on the curved path in the figure. It is easy to misinterpret that light as a reflection from (nonexistent) water on the road, particularly because it tends to shimmer (due to air convection currents). You know from experience that this illusion only appears in the distance, not up close.

e. The equation you found in (c) involves $\hat{\boldsymbol{x}} \cdot (\mathrm{d}\boldsymbol{\ell}/\mathrm{d}s)$; express that quantity in terms of $h' = \mathrm{d}h/\mathrm{d}x$. Thus show that the equation can be restated as

$$0 = \frac{\mathrm{d}}{\mathrm{d}x}\left[(1 + (h')^2)^{-1/2}n\right].$$

This formula says that a certain quantity does not change as we move along the trajectory. Reexpress it one more time, in terms of $\theta(x)$.

f. Suppose that $n(z)$ equals the index of air at 30°C everywhere except in a thin layer near the ground, but rises to the value appropriate for air at 50°C right at the ground. Look up the values of the two indices of refraction for visible light in Appendix C. Use those numbers to compute $\alpha = 1 - (n(0)/n_\infty)$, a small positive quantity. Then use your result in (d) to see how close θ_0 must be to $\pi/2$ in order for the stationary-phase path to level off before hitting the ground. Then estimate how far away the mirage will appear to be.

g. To find the complete trajectory, we must specify a specific temperature profile, or equivalently a profile for the index of refraction that changes from $n(0)$ to n_∞ over a small range of height. One such function is $n(z) = n_\infty(1 - \alpha e^{-z/L})$, where $L = 20\,\text{cm}$. Your eyes are about $z_0 = 2\,\text{m}$ off the ground.

h. In (c), you found an equation determining the entire curve. Use a computer to solve it for the situation discussed in (e) with the index profile just given. Use the

---

[19] Section 4.7.2 (page 164) introduced mirages.

smallest value of θ_0 for which you found that a mirage would be possible, and make a graph showing your solution.

Your computer will choose different scales for the x and z axes, so label them clearly.

PART II

Human and Superhuman Vision

Image formation by a spherical lens. View through a sphere of naturally occurring quartz with diameter 25 cm (Qing Dynasty, 19th century).

CHAPTER 6

Direct Image Formation

> Now, it is not too much to say that if an optician wanted to
> sell me an instrument which had all these defects, I should
> think myself quite justified in blaming his carelessness in the
> strongest terms and giving him back his instrument.
> — *Hermann von Helmholtz, on the human eye's optics*

6.1 SIGNPOST: *BRIGHT YET SHARP*

We have made a big investment in understanding how photons know where to go. It is
now time to put that effort to work to see how living organisms create visual images
in their eyes. In the world of technology, too, we often wish to focus images onto an
array of light detectors, to make a camera or microscope. Such devices can give us
visual acuity far beyond what any animal possesses.

The whole idea of "focusing" light seems to imply a level of control at odds
with the Light Hypothesis, which seems to say that "light does whatever it wants."
However, Chapter 4 introduced a way out of that dilemma: the stationary-phase
principle. This chapter will systematically exploit that insight, and upgrade it from
the isolated stationary-phase paths studied so far to whole ranges of paths that all
have the same phase. We will see that cameras, and animal eyes, use this trick to
evade a trade-off between having *bright* versus *sharp* images.

The Focus Question is

Biological question: How do swim goggles help you to see underwater?

Physical idea: The curvature of an air-water interface determines whether it will focus
light.

6.2 IMAGE FORMATION WITHOUT LENSES

In the next few sections, we will suppose that light travels exclusively along stationary-
phase paths (the "ray-optics approximation"). After we understand the main ideas
about image formation, Section 6.8 will examine this approximation's validity. In
particular, we will derive the limitation on the resolving power of a microscope due
to diffractive effects. Chapter 7 will discuss some recent breakthroughs that overcame
this limitation.

6.2.1 Shadow imaging

In the ray-optics approximation, light travels along straight lines when it passes
through vacuum or a uniform, transparent medium. If some opaque object intervenes
between a light source and an array of detectors, then a shadow will fall on the detecting

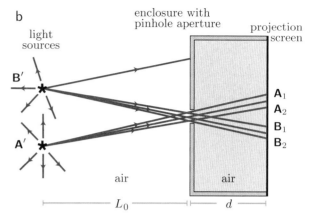

Figure 6.1: **Pinhole imaging system.** (a) [Schematic.] Camera obscura, from a seventeenth century manuscript of military designs. (b) [Ray diagram.] Light from distant sources (*left*) passes through a narrow aperture and lands on a projection screen (*right*). Light emerging from source **A**′ arrives at a range of locations on the screen (**A**$_1$–**A**$_2$), corresponding to all the different straight line segments from **A**′ that pass through the aperture, and similarly for another source **B**′. The two images can only be resolved if their separation is larger than this blurring.

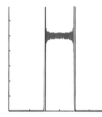

Fig. 4.14a1 (page 167)

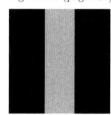

Fig. 4.14a2 (page 167)

surface. Suppose that the light source is approximately a distant point (for example, the Sun on a clear day), and its light lands on a projection screen perpendicular to the line to the source (the sidewalk, at noon). Then the shadow will have the same shape and size as the object's profile (Figure 4.14a). In the same way, ordinary x-ray imaging creates a full-size shadow image of our bones or teeth (which are opaque to x rays) on an array of detectors.

6.2.2 Pinhole imaging suffices for some animals

Shadows are silhouette images of objects lit by a distant point source of light. Animals have evolved systems to create more useful images, in more general situations. The simplest of these is the **pinhole imaging system**, or "camera obscura."

Figure 6.1 shows the idea: We place a barrier with a narrow aperture between the world and a projection screen. Suppose that the world contains just two distinct point sources of light, called **A**′ and **B**′ in the figure. The sources could either be light-emitting objects (two stars in the sky), or ordinary objects each reflecting light from the Sun. Ideally, we'd like each point source of light to illuminate a unique corresponding point on the screen, forming an image. The figure shows that, in the ray-optics limit, this situation is nearly realized, because in that limit, light follows straight-line paths.[1] The result is an inverted (upside-down) image, scaled by the geometric factor d/L_0.

A closer look at Figure 6.1b reveals a drawback with the pinhole camera, however. The multiple lines drawn from **A**′ illustrate that even a point source of light will show up as a blob on the screen, at least as large as the aperture. (In the figure, straight-line paths join **A**′ to a range of points, including **A**$_1$ and **A**$_2$.) Similarly, a continuous distribution of light sources will appear blurred, reducing the **visual acuity** of any creature that uses pinhole imaging. We could try to minimize this problem by

---

[1] The sketches in this section are cross sections, and our derivations, too, will usually neglect the third dimension, to simplify the math.

Figure 6.2: [Photograph.] **Head of *Nautilus pompilius*,** showing its pinhole eye. [Photo Hans Hillewaert.]

using a tiny aperture, but then we get a very dim image—most of the light is blocked. Moreover, when the aperture is too small diffraction takes over, and image sharpness starts to get *worse*.[2]

Partly compensating these disadvantages, pinhole imaging has the advantage that the distance L_0 to the light source is immaterial, as long as it's much bigger than the width of the aperture: There is no need to "focus" the system.

Snake sensory organ; nautilus eye

Despite its limitations, pinhole imaging is adequate for some animals. For example, some snakes have a "pit" lined with infrared sensors, with a narrow opening that crudely projects an image of their world as represented by the body heat of potential prey.

The cephalopod *Nautilus* also has a pinhole imaging system (Figure 6.2). Its eye even has an adjustable aperture size, so that it can take advantage of brighter conditions to achieve better visual acuity by narrowing its aperture, or sacrifice acuity when the illumination is poor.

Section 6.2 has surveyed some primitive yet effective strategies for image formation. We now ask what improvements on those schemes are possible, within the limitations of what real living tissues can do.

6.3 ADDITION OF A LENS ALLOWS FORMATION OF BRIGHT, YET SHARP, IMAGES

Pinhole imaging works, but it discards most of the available light. It would be better to find a way to collect light over a wide aperture, without the blurring that results from the pinhole scheme. That is, we'd like to arrange that all light entering the aperture

---

[2]Section 4.7.3 (page 165) analyzed this phenomenon.

Figure 6.3: [Path diagram.] **Lens-based imaging.** In this cross section, a point source of light, **A**′, lies on the centerline (*dotted*), that is, the line perpendicular to the opaque barrier (*middle*) and passing through the center of an aperture of width $W = 1\,\mathrm{cm}$. The aperture contains a thin, transparent element (the lens, represented by a *question mark*), which changes the phase of each light path passing through it. A typical path is shown, passing through the lens at a distance u from the center and ending at point **A** on the projection screen.

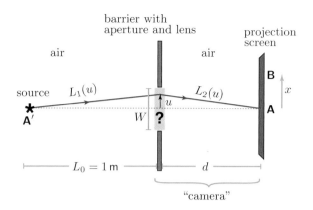

from **A**′ arrives at a *single* point **A**, and that light from a second source **B**′ all arrives at some *different* point **B**. This feat can be achieved with the help of a **lens system**. To develop our ideas, we'll consider a concrete, macroscopic example system that we'll call a "camera," as well as another example whose dimensions approximate those of our eyes.

6.3.1 The focusing criterion relates object and image distances to lens shape

Let's explore what we could achieve by filling a wide aperture with a device that manipulates the *phases* of photon paths passing through it (Figure 6.3). We'll call this device a **lens**, but won't describe it in detail yet; thus, it appears in the figure as a question mark. The distance d shown in the figure could be $\approx 1\,\mathrm{m}$ in a demonstration setup.

To get started, let's first consider a single light source **A**′ that lies on the line passing through the center of the aperture, as shown in the figure, and imagine placing a light detector at its opposite point **A**, also on the centerline. The goal is to arrange that

> *All the paths joining **A**′ to **A** have the same phase, despite having different geometrical lengths. But paths from **A**′ to any other point **B**, off the axis, all have differing phases.*

criterion
for focus

(6.1)

To understand Idea 6.1, recall the significance of the phase: The Light Hypothesis says that all the paths between two points contribute to the probability amplitude for photon arrival, and that these contributions will reinforce one another if the phases agree.[3] In the pinhole camera, Figure 6.1b shows that multiple points **A**$_1$, **A**$_2$, … each have a single stationary-phase path from **A**′; hence, they all share the total illumination, giving a spread-out image.[4] In contrast, satisfying the criterion in Idea 6.1 would imply that *every* path from **A**′ to **A** is a stationary-phase path, whereas there is *no* stationary-phase path from **A**′ to any other screen point. Then nearly all the light that gets through the aperture arrives at one spot, giving a bright, sharp image

---

[3] Idea 4.5 (page 154) states the Light Hypothesis part 2a.
[4] The same thing happened in Section 4.7.3 (page 165): In the ray-optics regime, a distant point source illuminates the screen in a region with the width of the aperture.

of **A**′ at **A**.[5] Thus, an imaging system satisfying Idea 6.1 would eliminate the trade-off we faced in the pinhole system between bright versus sharp images. Can this really be done?

As in Chapter 4, we consider only photon paths consisting of two straight segments. So we can again label the various paths by the distance from the centerline, u, at which each one passes through the aperture. This time, however, we will not assume that the object distance L_0 is much larger than d. Also, we are introducing the lens, which was not present in Chapter 4.

Because each of the paths has a different total length, the lens (symbolized by a question mark in Figure 6.3) needs to introduce a compensating *time delay*, $\Delta t(u)$, that depends on u. The total phase attributed to a path is then equal to $2\pi\nu\big(L_{\text{tot}}(u)/c + \Delta t(u)\big)$, where L_{tot} is the actual (geometric) length of the path. The prefactor $2\pi\nu$ is independent of u, and hence immaterial to finding the stationary-phase path, so we may drop it.

We want to know if a function $\Delta t(u)$ can be found such that the criterion in Idea 6.1 is satisfied, and if so, whether a physical device can be constructed that creates such delays. Writing an expression similar to Equation 5.9 (page 193), but with the extra delay, we compute the total phase and require that it be independent of u. That is, we want $\Delta t(u)$ to be a function with the property that

$$L_{\text{tot}}(u)/c + \Delta t(u) = c^{-1}\left(\sqrt{L_0{}^2 + u^2} + \sqrt{d^2 + u^2}\right) + \Delta t(u) = \text{const.} \qquad (6.2)$$

To simplify the math, suppose that L_0 and d are both much bigger than the aperture size, so that we can approximate the square root by using its Taylor series, as we did in Chapter 4.

Ex.   a. Show that the requirement that all paths from **A**′ to **A** have the same phase can be expressed as

$$\frac{u^2}{2L_0} + \cdots + \frac{u^2}{2d} + \cdots + c\Delta t(u) = \text{const.} \qquad (6.3)$$

Here the dots represent higher-order terms in u. To do this, follow steps like those in Section 4.7.3 (page 165).
b. Substitute values appropriate to our "camera" example. Is it really valid to neglect the higher-order terms in Equation 6.3?
Solution: a. We are making the simplifying approximations that $u \ll L_0$ and $u \ll d$. Applying Taylor's theorem to Equation 6.2 gives

$$L_0\Big(1 + \frac{u^2}{2L_0{}^2} + \cdots\Big) + d\Big(1 + \frac{u^2}{2d^2} + \cdots\Big) + c\Delta t = \text{const.}$$

Lumping all the constants together on the right side gives Equation 6.3.
b. The ellipses represent terms that are smaller than the ones retained by factors of $(u/L_0)^2$ or $(u/d)^2$, respectively. We are supposing that d is also roughly a meter, so these terms are of similar importance. Figure 6.3 shows an aperture width of $1\,\text{cm}$, so $|u| \le 5 \times 10^{-3}\,\text{m}$. Thus, for light of wavelength $500\,\text{nm}$ the contributions to the

---

[5]You'll make this implication more precise in Problem 6.12.

Figure 6.4: [Mathematical functions.] **Lens-based imaging.** *Solid curve:* The phase function corresponding to the case of a focused system, relative to $u = 0$. This function is nearly constant. The text uses an approximation that retains only terms up to order u^2; in that approximation, the function is perfectly constant. *Dashed curve:* Changing d defocuses the system; this time there is only an isolated stationary-phase path at $u = 0$. (Both curves used Equations 6.2 and 6.5 with $f = 0.25\,\mathrm{m}$, and $\lambda = 500\,\mathrm{nm}$. The distance to the screen was $d = 0.33\,\mathrm{m}$ or $0.4\,\mathrm{m}$ for the solid and dashed lines, respectively.)

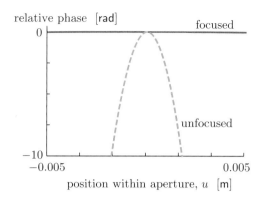

photon path's phase that we are neglecting are about

$$\frac{E_{\mathrm{ph}}}{\hbar} \frac{L_0}{c}\left(-\frac{1}{8}\right)\frac{u^4}{L_0{}^4} = -\frac{2\pi}{5\times 10^{-7}\,\mathrm{m}}\frac{1}{8}\frac{(5\times 10^{-3}\,\mathrm{m})^4}{(1\,\mathrm{m})^3} \approx 10^{-3}.$$

This is much smaller than 2π, so neglecting it in an exponential is not a big error.

Your Turn 6A

Repeat for the human eye, whose pupil diameter in bright light is about $2\,\mathrm{mm}$, whereas $d \approx 24\,\mathrm{mm}$.

Thus, a focusing device must delay light by the amount

$$\Delta t(u) = c^{-1}\left[\mathrm{const} - \frac{u^2}{2}\left(\frac{1}{L_0} + \frac{1}{d}\right)\right]. \tag{6.4}$$

Figure 6.4 shows the complete phase function (from Equation 6.2), contrasting cases where the condition Equation 6.4 is and is not satisfied. Even when it is satisfied, the phase is not perfectly constant: The solid line in the figure is not perfectly straight, because of the higher-order terms in u that we are neglecting. The Example showed that this neglect can be justified in some geometries (when the aperture is much smaller than L_0 and d).

What sort of object could introduce the required delay? We know that a transparent medium like glass slows light.[6] Suppose that part of a photon path, of length L_{g}, passes through the glass.[7] Then the increase in net transit time for that path due to the lens is $L_{\mathrm{g}}/(c/n) - L_{\mathrm{g}}/c$, or $L_{\mathrm{g}}(n-1)/c$. (Note that in the limit $n \to 1$ the effect disappears, as it should: In this case, there's effectively no lens!)

Equation 6.4 implies that a suitable lens could consist of a piece of glass that's thin at the ends, but thicker in the middle (that is, "lens-shaped"), creating a variable delay that is maximal when $u = 0$. Specifically, if its shape follows the profile $L_{\mathrm{g}}(u) = c\Delta t(u)/(n-1)$, where $\Delta t(u)$ is given by Equation 6.4, then all paths from an on-axis source ($\mathbf{A}'$) to $\mathbf{A}$ will make reinforcing contributions, as desired.[8]

---

[6]Section 1.7 (page 47) introduced this idea.

[7]As in earlier chapters, underlining the initial "g" in the word "glass" is a hint about the meaning of the notation L_{g}: The subscript refers to glass.

[8]You'll do more accurate calculations, for more realistic systems, in Problems 6.8 and 6.13.

We can restate Equation 6.4 in a more practical way. Suppose that our lens has some fixed shape that is symmetrical about the centerline. Then we can express the corresponding time-delay function in a Taylor series expansion with only even terms:

$$\Delta t(u) = c^{-1}\big(\text{const} - u^2/(2f) + \cdots\big), \tag{6.5}$$

where f is a constant with dimension $\mathbb{L}$ describing the lens. The dots denote terms of higher order in u. Equation 6.4 says that this lens will focus light from a source at L_0 only if the distance d to the screen satisfies

$$\boxed{\frac{1}{f} = \frac{1}{L_0} + \frac{1}{d}. \quad \text{lens formula}} \tag{6.6}$$

An important special case of the lens formula is when the source is distant ($L_0 \to \infty$); then d must equal f. For this reason, f is called the **focal length** of the lens.

Note that the focal length is an intrinsic property of the lens, not of the whole imaging system; its value depends only on the lens's shape and composition. It is *not* the distance d at which a focused image will form, except in the special case where L_0 is taken to be infinity. A lens with large f is called "weak," because it doesn't bend light very much. Instead of specifying f, we can equally well describe a lens by its **focusing power**, defined as $1/f$. In this context, the relevant unit m^{-1} is often renamed the **diopter**.

Ex. Assuming that the condition in Equation 6.6 is satisfied, what can we say about the phases of light paths that start at the same point $\mathbf{A}'$ as before, but that end at a different point $\mathbf{B}$?

Solution: Let $\mathbf{B}$ lie at a distance x from the centerline (see Figure 6.3), and suppose that this distance is much smaller than either d or L_0. We must modify Equation 6.2, replacing $d^2 + u^2$ by $d^2 + (u - x)^2$.

As before, we also take the aperture size to be much smaller than either d or L_0. Then the same derivation as in the previous Example, combined with Equation 6.5, gives the effective path length as

$$\approx L_0 + \frac{1}{2L_0}u^2 + d + \frac{1}{2d}(u - x)^2 - \frac{1}{2f}u^2 + \cdots$$

$$= \text{const} - \frac{ux}{d} + \frac{1}{2}\left(\frac{1}{L_0} + \frac{1}{d} - \frac{1}{f}\right)u^2 + \cdots.$$

The term $x^2/(2d)$ is independent of u, so it has been included in the initial constant.

Comparing to the on-axis case, the coefficient of the u^2 term is still zero, but the linear term is not if $x \neq 0$. Thus, the effective path length has *no* stationary-phase point at any point $\mathbf{B}$ different from $\mathbf{A}$, and so light from $\mathbf{A}'$ interferes destructively. In the ray-optics approximation, we won't see any light at $\mathbf{B}$ from a source at $\mathbf{A}'$: All the light from $\mathbf{A}'$ that passes through the aperture ends up at $\mathbf{A}$.[9]

Equation 6.6 and the following Example say that a given lens, in a given camera, will only focus objects at one particular distance L_0. In order to focus at various

---

[9]Section 6.8 will explore how diffraction effects modify this statement.

distances, a camera must adjust either its lens, or the distance d to the projection screen. The human eye uses the first option. Other animals, for example fish, use the second.[10]

Your Turn 6B

Think about why none of the following will focus light:
a. A flat plate of glass perpendicular to the centerline;
b. A flat plate of glass at an angle to the centerline;
c. A wedge of glass (cross section in the plane of Figure 6.3 is triangular).

6.3.2 A more general approach

Chapter 4 explained why, in a macroscopic apparatus, light seems to travel on stationary-phase paths: Near such a path in the summation over all paths, many contributions have similar phases, which are summed before we compute the modulus squared of the probability amplitude. In the present chapter, we just saw that an even stronger reinforcement can arise if we arrange for the phase function to approximate a constant more closely than just pausing at a point. Specifically, we found a condition for the phase function's dependence on u to vanish to *second* order in u about its stationary point, not just first order. The rest of this chapter will exploit this insight systematically: For the other systems we'll study, the approach will also be to

- Compute transit time as a function of u.
- Set the order-u term to zero to find the stationary-phase path, and hence the location of the image point.
- Set the order-u^2 term to zero to find the condition for focus.

We will continue to assume that u is constrained by an aperture to be small enough to justify neglecting higher-order terms. In addition, when studying object and screen points that are off the centerline, we will suppose that they are not very far off. More precisely, we will make the **paraxial approximation** that $\epsilon = x/d$ and $\epsilon' = x'/L_0$ are numerically small, or equivalently that the angles θ and θ' in Figure 6.5 are small. In that case, we can make Taylor series expansions in the *four* small quantities u/d, u/L_0 ϵ, and ϵ'. We'll see that generally we must keep second-order terms (order u^2/d, u^2/L_0, $u\epsilon$ and $u\epsilon'$), but that we may drop higher order terms than those.

6.3.3 Formation of a complete image

So far, we considered only a single point source of light, on the centerline (the line passing through the center of the lens). It would be nice if we could simultaneously focus light from *many* places, thus forming an *image*. Referring to Figure 6.5, is it possible to find a single lens that focuses all the light from **A′** to **A**, and *also* focuses all the light from **B′** to some *other* point **B**?

Ex. Assuming that the setup satisfies the lens formula (Equation 6.6), find a point **B** on the screen at which all of the paths arriving from **B′** have the same phase.
Solution: We modify the approach of the preceding Example to allow both source and observation points to be off-axis. So this time, we also replace $L_0{}^2 + u^2$ by $L_0{}^2 + (u - x')^2$ in Equation 6.2. We make the paraxial approximation, this time

---

[10]Section 6.4 will return to the designs of animal eyes.

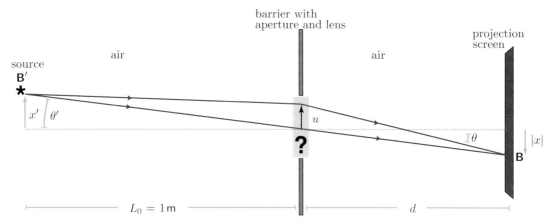

Figure 6.5: [Path diagram.] **Lens-based imaging of an off-axis light source.** In this cross sectional diagram, two illustrative photon paths are shown, both joining **B′** to **B**. One path has $u = 0$; the other has $u > 0$. Point **B′** lies a distance x' above the centerline; **B** lies a distance $|x|$ below that line (here x is a negative quantity). The *question mark* again reminds us that we have not yet specified the nature of the focusing element; one possibility is the realistic lens discussed in Section 6.4.2.

including that **B′** is close to the centerline. Grouping the resulting terms according to powers of u now yields the effective path length

$$\left[\text{const}\right] + \left[\frac{1}{2}\left(\frac{(x')^2}{L_0} + \frac{x^2}{d}\right)\right] + \left[u\left(-\frac{x'}{L_0} - \frac{x}{d}\right)\right] + \left[\frac{1}{2}u^2\left(\frac{1}{L_0} + \frac{1}{d} - \frac{1}{f}\right)\right] + \cdots . \quad (6.7)$$

This formula consists of four terms in square brackets. Of these, the first two are independent of u. The last is zero, regardless of the value of x', because we assumed that the lens formula was satisfied. Thus, the third term is the crucial one. This term can only be independent of u if it's *zero*, which occurs when $x = -x'(d/L_0)$.

The condition $x = -x'(d/L_0)$ just found has a simple meaning: It implies that the line segments from **B′** and **B** to the lens center have the same slope, or in other words that they form a single straight line (see Figure 6.5). In short,

- Light from an off-axis source all arrives at a single image point **B**.
- **B** varies with the source location.
- In fact, **B** is the point we get by drawing a straight line from **B′** through the center of the lens until it hits the screen.

Applying these conclusions to a scene containing *many* source points then shows that:

> *A single-lens system that satisfies the lens formula creates a bright, inverted image that is scaled relative to the actual scene by d/L_0.* (6.8)

Idea 6.8 sounds familiar—it's similar to the result we found for pinhole imaging (Section 6.2.2)! The difference is that the lens system can gather more light, via a wider aperture, and still give sharp images of objects that lie in a particular focal plane.

Your Turn 6C

Continuing the preceding Example, suppose that $|x'| < 5\,\text{mm}$ and discuss why we had to retain the terms of order ux and ux', but could drop terms of higher than second order in u, x and x' jointly.

This section only discussed focusing in a two-dimensional world. But it's straightforward to generalize the argument to the more realistic case, of lenses whose faces are not curves but rather curved *surfaces*. We also implicitly assumed a flat projection screen, unlike the curved retina in an animal eye. The appropriate calculations for a curved screen are similar in character to the one studied here.

6.3.4 Aberration degrades image formation outside the paraxial limit

Equation 6.7 has a crucial feature: The order-u^2 term does not depend on x or x'. That's why we found that canceling this term by an appropriate choice of L_0 and d creates an entire range of positions x' that are simultaneously in focus. However, this convenient circumstance only holds in the context of the paraxial approximation. It fails when higher-order terms in the expansion must be retained, a situation generically referred to as **aberration**. The loaded word "aberration" might seem to imply some departure from physical law; on the contrary, the phenomenon is entirely predictable. It's just undesirable, if we wish to form images. Section 6.5.3 will discuss a strategy that living organisms have evolved to mitigate this limitation on focusing.

Section 6.3 has shown how to reconcile the apparently conflicting desires to create bright yet sharp images. We now look at some details of how animals implement this idea.

6.4 THE VERTEBRATE EYE

Vertebrates see with an organ like the one shown in Figure 6.6. It took Western scientists a surprisingly long time to notice that this arrangement of optical elements focuses an image of the outer world onto the inner surface of the eye: Not until 1625 did C. Scheiner remove the coating from the back of an animal's eye and look at the translucent inner wall, where he saw projected a miniature, upside-down image of the world that was both sharp and bright. Let's explore how such image formation works, in more concrete detail than we have done so far. Thus, instead of representing a lens as an abstract time-delay function, and drawing it as a question mark, we'll imagine it as a solid chunk of material with index of refraction different from its surroundings, and having a definite shape.

[T₂] *Section 6.4′ (page 239) discusses the surprising arrangement of elements in Figure 6.6.*

6.4.1 Image formation with an air-water interface

The most significant optical element in our eye may not be immediately apparent in Figure 6.6. It's well known that, when our ancestors emerged from the seas, they needed to adapt to obtaining oxygen in gas form. But emerging into air also created an *air-water interface* at the eye's front surface (cornea). Let's examine a simplified version of this situation.

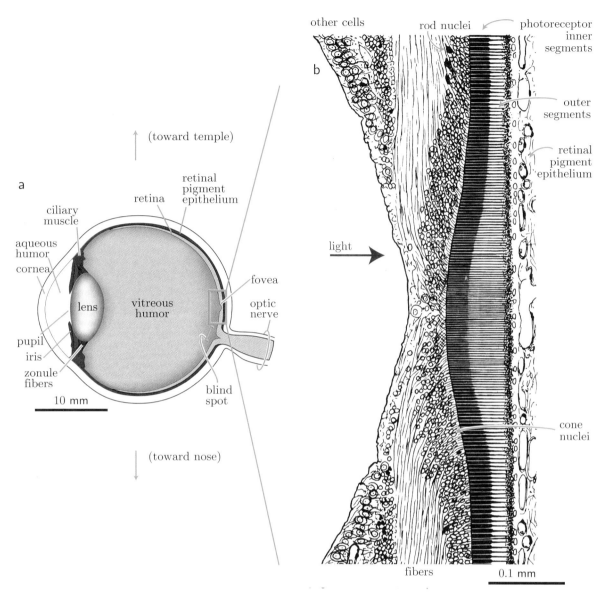

Figure 6.6: **Human eye.** (a) [Sketch.] Cross-section of the right eye (viewed from above). Incoming photons must pass through four transparent media (the cornea, aqueous humor, lens, and vitreous humor) before arriving at the retina. (b) [Anatomical drawing.] Enlarged view of the boxed region in (a), showing the central part of the human retina (the fovea), and its neighborhood. *Far right:* A layer of cells supports the photoreceptor (rod and cone) cells, whose outer segments are shown as *thin horizontal lines.* Thicker horizontal lines are cone cell inner segments, which supply the outer segments' metabolic requirements. The layer of *open ovals* represent nuclei of the cone cells, followed by a layer of nerve fibers that lead each cell's signal away from the retina to the next level of processing. *Solid ovals* at the periphery of the region shown are rod cell nuclei; the density of rod cells rises as we move away from the central region. Light from the pupil enters the retina from the left (*arrow*), and so must pass through multiple layers before arriving at the light-sensitive outer segments. Farther from the center (*top and bottom of figure*), even more layers intervene, including blood vessels and the bipolar and ganglion cells that relay signals from the photoreceptors to the optic nerve. When we look at the world, we do not notice these layers, partly because their shadows are not in focus at the layer containing photoreceptor outer segments. In addition, these layers are largely absent at the fovea. Away from the fovea, the distortion and scattering that they cause is less important, because our visual acuity is lower there. Other cells (the Müller glia, *not shown*) traverse the entire retinal thickness in the direction of the light and actually help guide photons to their destinations. [(b) From Polyak, 1957. ©1957 by University of Chicago.]

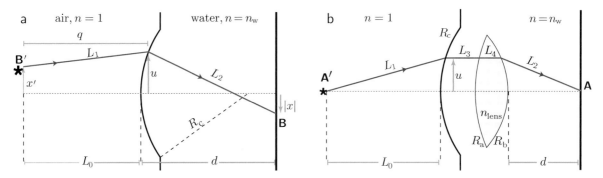

Figure 6.7: [Path diagrams.] **Curved air-water interface.** (a) Cross section of a circular interface between air and another medium, with radius of curvature R_c. This simplified optical system could represent the eye of a human whose eye lens has been removed by cataract surgery. It is also relevant for the scorpion, whose lens extends all the way back to its retina, if we understand the medium on the right to be that lens. The diagram is exaggerated; the analysis in the text actually assumes that x and x' are much smaller than L_0, d, or the radii of curvature. (b) A compound lens system consisting of an air-water interface followed by a lens. This system is similar to the normal vertebrate eye. Section 6.4.2 considers only paths like the one shown: They only bend at the first and last interfaces (thin lens approximation).

Figure 6.7a shows a curved boundary between two media with different refractive indices, in a notation familiar from previous figures. We can now repeat the analysis of Section 6.3.3 for this situation. To keep the formulas manageable, we again suppose that the displacements x', x, and u are all much smaller than the scale of the apparatus (L_0, R_c, and d), so that we can keep only low-order terms as before.[11] The key difference from previous sections is that the horizontal separation between the source and the kink in the path now depends on u, due to the curvature of the interface. This distance is

$$q = L_0 + R_c - \sqrt{R_c{}^2 - u^2} \approx L_0 + \frac{1}{2R_c}u^2. \tag{6.9}$$

Thus, the first part of the path shown has length

$$
\begin{aligned}
L_1 &= \sqrt{q^2 + (x' - u)^2} \approx \sqrt{L_0{}^2 + L_0 u^2/R_c + (x')^2 - 2ux' + u^2} \\
&\approx L_0 + \frac{1}{2L_0}\left((x')^2 - 2ux' + u^2\left(1 + \frac{L_0}{R_c}\right)\right).
\end{aligned} \tag{6.10}
$$

A similar derivation gives

$$L_2 \approx d + \frac{1}{2d}\left(x^2 - 2ux + u^2\left(1 - \frac{d}{R_c}\right)\right). \tag{6.11}$$

The total phase of the path shown is then $2\pi\nu/c$ times

$$L_1 + n_{\mathrm{w}}L_2 \approx \mathrm{const} - u\left(\frac{x'}{L_0} + \frac{n_{\mathrm{w}}x}{d}\right) + \frac{1}{2}u^2\left(\frac{1}{L_0} + \frac{1}{R_c} - \frac{n_{\mathrm{w}}}{R_c} + \frac{n_{\mathrm{w}}}{d}\right).$$

We can now interpret this result as in Section 6.3.3. In order to obtain a bright, focused image of **B′** at **B**, the phases of all the paths must be equal, at least up to second

---

[11]You'll do a more complete calculation in Problem 6.8.

order in powers of u. This means that the coefficients of u and u^2 must both equal zero, that is,

$$x = -x'd/(n_\mathrm{w} L_0) \tag{6.12}$$

and

$$\frac{1}{L_0} + \frac{n_\mathrm{w}}{d} = \frac{1}{f}, \text{ where } \quad f = \frac{R_c}{n_\mathrm{w} - 1}. \quad \text{lens formula for an air-water interface} \tag{6.13}$$

Just as in the single-lens system (Idea 6.8), a curved interface creates an inverted image at a distance d that is determined by a modified version of the lens formula (compare Equation 6.6). The modified lens formula again involves a quantity f that characterizes the interface, and we again call it the system's focal length. The image is also scaled, but this time by the factor in Equation 6.12.

Your Turn 6D

a. In the central part of your vision, the shape of your cornea roughly matches that of a sphere with radius of curvature $R_c = 7.8\,\mathrm{mm}$. The aqueous and vitreous humors are transparent media with indices of refraction similar to water, 1.34. Calculate the focal length, ignoring the eye lens.

b. Patients whose eye lens has been removed by cataract surgery cannot focus on even distant objects without supplementary lenses. Comment in the light of your result in (a).

Equation 6.13 shows that the focal length becomes longer (the focusing gets "weaker") if the difference in refraction indices is small, or if the curvature is small. In the limits $n \to 1$ or $R_c \to \infty$, there is no focusing at all. You may be familiar with an analogous fact: When you open your eyes underwater, your cornea becomes a water-water interface, so n is the same on each side, and everything looks very blurry, due to the loss of the cornea's focusing power. Animals that must see in *both* environments, such as seals, have flatter corneas than either land or sea creatures; the resulting large value of R_c reduces the focusing power of the interface in any medium, and hence the difference between air and water.[12]

Your Turn 6E

You can correct your underwater vision by wearing swim goggles. Why does that help? After all—you're still under water.

6.4.2 Focusing powers add in a compound lens system

Fish and octopus have no air-water interface, so they need some other kind of focusing element to create images on their retinas. The solution is an eye lens, made of a transparent protein mixture with index of refraction higher than that of water.[13] In evolutionary history, when aquatic animals moved to land, their eyes changed to adapt to the new focusing element at the cornea; yet they retained the eye lens. Thus,

---

[12]The surface-dwelling fish *Anableps* has evolved an even more remarkable adaptation in response to this problem: natural bifocals!

[13]Other creatures have evolved different focusing elements; see Problem 6.3.

to understand our own eyes, we need to consider an imaging system that consists of multiple elements. Figure 6.7b shows the situation schematically. A light path emerges from a source and encounters three interfaces: air-water, water-lens, and lens-water, each with its own radius of curvature (R_c, R_a, and R_b, respectively).

The mathematics of compound lens systems can get complicated, but a simplifying approximation is often justified: We will assume that all three of the interfaces are very close to one another (the **thin-lens approximation**). In such a situation, we may ignore the possibility of kinks in the intermediate parts of the photon path.[14] That is, we will consider only paths that include a single straight, horizontal segment between the first and third interfaces, as shown in Figure 6.7b. Such a path can be fully characterized by giving its distance u from the centerline. We will also simplify by considering only the case where the source $\mathbf{A'}$ and detector $\mathbf{A}$ are both on the centerline; one can generalize to other nearby points by adapting the analysis of Section 6.3.3.

Again, the condition for focus is that the total phase must be independent of u up to second order in the small quantities. You'll show in Problem 6.6 that this condition has the same general form as previous formulas (for example, Equation 6.13), but with focusing power given by[15]

$$\frac{1}{f} = \frac{1}{R_\mathrm{c}}(n_\mathrm{w} - 1) + \left(\frac{1}{R_\mathrm{a}} + \frac{1}{R_\mathrm{b}}\right)(n_\mathrm{lens} - n_\mathrm{w}). \quad \begin{array}{l}\text{compound lens system}\\\text{in Figure 6.7b}\end{array} \qquad (6.14)$$

The contributions to the focusing power from all three of the interfaces *add,* because $n_\mathrm{w} > 1$ and $n_\mathrm{lens} > n_\mathrm{w}$.

Some numbers

The human eye lens is not a uniform material; its index of refraction varies from the middle to the edges.[16] Nevertheless, the lens is sometimes approximated as a uniform material with $n_\mathrm{lens} = 1.42$. In a relaxed eye, the lens's surfaces are approximately spherical, with curvature $1/R_\mathrm{a} = 1/(10\,\mathrm{mm})$ at the front surface and $1/R_\mathrm{b} = 1/(6\,\mathrm{mm})$ at the back. Your Turn 6D gave $R_\mathrm{c} = 7.8\,\mathrm{mm}$. The distance from iris to retina is about $20\,\mathrm{mm}$.

Other transparent media in the eye include the aqueous and vitreous humors (see Figure 6.6a), which we'll model as uniform media with $n = 1.34$.

Your Turn 6F

a. Find the focal length in the thin-lens approximation.
b. Suppose that we look at a distant object, located at $L_0 \approx \infty$. Compute d by using Equation 6.14, and comment.

6.4.3 A deformable lens implements focal accommodation

One conclusion from Your Turn 6F is that we obtain most of the focusing power we need from the contribution of the air-water interface. The human eye lens's contribution is

---

[14] $\boxed{T2}$ You'll explore the validity of this claim in Problem 6.13.
[15] Some authors use a convention in which some of the R's are taken to be negative numbers. Our values are all true (positive) radii of curvature; the appropriate minus signs appear explicitly in Equation 6.14.
[16] See Section 6.5.3 (page 225).

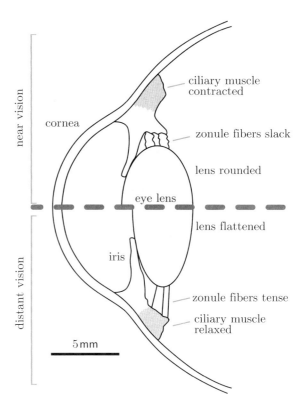

near vision

distant vision

cornea

ciliary muscle contracted

zonule fibers slack

lens rounded

eye lens

lens flattened

iris

zonule fibers tense

ciliary muscle relaxed

5 mm

Figure 6.8: [Anatomical sketch.] **Accommodation in the human eye.** *Top:* For near vision, a ring of muscle surrounding the eye lens contracts, releases tension on the zonule fibers, and hence allows the lens to adopt its preferred round shape. *Bottom:* For distance vision, the muscle relaxes and so moves radially outward; the zonule fibers then pull the lens into a flatter shape.

numerically smaller; however, we will now see that it has the crucial feature of being *adjustable.*

Although our relaxed eye is arranged to focus approximately on objects at infinity, we also need to focus on near objects. The lens formula, together with Equation 6.14, imply that a change in L_0 must be accompanied by compensating changes in either d, the R_i, or some combination of these. Indeed,

- Most fish have fixed lenses; they focus by adjusting the distance d, as do microscopes and many cameras.
- Humans instead adjust the curvature of their lens.[17]
- Some birds, such as eagles, can also adjust the curvature of their cornea.

Figure 6.8 depicts the human lens system in some detail, and shows the mechanism by which it **accommodates** to focus on near objects. The lens is an elastic body. It is held in place by filaments called **zonule fibers**, which attach to a surrounding ring of **ciliary muscle**. When that muscle is relaxed, the zonule fibers are tense, keeping the lens stretched radially, flattening it slightly, and giving rise to the values of R_a and R_b used in Section 6.4.2. When the ring of muscle tenses, its inner diameter decreases. The zonule fibers then go slack, and the eye lens adopts an unstretched, rounder shape, increasing its curvature (that is, *de*creasing its radii of curvature).

Very little deformation of the lens is needed for accommodation:

---

[17]The ubiquitous Thomas Young offered this explanation in 1793, even before his work on color vision. Prior to that point, others had believed that our eyes changed their overall shape.

Your Turn 6G

Suppose that both R_a and R_b decrease by the same, small fractional change ϵ. Using the numerical values previously given, calculate the value of ϵ needed to focus on an object at distance $L_0 = 25\,\text{cm}$, and comment.

Failures of accommodation

In a normal human child, the eye lens is arranged so that the combined system can focus objects on the retina from $L_0 \approx 5\,\text{cm}$ to infinity. For various reasons, however, some people cannot focus at infinity: They are "nearsighted." For example, their eye's overall shape may be such that the distance from lens to retina is different from normal (**axial myopia**). Or the cornea and/or lens may have surfaces with abnormal curvature, or one or more of the ocular media may have abnormal index of refraction (**refractive myopia**). Finally, after childhood the eye lens slowly begins to lose its elastic deformability, so that even when stretched it is not as flat as before (eventually leading to **presbyopia**, or age-related failure of accommodation). Each of these causes can give rise to images focused at a plane in front of the actual retina even when the eye lens is maximally stretched. The simplest treatment for myopia is to place another lens in front of the eye (eyeglasses or contact lenses), with a slight negative focusing power.

Conversely, some other people cannot focus on close objects: They are "farsighted." A similar set of causes as the ones just listed, but with opposite sign, can cause farsightedness (**hyperopia**). Again, the simplest treatment is to place an extra lens in front of the eye with *positive* focusing power. It is even normal for older subjects simultaneously to lose both ends of their focusing range; for them, Benjamin Franklin invented eyeglasses with two different sets of lenses (bifocals).

Section 6.4 has extended the idea of a lens-based focusing system to the sorts of arrangements actually invented by evolution (and later by scientists). Next, we will examine some extensions of this idea.

6.5 LIGHT MICROSCOPES AND RELATED INSTRUMENTS

6.5.1 "Rays of light" are a useful idealization in the ray-optics regime

So far, we have studied image formation from first principles, going all the way back to the Light Hypothesis and summing over paths. This approach quickly gets cumbersome, however, when we want to study a multi-element lens system, particularly when the thin-lens approximation is not applicable. A simpler, summary approach is needed.

Elementary physics texts generally present diagrams of a different sort from those used so far in this chapter: Instead of considering *all* possible photon paths, those diagrams draw special paths called "rays," such as those shown in Figure 6.9.[18] Each such line is a stationary-phase path with the given starting and ending points. To understand their significance, imagine inserting an opaque barrier just to the left of the leftmost lens in the figure.[19] The barrier has a hole that is small compared to the overall apparatus, though not so small as to create significant diffraction. If we now fill the spaces between the lenses with smoke or mist, a thin path in space will be illuminated. We think of this glowing path as a bundle of "rays." Moving the pinhole around lets us map out all the rays in the problem.

---

[18] Actually, Figure 6.1 was also a ray diagram.
[19] We imagined a similar procedure when we discussed refraction in Section 5.3.3.

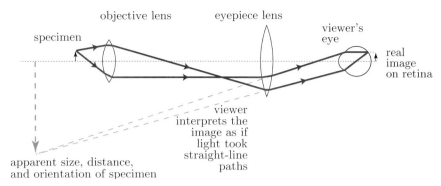

Figure 6.9: [Ray diagram.] **A basic microscope.** Two representative light rays are shown (*solid lines*); *dashed lines* are the viewer's interpretation of what she sees. The objective lens (*left*) forms an image of the specimen in a plane to the right of the objective (the "real intermediate image"). Instead of placing a viewing screen at this location, however, a second lens is added (the "eyepiece"), which bends the light back to a viewer's eye (or camera). The viewer interprets the resulting scene as arising from a magnified, inverted object, typically 25 cm in front of the eye (the "virtual image," *far left*).

A **ray diagram** is a sketch of an apparatus in which only rays of this sort are shown. They travel in straight lines and reflect or refract following the appropriate laws that we have previously found. The ray picture neglects diffraction. Nevertheless, it does help us to visualize roughly how light behaves in compound lens systems (Figure 6.9), and in other more complex contexts than those considered so far.

6.5.2 Real and virtual images

Conventional microscopy, as developed by Robert Hooke in the 17th century, involves illuminating a specimen, collecting the scattered light, and forming a magnified image. In the apparatus sketched in Figure 6.9, each ray bends at each air-glass interface according to the law of refraction. If there is a point at which many rays from the same part of the specimen reconverge, we say that the rays "focus" there. If we find a plane consisting of focal points for each point in a thin specimen (between the two lenses in the figure), we say that a **real image** is formed there. Up to now, this chapter has dealt exclusively with the formation of real images.[20]

Ultimately, a microscope must form a real image, either in our eye or in a camera. However, there is a second useful notion of image, distinct from this one. When we look through the compound lens system in Figure 6.9, our eye accommodates to bring the real image into focus. Because our brain expects light to travel on straight lines, it interprets the image as having come from an object located at a distance that *would have been* in focus with that much accommodation, had there been no microscope. The imagined object at this location is called the **virtual image** of the specimen.

6.5.3 Spherical aberration

We have seen that the key to focusing light is to introduce an element that adds a time delay to each photon path that passes through it. Section 6.3.1 mentioned that one way to accomplish this is via a curved object, of uniform index of refraction different

---

[20]For example, Section 6.3 (page 211) found in paraxial approximation that points at the distance d determined by the lens equation were in focus.

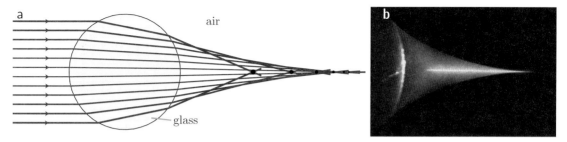

Figure 6.10: **Spherical aberration.** (a) [Ray diagram.] Parallel rays arriving at a lens with spherical surfaces, and passing close to its center (*thinner lines*), nearly coincide at a common focus (*smallest dot at far right*). However, rays initially farther from the axis (*heavier lines*) cross it in a spread-out array (*larger dots*). The rays shown were computed by using the law of refraction (Equation 5.10, page 193), for the case of a glass lens immersed in water. (b) [Photograph.] The spread-out focus is visible as the bright line in this photo. [(b) From Cagnet et al., 1962.]

from the surrounding medium, for example, an air-water interface or glass lens. But our analysis was restricted to light paths that stay close to the centerline of such an element. For example, at a curved interface we assumed that the distance u was much smaller than the radius of curvature, so that terms of order u^4/R^3 and higher were negligible compared to those of order u^2/R in Equation 6.9 (page 220) and elsewhere. Section 6.3.4 pointed out that more generally, terms like these can lead to aberrations.

If we wish to view a very dim sample in a microscope (perhaps a single fluorophore), then we need to collect a large fraction of the light it gives off. For that, we need our aperture to intercept a large angular range. A more detailed analysis is needed for this non-paraxial situation. For example, the ideal shape of a lens is not circular (in two dimensions) or spherical (in three dimensions), as we have been depicting lenses in the figures up to this point. Figure 6.10 shows the problem, for the case of a uniform, spherical lens. Light rays passing nearly through the center do focus to a point; however, more peripheral rays do not. Extending them in the figure beyond the point where they cross the centerline shows that there is *no* focal plane at which they all arrive at one point, a form of aberration called **spherical aberration**.[21]

Gradient-index lenses partially compensate for aberration in animal eyes

To address this aberration, some optical instruments have glass lenses with shapes that are more exotic than spherical. But animals long ago invented another approach: lenses made from a material whose index of refraction is not uniform.[22] For example, our eye lens's index of refraction varies continuously from 1.43 at the center to 1.32 at the edge. Figure 6.11 shows the improvement in spherical aberration that is possible with an appropriately graded index. Technology has caught up to Nature: Manmade **gradient-index lenses** can now focus light, despite in some cases being *flat*.

Land-dwelling animals further compensate for lens aberration by creating suitably nonspherical shapes for their cornea and eye lens. Despite all of these adaptations, however, optical aberration in our eyes still becomes significant when our pupils are fully open (see Figure 6.12).

---

[21]Spherical aberration is yet another phenomenon described by al-Haitham in the 11th century. In Problem 6.8, you'll study it without making the ray approximation.

[22] $\boxed{T_2}$ You'll explore gradient-index materials in Problems 5.6 and 6.14.

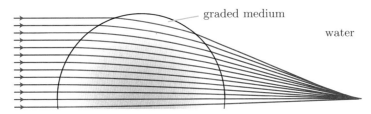

graded medium

water

Figure 6.11: [Ray diagram.] **Computed correction for spherical aberration by a continuously graded index of refraction.** A set of parallel incoming rays is shown, similarly to Figure 6.10a. However, in this case the rays curve inside the lens, because its index of refraction is greater in the center than at the periphery. The extra bending has been arranged to make all the rays nearly meet at a common focus. (Problem 6.14 describes the index function that was used to make this diagram.)

Figure 6.12: [Simulations based on experimental data.] **Point spread function of a normal human eye,** for several different pupil sizes. At 1 mm aperture diameter, diffraction spreads the spot of light; beyond 3 mm, aberrations become prominent. To create these simulated patterns, the subject's pupil was fully dilated and its aberration function was measured once; then, various pupil functions were applied to predict mathematically the illumination patterns on the retina. (Actually, the human pupil can only contract down to about 2 mm.) [Courtesy Austin Roorda.]

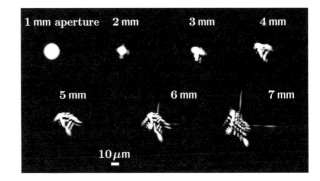

6.5.4   Dispersion gives rise to chromatic aberration

Section 5.3.5 (page 196) mentioned another phenomenon that limits image formation: The index of refraction of a medium, and hence the focusing power of a lens, in general depend on the frequency (or vacuum wavelength) of the light sent through it. Thus, the different colors in a visual scene will all have slightly different planes of focus. This chromatic aberration effect depends on the material composing the lens. In lab or photographic optical systems, it can be compensated by arranging two successive lenses, made from materials with opposite signs of the effect. Some species of animal eyes are also known to compensate for chromatic aberration.

The human visual system illustrates another approach to chromatic aberration. Our L- and M-type cone cells have peak sensitivities that are rather close to each other, but farther from that of the S cells.[23] Thus, we can simultaneously focus light imaged by the L and M cells, or that imaged by the S cells, but not all three at once. Accordingly, the central part of the fovea, where our visual acuity is greatest, is insensitive to blue light; it has only L and M cells. Moreover, the entire fovea has a filter, the **macula** or "yellow spot," that actually *removes* blue. Even outside the fovea, S cells are more coarsely spaced than the red and green ones (Figure 3.14). This disparity suggests that, although it's advantageous to have some S cones, there's no point in having too many: The blue light won't be well focused anyway when the red and green are.

Our brains unconsciously reconstruct blue in our central vision by constantly

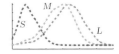

Fig. 3.11 (page 123)

Fig. 3.14 (page 130)

---

[23]Chapter 3 introduced cone cells; Figure 3.11 shows their spectral sensitivity functions.

Figure 6.13: [Ray diagrams.] **Epifluorescence versus confocal microscopy.** (a) Review of epifluorescence microscopy. The diagram has been simplified from Figure 2.4 (page 65), but here some more realistic light rays are shown. Illumination is sent in, with its rays converging in such a way as to flood the entire specimen; the resulting fluorescence light is then focused onto a camera. (b) Confocal fluorescence microscopy. This time, the illuminating light is arranged to focus to a spot, leaving most of the sample unilluminated. This spot lies in the focal plane of the imaging optics. Objects that are near to the illuminated spot, but at slightly the wrong depth, still receive some exciting light, but their fluorescence arrives at the detector out of focus. Fluorescence from objects that are near to the illuminated spot, but slightly offset in the transverse direction (*not shown*), arrives at the detector in focus but displaced, as in any lens system. Both of these unwanted light sources are rejected by a pinhole in front of the detector, which admits all light emitted from the desired region while blocking most of the light from nearby points.

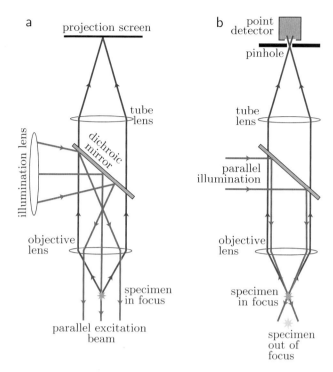

shifting our gaze, even when we think we are looking straight at one object; thus, we momentarily present that object to other parts of the retina, that do have S cells. One moment we are gathering high-resolution spatial information with our blue-blind central fovea; the next moment, we are gathering more complete color information with other parts of the fovea, and similarly with every other part of the scene. The brain assembles all these fragmented, rapidly changing messages into a perception that appears to us to be steady and complete.

6.5.5 Confocal microscopy suppresses out-of-focus background light

Earlier chapters have pointed out that much of the difficulty in microscopy comes from the need *not* to see the many unwanted, distracting objects present in a specimen.[24] Our own eyes use yet another scheme: We accommodate to bring only one range of distances into focus at a time, deemphasizing objects at other distances. (For example, when you look out through a screened window, you are generally unaware of the screen, even though you could easily focus on it if you chose.) Microscopes, too, focus at only one narrow range of depths in a sample.

Even out-of-focus objects still degrade an image, however, by contributing diffuse light, reducing image contrast. We have discussed total internal reflection fluorescence microscopy, which cuts down on this unwanted light by illuminating only selected regions.[25] But TIRF has a big limitation: It only sees objects very close to the coverslip, that is, to the boundary of a specimen. Another class of methods, generically called

---

[24]Section 2.3.2 (page 64) motivated fluorescence microscopy starting from this problem.
[25]Section 5.3.4 (page 194) introduced TIRF microscopy.

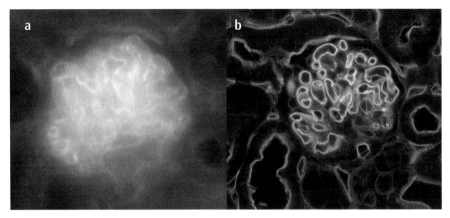

Figure 6.14: [Micrographs.] **Advantage of confocal microscopy.** Images of a $50 \, \mu m$ thick section of mouse kidney, taken using conventional epifluorescence microscopy (*left*) and confocal microscopy (*right*). The defocused light present in the conventional image reduces contrast and obscures details that are clear in the single cross section captured by the confocal microscope. [Courtesy Luke Fritzky; see also Fritzky & Lagunoff, 2013.]

confocal fluorescence microscopy, was invented earlier and has become a standard tool for imaging within thick samples.[26]

Figure 6.13b and its caption describe how confocal microscopy views a single point of a sample, rejecting light from other locations. The first step is to focus the excitation light down to only one spot. This technique is good at rejecting objects displaced transversely to the axis, but the excitation light must still pass through unwanted parts of the specimen on its way into, and beyond, the focused spot.[27] To reject fluorescence from those regions, we now make use of the fact that they will be out of focus. Placing a narrow aperture in the detector plane will then reject most of the light from the unwanted objects, while passing nearly all of the light from the in-focus region of interest. In order to form an entire image (Figure 6.14), the specimen can be moved, placing each volume element in turn into the focal region; other implementations instead sweep the illumination/imaging spot through a stationary sample.

Fig. 2.15 (page 80)

The confocal technique is not restricted to fluorescence mode; it can also be used in ordinary microscopy. Either way, it uses a single microscope objective lens both to focus the illuminating beam, and to image the returning light. Because it examines only a single spot at a time, we can use an inexpensive point light detector instead of an array of detectors to acquire the data. But for the same reason, it is slower than techniques that image an entire focal plane of the specimen at once.[28]

Section 6.5 has introduced ray diagrams as a conceptual aid to understanding optical systems (for example, confocal microscopes) and their aberrations.

---

[26]Two-photon excitation also allows deep imaging (Section 2.7, page 79), but the confocal approach described here is technically easier for some applications.

[27]See the lower beam in Figure 2.15 (page 80).

[28]Nor does the confocal method address the diffraction limit on resolution, introduced in Section 6.8 below.

6.6 DARWIN'S DIFFICULTY

We can perhaps imagine how a giraffe's long neck could have evolved gradually, with a sequence of small extensions each supplying enough fitness payoff to become fixed in the population. But the eye at first appears to be a very different case: It's an assembly of many precision parts, and it may seem hard to see how any of them would be of any use without the others. Darwin wrestled with this issue; in his words, "To suppose that the eye, with all its inimitable contrivances for adjusting the focus to different distances, for admitting different amounts of light, and for the correction of spherical and chromatic aberration, could have been formed by natural selection, seems, I freely confess, absurd in the highest possible degree."

Darwin continued, "Yet reason tells me, that if numerous gradations from a perfect and complex eye to one very imperfect and simple, each grade being useful to its possessor, can be shown to exist; if further, the eye does vary ever so slightly, and the variations be inherited, which is certainly the case; and if any variation or modification in the organ be ever useful to an animal under changing conditions of life, then the difficulty of believing that a perfect and complex eye could be formed by natural selection... can hardly be considered real." Although he had a very incomplete fossil record of our ancestors' eyes, nevertheless Darwin reasoned that we could examine a group of present-day species, in order to see what gradations are possible in viable organisms.

Figure 6.15 shows a series of light-sensing organs, all found in contemporary animals, in a drawing made shortly after Darwin's death. Each of these partial "eyes" confers enough fitness to enable an animal to prosper in its ecological niche; in fact, some are superior to more "advanced" eyes, for the specialized visual needs of their owners. Following Darwin's line of reasoning, then, each could be similar to one of the evolutionary steps to the full vertebrate eye.

What Darwin could not have guessed is that, despite the huge diversity of eye types, eye development always begins with the activation of a single gene that controls a genetic switch, and that is essentially the same throughout the animal kingdom. Incredibly, the version of this gene found in mammals, called *PAX6*, can be inserted into the genome of an insect, where it directs the growth of an *insect* eye, in any tissue where it is activated!

6.7 **BACKGROUND:** ANGLES AND ANGULAR AREA

The preceding sections have studied imaging systems using the ray picture. Section 6.8 will go beyond ray-optics approximation, uncovering a key limitation imposed on such systems by the physical character of light. First, however, this background section will review the measurement of angles.

6.7.1 Angles

A geometric **angle** is an example of a dimensionless measured quantity. For example, we can think of its value in terms of the fraction of a unit circle that the angle covers.

More commonly, we express an angle as the circumference of any circular arc spanning that angle (dimension $\mathbb{L}$), divided by its radius (dimension $\mathbb{L}$). When using that system, we often append the symbol rad (or the word "radians"), just to alert the reader that we're not using some other system. Similarly, when phrases like "cycles

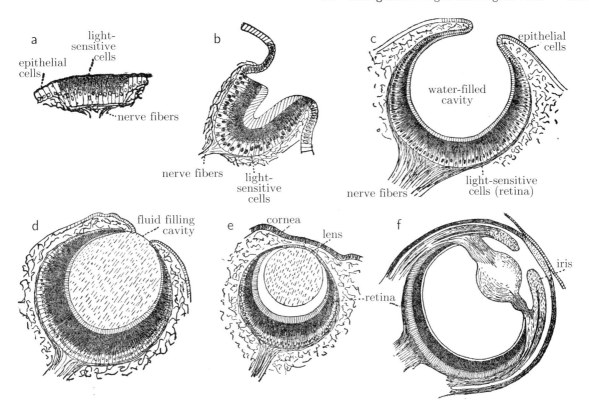

Figure 6.15: [Anatomical sketches.] **Light-sensing organs found in several marine animals.** Animals in this phy-
lum have differing needs for vision. The architectures are here arranged in a series of increasing complexity.
(a) Simple pigmented eyespot (starfish), consisting of a few light-sensitive cells (or even just one). (b) An
infolded pit lined with photoreceptors increases light collection (limpet, some crustaceans). (c) Pinhole eye
(*Nautilus*, abalone, ribbon worms). (d) Pinhole eye filled with secreted fluid (some giant clams). (e) A thin
film or transparent skin covers the entire eye apparatus, adding further protection. Also, some of the fluid
within the eye hardens into a convex lens that improves the focusing of light on the retina (marine snail *Litto-
rina*). (f) Full camera-type eye, with adjustable iris and focusing lens (squid). [From Conn, 1900.]

per second" or "revolutions per minute" are regarded as angular frequencies, we can
think of the words "cycles" and "revolutions" as dimensionless units (pure numbers),
both equal to 2π rad.

An older dimensionless unit of angle is called the "degree," defined as $1/360$ of
a full circle, or $(2\pi/360)$ rad. Related units that some authors consider convenient
for small angles include "arcminute" arcmin $= (1/60)$ deg, and "arcsecond" arcsec $=$
$(1/60)$ arcmin.

These Babylonian units (degree, arcminute, arcsecond) are sometimes abbreviated by a circle, a prime, and a double prime, respectively, as in GPS coordinates:
$42° 22' 42.29''$. This book will use the less confusing abbreviations deg, arcmin, and
arcsec in place of $°$, $'$, and $''$.

Physicists are more apt to express small angles in milliradians (mrad).

6.7.2 Angular area

Similarly, we can discuss **angular area** (which physicists often call by the confusing name "solid angle"). Suppose that some region in the sky is of interest, and we wish to find the angular area that it subtends (occupies). Imagine projecting it onto a sphere of radius r surrounding your head, and finding the area of the region of the sphere that its projection occupies. That area (with dimensions $\mathbb{L}^2$), divided by r^2 (with dimensions $\mathbb{L}^2$), is the angular area (dimensionless). The whole sky (a hemisphere) thus has angular area 2π. As with angles, we sometimes add the word **steradian** (abbreviated sr), just to indicate that we are using this system. Scientists who study vision often want to talk about very small fractions of a sphere, so they introduce other dimensionless units of angular area, for example, msr or $arcmin^2$ (see Problem 6.2).

6.8 DIFFRACTION LIMIT

We have seen at least three reasons why image formation by a lens system may not be perfect:

- The lens formula may not be satisfied. This problem can be corrected by "focusing," that is, by adjusting the lens shape or camera geometry (the distance d).
- Higher-order corrections to the phase, for example, the terms of order u^4 that we have been neglecting, may not be negligible, leading to aberration. This problem can sometimes be improved by making an arrangement where paraxial approximation holds,[29] and in other situations by using more complex lens systems designed to give such terms small coefficients.
- Different wavelengths may focus differently (chromatic aberration).

Even when these sources of focusing error are minimized, however, a fourth one remains: Diffraction, which we have so far neglected, also degrades focusing in an ordinary imaging system. We will now make that claim more precise, and at the same time gain insights into the design of natural imaging systems.

6.8.1 Even a perfect lens will not focus light perfectly

The analysis of pinhole imaging in Section 6.2.2 considered only the stationary-phase path from light source to observation point. But the stationary-phase approximation breaks down when the aperture is small, leading to diffractive effects.[30] A similar caveat applies to a lens system, and it imposes a fundamental limit to the sharpness of the resulting image.

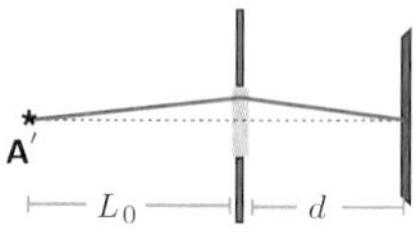

Let's return to the case of a single lens in air, with a point light source on the centerline (point **A**′, at position $x' = 0$, in Figure 6.3). We assume that the object distance L_0, and the projection screen distance d, are arranged to obey the lens formula. In this situation, all paths have the same phase when we look for light at the opposite

Fig. 6.3 (page 212)

---

[29]Section 6.3.1 (page 212) introduced this approximation.
[30]Section 4.7.3 (page 165) analyzed this phenomenon.

point **A**.[31] But is it really the case that there's *no* light *anywhere* else? Suppose that we move a tiny distance x off axis on the screen. Let's see how sharp the best-focused image of a point source can be, by asking how far away from $x = 0$ we can go on the screen before the total probability amplitude decreases significantly.

Equation 6.7 (page 217) gives the effective length for a photon path that leaves **A**$'$, passes through the aperture at height u, and arrives at screen position x: It's a constant minus ux/d. We now ask: If $x \neq 0$, will the corresponding phases sweep through at least a full circle? In that case, their contributions will interfere destructively, because $\int_0^{2\pi} \mathrm{d}\varphi\, \mathrm{e}^{\mathrm{i}\varphi} = 0$. Whether this happens or not depends on the width W of the aperture and lens, because W controls the range of u values. If the phases of paths change a lot over this range of u, then there will be cancellation and little light will be received at x. If on the contrary, all paths have nearly the same phase, then there will be reinforcement, and significant light will be received at x. Thus, we must compute

$$(2\pi\nu/c)\left[\frac{W}{2}\frac{x}{d} - \left(-\frac{W}{2}\frac{x}{d}\right)\right], \tag{6.15}$$

and determine whether it is much smaller than 2π. Rephrasing this result in terms of wavelength, destructive interference only begins to matter when $|x|$ exceeds x_{max}, where $x_{\mathrm{max}} \approx \lambda d/W$.

There's a useful restatement of the criterion just given. In the ray-optics approximation, light from **A**$'$ would all land on **A**. If any light lands elsewhere on the screen, we might misinterpret it as coming from some other point in the real scene, or as coming from an apparent angle $\tan^{-1}(x/d)$. We are assuming that all angles are small, so this angle is $\approx (x/d)\,\mathsf{rad}$.[32] We can therefore restate the preceding conclusion by saying that the apparent angular width of a point object is $\Delta\theta = 2\lambda/W$, or

> The apparent size of a point source of light corresponds to an angular width, in radians, of $\Delta\theta \approx 2\lambda/W$. (6.16)

Idea 6.16 implies that the blurring effect of diffraction gets more severe for longer wavelength light, or for narrower apertures.[33]

Extending this idea, suppose that a visual scene contains two or more bright objects on a dark background.[34]

Your Turn 6H

The two objects can't both be at the central point, so think about how the analysis leading to Equation 6.15 will change for a light source located a small but nonzero distance x' away from the axis. Show that the probability density for photon arrivals depends only on the distance between the screen observation point, x, and the location predicted in ray optics, $-x'd/L_0$.

Now suppose that there are many bright objects in the field of view. The "scene" consists of a probability distribution describing which source point will emit the next

---

[31]Section 6.3.1 (page 212) obtained this result, in the paraxial approximation.

[32]This follows from the Taylor series expansion of the tangent function (page 18).

[33]Why not just use a bigger aperture? In fact, enormous astronomical telescopes do just that. But there are practical drawbacks to this design, such as aberrations and the cost to a land- or air-based animal of having to carry heavy eyeballs. Marine animals are exempt from the latter cost; see Figure 9.1 (page 291).

[34]You'll see more explicitly what goes wrong when point sources are too close in Problem 6.9.

photon. Your result in Your Turn 6H implies that each detected photon will have a random displacement added to its idealized arrival point, which itself is a random variable. Thus, we can interpret this statement by saying that the resulting image will be the convolution of the scene with the imaging system's diffraction, which blurs each object.[35]

In short, if the actual distance between the objects gives them an angular separation less than about λ/W, then they will not appear separate in the image; we say that Idea 6.16 gives the **diffraction limit** for resolving distinct objects.

6.8.2 Three dimensions: The Rayleigh criterion

The analysis in the preceding section was appropriate for slit apertures. However, an analysis for a 3D lens with circular aperture of diameter W (and using a more precise definition of resolution) yields a similar result: The minimum resolvable separation of two independent point sources corresponds to the angular separation

$$\Delta\theta \approx 1.2\lambda/W. \tag{6.17}$$

Your Turn 6I

Consider a lens of diameter $2\,\mathrm{mm}$, arranged to focus on objects $10\,\mathrm{cm}$ away.
a. What's the minimum resolvable angular separation for light of wavelength $500\,\mathrm{nm}$? Express your answer in arcsec.
b. Suppose that the imaging screen is $5\,\mathrm{cm}$ away from the lens. What's the size on the screen corresponding to the diffraction limit you found in (a)?
c. What would be the corresponding spot size with *no* lens (pinhole camera), in the ray-optics approximation?

The criterion for resolving point sources can be generalized to handle situations where

- Some transparent medium other than air fills the space between the sample and the lens, and/or
- The aperture is not much smaller than the distance to the sample.

In such cases, we let n be the index of refraction of the medium and α_{max} be the angular radius of the aperture as viewed from the sample; then we define the **numerical aperture** as the quantity $\mathrm{NA} = n\sin\alpha_{\mathrm{max}}$. Two points in the sample can be resolved if they are separated by at least $\Delta x'$, where

$$\boxed{\Delta x' = 1.2\lambda/(2\mathrm{NA}).\quad \text{Rayleigh criterion}} \tag{6.18}$$

For the case of a narrow aperture, of width $W \ll L_0$, $\alpha_{\mathrm{max}} \approx W/(2L_0)$, and Equation 6.18 reduces to the previous form.

The Rayleigh criterion suggests that, for highest resolving power, the lens must be wide (large W) and close to the sample (small d). But this condition requires that light must emerge from the sample at a very large angle away from the centerline. At

---

[35]Section 0.5.5 (page 16) introduced convolution.

Figure 6.16: [Ray diagram.] **Oil-immersion objective.** A layer of oil between a microscope lens and its specimen increases the angular extent over which the lens can accept light. *Top:* For dry lenses, the numerical aperture is limited: Light emerging from the specimen at large angles never enters the lens, due to total internal reflection (*upper dashed path*). The practical limit for a dry lens corresponds to NA $\approx$ 0.95. *Bottom:* An oil immersion objective lens uses oil that matches the index of the glass coverslip ($n \approx 1.52$), eliminating the reflection. Such lenses can collect light from paths emerging at up to 67 deg (*lower dashed path*), which corresponds to NA = 1.4. Even higher values, up to 1.6, are possible with synthetic oils.

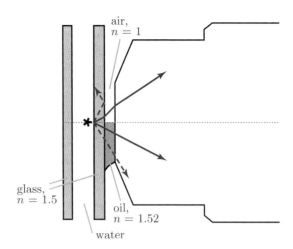

such large angles, total internal reflection can prevent light from escaping the sample (or its coverslip) at all.

Ex. Suggest a reason why some microscope objective lenses with high W/d are designed to be immersed in a drop of oil connecting them to the sample.

Solution: The oil ($n \approx 1.52$) matches the index of refraction of the glass coverslip better than does air. This in turn increases the critical angle beyond which we get total internal reflection, allowing more light to escape from the sample to fill the full lens aperture (see Figure 6.16). In addition, the factor of n in the denominator of the Rayleigh criterion implies an improvement in resolution if we use oil with $n > 1$.

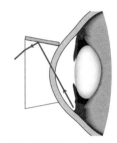

A similar idea underlies the gonioscope (Figure 5.10).

T2 *Section 6.8.2′ (page 239) discusses another resolution criterion that is more appropriate for some kinds of microscopy.*

Fig. 5.10 (page 194)

6.8.3 Animal eyes match their photoreceptor size to the diffraction limit

Human

The preceding section interpreted the diffraction limit in terms of an apparent displacement in the visual field that limits our visual acuity. An equivalent formulation of that result gives another insight. Diffraction spreads light from a point source on the projection screen (for example, our eye's retina), so there is little point in investing in a photoreceptor grid that is finer than that size.

In the central part of our visual field, angles are all small and we may approximate $\tan \theta \approx \theta$. Then Equation 6.17 states that two point sources whose images on the retina are closer than $1.2\lambda d/W$ will not appear distinct, regardless of how fine the photoreceptor grid may be, or how high the quality of the lens.

Your Turn 6J

a. In humans, the distance from iris to retina is about 20 mm. The diameter of the opening in our iris (the pupil) in dim light is about 8 mm. Find the diffraction limit on our ability to resolve two points that emit light with wavelength 500 nm. Convert it to an equivalent distance on the projection screen (retina), and compare with the actual diameter of cone cells in the human fovea (see Figure 3.9b).

b. In bright sunlight, the human pupil can contract down to 2 mm diameter. Repeat your estimate for this situation.

c. Translate your result to find the physical separation of two objects at arm's length, 50 cm from the eye, that could just barely be resolved in bright sunlight (if diffraction were the only limitation on visual acuity).

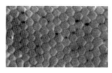

Fig. 3.9b (page 120)

Eagle

Birds of prey have evolved eyes with better acuity than our own. In part, this impressive performance reflects a larger pupil diameter in daylight, and hence a smaller angular size $\Delta\theta$ of their diffraction-limited spot. For example, the wedge-tailed eagle *Aquila audax* has a daylight pupil diameter of 5 mm, so the diffraction limit on its angular resolution $\Delta\theta$ is about 3/5 as large as humans'. The eagle's photoreceptor cells have finer spacing than our own, again roughly matching the diffraction limit in its main ecological niche (daylight hunting).

THE BIG PICTURE

We have seen how to understand the formation of images that are *bright yet sharp* in optical instruments such as the vertebrate eye, by starting from the Light Hypothesis. Although the analysis involved a fair amount of algebra, in every instance we were finding a path family with nearly constant phase. We also explored some ways in which image formation can be imperfect (aberrations). Going beyond ray-optics approximation, we then made quantitative estimates of diffractive spreading. Chapters 9–11 will continue the story by looking at what happens next when an image is formed on the array of photoreceptor cells in the eye's retina.

This chapter made little explicit mention of the lumpy character of light, but we have worked in a framework that accommodates that aspect. It was important to incorporate this much realism, because we will see that our rod cells are actually able to respond to a single photon absorption event. Moreover, the lessons we learned about diffraction also apply to the single-photon world, and set the stage for our discussion of localization microscopy in Chapter 7.

KEY FORMULAS

- *Lens in air:* We characterized a lens as a transparent optical element that imposes a delay $\Delta t(u)$ (extra transit time) on photon paths passing through it, depending on their distance u from its center. We considered only the small-angle case, where u is always much smaller than the size of the apparatus.
  We described the lens by writing Δt in the form $\left(\text{const} - u^2/(2fc) + \cdots\right)$, where f is a constant and the dots represent terms of higher order in u. Then we say

that the lens has focal length f. An object at distance L_0 will focus on a screen at distance d determined by the lens formula,

$$\frac{1}{L_0} + \frac{1}{d} = \frac{1}{f}.$$ [6.6, page 215]

We can adjust the distance to the objects that will be in focus, L_0, by changing either d (as in a camera, and the eyes of some fish) or f (as in human eyes). Although we only studied the two-dimensional case (cylindrical lens), the same formula also holds for the more familiar circular-symmetric lenses.

An image formed in this way is magnified or reduced by the factor d/L_0.

- *Air-water interface:* In this situation, the lens formula is modified to

$$\frac{1}{L_0} + \frac{n_w}{d} = \frac{1}{f} \quad , \text{ where } \quad f = \frac{R}{n_w - 1}$$ [6.13, page 221]

and R is the radius of curvature of the interface.

In this situation, an image formed on the water side is inverted and magnified or reduced by the factor

$$d/(n_w L_0).$$ [6.12, page 221]

- *Angles:* A degree of angle is defined as $1\,\mathsf{deg} = (\pi/180)\,\mathsf{rad}$. An arcminute ($\mathsf{arcmin}$) is $(1/60)\,\mathsf{deg}$. An arcsecond (arcsec) is $(1/3600)\,\mathsf{deg}$. Another small unit of angle is milliradian (mrad).
- *Angular area:* To find the angular area of a bundle of rays that all intersect at their end points, draw a spherical shell, centered on the common end point. The angular area in steradians (sr) is the ratio of the area spanned by the rays on the spherical shell to the square of the shell's radius. Thus, angular area, and the unit sr, are dimensionless.
- *Diffraction limit (Rayleigh criterion):* If light of wavelength λ emerges from a point source (zero size), passes through an aperture, and then gets focused onto a screen by a perfect lens obeying the above condition for focus, then the resulting image will nevertheless have a spread: It appears to be made by an object of nonzero angular size.

  Suppose that light passes through a circular aperture (hole) of diameter W, and is focused by a lens. Then in the small-angle (paraxial) approximation, the minimum resolvable angular separation of two independent point sources of light (for example, two fluorophores) is about $1.2\lambda/W$ radians. The smeared image of a point source is called the diffraction-limited spot, and its intensity profile is called the point spread function of the optical system.

FURTHER READING

Semipopular:
Saxby, 2002.
Optics of the human eye: McCall, 2010.
Evolution of eyes: Lane, 2009; Shubin, 2008; Carroll, 2006; Dawkins, 1996.

Intermediate:
Imaging and optics: Lipson et al., 2011; Hecht, 2002.
Animal eyes: Cronin et al., 2014; Land & Nilsson, 2012; Ahlborn, 2004.

Microscopy: Murphy & Davidson, 2013; Nadeau, 2012; Cox, 2012; Lanni & Keller, 2011; Mertz, 2010; Chandler & Roberson, 2009; `http://micro.magnet.fsu.edu/primer/` .

Confocal microscopy: Fine, 2011; Mertz, 2010; Pawley, 2006.

Other imaging methods not discussed in this book: Chandler & Roberson, 2009; Nolting, 2009.

Evolution of eyes: Zimmer & Emlen, 2013; Schwab, 2012.

Technical:

The Olympus Microscopy Resource Center: `http://www.olympusmicro.com/` .

Biological gradient-index lenses: Pierscionek, 2010.

Evolution of eyes: Lamb et al., 2007.

$\boxed{T_2}$ Track 2

6.4′ The retinal pigment epithelium

Figure 6.6 shows rod and cone cell outer segments (the parts that are actually sensitive to light) on the far right side, apparently blocked by many other things that light must traverse first. This may seem to be a perverse arrangement, and indeed squid and octopus have a camera-type eye much like ours (with lens, pupil, and so on), but with the rod cells inverted relative to ours. So our design choice was not the only possible way. But the vertebrate eye does place the outer segments in direct contact with the rear wall of the retina, whose **retinal pigment epithelium** (RPE) cells take up, recycle, and return used pigment molecules. The RPE cells also continually nibble off (phagocytose) the outermost (oldest) segments of the photoreceptor cells; new segments are generated from the other end, near the inner segment.[36]

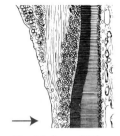

Fig. 6.6 (page 219)

Moreover, remarkably, the vertebrates' arrangement may actually have an optical advantage as well. The photoreceptor inner segments and other elements are transparent at visible wavelengths, so they do not appreciably steal light from the light-sensitive outer segments. And far from being in the way, these elements appear to function as light guides, funneling more photons into the outer segments than would otherwise be the case, and so improving the photon catch (Lakshminarayanan & Enoch, 2010; Franze et al., 2007).

$\boxed{T_2}$ Track 2

6.8.2′ The Abbe criterion

The discussion of the diffraction limit in the main text implicitly assumed that the two objects we wish to resolve are independent sources of light, like two stars in the sky, or two fluorophores in a specimen. Much of microscopy, however, is done in a situation where a *single* source of light illuminates an entire sample and we observe scattered or transmitted light. In this situation, two objects in the specimen will create two sets of paths from the source to the projection screen. As in our analysis of two-slit interference, the corresponding contributions to the probability amplitude must be added before computing $|\Psi|^2$, modifying calculations like the one in Section 6.8.1 and Problem 6.9.

E. Abbe studied the diffraction limit for such cases in 1873, using the wave model of light. He found that the attainable resolution depends not only on the nature of the sample and the optical system used to collect the light, but also on the character of the illumination. See Lipson et al., 2011, and Cox, 2012, for details on this resolution criterion, which in practice is often similar to the Rayleigh criterion: $\Delta x' = \lambda/(2\mathrm{NA})$.

---

[36]Section 10.7′c (page 348) describes the recycling scheme, or "retinoid cycle."

PROBLEMS

6.1 *Sundappled*

a. As you walk through a forest, sunlight passes through gaps between tree leaves, projecting small spots of light on the ground. When they land on a flat surface, the smaller spots are shaped as perfect ellipses, even though openings between leaves are not that shape. Explain.

b. When you make a 1 cm square hole in a piece of paper, hold it in front of a distant point source of light, and project to a screen a few centimeters away, you get a square spot of light. Why does this not contradict your answer to (a)?

6.2 *Angular areas*

When looking straight ahead, what is the angular area of the visual region that is within one degree of your line of sight (that is, your "central vision")? Express your answer in msr.

6.3 *Parabolic reflector*

Sea scallops of the genus *Pecten* use a focusing scheme quite different from our own (see Figure 6.17a). In this problem, you'll investigate focusing in a simplified version of the scallop's eye.

Monochromatic light of wavelength λ lands on a curved mirror shaped as shown in Figure 6.17b. The light comes from a point source that we will suppose is very far away on the horizontal dashed line. The figure shows the mirror's cross section. Work in two dimensions, as in the problems studied in the chapter.

The distance from the vertical dashed line to the mirror is au^2, where a is a constant and u is displacement from the dotted centerline. You might guess from

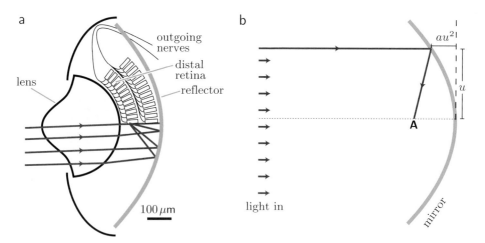

Figure 6.17: [Ray diagrams.] **Parabolic reflector imaging.** See Problem 6.3. (a) General layout of the sea scallop eye. The eye has a lens, but the lens has little focusing power; by itself, it would cast an image far behind the back of the eye. Instead, a curved reflector sends light back to a photoreceptor array located between it and the eye lens (the "distal retina"). For clarity, the photoreceptors are only shown in the top half of the figure, but they are present in the bottom half as well. A second layer of photoreceptors is shown (the "proximal retina"), but they do not receive a focused image. [Adapted from Land & Nilsson, 2006.]   (b) Cross section of a simplified optical system, with a reflector but no lens.

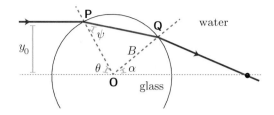

Figure 6.18: [Ray diagram.] **See Problem 6.7.**

symmetry that this mirror will focus the incoming light to a point somewhere on the centerline.

a. Find the location of that point and explain why light focuses there, using the methods developed in this chapter, but do not assume that u is small. [*Hint:* If there were no mirror, every incoming light path would have the same path length to arrive at the vertical dashed line. Draw a family of paths that each consist of two straight line segments, as shown. The first, horizontal segment ends on the mirror and is shorter than the distance to the vertical dashed line. The second segment connects to a point **A** on the dotted centerline. Adjust the location of that point until you get what you need for a focus.]

b. Suggest why a mirror with this shape (a parabola) might be superior to one with a circular cross section.

6.4 *Depth of focus*

a. Take a thin card and puncture it in the center with a thumbtack. Remove your eyeglasses, if you wear them, and hold the pinhole as close as possible to one eye (closing the other). Then look at a brightly lit scene and describe what you see.

b. Section 6.3.3 argued that, to obtain a focused image, it was necessary to satisfy the lens formula (Equation 6.6, page 215). Otherwise, we argued, light from a source point would not all arrive at a single image point. But what if the lens formula is *almost* satisfied, that is, $1/L_0 + 1/d$ is almost, but not quite, equal to $1/f$? Then the u^2 terms in Equation 6.7 (page 217) may still not matter, as long as u is constrained to very small excursions. Connect this idea to what you observed qualitatively in (a).

6.5 *Air-water interface*
Interpret Equation 6.12 (page 221) in the light of the law of refraction.

6.6 *Three curved interfaces*
Establish Equation 6.14 (page 222).

6.7 *Spherical aberration in ray optics*
In this problem, you'll make a figure like Figure 6.10a, but using index of refraction values appropriate for a human eye lens (index 1.42) in water. Use the ray-optics approximation for this problem.

Fig. 6.10a (page 226)

a. Consider a set of parallel incoming rays in a plane passing through the center of the sphere. An incoming ray arrives at distance y_0 from the centerline, as shown in Figure 6.18. Find the angle θ shown in terms of y_0 and the sphere radius B. This is also the ray's angle of incidence.

b. Use the law of refraction to find the angle ψ.

c. The triangle shown is isosceles. Use that fact to find the angle α, and hence the point **Q** of exit.

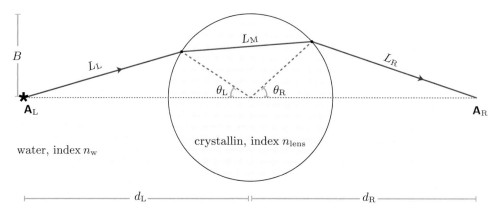

Figure 6.19: [Path diagram.] **See Problem 6.8.**

d. Use the law of refraction again to find the angle that the ray makes to the horizontal after exiting the glass. Then continue that ray until it hits the centerline.

e. Use a computer to draw the three segments of this ray, and repeat for other y_0 values. Comment on where the rays hit the centerline.

6.8   *Thick lens, including diffraction*

The lens in a fish eye is a roughly spherical ball of transparent proteins called **crystallins**.[37] In this problem, we will simplify by considering a lens that, unlike a real fish eye lens, has *uniform* index of refraction n_{lens}, larger than the index n_{w} of the surrounding water. We will also simplify by considering a two-dimensional geometry, as in other problems studied in the chapter.

Suppose that a point source of light is placed at $\mathbf{A}_{\text{L}}$ on the centerline (dotted line in Figure 6.19), at a distance d_{L} from the center of the lens. We would like to arrange for all the light that passes through the lens to arrive at another point $\mathbf{A}_{\text{R}}$, also on the centerline at distance d_{R}. Following Section 6.3, consider all light paths consisting of straight line segments in each region of uniform index. The figure shows that we can characterize each such path by the two angles θ_{L} and θ_{R}. The path that runs straight through the center of the sphere has $\theta_{\text{L}} = \theta_{\text{R}} = 0$.

a. To begin, consider the special case where $d_{\text{L}} = d_{\text{R}}$. In this case, the problem is symmetric under reflections exchanging left and right, so we might expect that the most important paths would be those with $\theta_{\text{L}} = \theta_{\text{R}}$. Assume that's true and write exact expressions for L_{L}, L_{M}, and L_{R} as functions of the common angle θ and distance d. [*Hint:* The law of cosines may be helpful (page 18).]

b. Use your answer in (a) to write an expression for the phase of the photon path as a function of θ. Then specialize your answer to the case of small θ, by writing a Taylor series expansion about $\theta = 0$.

c. What must the common distance d be in order for all the path phases to agree up to order θ^2? Interpret your result as a statement about the focal length of this lens system, using the lens formula (Equation 6.6, page 215).

d. The fish *Astatotilapia burtoni* has an eye lens with average index of refraction about 1.45. Evaluate your result from (c) for the focal length. Is it practical to

[37] Despite the name, the individual protein molecules are not arranged in a crystalline lattice; they are maintained in a disordered state, like window glass.

have an eye with the ratio of f to d that you found?

e. If the condition you found in (c) is met, will all the phases be *exactly* the same? Get a computer to make a graph supporting your answer, then interpret it.

f. How would your result change in a 3D geometry, with a spherical lens?

g. $\boxed{T_2}$ You may not be convinced that it's legitimate to consider only paths with $\theta_{\rm L} = \theta_{\rm R}$. So repeat your analysis, relaxing this condition, to find a phase function that depends on $\theta_{\rm L}$, $\theta_{\rm R}$, and the constant parameters d, B, $n_{\rm w}$, and $n_{\rm lens}$. Again expand this function in a Taylor series and truncate to a quadratic function of the two small angles, this time specified by a 2×2 matrix M that you are to find:

$$\text{const} + \tfrac{1}{2}[\theta_{\rm L}, \theta_{\rm R}]\mathsf{M}\begin{bmatrix}\theta_{\rm L}\\\theta_{\rm R}\end{bmatrix} + \cdots .$$

In order to have bright illumination at $\mathbf{A}_{\rm R}$, we require that one of the eigenvalues of this matrix be zero, so that a one-parameter family of paths will have nearly constant phase. Find the condition for this to be true, and compare it with your result in (c).

h. $\boxed{T_2}$ Finally, generalize your result in (g) to the case in which $d_{\rm L} \neq d_{\rm R}$. In this case, there is no reflection symmetry, so we cannot assume that the stationary-phase paths have $\theta_{\rm L} = \theta_{\rm R}$. Compare your result with the lens formula, where $d_{\rm L}$ was called L_0 and $d_{\rm R}$ was called d.

6.9   *Unresolvable point sources*

For this problem, work in two dimensions, as in other problems studied in the chapter. Section 6.8.1 argued qualitatively that two point sources cannot be resolved if they are too close together. See Figure 6.5.

Fig. 6.5 (page 217)

a. Specialize Equation 6.7 (page 217) to the case where a point source at $\mathbf{A}'$ is on the centerline and focused (that is, the lens formula is satisfied). Find the pattern of illumination on the screen (probability of photon arrival), assuming an aperture width W. Do the integral analytically, not numerically.

b. Graph the resulting function of x (you don't need to normalize it). [*Hint:* First define an appropriately rescaled, dimensionless position variable $\bar{x}$ on the screen.] Comment on the connection to Equation 6.16 (page 233).

c. Suppose that there is a second, independent source of light at $\mathbf{B}'$, at sideways distance $\Delta x' = L_0(\Delta\theta)$ away from $\mathbf{A}'$, where $\Delta\theta$ is given by Equation 6.16. Add the function you found in (b) to an appropriately shifted version of the same thing, make a graph of the resulting illumination pattern, and comment. Why add probabilities, not probability amplitudes?

d. Repeat for a separation of $0.42L_0(\Delta\theta)$ and comment.

6.10   *The fine print*

The goal of this problem is to estimate the smallest print that you can read from a distance of $30\,\text{cm}$, assuming that your visual acuity were limited only by diffraction. Take the diameter of your pupil to be $0.3\,\text{cm}$ and the wavelength of light to be $0.5\,\mu\text{m}$.

a. What is the approximate angular width of the diffraction-limited spot on your retina?

b. How far must two point sources, both $30\,\text{cm}$ from your eye, be separated in order for their respective images not to overlap?

c. How does your answer to (b) compare to the finest print that you are accustomed to reading? [*Note:* In typography, "n-point type" means that the letter "M" has width equal to $(n/72.27)$ inches, that is, n times about $0.35\,\text{mm}$.]

6.11   *Resolution*

Imagine a wire mesh (a piece of windowscreen), a grid of $0.5\,\text{mm}$ squares.

a. Suppose that a thin, circular lens of diameter $3\,\text{mm}$, at distance L from the mesh, focuses an image of the mesh onto a plane at distance $d = 24\,\text{mm}$ from the lens. What's the diffraction limit to the angular resolution for light of wavelength $500\,\text{nm}$? Express your answer in arcsec or in mrad.

b. You know from experience that when you observe such an object at a distance of $20\,\text{cm}$ you can see the squares, whereas at $5\,\text{m}$ it just looks gray. Comment quantitatively on this fact, in the light of your answer to (a). Specifically, use your answer to (a) to estimate the distance L at which the mesh crosses from resolvable to unresolvable.

6.12 $\boxed{T_2}$ *A more realistic point spread function*

The discussion of Section 6.3.3 made two approximations that limited the accuracy of our results. First, the various light paths from a source to the focus do not have exactly the same phase: The approximation of dropping terms of order u^3 and higher is not exact. Second, Section 6.8 (page 232) pointed out that the illumination on the projection screen will not be exactly zero away from the focus, because even though the phases partly cancel one another, that cancellation is not complete. Nevertheless, a large, high-quality lens really can concentrate quite a lot of sunlight into a small region.

In this problem, you'll address the shortcomings of the discussion in the main text by doing a numerical calculation. Idealize the Sun as a distant point source of light with wavelength $600\,\text{nm}$, and consider a two-dimensional world with a lens that modifies the phases of photon paths according to Equation 6.4 (page 214) with focal length $f = 2\,\text{cm}$ (see Figure 6.5). The lens focuses the source onto a screen.

Fig. 6.5 (page 217)

a. Perform a numerical integral over u in the range $|u| < 0.5\,\text{cm}$ to find the profile of light intensity on the projection screen located at a distance $d = 2\,\text{cm}$ from the lens. (We suppose that larger values of u are blocked by an opaque barrier.) Use the exact expression (involving square roots), not the approximate form (Equation 6.7). [*Hint:* The integral over u is complicated by the oscillatory character of the integrand. When performing integrals of this sort numerically, be sure to use a fine enough grid to get a stable answer.]

b. The Sun delivers roughly $1.4\,\text{kW/m}^2$ to Earth's surface. What is the corresponding intensity at the center of the illumination pattern you found in (a)? Can you imagine a three-dimensional version of this device being able to start a fire?

6.13 $\boxed{T_2}$ *Compound lens system*

Consider a compound lens system with two interfaces (Figure 6.20): Light travels through air, passes through a lens with index of refraction n_{lens}, then travels through water to a screen. (This system is a simplified version of the one in an animal eye, Figure 6.7b.) Each interface has a circular cross section, although their radii of curvature, R_{a} and R_{b}, may not be equal. In this problem, work in two dimensions, as was done in the main text.

Following ideas in the main text, characterize a light path by the distances from the axis, u_{L} and u_{R}, shown in the figure. Suppose that light comes in from a distant object $(L_0 \to \infty)$, takes the path shown, and arrives on a flat screen located a distance d from the inner surface.

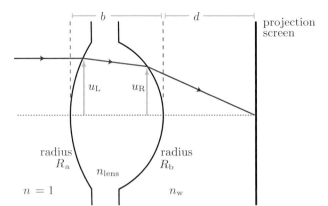

Figure 6.20: [Path diagram.] **See Problem 6.13.**

a. Find expressions for the lengths of the three path segments shown. Expand these in Taylor series in u_L and u_R, and keep only up to second order terms, as was done in the main text.

b. Find an expression for the total phase of this path, a quadratic function of u_L and u_R. Find a condition on d for there to be a constant-phase subfamily of paths, in terms of the constants b, R_a, R_b, and the indices n_lens and n_w. [*Hint:* See Problem 6.8.]

c. Specialize your result to the limiting case where b is much smaller than R_a and R_b (a thin-lens approximation). [*Hint:* This limit may seem paradoxical in the light of Figure 6.20, because as $b \to 0$ the lens thickness becomes negative in some parts of the diagram! But remember that we are also expanding in u_L and u_R, so we are only interested in a very small part of the figure near the centerline. Regardless how small we take b, there is always a narrow strip near the centerline where the lens thickness is positive.]

d. Comment on the claim made in Section 6.4.1 that, in the thin-lens approximation, we may take the middle path segment to be horizontal.

6.14 $\boxed{T_2}$ *Gradient-index eye lens*

Use the ray-optics approximation for this problem. If you haven't done Problems 5.6 and 6.7 yet, do them first. Problem 5.6 asked you to find light ray trajectories in a nonuniform medium whose index of refraction depends on only one Cartesian coordinate, the height. In the present problem, you'll generalize your results to a nonuniform medium (a "gradient-index lens") whose index of refraction depends only on *radius,* that is, the distance r to the center of the lens. Section 6.5.3 mentioned that this situation holds for the eye lenses of animals, and claimed that such nonuniformity can eliminate much of the aberration created by a uniform spherical lens (compare Figure 6.10a to Figure 6.11).

In this problem, you can scale all lengths by the radius a of the sphere, that is, work in terms of $\bar{r} = r/a$ and so on. Let $n_\mathrm{c} = n(0)$ be the index at the center, $n_\mathrm{p} = n(1)$ its value at the periphery, and $K = n_\mathrm{p}/n_\mathrm{c} - 1$. Fish eyes have $n_\mathrm{c} \approx 1.52$, $n_\mathrm{p} \approx 1.38$, and

Fig. 6.10a (page 226)

Fig. 6.11 (page 227)

$$n(\bar{r}) \approx n_\mathrm{c}\big(1 + K(0.82\bar{r}^2 + 0.30\bar{r}^6 - 0.12\bar{r}^8)\big),$$

and are immersed in media with $n_\mathrm{w} \approx 1.33$ on both sides. It will be convenient to define $g(\bar{r}) = n^{-1}(\mathrm{d}n/\mathrm{d}\bar{r})$.

a. Choose coordinates centered on the lens center, and a plane passing through that origin, say the xy plane. Write out both components of the ray equation (Equation 5.15, page 204), which determines the stationary-phase paths $\boldsymbol{\ell}(s)$. It's a pair of coupled, second-order ordinary differential equations in the two Cartesian coordinates of a path lying in the chosen plane, $\ell_x(s)$ and $\ell_y(s)$. Parameterize the curve by arclength s (Equation 5.14, page 204).

b. Now generate a picture similar to Figure 6.11, by constructing a series of solutions to the ray equation. Each ray initially starts outside the lens, traveling parallel to the x axis. Find the x and y values at which the incoming ray enters the lens, and the angle it makes relative to the perpendicular (the angle of incidence).

c. Use the law of refraction to find the tangent vector to the ray just *after* it enters the lens.

d. Use your results in (b,c) to get the required four initial conditions for the ray equation, then use a computer to solve it numerically.

e. Follow your solution to find the value $\bar{s}_{\text{exit}}$ at which $\bar{r}$ once again reaches the value 1.

f. The tangent vector $\mathrm{d}\bar{\boldsymbol{\ell}}/\mathrm{d}\bar{s}|_{s_{\text{exit}}}$ then tells you the angle of incidence as the ray crosses the lens→water interface. Use the law of refraction again to find its angle after it leaves the lens.

g. After leaving the lens, the ray is once again straight. Find the point where it hits the x axis, then have your computer draw all three segments (straight, curved, straight). Repeat for each ray that you wish to trace.

CHAPTER 7

Imaging as Inference

> You cannot depend on your eyes when your imagination is
> out of focus.
>
> — *Mark Twain*

7.1 SIGNPOST: *INFORMATION*

Chapter 6 described a fundamental limitation on direct image formation: Microscope images will be unavoidably blurred, due to diffraction. The Rayleigh criterion is one way to make this statement quantitative:[1] It says that two neighboring point sources of light, for example fluorescent molecules, will create images that are spread into overlapping blobs if they are closer than $1.2\lambda/(2\mathrm{NA})$. The numerical aperture, NA, is defined as the index of refraction of the medium times the sine of the angular radius of the aperture as viewed from the sources, so it cannot exceed $n\sin(\pi/2)$, or more precisely about 1.6 for an oil-immersion lens. Thus, the resolution limit for visible light, $\lambda \approx 500\,\mathrm{nm}$, is around $200\,\mathrm{nm}$, which is too big to resolve most of the biologically relevant structure inside cells.

For over a hundred years, the preceding paragraph was the last word on the subject, in part because light with wavelength shorter than visible is destructive to living cells. Investigations into molecular-scale structure mostly relied on electron microscopy (which also involves killing the cells), or even more unnatural procedures like crystallizing a protein of interest and subjecting it to x rays.[2] It seemed hopeless to watch the nanoscale choreography of a living cell.

The breakthrough to overcome the diffraction barrier came about gradually and involved many ideas and discoveries. One line of attack, now called localization microscopy, will be the main topic of this chapter.[3] Its pursuit culminated with three demonstrations in 2006. All three built on a much older program of reinterpreting imaging as a problem of optimally extracting *information* from data, so this chapter will begin with a review of some ideas from that discipline.

The Focus Question is

Biological question: If light is a wave, then how can you see objects smaller than its wavelength?

Physical idea: Light is *not* a wave. An image is really a probability density function for photon arrivals, and the center of a probability density function can be determined to much greater accuracy than its width, if you have enough samples.

---

[1]See Equation 6.18 (page 234).

[2]X-ray crystallography is still needed if we wish to resolve individual atoms (Chapter 8). Other indirect techniques include cryo-electron microscopy and nuclear magnetic resonance.

[3]Other superresolution methods, including stimulated emission depletion imaging (STED), involve different physical ideas.

7.2 **BACKGROUND:** ON INFERENCE

7.2.1 The Bayes formula tells how to update a probability estimate

This book's Prologue framed the concept of probability as an estimate of the degree to which we may believe a proposition, based on available evidence. In science, as in daily life, we constantly *update* those estimates, as more evidence becomes available. To see how to do this quantitatively, let's imagine a concrete situation that is relevant to imaging.

Suppose that we make repeated measurements of x, the apparent position of an object along some axis.[4] This position is a random variable, due to diffractive randomness in the arrival locations of photons on our camera's detector array. Suppose that we know x's probability density function, apart from its overall center; that is, we know that

$$\wp(x \mid x_\mathrm{t}) = f(x - x_\mathrm{t}), \tag{7.1}$$

where the point spread function f is known but the true position x_t is an unknown parameter.

The true position does not necessarily fluctuate like x. But still, it is subject to uncertainty. Thus, *it, too*, has a probability distribution. Initially, we know nothing about its value, other than that it lies within some range. Call that range from 0 to A; the constant A could correspond to the full field of view in our microscope. So prior to any observation, our best guess for the probability density function of x_t is that it is Uniform (constant over the given range). We call this distribution the **prior distribution** of x_t, often abbreviated as just "the prior," or $\wp(x_\mathrm{t})$.

Now we make one measurement, yielding an observed value x_1. It seems intuitively clear that now our best estimate of the center of the distribution is the value that we just observed. To justify and extend that intuition, we need to work out $\wp(x_\mathrm{t} \mid x_1)$, the probability distribution of x_t *given* the new information. Because this updated distribution only becomes available after a measurement, it's called the **posterior distribution** of x_t, often abbreviated as just "the posterior." The posterior distribution is a conditional probability, so we can express it by using the definition:

$$\wp(x_\mathrm{t} \mid x_1) = \frac{\wp(x_\mathrm{t} \text{ \textbf{and} } x_1)}{\wp(x_1)}. \qquad \text{[0.36, page 11]}$$

The numerator of this expression can be rewritten in terms of a different conditional probability: $\wp(x_1 \text{ \textbf{and} } x_\mathrm{t}) = \wp(x_1 \mid x_\mathrm{t})\wp(x_\mathrm{t})$. Combining the expressions gives

$$\boxed{\wp(x_\mathrm{t} \mid x_1) = \wp(x_1 \mid x_\mathrm{t})\frac{\wp(x_\mathrm{t})}{\wp(x_1)}. \quad \text{the \textbf{Bayes formula}}} \tag{7.2}$$

Equations 0.36 and 7.2 use a notation that, although widespread, can be confusing. It's important to understand that the second of these, for example, asserts a relation

---

[4]In preceding chapters, x referred to position on a detector plane and x' referred to position in the world. In this chapter, we will drop the prime because we won't need to write any formulas involving both quantities.

between *four different functions*. The generic symbol $\wp$ just means "probability density"; the symbols inside the parentheses specify *which* probability density is meant. Thus, Equation 7.2 expresses the posterior distribution (left side) as the product of the prior, $\wp(x_t)$, times two correction factors. The first of these, $\wp(x_1 \mid x_t)$, is called the **likelihood**. It was given to us; it's the function $f(x_1 - x_t)$. The other correction factor, $1/\wp(x_1)$, wasn't given, but it's independent of the quantity x_t that we are trying to characterize. We could obtain it by requiring that the expression in Equation 7.2 be a properly normalized probability density in its variable x_t, but we will see that often its value is not needed.

Discrete versus continuous

Equation 7.2 and its derivation assume that both x_1 and x_t are continuous variables. We can apply the same logic, however, if either or both is discrete. For example, suppose that the experimentally measured quantity K is discrete; then we have an analogous formula

$$\wp(x_t \mid K) = \mathcal{P}(K \mid x_t) \frac{\wp(x_t)}{\mathcal{P}(K)}. \tag{7.3}$$

7.2.2 Inference with a Uniform prior reduces to likelihood maximization

In the situation just outlined, all the physics we know about our system is encoded in Equation 7.1, which we'll call a **probabilistic model** for our measurement. We can now say that our best estimate of the object's true location, $x_{t,*}$, is the value of x_t that maximizes the posterior distribution (Equation 7.2), holding fixed the observed data x_1. That is, once our observation is done, its outcome x_1 is frozen while we entertain various hypotheses about the desired, but not directly observed, quantity x_t.

Equation 7.2 tells us the probabilities for various values of the unknown true position (x_t), given one measurement (x_1). The denominator of this formula is a constant, independent of x_t. Moreover, Section 7.2.1 pointed out that if we know nothing in advance about the value of x_t other than that it must lie in some range, then it may be reasonable to take the prior distribution, $\wp(x_t)$, to be Uniform, that is, a constant function. In that situation, everything in the Bayes formula is a constant in x_t except for the likelihood function, so finding the most probable value amounts to *maximizing the likelihood holding the experimental data fixed*. Suppose that f is known to be a simple bump function, for example, a Gaussian or Cauchy distribution centered on zero. Then the most probable value for x_t is just the value at which $f(x_1 - x_t)$ attains its maximum. That value is x_1, which agrees with our first intuition.

7.2.3 Inferring the center of a distribution

Inference via **likelihood maximization**, or "maximum likelihood estimation" (MLE), is a broadly applicable technique. For example, suppose that we made several independent measurements of x under identical conditions. The likelihood is then $\wp(x_1, \ldots, x_N \mid x_t)$, which is the product $f(x_1 - x_t) \cdots f(x_N - x_t)$. Substituting that expression into the right-hand side of Equation 7.2 gives us a more sharply peaked posterior distribution than the one with a single measurement. That sharpness means that we can have greater confidence in the estimate we obtain by maximizing it.

Your Turn 7A

Suppose specifically that we know that the function f is Gaussian, $f(x) = (2\pi\sigma^2)^{-1/2}e^{-x^2/(2\sigma^2)}$, with some known variance σ^2, and that every observed value x_i is far from the edges of the allowed region $[0, A]$. Find a formula for the value $x_{t,*}$ that maximizes the posterior distribution. Also find the variance of x_t, in terms of $x_1, \ldots, x_N$ and σ.

You may not find your result in Your Turn 7A to be very surprising. But you'll see in Problem 7.1 how well the procedure works in a more challenging situation.

Your Turn 7B

Why does the problem in Your Turn 7A get harder if some of the blip locations are close to the edge?

7.2.4 Parameter estimation and credible intervals

Maximizing the likelihood gives us the most probable value $x_{t,*}$ of an unknown parameter, but often we want to know more: We want to know *how good* that estimate is, or what *range* of values about $x_{t,*}$ probably contains the true value. If the posterior distribution is simple, as in Your Turn 7A, then it may suffice to state its variance. More generally, we may report a range of values around $x_{t,*}$ that contains, say, 95% of the area under the posterior distribution curve—a "95% credible interval" for the inferred value.

7.2.5 Binning data reduces its information content

Another approach to repeated measurements may be more familiar than the one suggested above. In this approach, we make a histogram of all the measured values of x, then find the peak and the width of that histogram. That approach may be the best we can do if we have no underlying probabilistic model, but it has drawbacks.

This book's Prologue defined a probability density function in principle as a limit obtained by making histograms of data frequencies with finer and finer divisions, $\Delta x \to 0$.[5] However, when we classify data into bins we destroy some of its information content: Each observed value of x may be measured to greater precision than Δx, but after binning, all we retain about that value is that it landed somewhere in its bin. Nor can we remove this problem by taking very small bins, because any real dataset contains only a finite number of measurements. If we take Δx to be too small, then the bin populations are small and therefore have large relative standard deviations.[6]

Likelihood maximization sidesteps this dilemma, because it doesn't require binning the data. If we have a probabilistic model with some unknown parameters, we just evaluate the likelihood function with the exact observed values and maximize it. Especially for small datasets, the resulting improvement in estimates can be significant.

This background section has proposed the posterior distribution as a summary of what we learn from new data, and the likelihood function as its surrogate in cases where we have no relevant prior information.

For further details on inference, see the references listed at the end of this chapter.

---

[5] Section 0.4.1 (page 10) introduced probability density functions.
[6] Section 0.5.3 discussed fluctuations in count data.

7.3 LOCALIZATION OF A SINGLE FLUOROPHORE

7.3.1 Localization is an inference problem

Section 7.1 stated that, to overcome the diffraction barrier, we must reimagine imaging as an inference problem: We observe the pattern of photon arrivals at a detector array, and wish to deduce what structures in the sample could have generated that pattern. When we look at a distant piece of paper using a telescope, the interpretation is straightforward: We are seeing a pattern of light that reproduces the structure of the paper's reflectivity. If conditions are not optimal, we may get some assistance by employing a prior expectation of what should be on the paper (perhaps words in a particular language).

Similarly, when we look at a microscope image, we are seeing a nonuniform distribution of photon arrival rates. The more prior knowledge we have about the object we're viewing, the better we can interpret that distribution. For example, we may know that the object is *static*, or moving only slowly. In that case, we can improve our vision by collecting light for a long time, especially if the objects we are looking for are very faint (for example, single fluorophores).

Localization microscopy begins with the realization that in some situations, we know a priori that there is *only one isolated fluorescent molecule* in the field of view, and it's *not moving* during some time interval. Once we have characterized our microscope, then we know everything about the distribution of photon arrivals *except* the location of the emitting molecule. We can express that small amount of unknown information as a pair of parameters (the x, y coordinates of the fluorophore) and proceed to estimate them by maximizing likelihood along the lines of Section 7.2.2. In fact, you found in Your Turn 7A that x and y could be determined to a precision far better than the width of the point spread function f, if enough individual measurements (photon arrivals) were obtained.

When will such a prior (isolated, static, point source) be applicable?

- Earlier chapters have outlined several strategies for reducing light generated by molecules other than the ones of interest. For example, tagging the important molecules with a fluorescent probe and filtering out wavelengths other than the probe's emission band reduces the background (Section 2.3.2). If the molecules of interest are themselves sparsely distributed, then we will indeed have isolated sources.

- We have also studied methods that further reduce stray light from regions of the sample other than the one of interest, such as confocal, two-photon, or total internal reflection microscopy (Sections 2.7, 5.3.4 and 6.5.5).

- A fluorescent probe will be fixed in space if it is bound to a static structure. Even a moving object, like a molecular motor, may in some cases make sudden discrete steps; in between steps, it remains nearly fixed in space.

Even if we could eliminate all light from every source other than our fluorescent probe molecules, we may still have many legitimate sources in our microscope's field of view. Section 7.4 will describe an approach to this problem that has yielded spectacular results. First, however, we will look at a biological context in which we are only interested in a single molecule at a time.

7.3.2 Formulation of a probabilistic model

A molecular motor is a macromolecule or complex that functions as an enzyme, repeatedly splitting molecules of adenosine triphosphate (ATP) into isolated phosphate groups and the remainder, adenosine diphosphate (ADP). The motor harnesses the chemical energy released by this hydrolysis reaction to take steps along a "track," a long straight polymer such as a microtubule. When anchored to "cargo," such as a vesicle, molecular motors can operate as the "trucks" in a cellular transport system, carrying their cargo from a production point to a distant demand point.[7]

When studying the operation of molecular motors, we typically view their progress individually. For example, a small set of motors can be fluorescently labeled; then we image them repeatedly, tracking each one's progress. To view individual steps of a motor requires that we find its locations on successive video frames and look for jumps. But typical molecular motors take steps of length in the range 6–70 nm, too short to be resolved by ordinary light microscopy.[8] We thus have a situation in which inference using the single-source prior is both applicable and needed. The situation is a bit more complicated than in Your Turn 7A, however; before proceeding, we must incorporate four real-world issues into our probabilistic model of the system.

Two dimensions

Instead of the one-dimensional likelihood function in Section 7.2.2, our fluorophore may lie anywhere in the microscope's focal plane. We will use a vector quantity $r = (x, y)$ to denote transverse position in the specimen; corresponding positions on the camera's detector array are scaled up by the microscope's magnification factor.

Realistic point spread function

Light from a point source appears on the camera's detector array as a blurred spot, a probability density function with a bump (the point spread function). However, that function is only roughly described by the Gaussian distribution assumed in Section 7.2.2. Problem 6.9 discussed the sort of patterns we can expect in a simple slit geometry. More generally, we will assume that, over the few micrometers of our field of view, the point spread function only depends on the vector $r - r_\mathrm{t}$; thus, we can empirically measure it just by imaging one immobilized fluorescent molecule. To simplify the analysis, we will also assume that this function is radially symmetric about its center (it only depends on the distance $\|r - r_\mathrm{t}\|$).

Stray light

Section 7.3 described the single-source prior as having absolutely no light from sources other than the one molecule being localized, but of course this is an approximation. In practice, there will be a faint, diffuse background light along with the signal of interest to us. Thus, our formula for the probability for a single observed photon to arrive from apparent location r is a generalization of Equation 7.1 (page 248):

$$\wp(r \mid r_\mathrm{t}) = \mathrm{const} \times \big(f(\|r - r_\mathrm{t}\|) + b\big). \tag{7.4}$$

---

[7]See Media 13 and 14.
[8]See Media 15.

In this formula, the constant b represents the strength of the background light; the normalization constant can be computed by integrating the expression over the entire field of view and requiring that the result must equal 1.[9]

Any real camera also generates false signals (electronic noise). We will model this noise by supposing that it can be regarded as another contribution to the constant b.[10] To determine the overall b empirically, we can look for a region of the sample with *no* fluorescent molecules and average the signal over that region (and over multiple video frames).

Pixelation

Unfortunately, most cameras do not literally tell us the arrival location of each photon on the detector array. Instead, detectors used in microscopy *bin* the photon positions. That is, the detector is a grid of pixels, and each pixel emits a signal whenever it captures a photon anywhere on its surface. Section 7.2.5 (page 250) pointed out that binning destroys some of the information that was potentially contained in the exact location of the electron that absorbed the photon, but this is a technological limitation we must live with.

In other words, the detector pixel grid converts the probability density function of photon arrivals, $\wp(\boldsymbol{r})$, to a *discrete* distribution, the probabilities $\mathcal{P}(i, j)$ that the next photon will land somewhere within the pixel located at column i and row j in the grid. If we observe N photon arrivals, then our instrument gives us N draws from this discrete distribution.

A typical setup uses a detector array called a "charge-coupled device" (CCD) with square pixels approximately $6\,\mu\mathrm{m}$ on an edge, and a magnification factor of 60. Thus, each pixel corresponds to a physical size of $6\,\mu\mathrm{m}/60 = 100\,\mathrm{nm}$ in the sample. It may seem that there is no hope of achieving resolution better than this bin size, but fortunately, we will see that this intuition can be very wrong.

$\boxed{T_2}$ *Section 7.3.2′ (page 262) mentions some fine points involving the point spread function.*

7.3.3 Maximum-likelihood analysis of image data

Figure 7.1a shows some simulated data. Four hundred draws were generated from a probability distribution of the form Equation 7.4, with point spread function f and background light level b corresponding to a typical experimental setup. Each of these simulated photon arrivals was assigned to a bin corresponding to a camera pixel. The photon arrivals for each pixel were counted and rendered on a gray scale. Because this is simulated data, we know the true location $\boldsymbol{r}_{\mathrm{t}}$; it is shown on the figure as a blue dot. Examining this simulated image by eye, we might feel confident that we could localize the molecule's position to within about 1.5 pixels, or 150 nm. We can do much better than this, however.

Section 7.3.2 formulated a probabilistic model for this system. To analyze the data, we generate a list of competing hypotheses for the true location and compute the likelihood of each from the data. The hypotheses are possible values for $\boldsymbol{r}_{\mathrm{t}}$. Although $\boldsymbol{r}_{\mathrm{t}}$ is a continuous quantity, we can only ask our computer to evaluate the posterior at a finite set of possible values. In the calculation shown, those values chosen for evaluation consisted of a grid that was 30 times finer than the pixel grid.

---

[9]Equation 0.33 (page 10) introduced this condition.
[10]Other, more realistic noise models are sometimes used in practice.

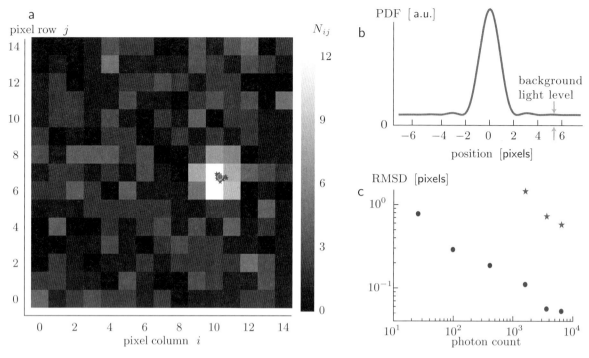

Figure 7.1: **Localization of a particle from a blurred, binned, and noisy image.** (a) [Simulated camera data.] A 15 × 15 grid of camera pixels with a total of 400 photons. The photon count in each pixel is indicated by gray level. The *blue circle* indicates the true location r_t of an emitter. Twenty *red stars* indicate the inferred source locations for this and 19 similar simulated video frames. (b) [Mathematical function.] The probability density function for photon arrivals used to generate the data in (a) includes a realistic point spread function superimposed on a Uniform background. (c) [Analysis of simulated camera data.] Log-log plot showing the performance of the maximum-likelihood localization algorithm. *Red dots* indicate the root-mean-square deviation of 20 determinations of source location from r_t, for various total numbers of photons collected. Collecting more photons is seen to lower the RMSD, implying greater accuracy of determining the true location. For comparison, *gray stars* indicate the performance of a simpler algorithm, which finds the mean location of the detected photons. [Simulated data from Dataset 8. See Problems 7.4 and 7.7.]

For each hypothesis, we want the posterior probability $\wp(r_t \mid \text{data})$, where "data" refers to the collection of observed photon counts in each pixel, $\{N_{ij}\}$. Choosing a Uniform prior and applying the Bayes formula (in the version Equation 7.3) gives this probability density function as a constant times the likelihood, so we wish to evaluate

$$\mathcal{P}(\text{data} \mid r_t).$$

Thus, for each r_t under consideration, we evaluate the continuous likelihood for r (Equation 7.4), then partition this probability among the pixels to find the discrete likelihood function[11] for a photon to be detected in the pixel at column i and row j:

$$\mathcal{P}(i, j \mid r_t) = \int_{\text{pixel } i,j} \mathrm{d}^2 r \, \wp(r \mid r_t). \tag{7.5}$$

In this formula, "pixel i, j" means that we are to integrate over all r values in a square centered at (i, j) in pixel coordinates.

---

[11]Equation 0.34 (page 11) shows how to obtain a probability from a PDF in this way.

Each photon's arrival is independent of the others, so we can now find the complete likelihood function $\mathcal{P}(\text{data} \mid \boldsymbol{r}_{\text{t}})$ by multiplying together one of the quantities in Equation 7.5 for each observed photon. In practice, it is better to compute the logarithm of the likelihood, by *summing* the logs of each of the factors:

$$\text{log-likelihood} = \sum_{ij} N_{ij} \ln \mathcal{P}(i, j \mid \boldsymbol{r}_{\text{t}}). \tag{7.6}$$

Because the logarithm function is strictly increasing, maximizing Equation 7.6 is equivalent to maximizing the likelihood. Given image data $\{N_{ij}\}$, then, we evaluate the quantities in Equation 7.6 (using Equation 7.5) and choose the hypothesis $\boldsymbol{r}_{\text{t},*}$ that gives the largest value.

Figure 7.1a displays the result in one way: Twenty successive "video frames" were simulated by drawing 20 sets of 400 photons from the distribution (Equation 7.4).[12] The procedure just described was applied to each frame, leading to 20 estimates of the fluorescent molecule's true location. These estimates are shown in the figure as red stars; they cluster tightly around the location actually used to generate the simulated data.

Figure 7.1c gives another representation of the results by displaying the spread of inferred $\boldsymbol{r}_{\text{t},*}$ values as a function of the total number of photons collected in each video frame. As might be expected, if we have only a few observations (photons), then the relative statistical fluctuations are large and we get a poor estimate. You may also find the general trend of the figure to be reasonable, because in Your Turn 7A you found that the estimate improves proportionally to $1/\sqrt{N}$. Remarkably, however,

- The improvement with increasing N persists down to a small fraction (here ≈ 0.05) of a camera pixel, and
- Under the conditions simulated here, the best result shown corresponds to an uncertainty (spread) of inferred location of $(0.05\ \text{pixel})(100\ \text{nm/pixel}) = 5\ \text{nm}$, far smaller than the wavelength of the light used!

$\boxed{T_2}$ *Section 7.3.3′ (page 262) discusses the advantage of the maximum-likelihood algorithm for localization microscopy.*

7.3.4 Results for molecular motor stepping

Figure 7.2 shows an application of these ideas to experimental data on molecular motor stepping. A. Yildiz and coauthors fluorescently labeled the single-molecule motor protein myosin V. This motor binds to, and walks along, actin filaments, which were themselves bound to a microscope coverslip. Thus, those motors that were ready to perform their function were automatically immobilized near the wall of their chamber, a setup suitable for total internal reflection fluorescence microscopy.[13] When ATP was added to the solution, the bound motors began to take steps of length about 76 nm.[14]

Several real-world issues complicate the analysis of real experimental data. In between steps, the imaged fluorophore may not be fully immobile, due to thermal motion; thus, the spread of its distribution may be wider than that predicted solely

---

[12] $\boxed{T_2}$ Problem 7.7 describes the procedure in more detail. Both the generation of simulated data and the analysis used the more realistic point spread function (not quite a Gaussian) shown in Figure 7.1b.
[13] Section 5.3.4 (page 194) introduced TIRF.
[14] See Media 15.

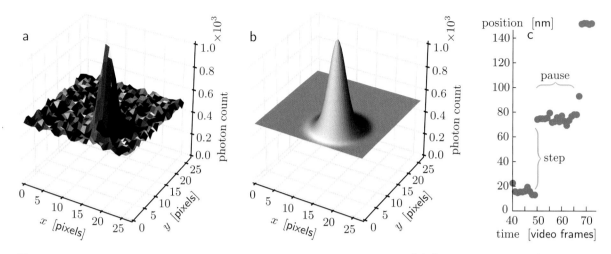

Figure 7.2: **Localization microscopy used to discern molecular motor steps.** (a) [Experimental data.] Typical video frame visualized as a surface plot: The height at each x, y value is proportional to the photon count at that location on the camera pixel grid. The pixel size corresponds to 86 nm in the sample. (b) [Fit to data.] Probabilistic model, consisting of a Gaussian distribution plus Uniform background light. (c) [Maximum-likelihood fits.] Inferred location of the fluorophore over the first few video frames, showing pauses interspersed with abrupt motor steps. [Data courtesy Ahmet Yildiz, available in Dataset 9 and Media 15.]

from diffraction. Nevertheless, Figure 7.2a,b show that the distribution can be modeled as a Gaussian plus Uniformly distributed background noise. Also, every aspect of the probabilistic model (level of background, width and height of the Gaussian distribution) may vary slowly over the time course of the trace, and must be fit to the actual data. With that procedure, however, panel (c) of the figure clearly shows sudden steps, executed in less than a video frame, of well-defined length that can be reliably determined to much better accuracy than the diffraction limit. Besides quantifying the lengths of such steps, the method gives precise timing information that the experimenters used to investigate the mechanism by which the motor transduces chemical energy into mechanical form.

Section 7.3 has outlined how to use likelihood maximization to extract position information from data that are partially obscured by noise, binning, and diffractive blur.

T_2 Section 7.3.4' (page 263) describes one way to obtain some parameters of the likelihood function.

7.4 LOCALIZATION MICROSCOPY

Section 7.3 showed that a dramatic improvement in localization is possible, if we have the prior knowledge that an image contains only a single pointlike emitter of light (a fluorophore). It's not difficult to generalize this result to images containing multiple sources, as long as they are sparse (well separated): We can then just crop out the bits of the image containing each one. Even if two emitters' point spread functions overlap somewhat, we could imagine a more elaborate set of hypotheses specifying the locations of each, and attempt to account for the observed data by maximizing likelihood over that larger hypothesis space.

Often, however, we want to image an extended structure, for example, a network of microtubules in a living cell. The objects of interest must then be labeled over all their extent, and so we must contend with many closely spaced light sources. The localization method discussed earlier is not directly applicable in this situation, so for some time it was considered to be useful only in specialized situations (such as motor stepping).

E. Betzig realized in the mid-1990s that the key to making progress must be to ensure that each emitter is somehow different from its neighbors, so that each *class* could form a sparse array, a subscene containing well-separated, point sources. Localizing each class of emitters separately, then combining the results, would then build up an entire image. Betzig initially proposed that the distinction between emitters could be in their emission spectra, but discoveries made soon after offered more practical implementations of his idea, based on **photoactivation**.[15]

Shortly after the discovery of green fluorescent protein, R. Dickson and coauthors found that GFP has a long-lived "dark" conformation that is unable to fluoresce. Unexpectedly, they found that they could pop individual molecules from this dark state to a fully fluorescent state by activation with light at 405 nm. Each molecule is either fully "on" or "off," and the choice of *which* ones turn "on" is random, reminiscent of photoisomerization.[16] Each photoactivated molecule fluoresces in response to 488 nm light until eventually it permanently photobleaches. Crucially, 488 nm light has no effect on the dark state; in particular, exposure to this light does not cause any change that would interfere with a molecule's ability to be photoactivated later. G. Patterson and J. Lippincott-Schwartz then engineered a modified GFP with similar behavior but greatly improved properties.

The implications of this work were profound. Instead of attempting to label a structure with many distinct kinds of probes, the discovery of photoactivatable proteins allowed the use of *one* probe molecule, which however had multiple *states*. Briefly irradiating a sample with the activating light converts just a few probes into their fluorescent state; those activated probes can be imaged by localization microscopy as usual until they photobleach. A second brief activating exposure then converts a few *other* probes, which can in turn be imaged, and the cycle is repeated many times. Each round of activation and imaging builds up a different subset of the labeled structure's points, until enough points have been accumulated to form an image.[17] In 2006, three research groups overcame the remaining technical obstacles, and the era of localization microscopy began in earnest. These first variants of the method were called photoactivated localization microscopy (PALM), stochastic optical reconstruction microscopy (STORM), and fluorescence photoactivated localization microscopy (FPALM).

Compared to other single-molecule imaging methods, such as atomic force microscopy and near-field scanning microscopy, localization microscopy is distinguished by working in the "far field" region: No part of the apparatus needs to be within a few nanometers of the objects being imaged. In particular, we can see deep inside living cells (a micrometer or more), without having to flatten them. Moreover, a camera with a large pixel array can simultaneously image many well-separated points of the object of interest in a single cycle, partly mitigating the need to image over many cycles. Later work extended the original idea to allow imaging in three dimensions.

Figure 7.3b shows an example of an architectural motif in neurons, a ladderlike arrangement of actin filaments, that was unknown prior to the discovery of localization

---

[15] $\boxed{T_2}$ Other related methods rely on similar phenomena called photoswitching and photoconversion.
[16] Section 1.6.4 (page 45) introduced photoisomerization.
[17] See Media 16.

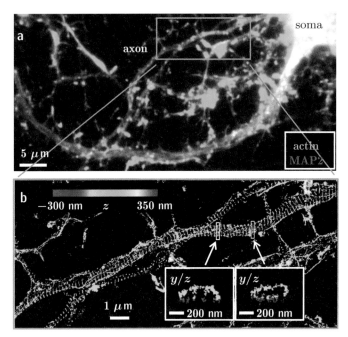

Figure 7.3: **Highly structured organization of actin filaments in nerve axons.** (a) [Conventional two-color fluorescence micrograph.] Actin filaments (*green*) and microtubule-associated protein 2 (*magenta*) in a neuron. (b) [Three-dimensional STORM image.] Organization of actin in a region containing axons [*box* in (a)]. Only actin is labeled; the color scale now denotes the height coordinate z (*top*). A periodic substructure, previously unsuspected, is clearly visible. *Insets* show cross-sections in the yz plane corresponding to the boxed regions. [From Fig. 1 D, E, p. 453 of Xu et al. Actin, spectrin, and associated proteins form a periodic cytoskeletal structure in axons. *Science* (2013) vol. 339 (6118) pp. 452–456. Adapted with permission from AAAS.]

microscopy.

$\boxed{T_2}$ *Section 7.4' (page 263) describes another extension of localization microscopy to three-dimensional imaging.*

7.5 DEFOCUSED ORIENTATION IMAGING

Fig. 6.5 (page 217)

This chapter has interpreted imaging as a process of inferring the spatial structure of an object from the pattern of photon arrivals at a detector. Chapter 6 considered cases where this inference was so straightforward that we didn't need to discuss it explicitly: A lens system, for example, the one in Figure 6.5, created a direct replica of the transverse placement of objects in the focal plane on a projection screen, albeit partly corrupted by diffraction and aberration. The present chapter addressed the diffraction problem by applying likelihood maximization.[18]

All of our discussion so far, however, has involved attempts to determine only the *locations* of pointlike sources of light. If we had perfect resolution, that might be enough. For example, we can at least imagine imaging an entire "leg" of a molecular motor, in order to see it swinging from one angular orientation to another as it walks

---

[18]Chapter 8 will discuss another technique, which however requires that the molecule in question be crystallized. That technique is therefore not suitable when we wish to observe the dynamics of a protein's normal function.

along its "track." In practice, the individual parts of a macromolecule are too small to resolve optically, and it is difficult to label them with enough distinct fluorescent tags to build up an image that would show their relative orientations in real time.

Remarkably, it is instead possible to infer the spatial orientation of a *single* fluorophore. If that fluorophore is bound rigidly to part of a macromolecule, for example the "leg" of a molecular motor, then it can report on the orientation of that entire domain. To see how this is possible, we must first acknowledge an implicit assumption made throughout Chapters 4–6: Our analysis of systems like the one in Figure 6.5 assumed that either

a. The light source is pointlike and far from the lens system (perhaps a star in the sky), or

b. If not distant, the light source is *effectively* distant because its photons are all traveling in the same direction (perhaps from a laser), or

c. If not distant nor parallel, the light source has a probability amplitude to emit photons that does not depend on direction (it is **isotropic**).

If either of the first two conditions holds, then we found that we may forget about the parts of each light path to the left of the lens: Each makes the same contribution to the probability amplitude, leading to a common factor that gets absorbed into the overall normalization.[19] Even if neither of **a,b** holds, so that we must account for both parts of each light path, we still found a condition for focus.[20] Suppose that we place light detectors at the focal plane. Then, neglecting the effects of diffraction, all light from the source arrives at one image point on the focal plane. Even if assumption **c** does not hold (the source is **anisotropic**), it won't matter: All light arrives at one point regardless of the source's orientation.

To find something interesting, now consider the case where *none* of conditions **a–c** holds, *and* the system is slightly out of focus. In particular, Chapter 13 will show that a single fluorophore generally emits photons with an anisotropic distribution of directions: An excited fluorophore has a special axis called its **transition dipole**. Photons are most likely to be emitted in a circle of directions perpendicular to the transition dipole (its "equatorial plane"). If the fluorophore's transition dipole points directly toward the lens, Figure 7.4a suggests that light will arrive in a *ring* at a defocused detector plane. If the transition dipole is tilted away from that direction, however, then different parts of the imaged ring will have different intensities. M. Böhmer and J. Enderlein showed that these conclusions also remain valid after we account for diffraction, and indeed, experiments later showed that this effect can be used to determine the orientation of a single fluorophore (Figure 7.4b).

If the fluorophore is attached to a macromolecule, the orientation can be read out in real time as the macromolecule does its job. Moreover, the loss of positional information from defocusing can largely be recouped by using localization microscopy, leading to a method named "<u>d</u>efocused <u>o</u>rientation and <u>p</u>osition <u>i</u>maging" (DOPI). [T_2] *Section 7.5′ (page 266) mentions orientation effects that appear even when a fluorophore is focused.*

---

[19]This was the situation considered in Section 4.7.3 (page 165).

[20]This was the situation considered in Sections 6.3.1–6.3.3 (page 212).

Figure 7.4: **Defocused orientation imaging.**
(a) [Ray diagram.] If a source does not give off photons equally in every direction, then its image in a defocused lens system will not be a simple blur. For example, a single fluorophore is characterized by an axis, called its transition dipole, along which it has zero probability to emit photons. If this direction points along the centerline, as shown, then in the ray-optics approximation the predicted illumination pattern on the screen shown will be a ring. Other orientations for the transition dipole yield different predicted patterns. In contrast, placing the screen at the focal plane brings all rays to a single point regardless of the transition dipole's orientation (*dashed red lines*). (b) [Experimental data and fits.] *Top:* Observed point spread functions for three single fluorophores. Lighter colors correspond to pixels with larger photon counts. *Bottom:* Corresponding physical predictions, after finding the best-fit value of the angle between the transition dipole and the centerline. From left to right, the fit values of this angle were 10 deg, 60 deg, and 90 deg. The in-plane orientation (azimuth) can also be obtained by fitting. [From Toprak et al., 2006.]

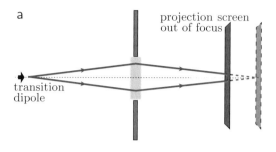

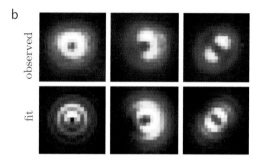

THE BIG PICTURE

The goal of optical microscopy is to gather *information* about an object that we cannot touch, by collecting and interpreting the light that it gives off (or scatters). Although light is a stream of photons, for many everyday uses we can ignore that fact, effectively performing all the necessary computations by focusing the light to a literal image on a detector with a lens system. In the diffractive regime, however, this hardware-based, analog computation loses some of the information potentially available in the photon stream.

This chapter has outlined one way to do better, in part by performing some of the computation numerically. The analysis built on Chapter 6, which explained how a lens system can create a (diffractively blurred) image. Equation 7.1 (page 248) summarized all of that physics; then we saw how to use the resulting information (and some pioneering discoveries in photochemistry) to improve upon the diffraction limit.

The next chapter will introduce a radically different imaging technique, which, however, can also be regarded as an inference problem: We will again consider various hypotheses about an unknown structure, see what diffraction pattern to expect from each one, and compare to data.

KEY FORMULAS

- *Bayes formula:* Suppose that we make a measurement x_1, and we wish to infer the value of a parameter x_t that is best supported by our measurement in the context of a probabilistic model of our system. We can write $\wp(x_t \mid x_1) = \wp(x_1 \mid x_t)\wp(x_t)/\wp(x_1)$, then maximize this expression over the parameter while holding fixed the measured data. We call $\wp(x_t)$ the prior distribution, $\wp(x_t \mid x_1)$ the posterior distribution in the light of new information x_1, and $\wp(x_1 \mid x_t)$ the likelihood function.
- *Isolated, point-source prior:*

$$\wp(\boldsymbol{r} \mid \boldsymbol{r}_t) = \mathrm{const} \times \big(f(\|\boldsymbol{r} - \boldsymbol{r}_t\|) + b\big). \qquad \text{[7.4, page 252]}$$

FURTHER READING

Semipopular:
Lippincott-Schwartz, 2015.

Intermediate:
Inference in physical and life science: Nelson, 2015; Woodworth, 2004.
Superresolution microscopy: Mertz, 2010.

Technical:
Fluorescence imaging at one nanometer accuracy (FIONA): Simonson & Selvin, 2011; Selvin et al., 2008; Toprak et al., 2010. Precursors to this method include Ober et al., 2004; Thompson et al., 2002; Cheezum et al., 2001; Lacoste et al., 2000; Gelles et al., 1988; Bobroff, 1986.
FIONA applied to molecular stepping: Yildiz et al., 2003.
Localization microscopy precursors: Lippincott-Schwartz & Patterson, 2003; Patterson & Lippincott-Schwartz, 2002; Dickson et al., 1997; Betzig, 1995.
Localization microscopy discovery: Betzig et al., 2006; Hess et al., 2006; Rust et al., 2006. A fourth article published in the same year used a different approach to obtain the sparse, stochastic labeling need for localization: Sharonov & Hochstrasser, 2006.
Localization microscopy without exogenous fluorescent dyes: Dong et al., 2016. Three-dimensional STORM applied to neurons: Xu et al., 2013.
Reviews, including other superresolution methods: Small & Parthasarathy, 2014; Bates et al., 2013; Zhong, 2011; Bates et al., 2011; Hell, 2009; Hinterdorfer & van Oijen, 2009, chapt. 4; Huang et al., 2009; Bates et al., 2008; Hell, 2007.
Defocused orientation imaging: Böhmer & Enderlein, 2003. Defocused orientation and position imaging: Toprak et al., 2006.
$\boxed{T_2}$ Interferometric PALM: Shtengel et al., 2014.

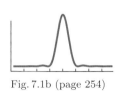

Fig. 7.1b (page 254)

Track 2

7.3.2′a Airy point spread function

One commonly used, idealized formula for a system's point spread function is the Airy function (see Figure 7.1b),

$$f(r) = \left[\frac{2J_1(ar)}{ar} \right]^2 . \tag{7.7}$$

In this formula, the constant $a = 2\pi \mathrm{NA}/\lambda$, NA denotes the system's numerical aperture, and J_1 is a Bessel function of the first kind. In practice, however, it is generally adequate to model the likelihood function as a Gaussian distribution (plus Uniform background), because the subtle bumps in the tail of the Airy function are generally washed out by pixelation and the added background. In any case, the Airy function itself is not a perfect representation of the point spread function in a real, high-NA optical system.

7.3.2′b Anisotropic point spread function

Chapter 13 will discuss the fact that a single fluorophore's emission of light is not uniform in all directions. If a fluorescent probe is free to rotate, then its thermal motion will average out this anisotropy, giving the sort of symmetric point spread function assumed in the main text. If it is not, then acknowledging its anisotropy can give us a useful additional measurement of the molecule's spatial orientation (Section 7.5; Backlund et al., 2012; Toprak et al., 2006).

7.3.2′c Other tacit assumptions

The analysis in the main text was based on Equation 7.4 (page 252), which contains some assumptions not mentioned there. First, the point spread function and background light were assumed not to vary across the field of view. In any particular experiment, these assumptions can be checked and if necessary replaced by a more accurate empirical model.

Second, the center of each distribution x_t was assumed to bear some simple relationship to the physical location of the fluorophore (it was assumed to be a linearly magnified version of the true location). There may instead be some nonlinear distortion of positions. This effect can be mapped out by mechanically sweeping an immobilized fluorophore over the field of view, determining its apparent position as a function of the known true position, and applying the resulting correction to the superresolution images of more interesting specimens.

Track 2

7.3.3′ Advantages of the maximum likelihood method

The main text describes a method for squeezing as much information as possible from data that are partially corrupted by diffractive blurring, binning (pixelation), and the randomness of photon arrivals in time (shot noise). Even if we have an accurate probabilistic model for our system, however, there is a limit to how much information we can obtain in this way. Information theory makes this notion precise in the form of a theorem, the "Cramér-Rao lower bound" on the variance of our estimate of emitter position (Abraham et al., 2010; Mortensen et al., 2010; Ober et al., 2004). The advantages of maximum likelihood estimation (MLE) include the facts that it actually achieves the lowest possible variance, is free from biases (for example, toward the center of the field of view, or toward the nearest pixel center), and gives equally accurate results regardless of where the emitter is located relative to the camera's pixel grid. Optimality is be important because we have only a limited number of photons available: Each fluorescent molecule photobleaches after emitting about 10^4–10^6 photons.

Despite the optimality of the maximum likelihood method, it must be used with care. Many cameras introduce additional electronic noise into their output, which may not have the Poisson character we assumed in our simplified discussion. If this effect is not properly incorporated into the probabilistic model, then MLE estimation can perform worse than a simpler approach, that of least-squares fitting. Also, it is important to use the correct, empirically determined point spread function in the probabilistic model; optical aberrations and other effects may make it significantly different from the ideal Airy function form (Equation 7.7).

T_2 Track 2

7.3.4′ Background estimation

The likelihood function, Equation 7.4 (page 252), includes both a point spread function f and a Uniformly distributed background b representing stray light. In practice, b must be determined from the data themselves, and it may be time dependent. One approach is to find a region of the image that does not contain any fluorophores over many video frames, and average the reported photon count over that region.

After we determine $b(t)$, we can subtract it from the data and fit the remaining photon counts to a proposed family of point spread functions, for example, a Gaussian distribution characterized by its spatial width and strength (which may also vary slowly over time). This procedure yields the empirical point spread function $f(\boldsymbol{r}, t)$ to be used in the likelihood analysis at time t.

The two steps just described were taken prior to the frame-by-frame likelihood maximization that resulted in Figure 7.2c.

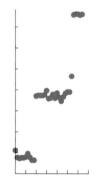

Fig. 7.2c (page 256)

T_2 Track 2

7.4′ Interferometric PALM imaging

The main text described methods that use visible light to obtain images with few-nanometer resolution in the directions transverse to the microscope's optical axis. But in the axial direction, such images have resolution no better than the ordinary point spread function of the microscope, generally hundreds of nanometers. This limitation is not serious when the objects of interest are confined to an extremely thin layer (for

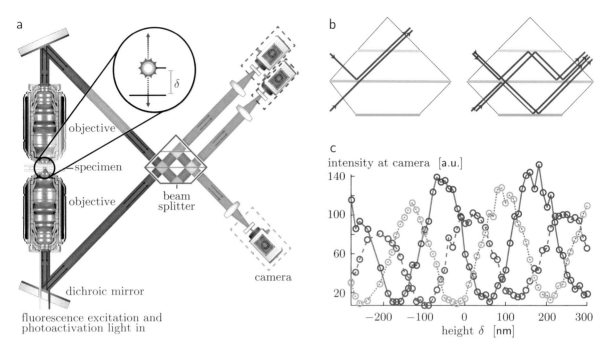

Figure 7.5: **Interferometric PALM microscopy.** The apparatus shown can determine axial position of many image elements, simultaneously and to high accuracy. (**a**) [Schematic.] A fluorophore at height δ emits a single photon. There are multiple paths that the photon can take. In the first stage of its journey, the photon can pass through either of two microscope objectives; then it enters a three-way beam splitter, and ultimately contributes to an image in one of three sensitive cameras (*right*). (**b**) Detail of the beam splitter. Photons can enter via the two paths on the *left*. They first encounter a layer (*middle*) that randomly reflects half of all arriving photons, transmitting the rest. Some then encounter a layer (*bottom*) that reflects all photons; others encounter a third layer (*top*), which reflects 1/3 of all arriving photons, transmitting the rest. The left subpanel shows two paths that end in the top camera in (a); the right subpanel shows four paths that end in the middle camera. (Four other paths end in the lower camera, *not shown.*) (**c**) [Experimental data.] When both incoming beams are open, interference partitions the outgoing photons unequally, in a way that depends on the source height, δ. To demonstrate this, a sample containing an immobile fluorophore was mechanically swept over 600 nm and its brightness in each of the three cameras was recorded. The data show that the relative brightnesses in the three cameras can be used to determine the height δ. The *lines* are not fits; they just join the data points. [Diagrams and data courtesy Pakorn Kanchanawong; see Shtengel et al., 2009.]

example, microtubules bound to the surface of a glass coverslip, as in motor protein assays, or the receptors in a cell membrane, mentioned in Section 5.3.4 on page 194). But many living cells are not so thin, and the three-dimensional architecture of specific molecules within them is of great interest.

G. Shtengel and coauthors addressed this problem, implementing a remarkable extension of localization microscopy based on ideas introduced in Chapter 4. They approached the challenge (simultaneous high-resolution axial and transverse localization) by exploiting the fact that a single photon can display interference phenomena.[21] More precisely, the interference is between potential *paths* that a photon can take from source to observer (Figure 4.2b). Figure 7.5a shows a schematic of their apparatus. The role of the two slits in Figure 4.2b is played by two microscope objective lenses, which together gather photons emitted in nearly any direction. A simple setup could

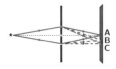

Fig. 4.2b (page 147)

_____

[21] Taylor's experiment; see Section 4.2 (page 146).

be to use mirrors to combine the light from those two path families onto a single light detector. Then each photon from a pointlike source (a fluorophore) in the specimen would either arrive at the detector, or not, with probability depending on the difference of lengths of the two available paths. In the arrangement shown at the left of the figure, a small vertical displacement of δ would change the path length difference by 2δ, so the intensity of the fluorescence received gives information about axial position.

Although the strategy just outlined would give some axial position information, it can be improved. For example, objects located at certain values of δ would all suffer destructive interference and hence be invisible! We could imagine moving the entire sample in the axial direction, to bring each object in turn to a visible position, but then we would need to compare the apparent brightness of the fluorophore at different moments in time—and a fluorophore's intrinsic brightness can vary over time.[22] Instead, Shtengel and coauthors constructed a more elaborate apparatus by adding a three-way beam splitter (center of Figure 7.5a). A photon entering this compound prism from either of the two incoming beams ultimately leads to a signal from one of the *three* cameras shown.

The beam splitter contains three parallel, reflecting surfaces: The top internal surface reflects one third of the photons incident on it, and transmits the rest. The middle internal surface reflects half of the photons incident from either side, and transmits the rest. The bottom surface reflects all of the incident photons. There is only one stationary-phase path from each input to the top camera, and hence no possibility of interference if one input is blocked. Thus, if either incoming beam is blocked, then each incoming photon from the other beam will have probability $1/3$ to be detected by the top camera, regardless of its position δ. The other two cameras are a bit more complicated, but they, too, give no δ dependence if one input is blocked.

When both incoming beams are unblocked, however, path interference effects will modify this outcome. To calculate the interference, the experimenters noted that thin internal reflecting films introduce an additional phase contribution $-\pi/2$ to the probability amplitudes of reflected photon paths, relative to the transmitted ones.[23] They also used the fact that the distance between the top two reflecting surfaces was slightly different from that between the middle and bottom surface. Adding all path contributions according to the Light Hypothesis showed that each camera gets an interference pattern as the height δ of the light source is changed, but *the three patterns are offset,* as shown in Figure 7.5c. Thus, regardless of the value of δ, one or more cameras will always be illuminated, and by comparing the strength of illumination among the three cameras the experimenters could fit to find the value of δ to $10\,\mathrm{nm}$ accuracy, far better than the width of the point spread function (about $500\,\mathrm{nm}$ in the axial direction). Moreover, because each camera was observing an image, the authors could still use localization microscopy in the two transverse directions, and so construct a truly three-dimensional superresolution image.

Figure 7.6 shows a biological application of this technique.

---

[22] "Blinking" is an extreme form of variation; see Section 1.6.4 (page 45).
[23] This behavior contrasts with that of a single interface (Idea 4.6, page 155).

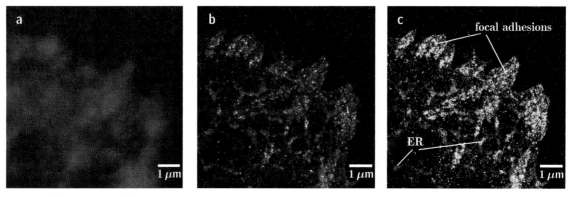

Figure 7.6: [Micrographs.] **Interferometric PALM.** Each panel shows the edge of a cell expressing fluorescently labeled integrin molecules. (a) Standard epifluorescence image. (b) A PALM image of the same cell shows greater detail, but is still a two-dimensional projection. (c) This interferometric PALM image has been false-colored according to the inferred height δ of each fluorophore. Protein complexes that attach the cell to its neighbors (focal adhesions) are a few tens of nanometers above the coverslip, and appear *yellow*; endoplasmic reticulum (ER) is higher, and appears in *blue and magenta*. The two extremes, coded yellow and magenta, correspond to an axial separation of about 200 nm. [Images courtesy Pakorn Kanchanawong; see also Shtengel et al., 2009, and Media 17 for a full 3D reconstruction.]

$\boxed{T_2}$ **Track 2**

7.5′ More about anisotropy

Even when a fluorophore is perfectly focused, the angular dependence of photon emission will affect its diffractive blur: Diffraction involves slight differences in path transit times, which can interact with the angular dependence of the probability amplitude for photon emission. Advanced methods of localization microscopy account for this phenomenon and use it to improve upon the precision of localizing the fluorophore, and to infer its orientation (Mortensen et al., 2010).

PROBLEMS

7.1 *Cauchy distribution estimation*

a. Use a computer to draw 600 random numbers from the Gaussian distribution with expectation 0 and variance 1. (Your computer math package may refer to this as the "normal" distribution.) Find the sample mean of the first N of these, for $N = 1, \ldots 600$, graph the result as a function of N, and comment.

b. Momentarily forget that you know where your distribution is centered, and try to infer its center from your data. That is, imagine that the numbers you drew in (a) were experimental observations of a quantity that you know follows a Gaussian distribution with variance 1 but unknown expectation. Evaluate the posterior distribution $\wp(x_t \mid x_1, \ldots, x_N)$ by using the Bayes formula (Equation 7.2, page 248) for the first N of these, plot it for $N = 1$, 10, 100, and 600, and comment. [*Hint:* It is better to compute, and plot, the log of the distribution. If you do this, then you needn't worry about normalizing the distribution—why not?] Run your code more than once to see what is different, and what is similar, each time.

c. Use a computer to draw 600 random numbers x_i from the Uniform distribution on the interval $[0, 1]$. Transform these into draws from the Cauchy distribution[24] by applying the function $\tan\big(\pi(x - 1/2)\big)$ to each of the x_i. Find the sample mean of the first N of these, for $N = 1, \ldots 600$, and graph the result as a function of N. What unpleasant surprise do you find, compared to (a)? Does it help to change 600 to, say, 10 000?

d. In fact, finding the sample mean is not an effective way to determine the center of a long-tail distribution, such as the one in (c), from samples. This time, instead adapt the procedure in (b) to handle Cauchy-distributed data, implement it, and comment.

e. Section 0.5.2 (page 15) claimed to prove that the sample mean gives us better and better estimates of the expectation as the sample gets large. What goes wrong in the case of the Cauchy distribution?

7.2 *Likelihood analysis of a Poisson process*

Suppose that you are studying the activity of an enzyme (for example, a molecular motor), which makes discrete steps at irregular time intervals. You believe that the waiting times between steps are distributed according to an Exponential distribution: $\wp(t_w) = Ce^{-\beta t_w}$, where C and β are constants. The enzyme only made seven steps before the experiment ended, leading to six measured waiting times $t_{w,1}, \ldots, t_{w,6}$. Your job is to find the values of the constant parameters in the model distribution that are best supported by the data.

a. Express C in terms of β.

b. Write an expression for the likelihood of any particular value of β, in terms of $t_{w,1}, \ldots, t_{w,6}$.

c. Find an expression for the maximally likely value of the parameter β.

d. In (c), it is important to account for the dependence of the normalization factor on the model parameter. But in Your Turn 7A (page 249), this wasn't necessary. What is the difference between these situations?

---

[24]Sections 0.4.2 (page 12) and 0.5.1 introduced and discussed this family of probability density functions. Section 0.5.1 (page 15) explains why applying a nonlinear function (such as tangent) to a random variable transforms its PDF.

7.3   *Power-law distribution*

Suppose that a continuous measured quantity lies in the range $1 < x < \infty$ and is distributed with probability density function $\wp(x) = Ax^{-\alpha}$. Here A and α are positive constants.

a. The constant A is determined once α is specified. Find this relation. Are there any limitations on the value of α?

b. Find the expectation and variance of x, and comment on when these quantities can be defined.

c. Suppose that we measure N independent values $x_1, \ldots, x_N$. Find a formula for the maximally likely value of α given these data.

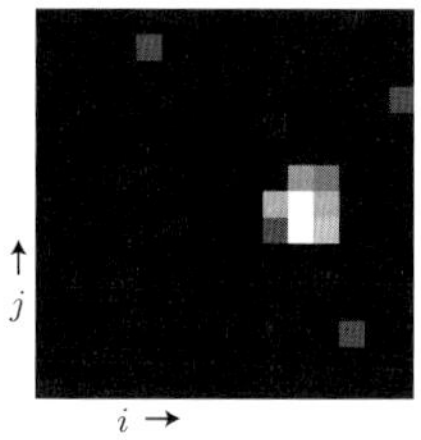

7.4   *Localization by finding the centroid*

Obtain Dataset 8, which contains simulated "video frames," that is, sets of numbers representing photon counts in a sensitive camera. For each of six different "exposures" (total photon counts), 20 such simulated frames are given, a total of 120 frames. In every case, the "scene" that is being simulated consists of a single point source of light somewhere in the field of view, plus some stray background light. Figure 7.1a displays one of these frames, using gray levels to represent photon counts. All frames have the same true source location and brightness, point spread function, and background level.

Fig. 7.1a (page 254)

Your Turn 7A (page 249) outlined a scheme for analyzing such data: Assume that light from the source has a Gaussian distribution in space and maximize likelihood.

a. To apply this scheme to the data, you must first estimate the background and subtract it. Find a region of the image where you're pretty sure the point source is *not* located, and, for each of the six "exposures," find the sample mean of the background photon count over all pixels in that region, also averaging over all 20 simulated video frames.

b. Now subtract the background you found in (a) from the data, and use them to compute the mean position:[25]

$$\bar{\boldsymbol{r}} = \sum_{ij} N'_{ij} \boldsymbol{r}_{ij} / N'_{\mathrm{tot}}.$$

In this formula, $\boldsymbol{r}_{ij}$ is a 2D vector giving the location of the center of the camera pixel at column i and row j. The value N'_{ij} is the photon count minus the background. The sum is over all pixels in any one video frame.

c. In this way, you can make an estimate of the source location in each frame, then compute the root-mean-square deviation of those estimates across the 20 samples for each simulated "exposure." How does the RMSD depend on the total photon count? Make a log-log plot of variance versus photon count and see if it behaves the way you expect in light of Your Turn 7A.

d. Also make an image like Figure 7.1a, showing one representative frame for one representative photon count. Decorate your image with stars indicating your 20 determinations of the weighted mean position.

7.5 $\boxed{T_2}$ *Inferring a Bernoulli trial parameter*

Suppose that an experiment yields a binary random variable ("success/fail"), and that we believe this variable is drawn from a Bernoulli trial distribution (each trial is

---

[25]This quantity is the vector sample mean of the data, also called the "centroid."

independent of the others). Thus, its distribution is described by a single parameter ξ, whose value is constant but unknown. We suppose no prior information about ξ, that is, ξ is itself a random variable with Uniform prior on the interval $[0, 1]$. Now we observe N trials and find that k came out "success." We want the best estimate of ξ, and also an estimate of how good our estimate is.

a. Use the Bayes formula to get an expression for the posterior distribution of ξ.

b. One approach to finding the best estimate is to compute the expectation of ξ in the posterior distribution, so compute that in terms of N and k, starting from your result in (a). For large values of N and k, your answer will reduce to something familiar.

c. Similarly, compute the variance of the posterior distribution, and again find a simpler formula in the limit of large N and k. The square root of this quantity is a reasonable estimate of the spread in inferred ξ values if we made many batches of N trials, and so of the reliability of our estimate based on one such batch.

7.6 $\boxed{T_2}$ *Blinking of a fluorophore*

An illuminated fluorophore emits individual photons according to a random process. Even if the illumination is of constant intensity, the emission process may change over time: Its mean rate can pop back and forth between zero and a nonzero value (or between several values). Generically, this phenomenon is known as "blinking"; it's both an experimental complication for fluorescence measurements, and also intriguing physically.

A rigorous approach to blinking would begin with the actual arrival times of each emitted photon. This problem instead studies a simpler problem: Assume that a colleague has already determined the times at which transitions between "on" and "off" episodes occur and created an estimated probability density function for the waiting times t_{w} (that is, the time intervals between transitions). Specifically, Dataset 10 contains estimates for

- $\wp_{\mathrm{off}}(t_{\mathrm{w}})$, where t_{w} is the waiting time before the fluorophore switches "off" (the lifetime of an "on" state), and

- $\wp_{\mathrm{on}}(t_{\mathrm{w}})$, where this time t_{w} is the time before it switches "on" (the lifetime of an "off" state).

In this problem, you may assume that the durations of successive "on" and "off" episodes are independent random variables drawn from these distributions.

a. The dataset contains estimates of $\wp_{\mathrm{off}}(t_{\mathrm{w}})$ and $\wp_{\mathrm{on}}(t_{\mathrm{w}})$ at selected values of t_{w}, based on experimental data. Graph these on linear, semilog, and log-log scales, and formulate a hypothesis for their form.

b. If you haven't worked Problem 7.3 yet, do it now. Then consider the hypothesis that

$$\wp_{\mathrm{on}}(t_{\mathrm{w}}) = \mathrm{const} \times (t_{\mathrm{w}})^{-\alpha} \quad \text{for} \quad t_{\mathrm{w}} > t_{\mathrm{min}}, \tag{7.8}$$

where $t_{\mathrm{min}} = 1.8\,\mathrm{ms}$ is the shortest interval that was measurable in the experiment. Adapt your result from Problem 7.3 to get an expression for the constant prefactor in terms of α.

You could estimate the best-fitting value of α graphically, by laying a ruler along the points in the appropriate plot. But the probability estimates at large t_{w} are not as reliable as the others, because they are based on many fewer instances. Such points should get less of a "vote" than others. Likelihood maximization is one approach that makes this intuition precise.

c. Suppose that N "off" episodes are observed, with durations (waiting times t_w) equal to $t_{w,1}, \ldots, t_{w,N}$. Again apply your result from Problem 7.3 to get an expression for the log likelihoods of the models in Equation 7.8 with various values of α, and find the value α_* that maximizes it.

d. The dataset does not contain the individual durations $t_{w,i}$. Instead, it contains only aggregate information, the estimated probability density function $\wp_{on}(t_w)$ at a discrete set of time values. Figure out how to compute α_* using this information, and implement your algorithm.

e. Plot Equation 7.8 with your best-fit value α_*, together with the experimental data.

f. Repeat (d–e) for the "on" episode lifetimes.

7.7 $\boxed{T_2}$ *Localization by maximizing likelihood*

If you haven't yet done Problem 7.4, it's a good warmup to this one. This problem explores the more elaborate maximum-likelihood algorithm described in Section 7.3.3 (page 253). Again obtain Dataset 8, which gives the number of photons observed in each camera pixel during 20 simulated "video frames" for each of six "exposures" that all observed the same "scene." Here is a summary of some integer quantities to be introduced in the following paragraphs:

$\boldsymbol{r}_t$	true location, a two-component vector, measured in pixels (dimensionless)
i, j	horizontal and vertical coordinates of the center of a camera pixel, in units of pixel size
K	factor by which we subdivide the pixel grid
I, J	offset from pixel center that specifies a hypothesis about the true location (Equation 7.9)

As in Problem 7.4, we are interested in how reproducibly we can estimate the true location of the point light source in the scene. To do this, the strategy is to consider various hypotheses about the true location $\boldsymbol{r}_t$ of the light source, evaluate each one's likelihood, and choose the winner.

Examining Figure 7.1a (page 254) shows that the source is somewhere near the point $(i = 10, j = 7)$. Let hypothesis (I,J) be the proposition that the true location, measured in pixels, is

$$\boldsymbol{r}_t = (x_t, y_t) = (10 + \mathtt{I}/K, \ 7 + \mathtt{J}/K), \tag{7.9}$$

where $K = 5$ specifies how finely we subdivide a pixel in each direction; I and J are integers from $-K$ to $+K$.

For each of these propositions, we need the probabilities assigned by the model for photons to land in the pixel centered at i, j, where i, j are integers from 0 to 14 (that is, a 15×15 pixel grid). Use the function in Equation 7.4 (page 252) with background level $b = 0.1$ and point spread function f given by Equation 7.7 (page 262) with $a = 16.6 \, \mu\mathrm{m}^{-1}$. Suppose that the camera pixel size corresponds to $0.10 \, \mu\mathrm{m}$ in the object being viewed; thus, a measured in inverse pixels equals 1.66, and you can do the problem with dimensionless quantities.

Equation 7.5 tells us to integrate the likelihood function over each pixel. In practice, however, it's easier to approximate the integral by a finite sum. Thus, we subdivide the pixel grid into $(15K)^2$ cells, and for each (i, j) sum the function evaluated over the K^2 cells belonging to the pixel centered at i, j.

a. Carry out those steps, and so tabulate the likelihood $\mathcal{P}(i, j \,|\, \text{hypothesis } \mathtt{I,J})$ for all values of i, j, I, and J (a total of 27 225 entries).

b. Evaluate the point spread function, Equation 7.7, on your fine grid and plot it. Assume for illustration that an isolated, point source is located at the center of the brightest pixel in the first frame. You may wish to confirm that your code works the way you want by summing this function over the camera pixels, plotting it as a function of i, j in gray scale, and comparing to a similar plot of the first frame of the simulated data. (For example, be sure the two plots have their brightness peaks at about the same location!)

c. Now that you have things aligned properly, it's time to find the best estimate of the true position. For each of the 20×6 frames in the dataset, evaluate the log likelihood (Equation 7.6, page 255). Find the winning values of I,J and interpret them via Equation 7.9. Find the expectation and variance of the 20 determinations of r_t that you obtain in this way for each of the six "exposures" (total photon counts) and comment.

d. If you think you can get results accurate to better than one fifth of a pixel, run your code again with a larger value of K.

e. As in Problem 7.4, make an image like Figure 7.1a, showing the first frame for one representative "exposure." Decorate your image with stars indicating your 20 determinations of the weighted mean position.

CHAPTER 8

Imaging by X-Ray Diffraction

> The beginning of modern science can be dated from the time
> when such general questions as, "How was the universe
> created? What is matter made of? What is the essence of
> life?" were replaced by such limited questions as "How does
> a stone fall? How does water flow in a tube? How does blood
> circulate in vessels?" This substitution had an amazing
> result. While asking general questions led to limited answers,
> asking limited questions turned out to provide more and
> more general answers.
>
> — *François Jacob*

8.1 SIGNPOST: *INVERSION*

A big theme of this book is the information carried by light of all kinds. The two
preceding chapters have specifically been concerned with extracting that information
by direct image formation, that is, focusing light to give a replica of a visual scene
on a detector screen. Later chapters will continue that discussion, describing what
happens on the particular detector arrays found in animal eyes. First, however, this
chapter will digress to consider a radically different approach.

As motivation, we can note that even localization microscopy is limited to a
resolution of about a nanometer, which is still 10 times the size of an atom. But
structural biologists need to know the internal structures of molecules, down to the
placement of individual atoms. X-ray diffraction imaging is a much older technique
that can give this kind of detailed information.

In earlier chapters, diffraction has been presented as something bad, a limitation
on direct imaging. So it may be surprising to find that diffraction can instead be our
friend, and in fact underlies the deduction of molecular structures at the atomic scale!
How, and when, this works is the subject of this chapter. We will see that an x-ray
diffraction pattern is a particular transformation of the spatial arrangement of atoms.
To find that arrangement, we must *invert* the transformation.

We will pursue one particularly significant case study, the discovery of the structure
of DNA. Remarkably, we'll see that the key architectural features of this molecule can
be read out from a single diffraction pattern obtained in 1952.

The Focus Question is

Biological question: How can we find the details of macromolecule architecture? That
substructure is too small to resolve!

Physical idea: X-ray diffraction patterns contain structural information.

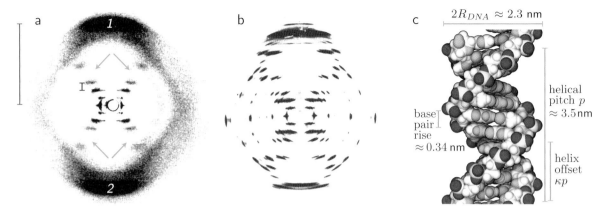

Figure 8.1: **Structure of DNA.** (a) [X-ray diffraction data.] R. Franklin and R. Gosling's historic x-ray diffraction pattern from a DNA fiber sample. Darker regions correspond to high photon arrival rate on a photographic plate. *Green arrows* point out the "missing" spots discussed in the text. The heavy blobs labeled *1, 2* are discussed in Section 8.4.1 (page 280). The *red marker* shows a screen separation that is one-tenth as large as that of the *blue marker;* Section 8.4.1 will explain why this ratio implies a structure with about 10 base pairs per helical turn. (b) A later, higher-resolution version of the same pattern. (c) [Mathematical reconstruction.] Structure of the DNA molecule, showing four key parameters (see also Media 18). Section 8.4.1 defines the parameters and shows how to obtain their values from the diffraction pattern. In this representation, the phosphorus atoms constituting the backbone are colored *greenish-yellow;* nitrogen atoms in the bases are *blue*. All three panels are for DNA in its standard "B" form. [(a) Reprinted by permission from Macmillan Publishers Ltd: Nature, Franklin and Gosling. Molecular configuration in sodium thymonucleate. *Nature* (1953) vol. 171 (4356) pp. 740 741, ©1953. (b) From Langridge et al., 1957. (c) Art by David S Goodsell.]

8.2 IT'S HARD TO SEE ATOMS

To get a sense of the problem, suppose that we want to resolve structures on the scale of atoms, 0.1 nm, by using visible light and a projection screen some macroscopic distance away, say 0.1 m.

Your Turn 8A

a. Show that such a setup is very far from the ray-optics limit needed for direct imaging. [*Hint:* Review Equation 4.20 (page 167).]
b. Show that even if we use x rays, with $\lambda \approx 0.1$ nm, we are still deep in the diffractive regime.

Conceivably we could try imaging with gamma rays, using some sort of very near detector, but many technical difficulties get in the way, including destruction of the sample by such high-energy photons.

Instead, this chapter will sketch how we can find molecular structures by working deep within the diffractive regime.[1] For example, Figure 8.1a shows an icon from 20th-century science, the pattern projected on a screen when R. Franklin and R. Gosling passed x rays through a fiber consisting of many aligned DNA molecules. Although it is hardly a literal picture, this single image nevertheless contained the information needed to obtain the crucial aspects of DNA structure.

---

[1]The method we will describe was first applied to a biomolecule in 1937, when D. Hodgkin used it in her study of the structure of cholesterol.

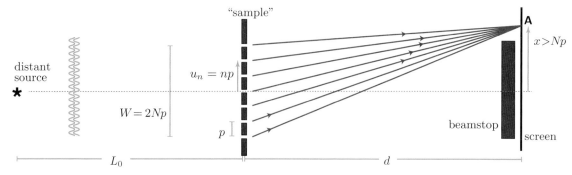

Figure 8.2: [Path diagram.] **First approach to understanding diffraction from a crystal sample.** Monochromatic light illuminates a "sample," in this case, a periodic array of narrow slits in an opaque barrier. (A similar arrangement was called a "transmission grating" in Section 5.3.2.) We observe the light intensity cast on a screen outside the sample's geometric shadow region. In practice, a "beamstop" (obstruction) is placed behind the sample, blocking the unscattered x-ray beam that would otherwise overwhelm the sensitive detector array. (The shadow of the beamstop appears as a blank circle in the middle of Figure 8.1a,b.) The schematic is not to scale; actually $L_0 \gg d$ (indicated by *wavy lines*) and $d \gg W$ and x, so that all paths are nearly parallel to the centerline.

The complete analysis of x-ray diffraction patterns is beyond the scope of our discussion. The rest of this chapter will build up just enough of the story to get a glimpse of how it can be done. We'll also see how the four key geometric parameters summarized in Figure 8.1c can all be inferred from the data in panel (a).

8.3 DIFFRACTION PATTERNS

8.3.1 A periodic array of narrow slits creates a diffraction pattern of sharp lines

Although our goal is to discuss patterns made by light scattered from atoms, we can warm up with a simpler problem that has a similar character. Instead of periodically arranged, localized scattering objects, we will consider light arriving at a periodic array of open slits in an opaque barrier. What's important is not the nature of the objects (slits versus atoms), but rather the fact that the objects are identical, identically oriented, and periodically placed.

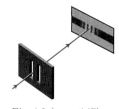

Fig. 4.2 (page 147)

We already have some experience with the case of one or two such slits; in particular, constraining light to pass through two narrow slits condenses the resulting screen illumination into an array of alternating light and dark bands.[2] Now consider a slightly more elaborate arrangement, in which the opaque barrier contains not two but K uniformly spaced, narrow slits, and for concreteness suppose that K is odd. Thus, we can write K as $K = 2N + 1$, where N is the number of slits above or below the one in the center. The slits are separated by a distance p, so the entire array runs from $u = -Np$ to $+Np$ (Figure 8.2).

We'll assume that

- The light source on the left is distant; and
- The projection screen is located at a distance d that is much larger than the "sample" size $W = 2Np$, and also much larger than the distance x on the screen.

---

[2]See Figure 4.2 (page 147).

This is a "small angle regime"; for example, the deflection angle $\tan^{-1}(x/d) \approx x/d$ is small.[3] (Earlier chapters also considered only this case.) We also assume that $|x| > Np$, that is, we look outside the region that would have been illuminated in ray optics.

Let $u_n = np$ denote the location of slit number n in the "sample." The probability amplitude Ψ at position x on the screen is then the discrete sum of $2N + 1$ terms:[4]

$$\Psi(x) = \text{const} \times \sum_{n=-N}^{N} e^{i\varphi_n}, \quad \text{where} \quad \varphi_n = (2\pi/\lambda)\sqrt{d^2 + (u_n - x)^2}. \tag{8.1}$$

We can simplify Equation 8.1 by using approximations that are valid in the assumed geometry, and by discarding three kinds of factors:

1. We are only interested in relative changes in illumination intensity on the screen. Thus, we can drop any overall multiplicative constant in Equation 8.1, where "constant" means independent of n and x.

2. Ultimately, we will only need the squared modulus of $\Psi(x)$, so we may also drop any overall factor (independent of n) whose modulus equals 1, even if it does depend on x. Equivalently, we may drop any real, additive term in φ_n if it's independent of n, because the exponential of (i times a real quantity) always has modulus equal to 1.[5]

3. Finally, any contribution to φ_n that is much smaller than 2π will make a negligible contribution, even if it depends on n and/or x. This observation lets us keep only terms up to first order in the small quantities x/d and u/d.

Rule **3** lets us approximate φ_n by the first terms of the Taylor series: $(2\pi/\lambda)\big(d + (u_n - x)^2/(2d)\big)$; rule **2** then lets us drop the first term of this expression. Expanding the square and using rule **2** again yields

$$\varphi_n \approx \frac{2\pi}{\lambda d}\Big(\frac{1}{2}(u_n)^2 - xu_n\Big).$$

We now introduce a second large-d approximation by noticing that, in practice, the sample size is much smaller than $\sqrt{\lambda d}$, so that $(u_n)^2/(\lambda d) \ll 1$. This approximation is the statement that we are far from the ray-optics regime; see Your Turn 8A. Together, the large-d limits give a regime sometimes called **Fraunhofer (far-field) diffraction**.[6]

Your Turn 8B

a. A typical situation in crystallography might have a sample size of $1\,\mu\text{m}$, x-ray wavelength $\lambda \approx 0.1\,\text{nm}$, and a screen located at $d = 1\,\text{m}$. Evaluate the last approximation made above.

b. A typical situation in a classroom demonstration might be "sample" size $\approx 1\,\text{mm}$, red visible light, and $d \approx 3\,\text{m}$. Is this far-field diffraction?

These approximations allow us to use less complicated math than the general case:

$$\varphi_n \approx \Big(-\frac{2\pi p x}{\lambda d}\Big)\, n. \tag{8.2}$$

---

[3] We used the Taylor series expansion for the tangent function (page 18).
[4] This expression is similar to the one appearing in Equation 4.16 (page 166).
[5] Section 4.4 and Appendix D review complex numbers.
[6] Note that the u^2 term was crucial for obtaining ray-optics behavior in Section 4.7.3, but now we are working very far from the ray-optics regime.

You'll complete the calculation of the probability amplitude in Problem 8.2. But already, we can understand the main qualitative consequences of Equation 8.2. First, notice that whenever the quantity $px/(\lambda d)$ is an *integer*, then all the phases are integer multiples of 2π, and so all the factors of $e^{i\varphi_n}$ are equal to one. Thus, in this case, all the contributions add: We expect uniformly spaced, bright bands on the viewing screen at the locations $x_\ell = \ell \lambda d/p$, where ℓ is any integer.[7] Notice that increasing p *decreases* the spacing of the pattern, so we get a **scale inversion**:

> *There is an inverse relation between the repeat distance in the sample and the spacing of the lines in the diffraction pattern.*   (8.3)

We may not seem to have found much that is new: Two-slit interference also led to a periodic illumination pattern.[8] But squaring Equation 8.1 shows that, at each maximum, the light intensity is proportional not to the number of slits K, but to K^2. The only way that this could happen is if each bright band is becoming *more concentrated* with larger K; you will confirm this prediction in Problem 8.2.[9] The increase in intensity is important because a macromolecule is not really an opaque barrier with an array of slits. Instead, each atom scatters light rather weakly, so the coherent amplification from many copies of a repeated motif (a crystal) is needed to get observable patterns with adequate intensity. Condensing the light into narrower bands (or spots) also helps an experimenter to resolve the more complex patterns that emerge from real macromolecules (for example, the one in Figure 8.1c).

Fig. 8.1c (page 273)

8.3.2 Generalizations to the setup needed to handle x-ray crystallography

Our analysis so far has given a key conclusion:

> *When a one-dimensional motif is periodically repeated, the usual diffractive blur condenses into isolated bright bands.*

Moreover, we found that a repeated linear object (an array of slits), with repeat period p, has a diffraction pattern that is also periodic. Measuring its period x_1 lets us find p via the formula $x_1 = \lambda d/p$, a determination of microstructure from a diffraction pattern.

We have not quite reached our destination, however, because

- We'd like to determine the *internal* structure of our objects, for example, DNA or protein molecules. We're less interested in how they are arranged into a periodic crystal, because generally in living contexts, macromolecules are not so arranged.
- A real crystal is a *three*-dimensional array of *atoms*, not an array of parallel slits in a plane.
- Moreover, in the original experiments of Franklin, Gosling, and M. Wilkins, the DNA molecules were *not* arranged in a perfect crystal.

The following sections will upgrade our first calculation to address these points.

---

[7] In classroom demonstrations, we generally illuminate with a narrow, circular beam of light, not a line source. In this case, the transmission grating expands the beam in the direction perpendicular to the slits, as above, but not in the direction parallel to the slits. The result is a line of bright *spots* on the screen, arranged perpendicular to the slits (see Figure 8.3b).

[8] Equation 4.10 (page 157) summarizes that result.

[9] Problem 4.4 (page 177) already showed a hint of this behavior.

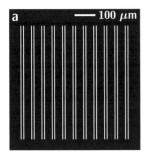

Figure 8.3: **Form factor of a 1D structure.** (a) [Drawing.] Detail of a transmission grating that has substructure. The arrangement shown corresponds to the fractional offset $\kappa = 1/4$ in Equation 8.5. (b) [Photograph.] Diffraction pattern when this grating is illuminated with a narrow beam of monochromatic, visible light. One out of every four spots is "missing"; that is, the spot intensities are modulated by a form factor arising from the double-bar pattern in (a), and that form factor is zero at certain spots. Had we used a wider illumination, then the spots would have been replaced by vertical bands. [(b) Courtesy William Berner.]

8.3.3 An array of slits with substructure gives a diffraction pattern modulated by a form factor

Suppose that we add to a strictly periodic array with spacing p a *second* array of slits, like the first but with a fractional offset given by a number κ between zero and one (Figure 8.3a). That is, the new slits are located at $u_n = (n+\kappa)p$. The same simplifying rules given in Section 8.3.1 now yield twice as many contributions:

$$\Psi(x) \approx \sum_{n=-N}^{N} \left[e^{-(i2\pi px/(\lambda d))n} + e^{-(i2\pi px/(\lambda d))(n+\kappa)} \right]. \tag{8.4}$$

Remarkably, the answer *factorizes* into the product of two simpler functions:

$$\Psi(x) \approx \left[\sum_{n=-N}^{N} e^{-(i2\pi px/(\lambda d))n} \right] \left[1 + e^{-(i2\pi px/(\lambda d))\kappa} \right]. \tag{8.5}$$

This factorized form is convenient, because

- The first factor in brackets, called the **structure factor**, is precisely the same as the corresponding expression for a simple array of slits;[10] for example, it does not involve the offset κ at all. It tells us about the repeating arrangement of objects (the "crystal structure") in our sample. In the example we are studying, the "objects" are pairs of slits and the structure factor once again has a set of sharp peaks located at integer multiples of $\lambda d/p$.
- The second factor, called the **form factor**, involves the *internal* arrangement of subelements in each individual object, in this case the value of κ.

As mentioned earlier, if the objects are macromolecules, then we are not very interested in the details of their periodic arrangement; the structure factor is then just a backdrop to the desired information, which is in the form factor.

Equation 8.5 gives $\Psi(x)$ as the product of these two functions of x. Thus, we can think of the form factor as *modulating* the intensities of the narrow spots generated

---

[10]Section 8.3.1 introduced this function; you'll work out some details in Problem 8.2.

by the structure factor. Figure 8.3b shows an example of this phenomenon. Because each repeated object must be narrower than the separation p between them, the scale inversion principle (Idea 8.3) implies that

> The fine structure of a diffraction pattern tells us about the repetitive arrangement of the constituent objects; the overall modulation tells us about the internal arrangement within each repeated object.

Your Turn 8C

a. To see how this works, note that we need only to evaluate the second factor in Equation 8.5 at the locations where the first factor is nonzero. So evaluate the form factor at $x_\ell = \ell \lambda d/p$, where ℓ is an integer (the "order" of the diffraction spot).

b. Comment on the cases where $\kappa = 1/4$ (shown in Figure 8.3b) and $\kappa = 3/8$.

You'll study the phenomenon just found in greater detail in Problem 8.2.

$\boxed{T_2}$ *Section 8.3.3′ (page 285) discusses Equation 8.5 from a broader viewpoint.*

8.3.4 A 2D "crystal" yields a 2D diffraction pattern

The next added element of realism is to consider a periodic *planar* array of small *holes* (not slits) in the opaque barrier. Now the viewing position $\boldsymbol{r} = (x, y)$ is a 2D vector, and similarly for the position $\boldsymbol{u} = (u, v)$ within the sample. For example, in a square lattice the hole labeled (n, m) is located at $\boldsymbol{u}_{nm} = (np, mp)$, where n, m are two integers, and Equation 8.2 is replaced by

$$\varphi_{nm} = (2\pi/\lambda)\sqrt{d^2 + (\boldsymbol{u}_{nm} - \boldsymbol{r})^2} \approx \text{const} + \left(-\frac{2\pi p}{\lambda d}\right)(nx + my). \qquad (8.6)$$

In order to get bright illumination on the screen, we must now require that φ_{nm} be a multiple of 2π for *every* (n, m) combination, or in other words, that *both* $px/(\lambda d)$ and $py/(\lambda d)$ be integers. That is, the diffraction pattern will be a square array of bright spots, with spacing $\lambda d/p$ in each direction.

Just as in Section 8.3.3, we can now generalize to more elaborate "crystals" consisting of regularly repeated motifs that have more structure than just small dots. Figure 8.4 shows the experimental diffraction pattern from a triangular lattice of rectangular objects. As before, the placement of individual bright spots is determined by the macrostructure of the sample (the triangular lattice); the *intensities* of those spots are modulated by the form factor, in a way determined by the *micro*structure (the rectangular objects).

8.3.5 3D crystals can be analyzed by similar methods

Section 8.3.2 pointed out that a real crystal is a *three*-dimensional array of objects, so we must generalize the analysis of the preceding section. We now suppose that the sample consists of small objects located on a cubic lattice, with positions $\boldsymbol{u}_{nmj} = (np, mp, jp)$. This case has a new feature: The objects are no longer all equidistant from the light source, so we may no longer drop the contribution to φ from the incoming part of

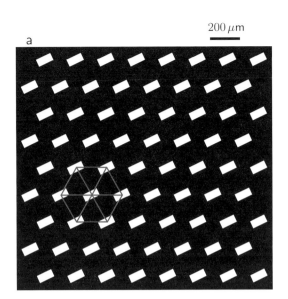

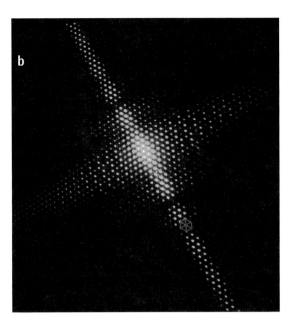

Figure 8.4: **Form factor of a 2D structure.** (a) [Drawing.] Detail of an array of rectangular objects, representing holes in an otherwise opaque barrier. The centers of the rectangles form a triangular lattice (*red lines*). (b) [Photograph.] Diffraction pattern formed on a screen when light with wavelength 543 nm was shone through the pattern in (a). Although the objects are far larger than molecules, and the light's wavelength is far longer than that of x rays, nevertheless the experimental geometry has been arranged to place the system in the diffractive regime, similar to that in x-ray diffraction experiments. The pattern is a triangular grid of dots (its microstructure, emphasized by *red lines*), corresponding to the macrostructure of (a). The dot intensities are modulated in a pattern with rectangular symmetry (the pattern's macrostructure), corresponding to the shape and orientation of the original rectangles in (a) (the microstructure). [(b) Courtesy William Berner.]

each path. Instead, the same rules given in Section 8.3.1 yield

$$\varphi_{nmj} = (2\pi/\lambda)\left(jp + \sqrt{(d - jp)^2 + (x - np)^2 + (y - mp)^2}\right)$$
$$\approx (2\pi/\lambda)\left(\cancel{jp} + d\left(1 + \tfrac{1}{2d^2}(\cancel{-2djp} + (jp)^2 + (x - np)^2 + (y - mp)^2)\right)\right) \quad (8.7)$$
$$\approx \text{const} - \frac{2\pi p}{\lambda d}(nx + my).$$

Here d is the distance from the plane $j = 0$ to the projection screen. That is,

> When we work in the limit of small angles and large d, the diffraction pattern of a 3D sample is the same as that for the 2D array obtained by   (8.8)
> squashing the sample into a single plane perpendicular to the beam.

Section 8.3 has built up a framework that predicts the x-ray diffraction pattern produced by a perfect crystal. The following section will show how we can get useful information even when perfect crystal samples are not available.

8.4 THE DIFFRACTION PATTERN OF DNA ENCODES
ITS DOUBLE-HELICAL CHARACTER

Earlier sections in this chapter pointed out that if we need atomic resolution, then diffraction will always be a dominant effect. And in the diffractive regime, photon arrivals do not reflect the structure of their source in a obvious way. Nevertheless, there is a mathematical relation between the source and the resulting photon arrivals; we worked out that relation in a series of increasingly realistic situations.

We can now apply some of these insights to DNA. Even a single molecule of DNA has the sort of repeating structure that we have been studying, due to its helical architecture.

Fig. 8.1c (page 273)

8.4.1 The helical pitch, base pair rise, helix offset, and diameter of DNA can be obtained from its diffraction pattern

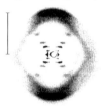

Fig. 8.1a (page 273)

Figure 8.1c shows two levels of repeating motifs. The smallest-scale periodic structure consists of the thin, horizontal, planar **base pairs** emerging out of the page. Imagine illuminating such a stack with monochromatic light coming from the direction perpendicular to the page. Then the scale inversion principle, Idea 8.3 (page 276), implies that the largest-scale part of the diffraction pattern will be a vertical array of spots; the first of these appear as the large blobs at the top and bottom of panel (a).[11] Measuring their separation (blue bar in Figure 8.1a) gives the **base-pair rise**, or vertical distance between base pairs, as 0.34 nm.

The longer-scale periodic structure will give rise to finer-scale features in the diffraction pattern. To understand this structure, note that the phosphate groups (greenish-yellow atoms in Figure 8.1c) are arranged on two helical paths. Each turn of the backbone helix repeats after it winds through a full circle around the central line, creating a periodic structure at longer spatial scale than the base-pair rise. Panels (a,b) show a subdivision of the space from the center to the large blobs into 10 horizontal levels, which gives the **helical pitch** $p \approx 0.34\,\mathrm{nm}/(1/10) \approx 3.5\,\mathrm{nm}$.[12]

To understand the "cross" structure of the diffraction pattern, recall that a 3D structure's low-angle diffraction pattern involves the projection of the structure into the plane perpendicular to the x-ray beam (Idea 8.8). The projection of a helix to a plane containing its axis is a sine wave.[13] A sine wave looks approximately like a series of left-leaning straight lines, or "zigs," alternating with a series of right-leaning lines, or "zags." Each of those repetitive series of lines makes a diffraction pattern that is a tilted row of spots, leading to the two arms of the cross seen in the diffraction pattern.

The fact that some spots are missing in the DNA diffraction pattern (Figure 8.1a,b) implies a form factor.[14] In fact, the pattern is consistent with the presence of *two* offset series of zigs and zags, with a relative displacement of 3/8 of a helical repeat.[15] The fact that the shift is not $p/2$ implies that the grooves between the two helical

---

[11] $\boxed{T_2}$ The beamstop has removed the zeroth-order spot that would have occupied the center of the pattern (along with unscattered x rays).

[12] Some of these equally spaced levels have their spots missing, but we count them anyway when measuring p. See below. More precise determinations give the helical pitch as slightly more than 10 base pairs, leading to the estimate given here.

[13] Figure 4.5a (page 151) illustrates this geometrical fact.

[14] Figure 8.3b (page 277) shows the missing-spot phenomenon for the case when the offset equals $p/4$.

[15] You'll establish this result in Problem 8.2.

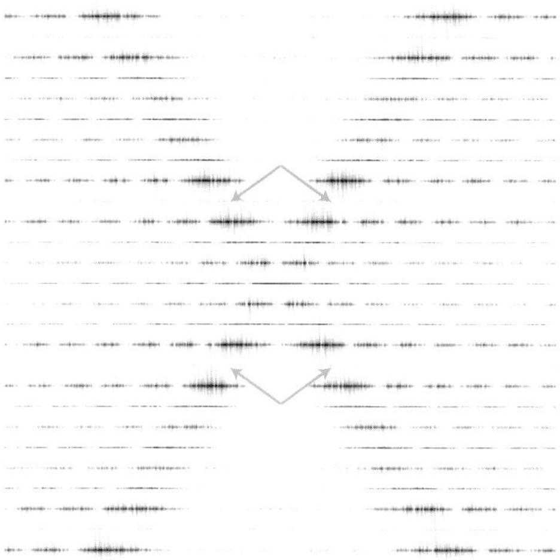

Figure 8.5: [Mathematical function.] **Calculated diffraction pattern of DNA.** The calculation made a simplified representation of a DNA fiber sample, showing the overall cross pattern and missing spots (*arrows*), analogous to those in the real experimental data (Figure 8.1a,b). In this figure, grayscale represents the probability density function for photon arrival at a detector array, as in experimental diffraction patterns. See Problem 8.3.

backbones have differing widths, with important consequences for DNA's function.

Figure 8.5 shows the result of a more detailed calculation of the x-ray diffraction pattern. The double helix was idealized as a pair of sine curves, and the base pairs themselves were omitted. As a result, Figure 8.5 shows the cross, including the absence of the fourth spot, but not the top and bottom blobs seen in the real diffraction pattern.[16]

Fig. 8.1c (page 273)

There is one additional architectural parameter shown in Figure 8.1c, namely, the radius R_{DNA} of each helix. X-ray diffraction gives us access to the slightly smaller distance R_p from the central axis to the center of the phosphate groups, because most of the electrons that can scatter are located there. To estimate it, first notice that we can construct a helix by wrapping a right triangle around a cylinder of radius R_p. Specifically, draw a right triangle on a piece of paper with one leg of length $2\pi R_p$, the other of length p. When the triangle is wrapped around a cylinder with the second leg parallel to the cylinder axis, then its hypotenuse becomes a helix with pitch p (see Figure 8.6a). If we slice the cylinder on any plane perpendicular to its axis, then the helix always makes the same angle Θ with the circular cross section, where $\tan\Theta = p/(2\pi R_p)$. This Θ is also the angle by which the zigs and zags of the projected helix are tilted away from the perpendicular to the axis, so the rows of spots making up the cross in the diffraction pattern are tilted from vertical by $\pm\Theta$. A more complete analysis shows that, more precisely, 2Θ should be measured as the angular width of the empty regions at the top and bottom of the diffraction pattern, that is, about $60\,\mathrm{deg}$.

In short,

> The two arms of the cross in the diffraction pattern have an angular separation of $2\Theta = 2\tan^{-1}(p/(2\pi R_p))$. $\qquad(8.9)$

Thus, knowing the helical pitch $p \approx 3.5\,\mathrm{nm}$ from the layer spacing in the diffraction pattern, and measuring $2\Theta \approx 60\,\mathrm{deg}$ from Figure 8.1a, allows us to compute $R_p \approx 0.96\,\mathrm{nm}$, the radius of the helical chain of phosphate groups. Watson and Crick's more accurate analysis yielded $R_p \approx 1.0\,\mathrm{nm}$.

---

[16]You'll create such a picture in Problem 8.3.

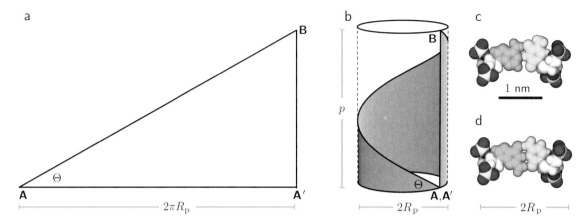

Figure 8.6: **DNA structural parameters.** (a,b) [Diagrams.] Illustration of the relation between the pitch p and radius R_{p} of a helix and the angle Θ. The right triangle **ABA**$'$ is rolled onto the surface of a cylinder so that **A**$'$ touches **A**. The diagram establishes a relation that lets us calculate R_{p} from measured values of p and Θ. (c) [Molecular structures inferred from x-ray crystallography.] Top view of a base pair of DNA. The complementary bases adenine and thymine, when positioned with their phosphate groups at a separation equal to the value in the DNA double helix, can be rotated to place all of their hydrogen-bonding sites into position for bond formation. Noncomplementary pairs of bases do not have this property, and so would create strained or weakly paired structures. Adenine and thymine form two such bonds. (d) Guanine and cytosine form three hydrogen bonds. [(c,d) Art by David S Goodsell.]

T2 *Section 8.4.1$'$ (page 285) discusses the final remaining item in the critique of Section 8.3.2, and some limitations on x-ray diffraction imaging.*

8.4.2 Accurate determination of size parameters led to a breakthrough on the puzzle of DNA structure and function

Watson and Crick went on to note that $2R_{\mathrm{p}}$ is just big enough for each chain's bases to sit inside the cage formed by the helical backbones, as shown in Figure 8.6c,d. If the base pairs are packed snugly inside the cage, however, then each base will touch its partner. It then becomes highly relevant that adenine and cytosine *fit onto* each other, and similarly guanine and thymine (see Figure 8.6c)—but adenine does not fit with adenine, and so on. Not only do the shapes of the molecules fit; also the "right" (complementary) partners bring together atoms that can form hydrogen bonds, whereas the "wrong" (noncomplementary) ones do not.

Watson and Crick therefore pointed out that

> *The detailed molecular dimensions of DNA suggest a base-pairing mechanism, in which each strand of the double helix contains a complementary duplicate of the other one's sequence.*

Strand complementarity implies that either strand can be used to direct the synthesis of its partner. Thus, to duplicate the genome, the cell "unzips" the two strands, then re-synthesizes a complementary strand for each one. No such elegant mechanism was possible in an earlier model of the molecule's structure, because that model placed the backbones in the center of a helix, with the bases directed outward, and hence unable to pair specifically with each other.

Section 8.4 has shown how even a noncrystalline sample can divulge structural information about its molecular constituents.

THE BIG PICTURE

This short chapter examined situations far from the validity of stationary-phase approximation, and developed another form of imaging as inference applicable to this case. Remarkably, here, too, we were able to extract structural information about objects from the pattern of illumination on a distant screen. Even though that pattern does not look like an "image" in any intuitive sense, we saw how to extract clues—a partial *inversion* of the transformation that generated the diffraction pattern.

The DNA story illustrates a new method for obtaining biological information by using light. It's also an outstanding example of the benefits of physical modeling for understanding living systems more generally. The riddle of the physical basis of heredity was solved by an imaginative approach to a biological problem (highly stable transmission of genetic traits), disciplined by the interplay of several kinds of constraints, including

- Biochemical facts (E. Chargaff's discovery that all organisms have equal quantities of adenine and thymine, as well as equal quantities of guanine and cytosine, in their DNA),
- Structural facts (the planar shapes and exact sizes of the individual base molecules),
- Geometrical facts (a stack of identical objects, with a uniform stacking rule, will generically form a helix), and
- X-ray diffraction, which yielded the key measurements (especially the helix radius and the existence and offset of the second helix deduced from the missing spots).

KEY FORMULAS

- *Diffraction pattern:* A 1D periodic array of narrow slits with spacing p, illuminated by a distant light source, will create a pattern of lines on a distant screen. At low angles, in the highly diffractive regime, the lines are located at positions x_ℓ that are integer multiples of $x_1 = \lambda d/p$. Here λ is wavelength and d is distance to the screen.
- *X-ray diffraction:* A 1D periodic array consisting of compound slits will create a pattern of lines at the same locations as for an array of single slits. But the lines' intensities will be modulated by a "form factor" determined by the substructure. For example, for two periodic arrays of slits with fractional offset κ, the form factor is

$$1 + \mathrm{e}^{-\mathrm{i}2\pi p x \kappa/(\lambda d)}. \qquad \text{[8.5, page 277]}$$

A 2D or 3D periodic array consisting of identical objects that selectively transmit (or scatter) light, illuminated by a distant light source, in the diffractive regime, will create a pattern of *spots* on a distant screen. If the objects have internal structure, then again the intensities of the diffraction spots will again be modulated by a form factor.

- *Determination of DNA radius:* The characteristic "cross" pattern of spots in the DNA diffraction pattern has opening angle equal to twice the tilt of the sugar-phosphate helix. That tilt is in turn given by $\Theta = \tan^{-1}(p/(2\pi R_\mathrm{p}))$. Thus, once p and Θ are measured by x-ray diffraction methods, the radius of the phosphate backbone may be determined by solving for R_p.

FURTHER READING

Semipopular:
Discovery of the structure of DNA: Judson, 1996.

Intermediate:
DNA: Nolting, 2009; Rhodes, 2006; Lucas & Lambin, 2005; Lucas et al., 1999.

Technical:
Rupp, 2010.

$\boxed{T_2}$ **Track 2**

8.3.3′ Factorization of amplitudes

When we studied a periodic array of repeated, complex objects (that is, a crystal) in the diffractive regime, we found a simplification of the probability amplitude for x-ray diffraction: Equation 8.5 (page 277) displayed the result as the *product* of two functions. One factor is determined by the crystal structure (it does not depend on the nature of the repeating objects). The other factor depends only on the repeating objects, not the crystal structure.

This result stems from a more general argument:

- The geometry of x-ray diffraction justifies some approximations, which led us to Equation 8.4 (page 277). This formula expresses the probability amplitude mathematically as the Fourier transform of the slit geometry.

- When a motif is repeated in space, we can regard the overall pattern as a convolution of two functions:[17] One function describes an individual instance of the motif; the other describes the spatial arrangement.

- A theorem from mathematics states that the Fourier transform of a convolution equals the product of the Fourier transforms of the individual functions. The derivation of Equation 8.5 proves a special case of this theorem.

This argument continues to hold in two or three dimensions. It also holds if we substitute more complex crystal structures, and if we use more realistic electron-density functions, in place of the simple ones discussed in the main text.

$\boxed{T_2}$ **Track 2**

8.4.1′a How to treat fiber samples of DNA

Section 8.3.2 mentioned that actually, the DNA sample Franklin and Gosling used to obtain Figure 8.1a was not perfectly organized into a single crystal. Instead, the experimenters prepared "fiber" samples by drawing out filaments consisting of many parallel strands from a concentrated solution of DNA, much as one can pull up a bundle of seaweed on the oar of a rowboat. The strands were all roughly parallel to one another, but were otherwise disorganized; nevertheless, they allowed Gosling and Wilkins to obtain the first x-ray diffraction patterns of DNA. Franklin and Gosling perfected the method, obtaining images that were good enough to allow structural deductions.

Thus, the sample consisted of many strands of DNA, each with its helical axis along a common direction, but otherwise randomly arranged. Despite this partially disordered arrangement, each *individual* DNA molecule has a great deal of regularity: It consists of many repeating helical turns. In Problem 8.3 you'll generate computer simulation images like the one in Figure 8.5 (page 281), and show that the features of the x-ray diffraction pattern that we found in Section 8.4.1 (page 280) are retained in the case of a fiber sample.

---

[17]Section 0.5.5 (page 16) introduced convolution.

Fig. 1.7a (page 40)

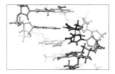

Fig. 1.7b (page 40)

Later researchers succeeded in making true crystals from short identical stretches of DNA. The structures shown in Figure 1.7 were obtained by x-ray crystallography on such samples.

8.4.1′b The phase problem

Although an x-ray diffraction pattern gives us many clues to a molecule's structure, in general those clues are not enough for us to determine that structure completely. That's because we cannot measure Ψ directly, and so lack the starting point for a mathematical inversion of the transform studied in the text. Instead, we only measure $|\Psi|^2$, which differs in that the phase of the probability amplitude has been discarded. Many clever techniques have been invented to get around this "phase problem," notably including a direct mathematical method (limited to small molecules) and multiwavelength anomalous diffraction (for proteins), in which sharp changes in diffraction pattern are observed as the x-ray wavelength is scanned past an absorption line for selenium atoms. (Selenium is introduced into the protein samples via a modified version of the methionine residue.)

This chapter instead took an indirect approach, which was to begin with a proposed structure, predict what diffraction pattern it would yield, and compare back to the data.

PROBLEMS

8.1 *Transmission grating*

Figure 8.2 represents a transmission grating with a monochromatic light source located far away along the center line. If there were a single, wide aperture of width W, then we wouldn't see much light at the screen point indicated. But with many narrow slits, it's possible. Get an approximate formula for the locations x that will be illuminated, in terms of the energy per photon and the distances d and p. You can assume that $p \ll W$ (there are many slits), and $W < x \ll d$. All you need to know about the distance to the light source is that it's much greater than any other distance in the problem.

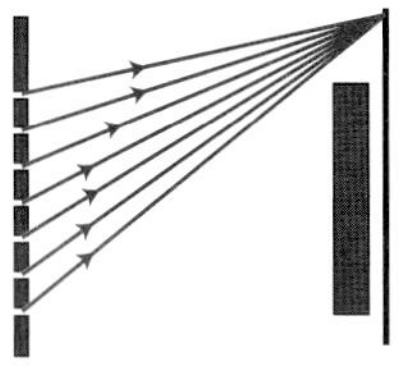

Fig. 8.2 (page 274)

8.2 *Form factor*

This problem illustrates the determination of the structure of a macromolecule from its x-ray diffraction pattern, in a simplified, two-dimensional context. First review Sections 8.3.1–8.3.3, particularly Your Turn 8C.

A real experiment involves a crystal consisting of many identical copies of the macromolecule of interest, all oriented the same way and regularly spaced at multiples of some distance p. To simplify the math, Section 8.4 (page 280) considered the diffraction of x rays from just *one* molecule. In this problem, you are to simplify still further, replacing the long, repeating molecule by an array of narrow slits in an opaque barrier. Specifically, consider a linear, regularly spaced array of K slits like the one in Section 8.3.1, characterized by a spacing p. Let $K = 2N + 1$ for some integer N.

The sample is illuminated by a distant light source, and the resulting interference pattern is projected on a screen at distance d from the sample (Figure 8.2). Let x be distance on the screen away from the centerline. Suppose that $d \approx 1\,\mathrm{m}$ is much larger than the size of the crystal (Np), and also much larger than the largest value of x that we will examine. We also make the additional large-d approximation discussed in Section 8.3.1 (far-field diffraction).

a. Find an analytic formula for the finite sum $\sum_{n=0}^{M} \xi^n$. [*Hint:* Your result from part (b) of Problem 0.4 will be useful.]

b. Write an analytic expression for the probability amplitude $\Psi(x)$ at a position x on the projection screen. Include only paths that consist of straight lines from some slit to the observation point on the screen, and treat each slit as infinitely narrow. Your answer should be expressed as a sum; don't evaluate it yet.

c. Simplify your answer by using the approximations made in Section 8.3.1 (page 274). Then use the identity in part (a) of this problem to evaluate your answer.

d. Although in general the Light Hypothesis gives the probability amplitude as a complex number, in this special case $\Psi(x)$ is a constant, times a factor that has modulus equal to 1, times a *real* function of x. Find that real function.

e. There are several dimensional parameters in the problem, but they appear in just one dimensionless group: $\bar{x} = xp/(\lambda d)$. Get a computer to graph $|\Psi|^2$ as a function of $\bar{x}$ over an interesting range. Make graphs, showing the cases with $K = 2, 3, 7$, and 9 slits. [*Hints:* The formulas in the text were only applicable for odd values of K, so make necessary changes for $K = 2$ (or recall Problem 4.3). You already worked out $K = 3$ in Problem 4.4.] Opening K slits allows K times as much light to pass as with a single slit; implement this by giving your functions appropriate normalizations.

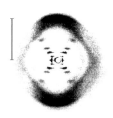

Fig. 8.1a (page 273)

Fig. 8.3b (page 277)

To make the calculation a bit more reminiscent of x-ray diffraction from macro-molecules, Section 8.3.3 (page 277) considered periodically repeated structures more complicated than a single slit. In addition to the slits located at np, we introduced an additional K slits located at $(n + \kappa)p$. Here κ is a fixed number between 0 and 1, a fractional offset between the two sets of slits.

f. Set $K = 7$ and make three graphs, showing the cases with $\kappa = 1/2$, $3/8$, and $1/4$. Comment on the three graphs and how their features relate to the values of κ. Discuss the relation to Figure 8.1a and Figure 8.3b.

g. $\boxed{T_2}$ Actually, the slits used to prepare Figure 8.3 were not infinitely thin. Rather, each had a thickness equal to $0.067p$. Add this additional element of realism to your model and make a plot of $|\Psi|^2$ for the case $K = 7$, $\kappa = 0.25$. [*Hint:* The needed modification to (f) is another overall multiplicative factor depending on $\bar{x}$.]

8.3 $\boxed{T_2}$ *DNA diffraction pattern*

In this problem, you'll predict the diffraction pattern expected from a simplified representation of the DNA molecule. You'll compute the pattern in a form appropriate for fiber samples like the ones originally used by Franklin and Gosling.

Section 8.3.5 argued that, in a certain approximation, the phases of photon paths for diffraction from a three-dimensional structure are given by Equation 8.7 (page 279). Thus, for a square lattice of identical objects, located at (na, ma, ja) within the sample, the resulting probability amplitude at the projection screen is

$$\Psi(x, y) = \sum_{n,m} \exp\left[-\frac{2\pi i}{\lambda d}(nax + may)\right].$$

(You can drop an overall multiplicative constant.)

Now adapt this derivation to a more realistic problem by imagining the sample to be a grid of points, some of which are occupied by "atoms." To do this, introduce a function $F(n, m)$ which equals zero if grid point (n, m) contains no object, or a nonzero value otherwise. In particular, we can limit to a finite-size sample by arranging for F to equal zero when $|n| > N_{max}/2$ or $|m| > N_{max}/2$.

a. Express the problem by setting up appropriate dimensionless quantities $\bar{x}$ and $\bar{y}$, in such a way that

$$\Psi = \sum_{n,m} F(n, m) \exp\left[-(2\pi i/N_{max})(n\bar{x} + m\bar{y})\right]. \tag{8.10}$$

b. Set up F as a square array of zeros, with size $N_{max} = 2048$, in your computer math package. Take the physical grid spacing to be $a = 0.04\,\text{nm}$. Set some of the entries of F equal to one, so that the nonzero entries are arranged on a sine curve, stretching horizontally across the array. [*Hint:* When a helix is projected down to a plane, it becomes a sine curve. Take the peak-to-valley height of this curve to be $2\,\text{nm}$, approximately the diameter of the DNA helix backbone. The peak-to-peak wavelength is the helical pitch of DNA, which is $p = 3.5\,\text{nm}$. Then define $\bar{p} = p/a$ and $\bar{R} = 2\,\text{nm}/a$, and use these dimensionless quantities when placing values into F.]

c. Create 100 such sine curves, each with the right amplitude and wavelength but each displaced by Uniformly distributed, random distances in n and m. Add up all the corresponding contributions to F, creating a simulated fiber sample.[18]

d. You may recognize Equation 8.10 as the discrete **Fourier transform** of F. Your computer math package probably has a fast algorithm for calculating this transformation, evaluated at a grid of integer values for $(\bar{x}, \bar{y})$. Compute this, and display its modulus squared as a grayscale image, similar to Figure 8.5. [*Hint:* Check carefully the exact definition of the Fourier transform used by your computer math package, which may be different from Equation 8.10. If so, you'll need to study the documentation to see how to use that built-in transform.]

e. If you don't like your image, it may help to make a histogram of the values in $|\Psi|^2$, and see if you can rescale them in a way that brings out the smaller ones prior to displaying your image.

f. Finally, upgrade your calculation from 100 single sine waves to 100 *pairs* of sine waves, to simulate *double*-helical molecules. Each pair has one copy shifted axially by 3/8 of a wavelength. Find the key qualitative change in the simulated diffraction pattern and comment.

---

[18]This problem can be done by using a separation into form factor and structure factor, analogously to what was done in Section 8.3.3, but it is easier not to bother with the associated manipulations, and instead to proceed directly.

CHAPTER 9

Vision in Dim Light

> The final goal of any visual system—biological, chemical, or electronic—[is] the ability to detect or count individual photons. A finite quantity of light means a finite number of photons, and a finite number of photons means a finite amount of information. Only by counting every photon can the total information in the light flux be extracted.
>
> — *Albert Rose*

9.1 SIGNPOST: *CONSTRUCTION*

Living creatures must gather information about their environment, *construct* an inferred model of the world, and then act appropriately. The ability to gather, sense, and interpret light gives an animal an especially rich source of information. Vision can report situations very far from the individual, and hence act as an early warning system—or even inspire us with the beauty of the night sky.

Previous chapters have touched on many aspects of vision: We studied the nature of light, its absorption by molecules, our visual system's partial ability to discriminate color, and the formation of images. We have also laid some groundwork by studying how to form optimal inferences from incomplete or noisy information. Chapters 9–11 will assemble these ideas into a more detailed picture of visual transduction. We'll also draw some more general lessons about other sensory systems.

This chapter begins the journey by documenting a remarkable aspect of our vision: The lumpy character of light is relevant, all the way to the point of influencing behavior. The Focus Question is
Biological question: What sets the absolute limit to night vision?
Physical idea: Light is granular—you can't see half a photon.

9.2 THE EDGE OF VISION

This chapter will focus on a set of questions that are critical for many animals: What sets the limit of visual sensitivity? How close are humans and other animals to that limit? Once we have characterized our visual system, we will be in a position to ask about the mechanisms that have evolved to implement its performance in Chapter 10.

9.2.1 Many ecological niches are dimly lit

Clearly, night vision confers a selective advantage. An animal with night vision can forage with less risk from predators; conversely, many predators developed night vision in order to exploit nocturnal prey.

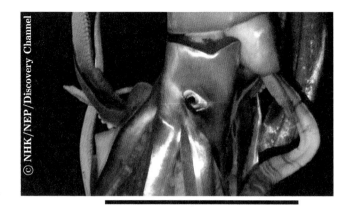

Figure 9.1: [Photograph.] **Detail of the giant squid *Architeuthis.*** This individual's total length was about 7 m. Others are much larger; some have been found with eyes (*center*) up to 37 cm in diameter. [See also Media 19.]

1 m

How dark does it get? Night levels of illumination range from about 10^{-6} times that of daylight (full Moon) to another 100 times smaller (starlight, no Moon). In other ecosystems, the availability of light is even more constrained: For example, even on a bright day the illumination that penetrates to 700 m below the ocean surface never exceeds that of starlight on the surface.

Beyond 1000 m depth, no useful sunlight is available at all, but even in this "bathypelagic zone" some creatures carry their own light sources to attract and illuminate prey. Still others rely on bioluminescent plankton, which disclose the presence of otherwise invisible predators by glowing when disturbed. For example, giant deep-sea squid apparently use vision in this way to spot sperm whales at great distances and take evasive action. To capture such faint signals, these squid have evolved eyes as large as basketballs (Figure 9.1)!

9.2.2 The single-photon challenge

Even before the photon concept was widely accepted, the physicist H. Lorentz realized around 1911 that, if light were indeed lumpy, that physical fact would impose an ultimate limit for visual sensitivity. Lorentz knew that physiologists had already roughly determined the energy delivered by the faintest flash that a subject can reliably detect. So he simply divided that value by the energy in a single visible photon.[1] Remarkably, the answer corresponded to roughly 25–150 photons arriving at the eye. Lorentz knew that some of these photons were unavoidably lost to absorption within the eye, reflection at its surfaces, and so on, so the actual number reaching the eye's retina must be even smaller. So clearly, animal photoreceptors must be rather close to the ultimate limit of performance: single-photon sensitivity. And yet, such performance seemed absurd: The energy of even 100 visible photons is many orders of magnitude smaller than the tiniest disturbance detectable by the sense of touch.[2] And even if Nature could design a photoreceptor with such exquisite sensitivity, wouldn't there be an enormous amount of noise from other sources? *How could anything like that possibly happen at all?*

---

[1] That energy is given by the Einstein relation (Equation 1.6, page 34).
[2] Recall the comparison in Problem 1.4 (page 59).

transduction module (parameters Q, μ_0 fixed)   decision module (parameter t flexible)

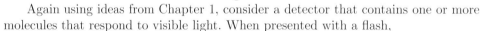

photons in

photoreceptors (rod cells)

relay (rod bipolar cells)

further processing

verbal report out

Figure 9.2: [Schematic.] **A model to characterize human visual performance in dim light.** The *dashed boxes* represent an abstract division of detection into two "modules," characterized by a quantum catch Q, false-positive parameter μ_0, and threshold t. The text defines these parameters more precisely. The *blobs* represent actual wetware elements to be discussed in this chapter and the two that follow. The rod bipolar cells straddle the boundary between abstract modules because processing at the first visual synapse affects the values of Q and μ_0, whereas the summation that bipolar cells perform will be considered part of the decision module.

9.2.3 Measures of detector performance

Before asking whether animal eyes really are ideal detectors, we should sharpen the notion of "ideal" into some quantitative figures of merit.

We might imagine an ideal detector as an instrument that, when presented with a flash of light, accurately reports the total energy that it contains. But confirming such performance, or measuring how a real detector falls short of the ideal, is not straightforward: Chapter 1 showed that the total energy of a monochromatic flash is a constant times the number of photons, but for ordinary light sources, the photon count varies uncontrollably, even if we try to make the flashes repeatable (recall Figure 1.2). Thus, a better statement of our goal is that we wish to describe how much *additional* randomness our eyes impose on the incoming light signal, beyond the minimum inherent in the nature of light.

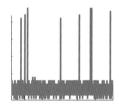

Fig. 1.2b (page 27)

Again using ideas from Chapter 1, consider a detector that contains one or more molecules that respond to visible light. When presented with a flash,

- The detector will randomly "lose" some of the arriving photons. For example, the lens of the human eye absorbs some photons before they even reach the retina. We can quantify this notion by stating the detector's **quantum catch Q**, the fraction of arriving photons that are productively absorbed. An ideal detector would have quantum catch equal to 100%.

- The detector may generate signals even in the *absence* of any light, analogously to the problem of false positives in other decision contexts. An ideal detector would have a rate of false-positive generation equal to zero.

- The right kind of molecule in the detector may absorb a photon, yet the detector may nevertheless fail to generate any signal unless some *minimal number* of photons was absorbed—a **threshold** requirement. An ideal detector would have threshold equal to one, that is, no threshold requirement.

Figure 9.2 imagines a dim-light detector as being logically divided into a "transduction module" (characterized by its quantum catch, false positive rate, and threshold), followed by a "decision module" (possibly introducing its own loss, noise, and/or

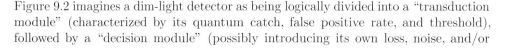

threshold).[3] With that viewpoint, it seems reasonable that each module can take various strategies with the signal it receives. It can choose to "clean up" a noisy input by imposing a high threshold, and in this way achieve a low false-positive rate at the expense of missing some real events.[4] Alternatively, each module could take the opposite strategy, and report every event, including the false positives. A really sophisticated detector may even be flexible enough to *change* strategies, and hence its performance, depending on the task at hand.

This chapter will present evidence that the rod cells in our eyes operate as transduction modules with high, though not perfect, quantum catch, and with "rod threshold" equal to one (no thresholding). Chapter 11 will go further, arguing that additional processing, downstream of the rod cells, suppresses false positives by imposing an *adjustable* criterion (a "network threshold") on the number of signals that must be generated by a collection of rod cells before the conscious mind registers that a flash of light has been seen.

Section 9.2 has introduced a framework for thinking about the detection of very dim light.

9.3 PSYCHOPHYSICAL MEASUREMENTS OF HUMAN VISION

Lorentz's observation led to many further attempts to quantify human visual sensitivity, culminating in key experiments by S. Hecht and coauthors, and by H. van der Velden, in the early 1940s. This section describes only the work of Hecht and coauthors; van der Velden's experiment was similar and was done independently.

Both groups performed **psychophysical** experiments, that is, experiments in which the input is characterized in precise physical terms, but the output is a verbal report by a human subject. Certainly human subjects are extraordinarily complex; their subjective reports of perception are potentially affected by a host of factors, including psychological effects. Yet these early experiments yielded a remarkably strong result, decades earlier than would have been possible had the researchers waited for single-cell techniques to become available. Their pioneering work forms a compelling case study in the value of indirect, probabilistic reasoning.

9.3.1 The probabilistic character of vision is most evident under dim-light conditions

In order to quantify dim-light vision at its absolute limit, Hecht and coauthors were careful to work under ideal conditions.

Optimal conditions: Dark adaptation

Immediately after turning off the lights, we experience poor night vision for a few minutes. In fact, our eyes' sensitivity increases rapidly in the first second as our pupils enlarge, but then other mechanisms of **dark adaptation** continue, until after 20 minutes our visual sensitivity is more than 2000 times greater than initially. The experimenters had their subjects wait 40 minutes in the dark prior to beginning the experiment, in order to get all the way to maximum sensitivity.

---

[3] $\boxed{T_2}$ We will use this model to understand dim-light vision only. Our visual system has different circuits, in parallel with this one, for vision in brighter light.

[4] Other cellular reaction networks also display threshold behavior, in some cases for a similar advantage.

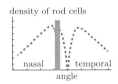

density of rod cells

nasal   temporal

angle

Fig. 3.9c (page 120)

Optimal conditions: Location on the retina

There are big differences in visual response, depending on what part of the retina receives the light. For example, the center of our visual field (the fovea) is devoid of the most sensitive photoreceptors, the rod cells (Figure 3.9c). The experimenters focused their flash on a small area (a "spot") about 20 deg away from the fovea.[5]

Optimal conditions: Wavelength

Dim-light sensitivity depends on the wavelength of the light: Rod cells have a spectral sensitivity that peaks in the blue-green part of the visible spectrum (wavelength around 507 nm). The experimenters chose to present flashes with that wavelength, in order to maximize their subjects' visual sensitivity.

Optimal conditions: Flash duration and spot size

Imagine several series of flashes, each delivering the same mean photon number but over a different time interval for each series. When the photons are spread out over too long a time, then the flashes become harder to see. Experiments prior to Hecht's had shown, however, that visual sensitivity is unaffected by flash duration, as long as it is shorter than about 200 ms (the rod cell's **integration time**). The experimenters chose a flash duration of 1 ms, well below the integration time.

Similarly, imagine several series of flashes, each delivering the same mean photon number but spread over a different spot size for each series. When the photons are spread out over too great an area on the retina, then the flashes become harder to see; again, however, earlier work had shown that visual sensitivity is unaffected by spot size, as long as the spot is no larger than about 10 arcmin in angular diameter. The experimenters accordingly used a spot angular diameter equal to 10 arcmin.

Probability of seeing

Even under optimal conditions, human subjects shown dim flashes were found to give partially random responses. Physiologists were well acquainted with randomness in perception, but had previously attributed it to the complex network of processes occurring in a subject's sensory apparatus, the subject's wandering attention, and so on. Following logic like that of Section 9.2.3, Hecht and coauthors suspected that on the contrary, the observed randomness in human responses was mainly due to the physical character of light.

Earlier chapters have given us some experience with hypotheses of this sort: Testing them requires *multiple trials.* As we make more trials, we gain greater conviction that the hypothesis is right (or wrong). Hecht and coauthors found that, for any particular flash strength, their subjects' responses were independent Bernoulli trials with a particular probability, which they called the **probability of seeing**, $\mathcal{P}_{\text{see}}$. Specifically, the experimenters trained their subjects to reply "yes" in a trial only if they were sure they saw a flash, that is, to minimize false-positive errors. After this training, the subjects were shown 50 flashes at each of several strengths. For each strength, $\mathcal{P}_{\text{see}}$ could then be estimated as the fraction of flashes that elicited a "yes" response.

Figure 9.3 shows the results for one subject. As expected, the curve does not correspond to decisively flipping a switch at some particular flash strength. Instead it is a sigmoid.

---

[5] Viewing with the right eye, the flash appeared 20 deg to the left of center. Thus, its inverted image on the retina was displaced horizontally from the fovea in the "temporal" direction (toward the temple, away from the nose). Angular displacement from the center is sometimes called "eccentricity."

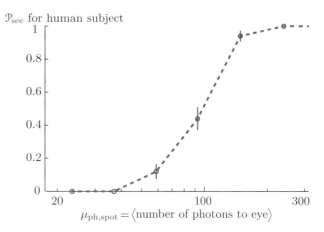

Figure 9.3: [Experimental data.] **Probability of seeing for a human subject.** Semilog plot of $\mathcal{P}_{\text{see}}$, estimated as the fraction of 50 trials in which a human subject reported seeing a faint flash of light, versus the expectation of the number of photons delivered to the subject's eye in each flash. The *dashed line* is not a fit; it just joins the points. The error bars represent the standard deviation of the posterior distribution for the parameter of a Bernoulli trial that was repeated 50 times (Problem 7.5). [Data from Hecht et al., 1942.]

9.3.2 Rod cells must be able to respond to individual photon absorptions

Hecht and coauthors also made a remarkable deduction by combining their psychophysical results with known physical facts about eyes. From optical measurements of the absorption of light by the eye lens and other elements, they estimated that only about half of the incoming photons actually arrived at the retina. From measurements of the absorption of light by the pigment molecule present in rod cells, they further estimated that about 20% of the surviving photons were absorbed by such molecules, for a net capture of about $50\% \times 20\% \approx 10\%$ of incoming photons. Not every absorbed photon is *productively* absorbed, that is, absorbed in a way that leads to a rod signaling event, just as not every photon absorbed by a fluorophore leads to fluorescence. Nevertheless, we can say that no *more* than about 15 are productively absorbed, on average, when 150 photons arrive at the eye.[6] Figure 9.3 suggests that the subject shown could reliably see flashes that delivered an average of 150 photons to the eye. Such flashes are very likely to create at least 10 productive photon absorptions, so that number is a rough estimate for the overall threshold of the human dim-light visual system.

Next, the experimenters noted that the spot size used in their experiment was large enough to spread this handful of events over several hundred rod cells. It was therefore very unlikely that any one rod cell would capture more than *one* photon.[7]

Your Turn 9A

a. Use Figure 3.9c (page 120) to estimate the surface density of rod cells at $20\,\text{deg}$ displacement in the temporal direction.

b. Then estimate the number of rods covered by a circular spot created by a light source with angular diameter $10\,\text{arcmin}$. [*Hints:* See Section 6.7 (page 230) for angular units. Approximate the eye as just a curved air-water interface located $24\,\text{mm}$ from the retina and use Equation 6.12 (page 221).]

Thus, the eye can reliably detect flashes of light so faint that no rod cell productively absorbs more than one photon. Assuming that each rod cell operates independently of the others, Hecht and coauthors were therefore led to conclude that

---

[6] $\boxed{T_2}$ Section 9.4.2′ (page 311) obtains more modern estimates of these numbers.

[7] You'll develop this argument in Problem 9.1.

- A rod cell must be capable of generating a signal based on a *single productive photon absorption* (no rod threshold).
- Some neural mechanism must combine the output signals of many rod cells prior to applying any other network threshold.

9.3.3 The eigengrau hypothesis states that true photon signals are merged with a background of spontaneous events

The experiment of Hecht and coauthors implied that a rod cell can respond to just a single productive photon absorption. Why, then, does the whole pathway demand a higher network threshold (roughly 10 photon absorption events, each exciting one rod cell) before it reports a definite flash to the subject's conscious mind?

H. Barlow proposed a physical model to answer this puzzle in 1956, using concepts introduced in Section 9.2.3:

Photoreceptor cells generate a random trickle of false-positive signals. Although individual cells impose no rod threshold, some level of processing downstream from them does impose a network threshold to eliminate distractions from these spurious events, when low false-positive reporting is required.

eigengrau
hypothesis

(9.1)

Barlow's hypothesis was rooted in other facts that were known at the time. For example, it was known that rod cells contain a pigment called **rhodopsin**, and that its absorption spectrum matches the spectral sensitivity of our dim-light vision. That result strongly suggested that rhodopsin was responsible for the first step of visual transduction.[8] It was also known that after exposure to light, rhodopsin photobleaches (loses its ability to absorb visible light). This observation suggested that the photobleaching could be the result of a light-induced change in the structure of the rhodopsin molecule (a photoisomerization[9]), somehow related to its signaling function. Barlow pointed out that the exact same conformational change could also occur from some other cause, for example, ordinary thermal collisions of the rhodopsin molecule with its neighbors. Thus, there would be some probability per unit time of spontaneously generating signals that were indistinguishable from those produced by real photons. We will refer to such baseline activity as the **eigengrau**, from German words meaning "one's own gray."[10]

Following Barlow, we can now sharpen Idea 9.1, with the ultimate goal of obtaining testable, quantitative predictions in Chapter 11.

Any particular rhodopsin molecule has a very small probability of absorbing any particular photon, or of spontaneously isomerizing, but each rod cell contains an enormous number of rhodopsins (about 1.4×10^8 in humans). Thus,

- In any short time interval, a rod cell can be thought of as either being presented with a photon or not, with a probability that depends on the mean photon arrival rate.[11]

---

[8]Cone cells contain closely related rhodopsin-like molecules with similar function but different absorption spectra (Figure 3.11, page 123).

[9]Section 1.6.4 (page 45) introduced photoisomerization.

[10]There are other potential sources for such activity besides spontaneous thermal isomerization. Section 11.2.2 discusses one example.

[11]Section 1.4.2 (page 36) connected mean rate to probability per time.

- When a rod cell is presented with a photon, at most one of its many rhodopsin molecules can absorb it and photoisomerize. The number of photoisomerizations is therefore either zero or one. This two-outcome random variable is a Bernoulli trial, which thins the initial Poisson process of photon arrivals.

- Some isomerizations may not trigger any signal from the rod (they may be "missed"). We model this effect by supposing that each isomerization is subjected to a further Bernoulli trial that determines whether it is missed or not.

- The thinning property of Poisson processes[12] then implies that, in dim light, rod signals follow a Poisson process, with mean rate proportional to the mean photon arrival rate.

- Each rhodopsin also has a small probability per unit time to isomerize spontaneously, for example, due to thermal agitation. Applying the merging property,[13] we finally predict that rod signals from all causes again constitute a Poisson process, whose mean rate is a constant plus the mean rate of photon-induced signals.

Each flash has a limited duration, so another property of Poisson processes lets us rephrase our prediction:[14]

> The total number of signals, ℓ, produced by a rod cell during a brief flash of light is Poisson distributed, with expectation given by the formula $\langle \ell \rangle = Q_{\mathrm{rod}}\, \mu_{\mathrm{ph,rod}} + \mu_{0,\mathrm{rod}}$.   (one rod cell)

$$(9.2)$$

In Idea 9.2, $\mu_{\mathrm{ph,rod}}$ denotes the expectation of the number of photons delivered to the rod cell. We can interpret Q_{rod} as the rod's quantum catch, and $\mu_{0,\mathrm{rod}}$ as the expected number of spontaneous events that occur during a flash (or near enough in time to be confused with it).[15]

Idea 9.2 may be reasonable, but it is not directly applicable to psychophysical experiments, because they do not record the output of individual rod cells. Moreover, the experimental methods available at the time could not illuminate individual rods. Barlow made some additional reasonable assumptions to connect his physical model to experiments:

- Psychophysical experiments supply light to the front of the eye, not directly to the retina. Each incoming photon thus has some probability of getting lost along the way, for example, by reflection from the cornea, absorption in the eye lens, and so on. That is, each suffers a series of independent Bernoulli trials, involving the probabilities of passing each of these sequential hurdles. By the thinning property, however, the net effect is still to supply photons to the retina in a Poisson process, albeit with reduced mean rate.

- Each photon arriving at the retina has some probability of hitting something other than a rod cell, further thinning the useful stream but preserving its Poisson character.

- The neural circuitry of the visual system *pools* signals from many rods in and near the stimulated area. Barlow made the simplest assumption, that this pooling

---

[12]Section 3.4.1 (page 111) introduced this property.
[13]Section 3.4.2 (page 111) introduced this property.
[14]Section 1.4.2 (page 36) found the distribution of counts over a fixed time interval.
[15] $\boxed{T_2}$ More precisely, the constant $\mu_{0,\mathrm{rod}}$ refers to spontaneous signals occurring during the rod cell's integration time, which may be longer than the actual duration of the flash.

consisted of evaluating the sum ℓ_{tot} of the individual rods' ℓ values in each integration time window.

Again applying the thinning and merging properties, Barlow concluded that the total count of rod signals must again be Poisson distributed. Its expectation must again be given by a formula similar to the one in Idea 9.2, but with different effective values for the quantum catch and eigengrau parameter:

> The total number, ℓ_{tot}, of rod signals during a brief flash is Poisson distributed. Its expectation depends linearly on the flash strength: $\langle \ell_{\mathrm{tot}} \rangle = Q_{\mathrm{tot}}\, \mu_{\mathrm{ph,spot}} + \mu_{0,\mathrm{sum}}$.

(whole eye)

(9.3)

This time, $\mu_{\mathrm{ph,spot}}$ denotes the expectation of the number of photons in the entire spot of light delivered to the eye, that is, the product of the photon arrival rate times flash duration. The quantum yield Q_{tot} is a new constant characterizing the chosen retinal region, not an individual rod cell. The eigengrau parameter $\mu_{0,\mathrm{sum}}$ again gives the expectation of the number of false signals occurring close enough in time to be confused with the flash; this time, it is also extended to include all rod cells close enough in *space* to be confused with the flash (the summation region).

Barlow proposed that ℓ_{tot} is the quantity submitted to the decision module. He also suggested a specific model, in which the decision module could impose any network threshold on ℓ_{tot}, bringing the false-positive rate down to whatever level was required for a particular task. If that hypothesis is correct, he argued, then subjects should be able to *program* their decision module to respond to even fewer than 10 productive photon absorptions, if their training permits them to make some false-positive reports. In contrast, the parameters Q_{tot} and $\mu_{0,\mathrm{sum}}$ belong to the lower-level transduction module (Figure 9.2); Barlow's model assumes that they are *fixed* for each human subject (after dark-adaptation).

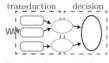

transduction   decision

Fig. 9.2 (page 292)

9.3.4 Forced-choice experiments characterize the dim-light response

Barlow performed some psychophysical experiments to test his hypothesis; a more systematic version was later done by B. Sakitt. Instead of Hecht and coauthors' somewhat nebulous instruction to avoid false positives altogether, Sakitt asked her subjects to rate flashes on a scale from 0 ("did not see anything"), 1 ("very doubtful that a light was seen"), ..., to 6 ("very bright flash was seen"). The subjects were shown flashes of two different strengths, and also "blanks" (no flash at all), all randomly mixed, and they were simply asked to rate each stimulus on the seven-alternative scale. As in Hecht's experiment, the subjects gave variable answers in response to repeated stimuli of the same nominal strength. But instead of a Bernoulli trial (a "coin flip" representing seen/not seen), each subject's responses were like rolls of a seven-sided die: For any choice of flash strength, the responses had unequal but definite probabilities for each of the seven allowed outcomes. The experimenter did know the strength of each flash, so those probabilities could be estimated as the fraction of trials for each flash strength that elicited each rating.

Figure 9.4 shows that the mean reported rating increased with increasing flash strength, as expected. Even at zero strength, the mean rating was nonzero, supporting the eigengrau hypothesis. However, the subject shown in the graph gave a "2" rating far more often when 55 photons were presented than when no flash was shown. This result, combined with the estimate made by Hecht and coauthors that the quantum

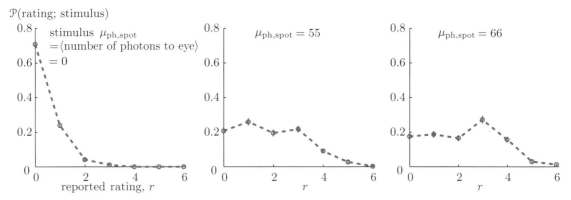

Figure 9.4: [Experimental data.] **Sakitt's results for one human subject.** Flashes of 500 nm light with three different strengths (mean photon count $\mu_{\mathrm{ph,spot}} = 0$, 55, or 66) were presented to the eye in random sequence, and the subject was asked to report their brightness as a "rating" $r = 0, \ldots, 6$. For each flash strength $\mu_{\mathrm{ph,spot}}$, the probability of each rating was estimated from the frequency of the subject's reporting that value of r. The error bars represent the standard deviation of the posterior distribution for the parameter of a Bernoulli trial that was repeated 400 times (for $\mu_{\mathrm{ph,spot}} = 55$ or 66 photons), or 800 times (for $\mu_{\mathrm{ph,spot}} = 0$). The *dashed lines* are not a fit; they just join the points. Other subjects gave similar results (*not shown*). (Figure 11.5 on page 359 will give a fit to data of this sort.) [Data from Sakitt, 1972, available in Dataset 11; see Problem 11.3.]

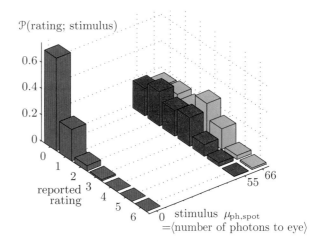

Figure 9.5: [Experimental data.] **Another representation of Sakitt's data.** This plot emphasizes that she measured a function of two variables: $\mathcal{P}(\text{rating}; \text{stimulus})$. [Data from Sakitt, 1972.]

catch is at most 10%, implies that as few as about *six* photoisomerizations suffice to generate a statistically measurable behavioral response. That statement is more precise than saying "a human can see a single photon"; Chapter 11 will sharpen it still further.

Extending Barlow's proposal, Sakitt suggested that

- *The number of rod signals during a flash is distributed according to Idea 9.3.*
- *A subject reports a rating r determined by applying a set of network thresholds to the rod signal count. Each allowed response has its own threshold, which the subject has unconsciously set during the course of prior training for this task.*   (9.4)

It is tempting to try to confirm or disprove this hypothesis by fitting its predictions to data such as those shown in Figures 9.4–9.5. Unfortunately, however, many different fits are possible. For example, we have only made a rough estimate of the quantum catch Q_{tot}. It could be high, leading to lots of signals per flash, but the eigengrau parameter could also be large, requiring large network thresholds to remove the false positives. Assuming smaller values for all of these unknown parameters leads to a prediction of similar performance—an equally good fit to the data. Psychophysical data alone cannot resolve this ambiguity, so we must postpone completing the analysis until we have found independent determinations of the constants Q_{tot} and $\mu_{0,sum}$ (see Section 11.3.4).

9.3.5 Questions raised by psychophysical experiments

Although the psychophysical experiments could not fully characterize the performance of photoreceptors, they did establish two key features indirectly:

1. The experiments gave indirect evidence that a dark-adapted human rod cell can generate a signal based on a single photon absorption.
2. Together with basic facts about photochemistry, the experiments suggested that some spurious signals must also be generated.

The rest of this chapter will describe how later experiments, motivated by the pioneering work, characterized dim-light vision at the level of single photoreceptor cells. Some questions related to the points just made include

1′. Can we confirm the absence of a rod threshold directly? What is the value of the quantum catch? What cellular machinery could implement that performance?
2′. Can we measure the false-positive rate directly? Are there other sources of noise?

The next section and Chapter 10 will address these questions.

9.4 SINGLE-CELL MEASUREMENTS

9.4.1 Vertebrate photoreceptors can be monitored via the suction pipette method

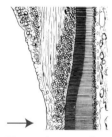

Fig. 6.6 (page 219)

Fig. 9.3 (page 295)

Anatomists had found in the early 20th century that the vertebrate retina[16] includes a tightly packed carpet of long cells, aligned parallel to the incoming light rays (Figure 6.6). Signals generated by these photoreceptors were difficult to measure, however. Accordingly, the first attempts at photoreceptor recording were made starting in the 1930s with invertebrates such as insects and the horseshoe crab *Limulus* (Figure 9.6). These efforts were rewarded in 1977, when P. Lillywhite found that single cells generated signals probabilistically, giving response curves similar to those found in psychophysics experiments (Figure 9.3). Another set of experiments used vertebrates, but measured signals sent to the brain from the retina's output layer (the retinal **ganglion cells**), which could be conveniently recorded. However, ganglion cell signals are already some steps removed from the initial phototransduction.

---

[16]See also the image on page 21.

Figure 9.6: [Photographs.] **The Atlantic horseshoe crab, *Limulus polyphemus*.** *Left:* Whole animal. *Right:* Closeup of the compound lateral eye. [Courtesy Lisa R Wright/Virginia Living Museum.]

The breakthrough needed to monitor an individual vertebrate rod cell involved electric currents set up by the motions of ions in the cell's cytosol and in the surrounding fluid.[17] All animal cells generate an electric potential drop across their outer membranes; neurons (nerve cells) are special in that they use these membrane potentials to transmit signals along their length. Two general techniques are commonly used to record signals from neurons, but each has drawbacks when used with photoreceptors:

- The experimenter can gently puncture the neuron with a micropipette, gaining direct access to measure the electric potential inside it ("intracellular recording"). This procedure works well on most neurons, but the change in membrane potential associated with single photon reception in vertebrate photoreceptor cells is small and hence was initially hard to measure.

- Alternatively, in some cases the experimenter can simply place an electrode close to a neuron's output transmission line (its axon), and pick up small extracellular potential changes when nerve impulses (action potentials) travel down the axon. But vertebrate photoreceptors were found not to generate action potentials, nor do they have any convenient long axon.

To overcome these obstacles, D. Baylor and coauthors developed a third measurement scheme around 1979 (Figure 9.7). Each rod cell has a cylindrical projection called its "outer segment"; this is the part of the cell that is sensitive to light. The experimenters gently pulled the outer segment of a single rod cell into a micropipette by suction. The rest of the cell remained outside the micropipette, which therefore interrupted an extracellular pathway for the flow of electrically charged atoms (ions). Normally, the ions emerge from the cell body, then reenter it via the outer segment. Pulling the rod partway into the micropipette forced this circulating electric current to detour through a measuring apparatus, which recorded it (Figure 9.8a).[18]

The experiment showed that a current of about 20 picoamperes (abbreviated pA) flows in the neighborhood of an unstimulated rod cell. This "dark current" fluctuates, with variations of amplitude around $0.2\,\mathrm{pA}$ called the **continuous dark noise**. When the

---

[17]Section 2.4 introduced membrane potentials; Chapter 10 will give more details about ion currents.
[18]Figure 10.8 (page 332) gives a more detailed depiction of this current loop. Continuous extracellular currents in the dark were first documented by W. Hagins and coauthors.

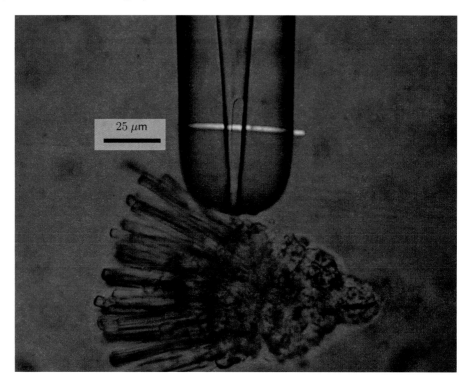

Figure 9.7: [Micrograph.] **Suction pipette recording.** *Bottom:* a patch of retina from the toad *Bufo marinus*, containing many long rod photoreceptor cells. *Center:* A single rod cell's outer segment was drawn into a micropipette in order to record its extracellular electric current. The rod was stimulated from the side by 500 nm light as shown (*yellow-green streak*). (In other experiments, the entire outer segment was illuminated.) To preserve the adaptation of the cell to the dark, pipette placement was done under infrared light, which was viewed on a video monitor. The photo shown here was taken with visible light after the end of the experiment. [Courtesy King-Wai Yau; see also Baylor et al., 1979.]

rod is exposed to light, however, the current drops significantly. For example, a bright flash can stop it altogether. Dim light brings about smaller, discrete current drops; in mammalian rods, their amplitude is 1–2 pA and they last about 300 ms (Figure 9.8b).

9.4.2 Determination of threshold, quantum catch, and spontaneous signaling rate

Having electrically isolated the photoreceptor, Baylor and coauthors proceeded to characterize its performance in the framework of Section 9.2.3, and specifically to find the single-cell quantum catch.[19] Because the blips observed in the dark current were larger in magnitude than the continuous dark noise, they were easy to identify. Moreover, although the current occasionally had such blips even in darkness, consistent with the eigengrau hypothesis, the chance of this happening in synchrony with an incoming light flash was small.[20] Thus, the response of a rod to a flash (blip or no blip) was nearly unambiguous, allowing the experimenters to construct a response

---

[19] The quantum catch appeared as a parameter in Idea 9.2 (page 297).

[20] T_2 Section 11.2.2 will point out that the pooling of many rod cells' signals makes spontaneous signaling significant in intact retina. However, no such pooling occurred in Baylor's single-cell preparation.

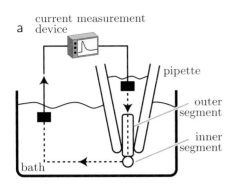

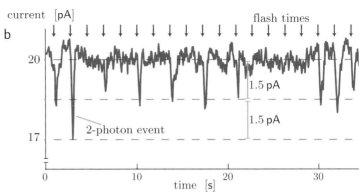

Figure 9.8: **Suction pipette data.** (a) [Schematic.] Electric current path in experiments of the sort shown in Figure 9.7. Normally, current carried by the movement of ions in solution (*dashed lines*) flows out of the rod inner segment and back into the outer segment. The suction pipette, however, provides a barrier which prevents the current loop from being completed in the solution just outside the cell. Instead, current must flow to an electrode placed in the bath, through a current-measuring amplifier, and finally into the pipette through another electrode (*solid lines and rectangles*). (b) [Experimental data.] Extracellular current of a primate rod cell in response to a series of fixed-strength flashes. Flash arrival times are indicated by *arrows* above the main trace. The trace shows light-induced signals rising out of a "continuous dark noise" background. The individual responses to such dim flashes fall into discrete categories by strength, corresponding to the productive absorption of zero, one, or occasionally two or more photons (see also Figure 9.12, page 307). [Data courtesy Greg D Field; see also Rieke, 2008.]

curve for single rod cells, analogous to the "frequency of seeing" plot (Figure 9.3).

Baylor and coauthors used the Poisson distribution to predict the probability for ℓ productive photon absorptions. To make this statement precise, let $\mu_{\mathrm{ph,rod}}$ denote the mean number of photons delivered to the rod cell, $Q_{\mathrm{rod,side}}$ the quantum catch for the sideways illumination used in the experiment (Figure 9.8), and $\mu = Q_{\mathrm{rod,side}}\,\mu_{\mathrm{ph,rod}}$. Then the probability for zero productive absorptions[21] is $\mathcal{P}_{\mathrm{pois}}(0;\mu) = \exp(-\mu)$, and so the predicted probability of a *nonzero* number is

$$\mathcal{P}_{\mathrm{see}}(\mu_{\mathrm{ph,rod}}) = 1 - \mathrm{e}^{-Q_{\mathrm{rod,side}}\,\mu_{\mathrm{ph,rod}}}. \quad \text{(single cell, if rod threshold } = 1\text{)} \qquad (9.5)$$

That is, if a rod cell imposes no threshold (as implied indirectly by Hecht's experiment), then its probability of responding to a flash should be given by Equation 9.5. If the rod threshold were instead *two* photons, then $\mathrm{e}^{-\mu}$ in the above formula would be replaced by the probability to catch zero *or one* photon, and so on.

Equation 9.5 embodies a physical model. Although this formula contains an unknown parameter $Q_{\mathrm{rod,side}}$, we can test the model by attempting to fit it to experimental data; if no value for the parameter gives a fit, then the model is wrong or oversimplified. The model is falsifiable because varying the parameter has only a limited ability to make it fit the data.

Fig. 9.3 (page 295)

---

[21]Section 0.2.5 (page 9) gives the full Poisson distribution.

Figure 9.9: [Experimental data with fits.] **Determination of the rod threshold.** *Circles:* Probability that a single rod cell will respond to a faint flash of 500 nm light, for four individual cells from a macaque monkey (distinguished by colors). The experimental arrangement was similar to the one in Figure 9.7. The rod outer segment was illuminated transversely to its long axis ("sideways"). The logarithmic horizontal axis gives the mean number of photons presented, divided by the area of the illuminated region (about $50\,\mu\mathrm{m}^2$). The *solid curve* shows a fit of all the data to a physical model that assumes that the productive absorption of a single photon can give rise to a signal; *dashed curves* show attempted fits to alternative hypotheses, in which two or three photons must be absorbed (see Problem 9.3). Error bars represent the standard deviation of the posterior distribution for the parameter of a Bernoulli trial that was repeated 65 times (Equation 7.2, page 248). (*Not shown:* Human rod cells behave similarly.) [Data from Baylor et al., 1984, available in Dataset 12.]

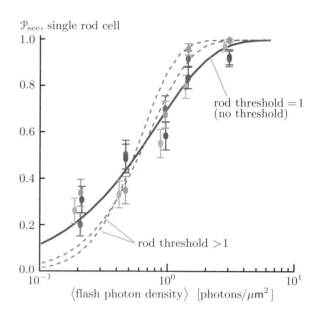

Your Turn 9B

a. Figure 9.9 plots $\mu_{\mathrm{ph,rod}}$ on a logarithmic scale. Show that changing the value of $Q_{\mathrm{rod,side}}$ just slides the graph of Equation 9.5 to the left or right, without changing its shape. [*Hint:* First write $\mathcal{P}_{\mathrm{see}}$ as $f(y)$, where $y = \ln Q_{\mathrm{rod,side}} + \ln \mu_{\mathrm{ph,rod}}$ and f is a function only of y.]

b. Also show the same thing for the versions of this formula appropriate for rod threshold of two or more photons.

Thus, if our model predicts a probability-of-seeing curve that is too steep (or too shallow), then adjusting the value of the fit parameter $Q_{\mathrm{rod,side}}$ *won't help.*[22] Figure 9.9 shows that, in fact, a fit can be found under the hypothesis that the rod threshold has the ideal value (that is, one); higher values are ruled out. This conclusion confirms the indirect inference made much earlier by Hecht and coauthors (Section 9.3.1). The fit also provides a numerical estimate for the quantum catch.[23]

K. Yau, G. Matthews, and Baylor also found the mean rate of spurious signals generated by a rod cell in the dark. A more recent estimate, for macaque monkey rods, is about $0.0037\,\mathrm{s}^{-1}$.

Your Turn 9C

To get the implication of this result, note that each macaque rod contains about 1.2×10^8 rhodopsin molecules. Find an upper bound on the probability per unit time that any particular rhodopsin will spontaneously isomerize.

Your result shows one of the features of rhodopsin that make it a good light detector:

---

[22] You'll explore this claim in Problems 9.3 and 9.5.

[23] You'll perform this fit in Problem 9.7.

It is thermally very stable. This stability is what allows the rod cell to carry so many copies of the molecule, achieving a reasonably high quantum catch, without too many false-positive signals.

$\boxed{T_2}$ *Section 9.3.4 ended by saying that we need to know the quantum catch Q_{rod} for a rod illuminated axially. The present section, however, outlined an experimental method to determine the quantum catch $Q_{\text{rod,side}}$ for sideways illumination. Section 9.4.2' (page 311) describes how to obtain Q_{rod}, and then the quantum catch for the entire retina, $Q_{\text{rod,ret}}$, starting from the measured value of $Q_{\text{rod,side}}$.*

9.4.3 Direct confirmation that the rod cell imposes no threshold

Earlier sections have discussed classic demonstrations that rod cells impose no threshold; that is, they can signal after productively absorbing a single photon. These experiments relied on indirect, probabilistic reasoning, however, because light sources available at the time could not produce flashes containing exactly one photon. One could reduce the flash intensity or duration to the point where the flashes have mean number of photons μ_{ph} smaller than one. Then, each "flash" is unlikely to deliver more than one photon; but most contain none at all. When such stimuli are presented to an isolated rod cell, it does sometimes respond, but this does not directly prove that the rod can respond to single productive absorptions—instead, it is possible that some of the observed rod signals were spontaneous, false-positive events that would have happened with *no* photons. Comparing the rod signals with and without the flashes is difficult, because a small difference of two noisy quantities has high relative standard deviation.[24]

More recent techniques, however, allow the creation of states known to contain exactly one photon. N. Phan and coauthors used one such method to revisit the question of a possible rod threshold (Figure 9.10). To overcome the difficulty of many zero-photon states, Phan and coauthors passed light flashes through a crystal of β-barium borate. The crystal let most photons pass through unchanged, but converted a small fraction into *pairs* of photons.[25] Each photon in the pair emerged at the same time and had the same wavelength, which was arranged to match the sensitivity peak of a rod cell from a frog. One of the pair was directed to a rod cell in a suction pipette apparatus. The other was directed to an electronic detector known to have high quantum catch, no threshold, and low false-positive rate. Signals from this detector were then used to trigger recording from the rod cell. Thus, data were collected only when a single photon was known to be arriving at the rod cell.

Figure 9.11 shows the distribution of rod peak current values observed after zero-photon and one-photon "flashes." The fact that these histograms differ is direct evidence that, some fraction of the time, rod cells do respond to stimulation by single photons. In Problem 9.6, you'll use these data to determine the numerical value of that fraction, and hence estimate the quantum catch of the rod directly.

9.4.4 Additional single-cell results

Multiple photon absorptions

We now return to experiments done with traditional light flashes. Figures 9.8b and 9.12 display another key feature of the response of rod cells to dim light. Although most signals peak at around 1.8 pA, a few are about twice as large. Baylor and coauthors proposed the natural interpretation, which was that

Fig. 9.8b (page 303)

---

[24]Section 0.5.4 (page 16) discusses the RSD of a difference.
[25] $\boxed{T_2}$ Section 1.3.3'a (page 50) describes the physics behind this splitting.

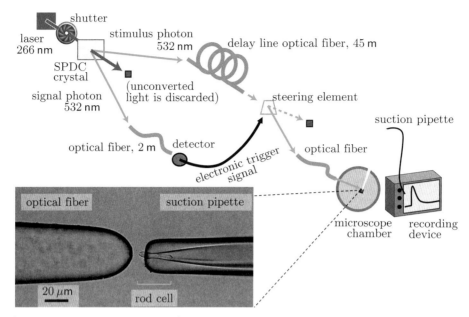

Figure 9.10: [Schematic; optical micrograph.] **Direct determination of the rod threshold.** Photon pairs are produced via spontaneous parametric down-conversion from an ultraviolet laser. One of them, designated the "signal photon," is detected by an electronic light detector, and its output triggers a beam-steering element (acousto-optic modulator). Each time a signal photon arrives, the steering element directs the other "stimulus photon" to an optical fiber and ultimately to a living rod cell. (A long delay line for the stimulus photon gives the steering element enough time to respond.) The trigger signal also starts recording of electric currents near the rod cell, measured by the suction pipette method. *Inset:* A microscope image shows the rod cell in the suction pipette and the tip of the optical fiber. [Courtesy Leonid A Krivitsky; see also Phan et al., 2014.]

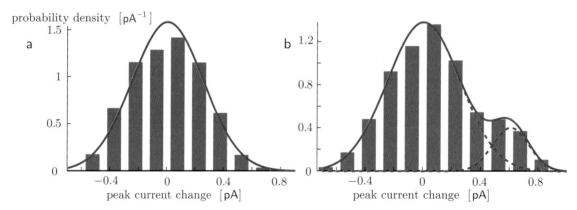

Figure 9.11: [Experimental data.] **Responses of a rod cell to single photon stimuli.** Observed current changes for a single rod cell from the frog *Xenopus laevis*, at room temperature. (a) Control experiment in which a rod cell was presented with zero photons and observed over 157 trials. The *curve* shows the maximally likely Gaussian distribution representing the data. (b) Currents from the same cell for 195 flashes consisting of exactly one photon each. The *left dashed curve* is the same distribution as in (a), apart from normalization; the deviation from this curve shows the effect of photon absorptions on the current. The *right dashed curve* is another Gaussian fit to the data (see Problem 9.6). The *solid curve* is the maximally likely weighted sum of those distributions, representing a mixture of events in which the incoming photon was or was not productively absorbed. [Data from Phan et al., 2014.]

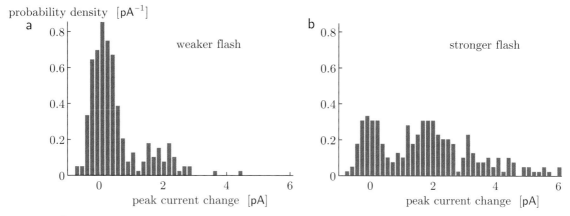

Figure 9.12: [Experimental data.] **Histograms of currents recorded in primate rod cells.** Poisson-distributed light flashes were presented to single cells, similarly to Figure 9.9 (but unlike Figure 9.11). However, instead of coarsely classifying each response as "see/no see," as in Figure 9.9, the actual response was recorded. In each panel, the bump on the *left* corresponds to trials in which the cell failed to respond; the bump centered near 1.8 pA corresponds to single photon responses. The second histogram shows responses to flashes containing on average four times as many photons as those in the first one. In (b), there is a third, broad peak centered near 3.6 pA, corresponding to flashes in which two photons were productively absorbed. Compare Figure 2.8 (page 72), where a similar distribution arose in a different context, and see Problem 10.3. [Data courtesy Greg D Field, available in Dataset 13; see also Rieke, 2008.]

- At the flash strengths they used, sometimes *two* photons (or more) are productively absorbed in response to one flash, and
- The response of a rod cell reflects the number of photons absorbed;[26] this additional information is therefore available to the next level of processing in the retina.

In fact, the histogram in Figure 9.12 suggests that different photon counts appear in distinct peaks, and the peak currents are roughly linear functions of the number of absorptions (at least when that number is small).

Univariance

The analysis of color vision in Chapter 3 made a key assumption, which was that the signal sent from any photoreceptor cell in response to light had very limited dependence on the color of that light. Specifically, the univariance principle states that a photon's wavelength affects only the probability of its being absorbed by a photoreceptor cell, and not the nature of the response if it is absorbed.[27] Baylor and coauthors tested the corresponding hypothesis in their study of amphibian rod cells. For example, they found that the peak current of the single-photon response was independent of the wavelength of light, even though the cell's sensitivity varied over a range of 10^5 as wavelength was scanned from 420 to 700 nm. Later experiments extended this result to mammals.

---

[26]You'll test this claim in Problem 10.3.
[27]Idea 3.8 (page 124) introduced the univariance principle.

9.4.5 Questions raised by the single-cell measurements

The previous sections have documented how a tiny lump of light energy can induce responses in rod cells with some remarkable properties:

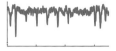

Fig. 9.8b (page 303)

- Compared with the energy of a photon, the rod cell response is *large*. One way to quantify this statement is to observe that the rod's response is an alteration in electric current of about 1–2 pA, which lasts for roughly 300 ms (see Figure 9.8b).

Your Turn 9D

a. Estimate the total change in electric charge passing through the rod membrane during such an event.

b. There is an electric potential drop across the rod cell's membrane of about 40 mV. Estimate the electric potential energy change when the charge you found in (a) crosses the membrane.

> The energy you just found is small compared with, say, lifting a brick by one meter. But it's vastly larger than the energy of the photon whose absorption triggered it.[28] How can the rod cell achieve such great amplification, without introducing so much noise that the signal is lost?

- Section 9.3.2 mentioned that roughly 20% of photons arriving at the retina are absorbed by rhodopsin. This is surprising when we consider that any one chromophore is quite unlikely to catch any incoming photon.
- The rod rarely emits photon-like signals in the *absence* of real photons. For example, every signal in Figure 9.8b appears to be a response to a real light flash. How can such a sensitive detector give so few false-positive signals?
- The rod cell's responses to single photon absorptions are fairly *uniform* (Figures 9.8b and 9.12, page 307). We might have expected great variation in the signals, due to the random character of events involving few molecules. So this observation, too, is surprising.

Keep these questions in mind as we explore in the next chapter how photoreceptors actually transduce light into signals suitable for transmission to the brain.

THE BIG PICTURE

This book has described a number of single-molecule detection methods, for example, in the realm of fluorescence microscopy. But now we know that each of us (and even the lowly dung beetle) has been a single-molecule biophysicist all along: Our photoreceptors report single photoisomerization events. Even our conscious minds can use just a handful of such events to *construct* a model of the world, a level of performance that brings great fitness benefits. Our story began with pioneering psychophysics experiments in the 1940s; we then surveyed some details that became available much later, via single-cell recording. Chapter 10 will begin to address the question of how such spectacularly good performance is possible.

---

[28] See Problem 1.4 (page 59).

KEY FORMULAS

- *Dim-light response of rod cell:* Section 9.3.3 introduced Barlow's proposal that the response of a single rod cell to a dim flash of light is to generate ℓ "signals," where ℓ is a Poisson random variable with $\langle\ell\rangle = Q_{\mathrm{rod}}\,\mu_{\mathrm{ph,rod}} + \mu_{0,\mathrm{rod}}$. In this formula, Q_{rod} denotes the rod's quantum catch, $\mu_{\mathrm{ph,rod}}$ the expectation of the number of photons delivered to it in a flash, and $\mu_{0,\mathrm{rod}}$ the expectation of the number of false positive signals it generates during an integration time.

  In experiments where an isolated rod cell is illuminated from the side, we again expect a Poisson distribution, but the rod cell quantum catch has a different value $Q_{\mathrm{rod,side}}$, and if we only "listen" to the rod during the very brief illuminating flash, then the number of spontaneous events will be negligibly small. In that situation, we modeled the probability of getting one or more signaling events in response to a flash as $\mathcal{P}_{\mathrm{see}}(\mu_{\mathrm{ph,rod}}) = 1 - \exp(-Q_{\mathrm{rod,side}}\,\mu_{\mathrm{ph,rod}})$.

- *Dim-light response of entire retina:* Section 9.3.3 argued that the response of the entire retina to a dim, short, localized flash is also a discrete random variable ℓ_{tot}, obtained by pooling many single-rod signals; thus it, too, is Poisson distributed, with expectation $\langle\ell_{\mathrm{tot}}\rangle$ that depends on flash strength as $Q_{\mathrm{tot}}\,\mu_{\mathrm{ph,spot}} + \mu_{0,\mathrm{sum}}$. In this formula, $\mu_{\mathrm{ph,spot}}$ is the expectation of the total number of photons in the flash. $\boxed{T_2}$ Chapter 11 will express the quantum catch parameter Q_{tot} in terms of its single-cell counterpart Q_{rod} by combining the derivation in Section 9.4.2′ with an estimate of loss at the first synapse. Similarly, we will estimate the eigengrau parameter $\mu_{0,\mathrm{sum}}$ from the measured value of $\mu_{0,\mathrm{rod}}$.

- $\boxed{T_2}$ *Cross section, absorptivity, extinction:* The absorption cross section of a chromophore is $a_1 = f/\sigma$, where f is the fraction of photons that are absorbed when they impinge on a thin layer of chromophores with area density σ. The molar absorptivity (also called extinction coefficient) is defined as $a_1/(\ln 10)$, traditionally expressed in $\mathrm{M}^{-1}\mathrm{cm}^{-1}$. The absorption coefficient of a bulk solution of this chromophore is the product of the molar absorptivity times the concentration.

- $\boxed{T_2}$ *Quantum yield versus quantum catch:* The quantum yield ϕ_{sig} for rod signaling is the probability that, if a photon is absorbed, it will trigger a rod response. In contrast, the quantum catch of a rod cell under sideways illumination, $Q_{\mathrm{rod,side}}$, is the probability that, if a photon is presented to the rod, it will trigger a response. Not all photons are absorbed, so the quantum catch is smaller than the quantum yield:

$$Q_{\mathrm{rod,side}} = \phi_{\mathrm{sig}}a_1 c\pi d_{\mathrm{rod}}/4. \qquad [\text{9.11, page 313}]$$

In this formula, c denotes number density (concentration) of chromophores and d_{rod} the rod diameter.

FURTHER READING

Semipopular:
Henshaw, 2012.

Intermediate:
Cronin et al., 2014; Bialek, 2012, chapt. 2; Land & Nilsson, 2012; Packer & Williams, 2003.
Hecht experiment: Benedek & Villars, 2000.

Threshold behavior in other cellular reaction networks: Nelson, 2015, chapt. 10.

Technical:

Single-photoisomerization detection from psychophysics: Rieke, 2008; van der Velden, 1944; Hecht et al., 1942.

Giant squid eye: Nilsson et al., 2012.

Eigengrau hypothesis: Sakitt, 1972; Barlow, 1956.

Extracellular current: Hagins et al., 1970.

Single-photoisomerization detection by single rod cells: Baylor et al., 1984; Lillywhite, 1977.

Spontaneous photon-like responses in darkness: Yau et al., 1979.

T_2 Track 2

9.4.2′a The fraction of light absorbed by a sample depends exponentially on thickness

Baylor and coauthors subjected single rod cells to light flashes of various strengths, obtaining "probability of seeing" data such as those in Figure 9.9. Although the curve in the figure looks superficially similar to the psychophysical data (Figure 9.3), it's easier to analyze, because no further neural processing is being done to the rod signals. You'll analyze the data in Problem 9.3 and show that they can be fit to a function of the form Equation 9.5 (page 303).

This is not quite the end of the story however, because Baylor's experiment exposed the rod to light directed *sideways* (transverse to its axis). We'd like to deduce the quantum catch of a rod cell for light that passes through it *axially*, as it does in an intact eye. To make the connection, we'll need to think a bit about how an optically dense medium absorbs light.[29]

To warm up, first imagine a thin sheet of transparent material with area A, illuminated perpendicularly to its surface. All the light passes through.[30] Now decorate the sheet with N perfectly absorbing spots, each of area a_1 (Figure 9.13a). The area density of the spots is $\sigma = N/A$. If none of them overlap, then they block a fraction $f = a_1 \sigma$ of the light from passing through.

Next consider a more realistic situation, in which some light-absorbing molecules are suspended in a thin layer of water, with surface density σ. We can measure the fraction f of incident light that they block. This fraction can be interpreted as the probability that any incoming photon with the wavelength we chose will be absorbed. The chromophores present a mutually exclusive set of alternatives to the photon: At most one of them can absorb it. Thus, the total probability to be absorbed is proportional to how many are present.[31] Moreover, imagine a series of experiments in which a fixed number N of chromophores are spread over larger and larger areas A, and the incoming light beam is also expanded to always fill A. Clearly the gaps between molecules grow larger, and each incoming photon is less likely to "find" a chromophore. Taken together, these statements imply that f is some constant times the area density of chromophores. We call the constant of proportionality the **absorption cross section** a_1 of the chosen chromophore because, although it may not equal the geometrical size of the molecule, it does have dimensions $\mathbb{L}^2$, and its definition $a_1 = f/\sigma$ is the same as in the macroscopic example in the preceding paragraph.[32]

Now consider a sample whose thickness b is *not* necessarily small. The sample contains chromophores at volume density (concentration) c. Imagine subdividing the sample into a stack of M thin layers. Each layer, of thickness b/M, contains chromophores with area density $\sigma = cb/M$. Light enters the sample, coming from a direction perpendicular to the imagined layers. When a single photon encounters the first layer, its probability to be absorbed is $f = a_1(cb/M)$. The probability of

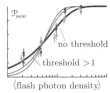

〈flash photon density〉

Fig. 9.9 (page 304)

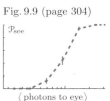

〈 photons to eye 〉

Fig. 9.3 (page 295)

---

[29] There is also a subtle point relating to the polarization of light, which we defer to Section 13.7.4 (page 412).

[30] We neglect any reflection at the surfaces.

[31] We neglect the possibility of quantum interference effects; thus we may use the rules of probability, in this case the addition rule (Equation 0.3, page 4).

[32] Chemists traditionally express the absorption cross section in units $\text{M}^{-1}\text{cm}^{-1}$. When a_1 is converted to those units, then $a_1/(\ln 10)$ is called the "molar absorptivity" or "extinction coefficient" of the chromophore. Cross sections can be defined for more specific processes; for example, absorption followed by fluorescence emission. The excitation spectrum discussed in Chapter 1 essentially such a cross section, as a function of wavelength.

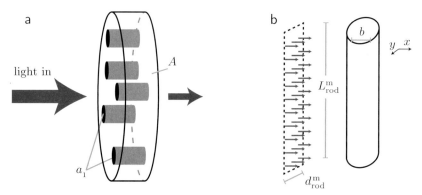

Figure 9.13: **Absorption by chromophores.** (a) [Metaphor.] A collection of N macroscopic, perfect absorbers in a thin layer, each with cross-sectional area a_1, removes a fraction Na_1/A of light traveling perpendicular to that layer. (b) [Schematic.] A rectangular beam of uniform intensity impinges from the side on a cylinder containing chromophores. The probability of absorption depends on the path length b, which in turn depends on y; Equation 9.9 evaluates the average over y.

passing through the entire sample is then the product of the probabilities of *not* being absorbed in *any* of the layers, or $\mathcal{P}(\text{transmitted}) = (1 - f)^M$. Taking M to be large, the compound interest formula (Equation 0.50, page 18) lets us rephrase this result as:[33]

$$\boxed{\mathcal{P}(\text{transmitted}) = \mathrm{e}^{-a_1 cb}. \quad \text{thick sample}} \tag{9.6}$$

Some authors rewrite this formula as $10^{-\zeta b}$, where the **absorption coefficient** is defined as $\zeta = a_1 c/(\ln 10)$. It depends on both the absorption cross section a_1, a molecular characteristic, and the number density (concentration) of chromophores c. If $a_1 cb \ll 1$, then we may approximate the probability that the photon *will* be absorbed as

$$\mathcal{P}(\text{absorbed}) = 1 - \mathrm{e}^{-a_1 cb} \tag{9.7}$$

$$\approx a_1 cb. \quad \text{(thin sample approximation)} \tag{9.8}$$

Baylor and coauthors used rod cells from macaque monkeys. Their outer segments are cylindrical, with diameter $d_{\text{rod}}^{\text{m}} = 2\,\mu\text{m}$ and length $L_{\text{rod}}^{\text{m}} = 25\,\mu\text{m}$. For rhodopsin at the concentration found in rod cells, the measured value of $a_1 c$ is $0.044\,\mu\text{m}^{-1}$. Thus, for sideways illumination of rod cells, Equation 9.8 is a good approximation, because any sideways path through the rod has length less than or equal to $d_{\text{rod}}^{\text{m}}$, and $d_{\text{rod}}^{\text{m}} a_1 c = 0.09 \ll 1$. In Baylor's experiments, the light beam had uniform intensity over a rectangle of size $d_{\text{rod}}^{\text{m}} \times L_{\text{rod}}^{\text{m}}$ (just covering the outer segment). Averaging the absorption probability over the variable thickness of a cylinder (Figure 9.13b) gives

$$\mathcal{P}\big(\text{absorbed (sideways)}\big) = a_1 c(\pi d_{\text{rod}}^{\text{m}}/4). \tag{9.9}$$

---

[33]This short derivation should sound familiar; see Section 1.4.1 (page 35). It was discovered in this context by P. Bouguer before 1729, then popularized by J. Lambert and extended by A. Beer.

9.4.2′b The quantum yield for rod signaling

We can now find the probability ϕ_{sig} that an absorbed photon actually triggers a rod response (the "quantum yield for signaling"[34]), by combining Equation 9.9 with the general product rule:[35]

$$
\begin{aligned}
\phi_{\text{sig}} &= \mathcal{P}\big(\text{rod signal} \mid \text{absorbed (sideways)}\big) \\
&= \mathcal{P}\big(\text{rod signal } \textbf{and} \text{ absorbed (sideways)}\big)/\mathcal{P}\big(\text{absorbed (sideways)}\big).
\end{aligned}
\tag{9.10}
$$

The numerator of this expression is the quantum catch (Section 9.2.3), so we find

$$
\phi_{\text{sig}} = \frac{Q_{\text{rod,side}}}{a_1 c \pi d_{\text{rod}}^{\text{m}}/4}.
\tag{9.11}
$$

   There is an interesting comparison to be made. The probability that a rhodopsin will isomerize, given that it has absorbed a photon, can be measured by purely physical means, because an isomerized rhodopsin absorbs light differently when a *second* flash is presented. This "quantum yield for photoisomerization" was found to be about 0.67: Most absorptions do trigger isomerization instead of just turning the photon into heat. Comparing to the value that you'll find for ϕ_{sig} in Problem 9.3 or 9.7 then shows that the rod cell has a remarkably high, though not perfect, probability to record a single isomerization anywhere in its farm of about 1.4×10^8 rhodopsin molecules.

   For us, the interest of ϕ_{sig} lies in the fact that its value should be *the same regardless of how light has been presented* to the rod cell. Once a photon has been absorbed by rhodopsin, its original direction no longer matters:

$$
\mathcal{P}\big(\text{rod signal} \mid \text{absorbed (axial)}\big) = \mathcal{P}\big(\text{rod signal} \mid \text{absorbed (sideways)}\big) = \phi_{\text{sig}}.
$$

We will now use this fact to find the rod's quantum catch for axial illumination, which is the situation relevant for normal visual function, and in particular for psychophysics.

9.4.2′c Quantum catch for a single human rod cell under axial illumination

To estimate Q_{rod}, we start with the probability for a photon to be absorbed when presented axially to a human rod cell, whose outer segment length is roughly $L_{\text{rod}}^{\text{h}} \approx 42\,\mu\text{m}$, then multiply it by ϕ_{sig} (see Equations 9.7 and 9.11):

$$
\begin{aligned}
Q_{\text{rod}} &= \mathcal{P}\big(\text{rod signal (axial)}\big) = \mathcal{P}\big(\text{rod signal} \mid \text{absorbed (axial)}\big)\mathcal{P}\big(\text{absorbed (axial)}\big) \\
&= \phi_{\text{sig}}\big(1 - \exp(-a_1 c L_{\text{rod}}^{\text{h}})\big).
\end{aligned}
\tag{9.12}
$$

Note that the thin-sample approximation (Equation 9.8) is *not* valid. We might have expected that: For good night vision, rod cells need to catch a large fraction of the photons that arrive on them, so $\exp(-a_1 c L_{\text{rod}}^{\text{h}})$ should not be close to 1.

9.4.2′d The whole-retina quantum catch is the product of several factors

We have now obtained a numerical value for the quantum catch of one rod cell—the quantity appearing in Equation 9.2 (page 297). Before we can interpret behavioral

---

[34] This terminology is consistent with that of Section 2.9.2, where the quantum yield for photosynthesis was defined as the probability that, if absorbed, a photon would give rise to the production of an oxygen molecule. You'll evaluate ϕ_{sig} numerically in Problem 9.3.
[35] Section 0.2.2 (page 4) introduced conditional probability. Equation 0.7 (page 4) gives the product rule.

experiments, however, we need the probability Q_{tot} that the *entire* retina will respond to an incoming photon. Photons must pass several hurdles before they arrive at a rod cell; each hurdle multiplies the probability of productive absorption by a number smaller than 1.[36] All of these factors vary considerably between individuals, so any calculations we make with them will be approximate.

The first factor arises because about 4% of the incoming light is reflected from the cornea—it never enters the eye at all. There is also loss due to absorption and scattering in the cornea, eye lens, and fluids prior to the retina. The fraction of light *not* lost in this way depends on the subject's age; a typical value, appropriate for a 35-year-old subject viewing 507 nm light, is 47% (van de Kraats & van Norren, 2007). Multiplying by 0.96 for corneal reflection gives the "ocular media factor":

$$[\text{ocular media}] \approx 0.45.$$

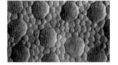

Fig. 3.9a (page 120)

Next, we must also account for the fact that rod cells do not completely fill the retina's surface (Figure 3.9a), so some of the photons that arrive at the retina don't enter any rod. This "tiling factor" depends on where on the retina we choose to direct the flash. At 7 deg away from central vision, the value used in Sakitt's experiments, it is roughly (Donner, 1992)

$$[\text{tiling}] \approx 0.56.$$

We can now combine these factors to estimate the effective quantum catch of rods over a region of the retina:

$$Q_{\text{rod,ret}} = [\text{ocular media}] \times [\text{tiling}] \times Q_{\text{rod}}. \tag{9.13}$$

In Problem 9.7, you'll evaluate the probability that a photon arriving at the eye will trigger a rod signal. We can compare it to Hecht and coauthors' estimate of 10%, which was based solely on optical effects in the eye and the absorption cross section of rhodopsin. Those authors did not know the quantum yield for rod signaling, but still their estimate was remarkably good, considering it was obtained over 40 years prior to the single-cell recordings!

The quantity $Q_{\text{rod,ret}}$ is still not quite our prediction for the psychophysical quantum catch Q_{tot}, however: Yet another loss mechanism will be introduced in Chapter 11, involving discrimination at the first synapse, which discards some true photon signals.

---

[36]Section 3.4.1 (page 111) discusses the thinning property of Poisson processes.

PROBLEMS

9.1 Rod threshold from psychophysics

Suppose that flashes of light are spread over 300 detectors (rod cells), each of which contains many light-sensitive molecules (chromophores). Each chromophore has a small chance to capture any incoming photon, but there are so many of them that each flash triggers a total of 10 productive absorptions, throughout the retina, on average. For this problem, you may neglect the possibility of spontaneous isomerization.

a. Write the probability distribution $\mathcal{P}(\ell)$, where the random variable ℓ is the number of productive photon absorptions in any one detector.

b. Evaluate the probability that any chosen detector will catch either 0 or 1 photon.

c. Evaluate the probability that *every* detector catches either 0 or 1 photon.

d. Evaluate the probability that *any* detectors (one or more) catch *more than* 1 photon.

e. Hecht and coauthors found about 50% probability of seeing for flashes of this strength. Draw a conclusion about the rod threshold.

9.2 Rod threshold from behavior

Imagine two hypothetical animal eye designs. Design A uses photoreceptors that are able to respond to one absorbed photon. Design B has photoreceptors that don't signal unless they get two productive photon absorptions within an integration time window of 200 ms. In other respects, the designs are identical; in particular, both photoreceptors have the same quantum catch $Q_{\mathrm{rod}} = 0.30$. Both are embedded in a visual system that, like our own, can usefully see the world if each photoreceptor emits signals with a mean rate of at least one photon every 1.4 hours ($0.0002\,\mathrm{s}^{-1}$).

Compare the mean photon arrival rates needed in each design to achieve this signaling rate and comment.

9.3 Probability of "seeing," single rod

Baylor and coauthors recorded from individual rod cells from macaque monkeys, and measured the "probability of seeing," $\mathcal{P}_{\mathrm{see}} = \mathcal{P}\big(\text{signal (sideways)}; \mu_{\mathrm{ph,rod}}\big)$ (Figure 9.9). Obtain Dataset 12, which contains the data shown on the plot. Convert the flash strength from photons per area to total photons, using the fact that the light was confined to a rectangular beam of size $2\,\mu\mathrm{m} \times 25\,\mu\mathrm{m}$.

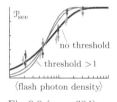

$\langle$flash photon density$\rangle$

Fig. 9.9 (page 304)

a. See if you can fit the data to the functional form expected for no rod threshold (Equation 9.5, page 303), by choosing an appropriate value for the fit parameter $Q_{\mathrm{rod,side}}$ (the rod quantum catch). A fit that looks good to your eye is enough.

b. Try part (a) again, this time using the functional form expected if the rod only signals upon catching *two* or more of the photons in a flash, and comment.

c. $\boxed{T_2}$ Use Equation 9.11 (page 313) and your result from (a) to get a numerical value for ϕ_{sig}, the quantum yield for rod signaling. (Other needed values are given elsewhere in Section 9.4.2.) Then use Equations 9.12 and 9.13 to get the quantum catch for rod cells, Q_{rod}, and for the whole retina, $Q_{\mathrm{rod,ret}}$, respectively.

9.4 Alternative presentation of single-rod response

Continue Problem 9.3:

a. Again obtain Dataset 12, but instead of plotting $\mathcal{P}_{\mathrm{see}}$, as in Figure 9.9 (page 304), make graphs of the quantity $-\ln(1 - \mathcal{P}_{\mathrm{see}})$ as a function of the mean number of photons in each flash. This is a potentially interesting presentation because, for our model with no rod threshold requirement (rod threshold equal to one photoisomerization), its graph will have a special property. The corresponding graphs with higher rod threshold don't have that property—what is it?

As in Problem 9.3, the goal is now to superimpose the experimental data points on your graphs and see which version of the model can be most successfully fit to the data. But there is a difficulty with this approach: After transforming from $\mathcal{P}_{\text{see}}$ to $y = -\ln(1 - \mathcal{P}_{\text{see}})$, the error bars on some points will be much larger than those on other points. The experimental data give the fraction of the approximately 66 trials that led to an observed rod signal, that is, an estimate for $\mathcal{P}_{\text{see}}$.

b. Using what you know about Bernoulli trials, find the standard deviation of this quantity for each trial and each flash strength chosen. Add and subtract that to the estimated $\mathcal{P}_{\text{see}}$ to get a range of values, compute the corresponding ranges of the y values, and place error bars on your graph of the experimental data to reflect this statistical uncertainty.

c. Now comment on which model appears best able to account for the data.

9.5   *Effect of a threshold requirement on single-rod response*
The text stated that raising the rod threshold steepens the probability-of-seeing curve for a single rod cell. In this problem, you'll investigate that claim.

Calculate analytically the maximum of the slope function $d\mathcal{P}_{\text{see}}/d(\ln \mu_{\text{ph,rod}})$, both under the hypothesis of no threshold, and under the alternative hypothesis that a rod cell sets a threshold of two photoisomerizations before it signals (see Equation 9.5, page 303, and the following text). Show that the maximum slope depends only on the assumed value of the threshold (not on the unknown quantum catch parameter). Comment in the light of Your Turn 9B (page 303).

9.6   *One-photon stimulus*
Section 9.4.3 (page 305) outlined how N. Phan and coauthors determined the quantum catch of a single rod cell.

a. Obtain Dataset 14. The first pair of arrays defined there describe the measured current noise from unstimulated rod cells. Use the method in Section 7.2.3 (page 249) to obtain the maximally likely Gaussian distribution given these data. Plot this probability density function superimposed on a bar graph representing the PDF estimated from the experimental data.

b. The other pair of arrays defined in Dataset 14 describe the measured current from rod cells that were presented with exactly one photon. Construct a physical model by supposing that (i) the cell productively absorbs a fraction ξ of incoming photons; and (ii) the response of the cell to a productive absorption is a peak current I drawn from the Gaussian distribution with expectation I_0 and variance σ^2. If no productive absorption occurred, then the cell's peak current is drawn from the distribution you found in (a).

c. Find the best fitting (maximally likely) values of the parameters ξ, I_0, and σ and make a graph analogous to the one you found in (a). To get started, make some estimates of the best values by examining the graph; then search near those values. Does the resulting graph seem to describe the observed distribution well?

d. $\boxed{T_2}$ The value of ξ that you find is the number of times that the rod cell responds, as a fraction of all photon pairs produced by the apparatus. However, in each such pair the photon that is headed to the rod cell can be lost along the way, reducing

ξ below the true quantum catch of the cell.[37] The experimenters estimated that 79% of photons were lost in this way. Divide your answer in (c) by $(1 - 0.79)$ to find an estimate for the quantum catch of the rod cell being studied. (Why is that a reasonable thing to do?)

9.7 $\boxed{T_2}$ *More about single rod response*

a. Instead of the informal fit you did in Problems 9.3 or 9.4, write a likelihood function, apply it to Dataset 12, and optimize it to find the best estimate for the rod quantum catch, $Q_{\text{rod,side}}$.

b. Use your result from (a) and Equations 9.12 and 9.13 to evaluate $Q_{\text{rod,ret}}$ numerically, for the retinal region and human subject studied in the experiment. When 100 photons strike the eye, about how many signals get generated by rod cells?

c. To choose objectively between alternative models, repeat (a) under the assumption that the rod threshold equals two productive absorptions. Compute the ratio of the best-fitting likelihood value for this model to the one in (a) and comment.

9.8 $\boxed{T_2}$ *Twinkle, twinkle*

In this problem, you'll investigate whether the human eye can detect a single fluorescent dye molecule (fluorophore), illuminated by light with an intensity typical of laboratory setups.

a. Chemical supply catalogs quote the extinction coefficient (molar absorptivity) of each of their fluorophores. Suppose that yours has the typical value $\epsilon = 10^5 \, \text{M}^{-1}\text{cm}^{-1}$. Convert this to a cross section, a_1, by the method in Section 9.4.2'a (page 311), or equivalently by comparing Equation 9.6 (page 312) with the similar formula, $\mathcal{P}(\text{transmitted}) = 10^{-\epsilon\,cb}$. Express your answer in m^2.

b. Now imagine exciting this fluorophore with a source that emits light with vacuum wavelength $\lambda = 510 \, \text{nm}$ at intensity $I = 100 \, \text{W cm}^{-2}$. Calculate the mean photon arrival rate from this source, per unit cross sectional area of the beam.

c. Suppose a typical quantum yield of 50%; that is, half of the absorbed photons lead to emission of a fluorescent photon. Suppose that a microscope objective captures half of these, and directs them to your eye. Suppose that about 10% of the photons arriving at your eye are productively absorbed. Find the mean rate of those absorptions.

d. Multiply your answer to (c) by the integration time of a rod cell, $\approx 200 \, \text{ms}$. Can you see a single fluorescent molecule with your dark-adapted eye?

---

[37] There is another correction, which you may neglect: A spontaneous rod event can coincide with a true photon stimulus, inflating the value of ξ because a rod response was measured even though the actual incoming photon was not productively absorbed.

CHAPTER 10

The Mechanism of Visual Transduction

> If the Lord Almighty had consulted me before embarking on creation thus, I should have recommended something simpler.
> — *Attributed to Alfonso X ("Alfonso the Wise"), 1221–1284*

10.1 SIGNPOST: *DYNAMIC RANGE*

The preceding chapter documented our eyes' remarkable performance. Now it's time to understand what mechanism inside photoreceptor cells can deliver such high sensitivity and low noise at body temperature. We'll sketch just a few of the ingenious experiments that underpin our understanding, to give a flavor of the heroic efforts that delivered that knowledge. In addition, we'll see how our photoreceptors give us useful information over a staggering *dynamic range* of overall illumination intensities (Figure 3.10).

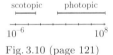

Fig. 3.10 (page 121)

The Focus Question is

Biological question: What mechanism can monitor one hundred million chromophores and reliably report when any one of them absorbs a photon?

Physical idea: A cascade of chemical and electrical transactions minimizes the small-number randomness inherent in single-molecule reactions.

10.2 PHOTORECEPTORS

10.2.1 Photoreceptors are a specialized class of neurons

The next few sections will outline the chain of events that occurs inside a single rod cell in response to light. It's an intricate story, drawing upon many themes in earlier chapters.

Fig. 9.7 (page 302)

To get started, let's consider the spatial organization of the human rod cell. Earlier images have shown a generally cylindrical form (Figure 9.7), with light traversing the cell along its length (Figure 6.6). Figure 10.1 shows the anatomy of the cell in greater detail. The long **outer segment** is packed with about 1000 organelles appropriately called **disks**.[1] Each disk is a flattened, closed surface made of bilayer membrane, about 16 nm thick and about $2\,\mu$m in diameter. Rhodopsin molecules[2] and other proteins are embedded in, or attached to, the disk's bounding membrane. Each disk exists for 8–10 days before being replaced; then it is shed by the rod cell and its constituents are recycled by neighboring epithelial cells.

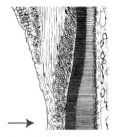

Fig. 6.6 (page 219)

The outer segment joins an **inner segment**.[3] The inner segment contains many mitochondria, the organelles that generate the cell's internal energy molecule ATP—a

---

[1] $\boxed{T_2}$ Cone cells have a different organization for the outer segment; see Section 10.4.1′ (page 344).

[2] Section 9.3.3 (page 296) introduced rhodopsin.

[3] Don't confuse the outer and inner *segments* with the outside and inside of the *cell*. Figure 10.1a shows that the outer and inner segments are joined; they share a common interior volume.

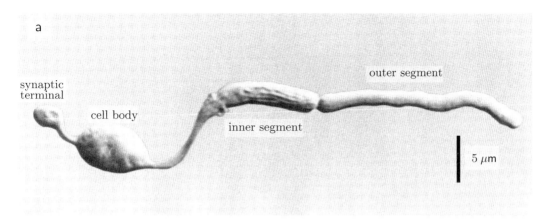

a

synaptic
terminal

cell body

inner segment

outer segment

5 µm

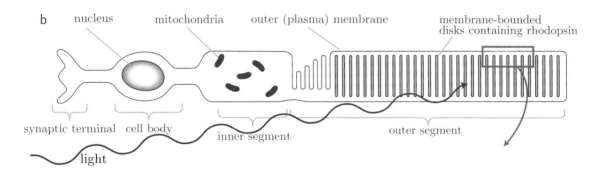

b nucleus   mitochondria   outer (plasma) membrane   membrane-bounded
disks containing rhodopsin

synaptic terminal   cell body   inner segment   outer segment

light

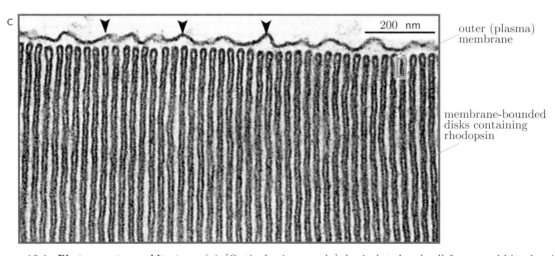

c   200 nm   outer (plasma)
membrane

membrane-bounded
disks containing
rhodopsin

Figure 10.1: **Photoreceptor architecture.** (a) [Optical micrograph.] An isolated rod cell from a rabbit, showing its major structural features. (b) [Cartoon.] Schematic representation of these structures and their contents. (c) [Electron micrograph.] Detail of the outer segment, corresponding to the *blue box* in (b), showing regularly spaced disks. Each disk is bounded by its own bilayer membrane; all are enclosed by the cell's outer (plasma) membrane (*arrowheads*). The region in the *orange box* appears further magnified in Figure 10.5a. [(a) From Townes-Anderson et al., 1988; (b) from Townes-Anderson et al., 1985.]

clue that the rod cell uses more energy than other cell types. Next comes a **cell body**, containing the nucleus. Finally, at the opposite end from the outer segment, there is a specialized projection, similar to the output end of any neuron. This **synaptic terminal** transmits signals to a specialized type of neuron called the **rod bipolar cell** (to be discussed later).[4]

10.2.2 Each rod cell simultaneously monitors one hundred million rhodopsin molecules

The large number of rhodopsin molecules present in the rod cell outer segment can explain its high quantum catch, because every incoming photon gets many opportunities to be productively absorbed.[5] But this strategy brings with it a new challenge: The rod cell must constantly monitor every one of those many rhodopsins. Thus, we can summarize the rod cell's performance by saying[6]

> *Each human rod cell maintains a total of about 1.4×10^8 rhodopsin molecules in its disks. A dark-adapted rod cell generates a well-defined signal with high probability whenever **any one** of its rhodopsins photoisomerizes.*   (10.1)

The following sections outline how the rod cell implements this performance.

Cat eye

Cats and many other nocturnal mammals have eyes that seem to "shine in the dark." More accurately, light from a point source, such as a flashlight, is brightly reflected from the eye back toward its origin. The effect arises from a layer in the retina that lies behind the photoreceptors, called the **tapetum lucidum**. This layer is highly reflective, in some cases with a specific wavelength dependence, because it has a layered structure similar to the ones studied in Chapter 5. Light from the source is focused to a point on the retina. Light not captured by the photoreceptors reflects from the tapetum lucidum, then gets redirected by the eye's optical elements back to its point of origin.

Nocturnal mammals do not ordinarily encounter bright point sources of light, so the phenomenon just mentioned does not confer any fitness advantage (or cost). Instead, the tapetum lucidum is an adaptation to improve low-light vision: Its functional role is to ensure that any photons missed when passing through the photoreceptor layer reflect, and hence get a second chance at being detected, on their way back out of the eye. Thus, the cat eye effectively doubles its photoreceptor cell length, improving its quantum catch, without the costs associated to actually lengthening the cells.

There is a trade-off, however, which explains why not all animals share this design: Light reflected from the tapetum lucidum can be absorbed by multiple different photoreceptors near to the one that would have captured it on its way in. That is, it creates a blurred image, reducing visual acuity.

Section 10.2 has sketched the general structure of a photoreceptor cell. Before attempting to describe its mechanism, we first digress to discuss some simpler ways in which cells can respond to their environment.

$\boxed{T_2}$ *Section 10.2.2′ (page 341) discusses rod cell response at higher photon arrival rates.*

---

[4] $\boxed{T_2}$ More precisely, each rod cell connects to 2–4 rod bipolar cells (as well as two cells from another class called "horizontal," discussed in Section 11.4.1′, page 369).

[5] $\boxed{T_2}$ You'll make this expectation precise in Problem 10.5.

[6] Section 9.4.5 (page 308) introduced this characterization of rod cells.

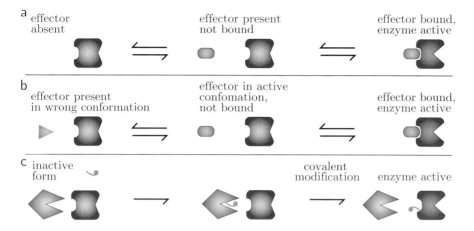

Figure 10.2: [Cartoons.] **A few mechanisms of enzyme control.** (a) An effector is a small molecule that binds to an enzyme and regulates its activity, typically via an allosteric interaction. (b) An effector may have two states, only one of which can bind and activate an enzyme. (c) A second enzyme (*left, blue*) may modify the first one, in this case activating it by covalently attaching a chemical group.

10.3 **BACKGROUND:** CELLULAR CONTROL AND TRANSDUCTION NETWORKS

We will soon see that each photoreceptor cell contains an elaborate signal processing apparatus that accomplishes the task of converting photoisomerization events to electrical signals sent to the brain. Before we can describe this apparatus, however, we first need to establish a graphical language that is useful for representing control systems in general, and intracellular circuits in particular.

10.3.1 Cells can control enzyme activities via allosteric modulation

Simple electrical circuits consist of discrete elements that communicate with each other by the guided flow of electrons. Electrons pass through metal wires that physically connect points that ought to be connected; other points are insulated from one another by air, plastic, or some other nonconducting material. Cells also contain discrete elements that must communicate, but they do so with two major differences:

- Instead of a single currency (electrons), there are many species of small molecules carrying information and energy.
- There is little compartmentation, especially in bacteria. One component can nevertheless speak specifically to another by employing a particular "messenger" molecule; the individual components generate and act on only the molecular species relevant to them.

Figure 10.2 lists three representative mechanisms for a small molecule (called an "effector," a particular kind of ligand) to influence a molecular actor (for example, an enzyme). In the top row, the enzyme is in an inactive state, unable to perform its biochemical function, until an effector binds to it. Binding a small molecule at one location can make the entire enzyme flex slightly, altering its shape at another site. Such **allosteric interactions** between binding sites on the enzyme can alter the "fit" between the second site and the **substrate** molecule on which the enzyme acts. If the fit improves, then the enzyme is more likely to bind and act on its substrate—it becomes "activated." If the fit worsens upon binding the effector, the enzyme is "inactivated."

Figure 10.2b symbolizes a situation in which the effector itself can have more than one conformational state; only one "active" form can bind to the enzyme and in turn activate it. In the bottom row of the figure, a second enzyme catalyzes a modification of the first one, for example, by covalently adding extra atoms such as a phosphate group, and in that way permanently modifies the activity of the enzyme (at least until yet another enzyme removes the group).

10.3.2 Single-cell organisms can alter their behavior in response to environmental cues, including light

Most animals will migrate from one habitat to another with a more comfortable climate, better food, or even the right amount of sunlight. Remarkably, even single-cell organisms can do all these things as well, a phenomenon generically called cell motility, or taxis. Thus, bacteria such as *Escherichia coli* or *Salmonella typhimurium* will move up a gradient of food concentration (**chemotaxis**); some photosynthetic bacteria will also move up a gradient of illumination (phototaxis, a primitive form of vision), and so on. There is also negative chemotaxis (avoidance of noxious chemicals).

To understand how these organisms make decisions like this, with *no brain,* we must first understand how they move in the first place. *E. coli* has several long, thin filaments protruding from its surface, called **flagella**. Each flagellum is fairly rigid, with a helical shape.[7] Each is anchored in the bacterial cell wall by a rotary motor. When all the cell's motors twirl their flagella in one direction, conventionally called "counterclockwise," then the flagella bundle together and create a net propulsive force directed along the long axis of the cell. When one or more of the cell's motors reverse, twirling their flagella in the "clockwise" direction, then the flagella unbundle and the cell flails about.[8]

Under neutral conditions, the cell's motors spin counterclockwise for about one second before reversing to clockwise rotation for a mean time of about 0.1 s, reverse again, and so on. Thus, the cell executes long **runs** of directed motion, punctuated by **tumbles** (episodes that randomize its orientation). The net effect is that each cell performs a **random walk** in space, mathematically similar to molecular diffusion although on much longer spatial and temporal scales.

If the environment is not uniform, however, something more interesting can happen. If the cell blunders into a direction where conditions are better (more food, less noxious chemicals, better temperature or illumination), then its motors get biased toward running; tumbles become less frequent. Similarly, if conditions worsen, the balance shifts the opposite way. That is, the cells execute a simple, yet effective algorithm:

> *If things are getting better, keep going. If things are getting worse, make a random change of direction.*   (10.2)

Continuously sampling the environment, and applying this algorithm, suffices to send most cells to the best environment available for them. That is, over time they move

---

[7]Single-cell eukaryotes like *Paramecium* bear larger cilia, which are flexible and execute whiplike motions.

[8]See Media 20. The difference in these two behaviors arises because the flagella all have the same helical sense, breaking the equivalence between clockwise and counterclockwise rotation.

up the gradient of chemoattractant concentration (or down the gradient of repellent concentration).

10.3.3 The two-component signaling pathway motif

How does a single cell connect its sensory input to its behavior ("output")? Remarkably, a single, ancient control scheme implements many of the environmental responses displayed by bacteria, and even some of those in eukaryotes such as plants, fungi, and protozoa.

Figure 10.3a shows this signaling pathway. A sensory apparatus, typically a collection of molecules jointly called the **receptor complex**, is embedded in, and extends just inside, the cell's membrane.[9] Each *E. coli* cell has several thousand copies of this complex. Each receptor receives information, for example by binding an attractant molecule to its extracellular end, and responds by changing its normal activity via an allosteric interaction. That activity involves binding an ATP molecule from the cell's interior fluid, removing one of that molecule's three phosphate groups, and reattaching the phosphate to itself. Binding an attractant molecule decreases the mean rate of this **autophosphorylation**, and so alters the balance between the phosphorylated and dephosphorylated states. (Binding a repellant has the opposite effect.) Enzymes whose job is to add a phosphate to something are generically called **kinases**; the receptor complex more specifically contains a "histidine kinase," because it phosphorylates a particular histidine residue on its substrate (which in this case is itself). In the chemotactic network of *E. coli*, the kinase in the receptor complex is abbreviated CheA.

The receptors are confined to a cluster near the front of the cell. How, then, do they communicate with the flagellar motors, most of which are at the back end? A second protein present in the cell, called the **response regulator**, couples the histidine kinase to the output. It removes the phosphate from a CheA molecule and reattaches it to itself. Unlike the kinase, however, the response regulator is mobile, and can diffuse throughout the cell. The phosphorylated form can then bind to some target and alter its activity (Figure 10.2c). In some two-component networks, the response regulator is a transcription factor, that is, it controls the expression of a gene. In the case of chemotaxis, it controls the flagellar motor and is abbreviated CheY.

The multistage transduction pathway used in two-component signaling confers advantages over a simpler design:

- Each catalytic stage gives the possibility of amplification. For example, after binding a single chemoattractant molecule, the receptor can prevent many phosphorylations of CheY that would otherwise have taken place.

- The constant shuffling back and forth between phosphorylated and dephosphorylated states gives the network a faster response to changes in chemoattractant concentration than it would otherwise have (for example, if the only way to reset the system was to wait for the phosphorylated forms to be broken down and recycled by the cell or diluted by cell growth).

- Several different types of receptor can all supply input to the network. For example, chemotaxis in *E. coli* integrates signals for multiple food molecules and for temperature.[10]

---

[9] $\boxed{T_2}$ More precisely, the receptors are embedded in the plasma membrane. Outside that membrane, *E. coli* has a "periplasmic space," bounded by a second, outer membrane. Attractant molecules enter the periplasmic space via pores in the outer membrane.

[10] $\boxed{T_2}$ Bacteria can also sense and act on gradients in pH and osmolarity.

Figure 10.3: [Network diagrams.] **Two-component signaling.** (a) The generic scheme involves a stimulus (presence of a ligand, mechanical stress, or even light) that can raise or lower the mean rate at which an enzyme (a histidine kinase, abbreviated "K") attaches phosphate groups ("P") to itself. The kinase transfers its phosphate to a response regulator ("RR"), whose phosphorylated state in this case modifies the action of an "output" molecule ("O"). The response regulator can spontaneously lose its phosphate group, although generally another enzyme ("Z") is present to speed up this process; that enzyme's activity can itself be regulated. (b) Bacterial chemotaxis and phototaxis in *E. coli* and related organisms involve a version of this circuit, in which a chemoattractant (or light) represses the autophosphorylation of a kinase called CheA. Although CheA is confined to a specific region on the cell membrane, it transfers its phosphate group to an intermediary called CheY, which can diffuse throughout the cell. When the phosphorylated molecule CheY-P encounters one of the rotary motors that drive the cell's flagella, it can bind to the motor, increasing the mean rate at which the motor switches from counterclockwise to clockwise rotation. The net effect is to increase the probability per time to switch to a "tumbling" episode. When chemoattractant is present, CheA and CheY remain dephosphorylated and the flagellar motor runs counterclockwise, driving directed motion (a "run"). A similar cascade implements phototaxis, with a different transducer supplying the stimulus.

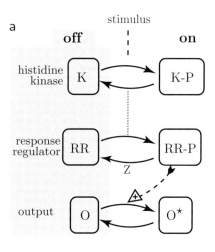

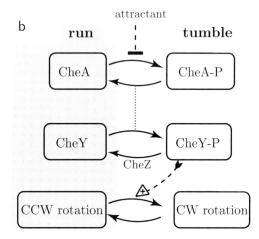

- Each receptor type can exist in thousands of copies. The balance between phosphorylated and dephosphorylated CheY reflects the sum of the activities of all receptors, a strategy that reduces noise.[11]
- Finally, each of the backward arrows in Figure 10.3 can also serve as a potential target for regulation.

T₂ *Section 10.3.3′ (page 341) gives more details about two-component signaling, and about phototaxis.*

10.3.4 Network diagrams summarize complex reaction networks

The preceding section involved a lot of long words. It is good to have a graphical shorthand that is simple, yet flexible enough to express ideas like these, including the variations that arise in each instance. Figure 10.3 illustrates one approach, which we will call the language of **network diagrams**.

---

[11]Section 0.5.2 (page 15) related relative standard deviation to sample size.

Chemical reactions depend on the availability of the reactants, so we will think of a cell as having states described by the total number (inventory) of each relevant chemical species. To begin making a network diagram, then, we draw a box to represent each inventory. Then we represent interconnections with some conventional symbols:

- An incoming solid arrow represents production of a species, for example, by conversion from something else. Outgoing solid arrows represent loss mechanisms.

- If a process transforms one species to another, and both are of interest, then we draw a solid arrow joining the two species' boxes, pointing away from the "input" and toward the "output" of the process.

- But if a precursor species is not of interest to us, for example, because its inventory is held constant by some other mechanism, then we can omit it, and similarly when the destruction of a particular species creates something not of interest to us. Thus, the phosphate groups that appear and disappear in Figure 10.3 come from ATP molecules in the cytosol, which are not shown. They leave as phosphate groups to be recycled by a mechanism that is also not shown.

- To describe how the population of one species affects another one's transformation, we draw a dashed "influence line" from the former to the appropriate solid arrow, terminating with a symbol: A blunt end, ------|, indicates suppression of a process, whereas an open arrowhead, ----▷, indicates its enhancement.

- A dotted line joining two processes indicates that they are coupled: For example, phosphate transfer simultaneously transforms one species from its phosphorylated to unphosphorylated state and has the opposite effect on another species (Figure 10.3).

Many variants of network diagrams exist in the literature.

Using the conventions just given, Figure 10.3b shows that chemoattractant molecules reduce the rate at which the kinase CheA autophosphorylates, and hence the rate at which CheY can find and acquire phosphate groups. Because CheY is constantly dephosphorylating itself (and being dephosphorylated by another enzyme called CheZ), the concentration of CheY-P drops in response to rising chemoattractant level, biasing the flagellar motor to remain longer in its counterclockwise, or "run," mode. In this way, the two-component signaling cascade implements the chemotactic algorithm (Idea 10.2).

$\boxed{T_2}$ *Actually, the simplified signaling network described up to this point has a defect: It responds to the* absolute *concentration of chemoattractant, not to its change over time. Section 10.3.4′ (page 341) describes a more elaborate mechanism ("adaptation") used by cells to estimate the time derivative implied by Idea 10.2.*

10.3.5 Cooperativity can increase the sensitivity of a network element

Although network diagrams can clarify the relationships between elements in a cellular control network, and even allow us to guess its qualitative behavior, still they leave much unsaid. For example, an influence line from one species to a process implies that the mean rate of the process depends on the concentration of the species, via some function. A blunt or pointed arrow on that influence line can indicate that the slope of that function's graph is negative or positive, respectively, but often we need greater detail.

For example, consider the binding of a small effector molecule to a protein. The protein's state (bound or unbound) will fluctuate as effectors bind to it, fall off, and

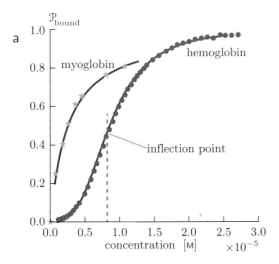
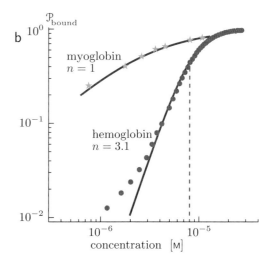

Figure 10.4: [Experimental data with fits.] **Binding curves.** (a) The *symbols* are experimental data for the binding of a ligand (in this case, oxygen) to each of two different proteins. For each protein, increasing oxygen concentration raises the probability of binding, but there is a qualitative difference: One case has an inflection point away from zero concentration (*dashed line*); the other does not. The *curves* are functions of the form given in Equation 10.3 with n chosen to fit the data. (b) The same functions in (a), displayed as log-log plots. In this presentation, the value of n can be read off from the slope of the low-concentration part of each curve. (The deviation of the data from the model at low concentration shows the limitation of the simplified cooperative model described in the text.) [Data from Mills et al., 1976, and Rossi-Fanelli & Antonini, 1958.]

bind again. It seems reasonable that the probability of being bound would increase with increasing effector concentration, but there are many possible increasing functions. For example, some receptors are **multimeric**: They have two or more identical binding sites. Often, binding an effector at one site creates an allosteric change in the ability of the other sites to bind. For this reason, some receptors are most likely to have either no sites bound, or else all of them, but nothing in between, a property called **cooperativity**. For such a receptor, a small increase in effector concentration c from zero has almost no effect on the probability of binding. Instead, the binding probability rises sharply near a critical concentration level called the **inflection point** of the binding curve.

The possibility to have a sharp switching behavior can help a receptor to ignore insignificant fluctuations of effector concentration (those occurring near $c = 0$). Figure 10.4a shows binding data for two famous oxygen-binding molecules. One of them (myoglobin) is monomeric (one binding site), and has a steep increase of binding probability, starting all the way down at zero concentration. The other (hemoglobin) is tetrameric (four binding sites), and instead has its steepest increase at $c \approx 8\,\mu\mathrm{M}$. Frequently, data like these are summarized empirically by selecting from a family called the **Hill functions**:

$$\mathcal{P}(\text{bound}; c) = \frac{1}{1 + (K_\mathrm{d}/c)^n}. \quad \text{cooperative binding curve} \qquad (10.3)$$

The parameter K_d is called the **<u>d</u>issociation equilibrium constant**.[12] The **cooperativ-**

---

[12] You'll find the relation between the inflection point and the parameters K_d and n in Problem 10.2.

ity parameter (or "Hill parameter") n equals one for monomeric or noncooperative receptors, but is greater than one in the cooperative case. Figure 10.4b shows that one way to find its value is by making a log-log plot of the binding data, then finding the slope of the part of the curve below the critical concentration.

Your Turn 10A

Why does that procedure yield the approximate value of the Hill parameter? What procedure might give a more accurate answer?

Figure 10.4a illustrates a key advantage that a sensory system can get from cooperativity: The steepest part of the binding curve corresponds to the greatest sensitivity to changes in concentration. In order for effector binding to be part of a transduction strategy, either at the initial point (chemosensing) or at some intermediate point of a cascade, the system should be arranged to have greatest sensitivity at some biologically significant value of concentration. Noncooperative binding always has greatest sensitivity at *zero* concentration, but cooperativity can move that ideal operating point elsewhere. Partly for this reason, *E. coli* receptors cluster into assemblies that sense food and other conditions cooperatively.

A second advantage of cooperativity becomes apparent when we consider the *fractional* change of binding probability, $\mathrm{d}\mathcal{P}/\mathcal{P}$, brought about by a given *fractional* change of effector concentration, $\mathrm{d}c/c$:

$$\frac{\mathrm{d}\mathcal{P}/\mathcal{P}}{\mathrm{d}c/c}. \tag{10.4}$$

Your Turn 10B

a. Relate this quantity to a feature in Figure 10.4b.
b. From the graph, make a guess about how the maximum value of the quantity (10.4) along the binding curve is related to the value of n.
c. Now work out this quantity starting from Equation 10.3 and comment.

This background section has developed a graphical language useful for describing the signaling cascades inside single cells. We will see later that in vision, too, an intermediate stage of transduction utilizes cooperativity.[13]
For further details on control networks in cells, see the references listed at the end of this chapter.

10.4 PHOTON RESPONSE EVENTS LOCALIZED TO ONE DISK

We can now return to vision. The first step in seeing amounts to a cellular signal transduction problem. Section 10.3 pointed out that cells have invented many sorts of biochemical cascades to solve problems of this sort, but we also noted that vision is different from other kinds of sensory problems, due to the small probability for any

---

[13]See Figure 10.9 (page 333).

Figure 10.5: [Artist's reconstructions based on structural data.] **The role of rhodopsin.** (a) Rhodopsin is a complex of the chromophore retinal (*blue*) with the protein opsin (*pink*). The complex embeds itself in the membranes of the disks in the rod cell outer segment. Although rhodopsin is constrained to stay embedded in the membrane, nevertheless it is free to wander laterally. This mobility enables one excited rhodopsin to encounter many transducin molecules during its active lifetime. (b) Prior to photoisomerization, retinal embedded in its opsin has a bent ("11-*cis*") conformation. Absorption of a photon, however, can change its conformation to the straight ("all-*trans*") form. [Art by David S Goodsell.]

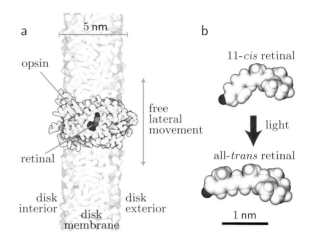

one chromophore to capture photons. Thus, instead of a few thousand receptors on the membrane, a photoreceptor cell has 10^8 chromophores, a massive parallelism that requires special adaptations.[14]

Because the rhodopsin molecules are confined to the disk membranes, we begin the story at the level of one such disk.

10.4.1 Step 1: photoisomerization of rhodopsin in the disk membrane

Rhodopsin consists of a small chromophore called **retinal**,[15] which is a modified form of vitamin A. The retinal molecule is embedded in a large protein called **opsin**, which in turn embeds itself into the disk's bounding membrane (Figure 10.5a). The opsin wrapper plays at least two roles in vision.

First, opsin tunes the absorption spectrum of retinal. Retinal absorbs light strongly because it has a few highly mobile electrons that can readily be shaken by the action of light.[16] But by itself, retinal absorbs light mostly in the ultraviolet region, which is not where we see. By surrounding the retinal and altering its immediate environment, opsin shifts its absorption band to peak around 500 nm.

Prior to photoisomerization, the retinal has a bent shape. This conformation is called 11-*cis*, because the carbon atoms flanking the one at position 11 follow a bent ("*cis*") path (Figure 10.5b). Chapter 1 outlined how absorbing a photon can knock a molecule into an alternate conformation, by enabling it to surmount an energy barrier.[17] In this way, light can isomerize 11-*cis* retinal to an unbent form called all-*trans* retinal. The energy barrier prevents the new conformation from spontaneously jumping back to the 11-*cis* form.

The second role of opsin is to read out the state of the retinal embedded in it. After photoisomerization, the retinal is still chemically bound to its opsin wrapper,

---

[14]Photosynthesis faces a similar problem, but solves it in a different manner (Section 2.9, page 93).
[15]We already encountered this chromophore in the context of optogenetics (Section 2.5, page 72). It is pronounced with the accent on the last syllable; its full name is retinaldehyde. When the word "retinal" is used as an adjective, meaning "concerning the retina," the accent is on the first syllable.
[16] $\boxed{T_2}$ More precisely, retinal has alternating single and double bonds. A similar property accounts for the chromophoric properties of chlorophyll (see Section 12.2.7, page 387). The amino acids that make up generic proteins absorb visible light much less strongly.
[17]See Section 1.6.4 (page 45).

but it no longer fits precisely into its usual binding pocket; instead it pushes on the surrounding opsin, deforming it slightly. Section 10.3.1 (page 321) outlined how a small conformational change in one part of a protein can trigger a rearrangement, affecting the protein's shape at distant places. Such allosteric interactions can alter the protein's ability to bind some other substrate. The affected protein can be an enzyme; allosteric interactions can then greatly alter its activity. We are interested in such a scenario (Figure 10.2b), but with a small twist: In vision, the opsin molecule carries its retinal with it at all times. Light converts the retinal to a new (unbent) form, which acts as opsin's effector.

Fig. 10.2b (page 321)

Immediately after absorbing light, the rhodopsin complex rapidly passes through some intermediate states, ending after about a millisecond in a long-lived state called metarhodopsin II. We will abbreviate this activated molecule as Rh⋆. (The original complex is abbreviated as Rh.)

$\boxed{T_2}$ Section 10.4.1′ (page 344) describes the modified opsins present in cone cells.

10.4.2 Step 2: activation of transducin in the disk membrane

The rhodopsin complex is embedded in its disk membrane, but it can still wander around the disk and bump into other membrane proteins (lateral motion). In its activated form, Rh⋆, rhodopsin transfers information about photon absorption by catalyzing a transformation of another membrane-bound complex called **transducin** (see Figure 10.6). Transducin belongs to a family called **G proteins**, which perform various signaling tasks in cells. The "G" in the name arises because these macromolecules bind the nucleotides GTP or GDP.[18] In the case of vision, transducin has an inactive state with bound GDP, but when influenced by Rh⋆ it exchanges the GDP for a GTP molecule from solution, triggering the separation of one of its subunits from the other two (see Figure 10.6). The separated "α subunit" of transducin, with its bound GTP, is the activated form, which can then participate in the next step of the visual cascade.[19]

Once it has activated a transducin, the activated rhodopsin detaches from it and continues to find and activate other Ts. In fact, within its lifetime one Rh⋆ activates an average of 10–20 transducins. This increase in the number of independent, information-carrying molecules accounts for some of the visual cascade's amplification.

10.4.3 Steps 3–4: activation of phosphodiesterase in the disk membrane, and hydrolysis of cyclic GMP in the cytosol

The activated transducin, T⋆, like Rh⋆, can wander laterally on a single disk membrane. But the signaling event that we wish to understand involves electric current through the rod cell's *outer* (plasma) membrane, which does not connect to the disk (see Figure 10.6)! There must be some mechanism that sends the information about light from the disk membrane to the plasma membrane. This job is done by a small molecule, or **second messenger**, in this case cyclic GMP, abbreviated **cGMP**.[20] cGMP differs

---

[18]The abbreviations refer to <u>g</u>uanosine <u>tri</u>- and <u>di</u>phosphate, small molecules analogous to the energy carrier ATP and its partner ADP. Some authors abbreviate transducin as G_t; we use the symbol T.
[19]Rhodopsin is just one entry in a huge list of "G protein coupled receptors." Section 11.5.2 will discuss other members of this family.
[20]The "first messenger" is the incoming photon. Section 10.3.3 (page 323) pointed out that *E. coli* faces a similar problem, and solves it in a similar way: The response regulator molecule CheY has two states (with and without a phosphate), and it can diffuse from the receptor to the flagellar motor.

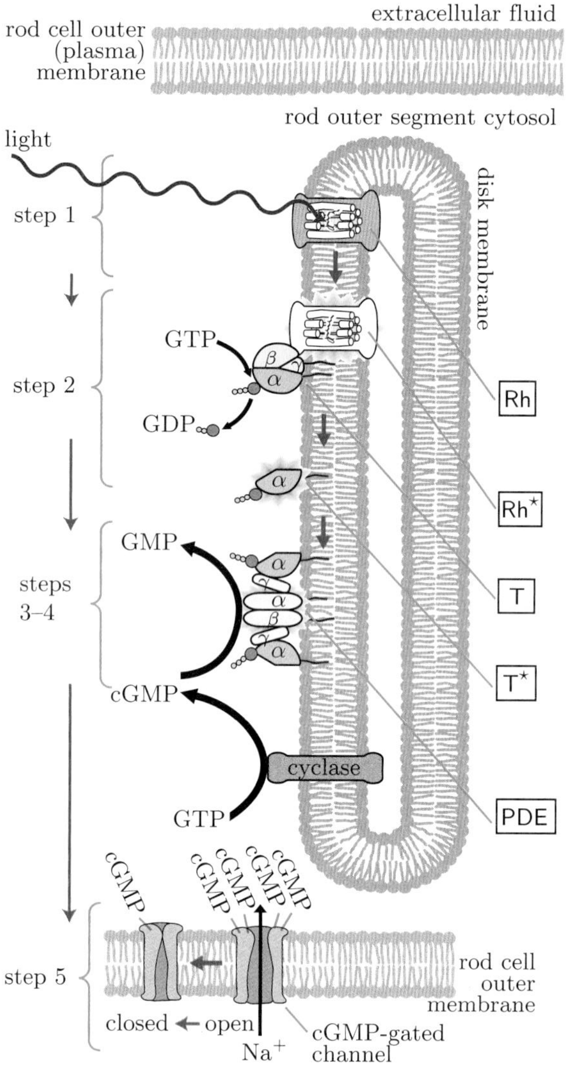

Figure 10.6: [Cartoon.] **The cascade of vertebrate phototransduction in the rod cell outer segment.** See text for descriptions of the five sequential steps, involving an incoming photon, rhodopsin (Rh, Rh*), transducin (T, T*), phosphodiesterase (PDE), guanylate cyclase, and cGMP-gated channels. Only the activated form of transducin, T*, can bind to and activate the phosphodiesterase, a situation reminiscent of the theme represented in Figure 10.2a (page 321). Distant ion pumps and ion exchangers (*not shown*) are discussed in Figure 10.8 and Section 10.7′b (page 346), respectively. They are not directly modulated in the light response; instead, they operate continuously. Other processes *not shown* continually reattach phosphate groups to GMP and GDP, replenishing GTP. The disk is actually symmetrical; all the membrane-bound equipment shown here on the left side is equally present on the right side as well. [Courtesy Trevor D Lamb, adapted from Lamb & Pugh, 2006.]

from guanosine monophosphate by an additional chemical bond that forms a ring structure; cells use it (and its cousin cAMP) for a variety of mostly intracellular signaling tasks.

Cyclic GMP is constantly being created by an enzyme in the disk membrane called **guanylate cyclase**, which we'll abbreviate as "cyclase." It forges GTP into a ring, by clipping off two of its phosphate groups and using the energy thus liberated to add a chemical bond. At the same time, however, another enzyme called **phosphodiesterase** binds any cGMP that diffuses to it, and cuts (hydrolyzes) the bond, converting the cGMP back to ordinary GMP.[21] This arrangement is able to transmit information because the phosphodiesterase is *modulated:* Activated transducin binds to it and raises its catalytic rate. The cyclase is not directly modulated by light, so raising the phosphodiesterase activity has the net effect of lowering the cGMP concentration inside

---

[21] A third enzyme replenishes the cell's GTP store by reattaching phosphates to GMP.

Figure 10.7: [Network diagram.] **Summary of events in a single disk occurring in response to light absorption.** In this figure, information flows downward from an initial photon capture at the top; labels *1–4* refer to the steps introduced in the text and depicted in Figure 10.6. The cascading effects of a flash of light correspond to population shifts from the boxes on the left to those on the right. The graphical conventions are the same as in Section 10.3.4: Each box represents the amount of some molecular species present, *solid arrows* represent processes that increase or decrease those numbers, and each *dashed line* represents the influence of one molecule on another process. The notation T*·PDE indicates a complex (transiently bound state). *Wavy arrows* correspond to recovery events that "reset" the rod cell after a short flash (see Section 10.7$'$c, page 348). Some occur in just a few hundred milliseconds; others, like the multi-stage reconstitution of rhodopsin, take much longer.

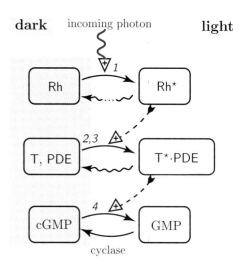

the rod cell.[22] That reduction of concentration encodes the information transmitted from the disk to the plasma membrane.

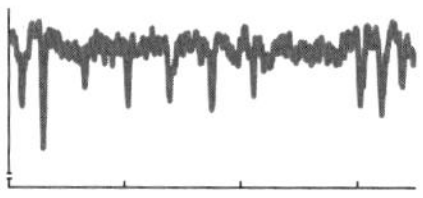

The visual signaling story has many steps. To make the global picture easier to grasp, Figure 10.7 replaces detailed cartoons like Figure 10.6 by the more schematic network diagram notation introduced in Section 10.3. Later sections will add more elements to this diagram.

Section 10.4 has sketched the first steps in photoreceptor response, which involve photochemistry and enzyme activity. We must now see how the rod cell transduces the output of these initial steps into electrochemical signals that can be transmitted to the brain.

[T2] *Section 10.4.3$'$ (page 345) discusses other classes of vertebrate photoreceptors.*

10.5   EVENTS ELSEWHERE IN THE ROD OUTER SEGMENT

After photon absorption, the local change of cGMP concentration spreads rapidly by diffusion to nearby regions of the rod cell's plasma membrane. To understand what happens there, we now outline the motions of ions in and near the rod cell, using ideas introduced in Section 2.4 (page 66).

10.5.1   Ion pumps in the rod cell plasma membrane
maintain nonequilibrium ion concentrations

Like any animal cell, a photoreceptor maintains an electric potential drop across its plasma membrane by actively pumping specific ions in specific directions across it.[23]

---

[22] [T2] Spontaneous thermal activation of PDE, that is, activation not induced by a T*, is thought to be the source of most of the continuous dark noise seen between the large events in Figure 9.8b (page 303). Section 10.7$'$ (page 346) discusses additional modulations of cyclase activity.

[23] Section 2.4 (page 66) introduced this concept.

Fig. 9.8b (page 303)

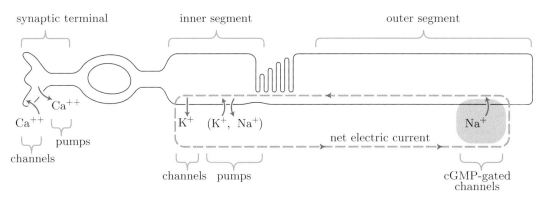

Figure 10.8: [Cartoon.] **Ion currents in and near a vertebrate rod cell, in the absence of light.** At all times, ion pumps (*pale blue blob*) use the cell's metabolic energy supply to push specific ion species in specific directions across the inner segment's membrane. The net effect is to remove positive charge from the cell, and hence to drive the cell's interior electric potential negative. In the dark, sodium ions (Na^+) can also pass through gated channels (*pink, right*) in the rod outer segment, replacing ions exported by the pumps and partially depolarizing the cell. Also, channels permeable only to potassium ions (K^+) constantly let them escape (*yellow*), so that overall the cell maintains a steady state. In the presence of light, the cascade shown in Figure 10.6 closes some of the cGMP-gated channels, reducing ion entry at the outer segment. The continued action of the pumps, and outflow of K^+ ions, then drives the interior of the cell to a more negative potential than in the dark. This hyperpolarization spreads rapidly across the entire cell membrane, and ultimately modulates neurotransmitter release at the synaptic terminal. *Dashed lines* represent the net electric current between outer and inner segments. (The calcium currents at the synaptic terminal are discussed in Section 10.6; other calcium currents in the outer segment are discussed in Section 10.7′.)

Figure 10.8 illustrates the action of one such pump; on each working cycle, it consumes ATP, pushes three sodium ions out of the cell, and pulls in two potassium ions. The figure also shows a class of channels that only allow potassium ions to pass. These K^+ ions leave the inner segment under the effect of their excess interior concentration (built up by the ion pumps), which overcomes a smaller electric force (inward, because the cell interior is more negative than the exterior).

Like any sensory neuron, a photoreceptor also has some channels that are controlled by prevailing conditions. Figure 10.8 shows one such class of channels in the rod outer segment; when open, they allow sodium ions to pass. These Na^+ ions move inward due to the combined effect of lower interior concentration (depleted by the pumps) and the same inward electric force as before. In the dark, the channels are open, and the net effect of the pumps and channels is to maintain a steady state with the interior electric potential about $40\,mV$ more negative than the exterior.

10.5.2 Step 5: ion channel closing in the plasma membrane

The story so far explains the existence of a steady, nonequilibrium membrane potential and circulating exterior electric current. But T. Tomita, A. Kaneko, and coauthors also found in the mid-1960s that a flash of light changes the cell's membrane potential, driving the interior even more negative than in the dark state. To see how this hyperpolarization comes about, recall that Steps 1–4 of the visual cascade[24] explain how a flash of light disturbs the intracellular concentration of the second messenger molecule cGMP. The connection between cGMP and electrical signaling was demonstrated

---

[24]Sections 10.4.1–10.4.3 described these steps.

Figure 10.9: [Experimental data with fit.] **The dose-response relation for opening of cyclic nucleotide-gated channels by cGMP.** The vertical axis represents ionic conductance relative to its maximal value, a proxy for the fraction of channels that are open. *Dots* on this log-log plot represent the averages of 20 experiments; error bars represent ± one standard deviation. The results were obtained from toad rod outer segments. The *curve* is the function $1/(1 + (57\mu\text{M}/c)^2)$; that is, a cooperativity parameter (Hill coefficient) value $n = 2$. The *asterisk* indicates the level of cGMP inside a normal rod cell in the absence of light. The steep slope of the curve near this point means that small relative changes in concentration are amplified to larger relative changes in the probability that the channel will be open, and hence also to changes in ion current (see Figure 10.4, page 326). [Data from Nakatani & Yau, 1988, available in Dataset 15.]

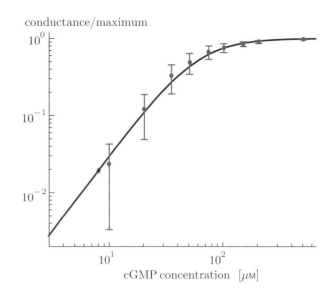

directly by isolating a small patch of outer-segment plasma membrane on the tip of a tiny pipette. Exposing either side of the membrane to various chemicals, and then monitoring its conductances to various ions, showed that exposure to cGMP on the intracellular face of the membrane raises its conductance for sodium ions. The interpretation of this change is that *cGMP biases the ion channels toward their open conformation;* accordingly, they are called **cyclic nucleotide-gated channels.**[25]

Fig. 9.8b (page 303)

Fig. 9.12a (page 307)

This result supplied the missing link between biochemical events in the disk and electrical events in the rod cell membrane. In the dark, there is plenty of cGMP, and some of the channels are open. In response to light, however, the cGMP level suddenly falls, either in a local region of the outer segment (for a dim flash) or everywhere (for a bright flash). For a single-photon event, the net effect is to close about a hundred channels. The rate of Na^+ entry then falls, and the circulating current drops. The reduction in the cell's extracellular current is the signal measured experimentally in single-cell recordings (Figure 9.8b). The magnitude of the change reflects the strength of the flash, as seen in experiments like the one in Figure 9.12.

The story of phototransduction up to this point has a puzzling feature, which only emerges when we examine it quantitatively. The amplification in previous steps amounts to a single Rh* molecule reducing the number of cGMP molecules by about 1400. This number is less than 1% of the total number of free cGMP in the outer segment. The gated channels need to be very sensitive in order to respond to such a small relative change. They achieve this sensitivity in part because their closing is cooperative.[26] Section 10.3.5 (page 325) discussed cooperativity in the context of a molecule (such as a gene's transcription factor, or hemoglobin) that can bind either n ligand molecules, or none. Similar remarks apply to an ion channel that remains closed unless it has bound n ligand molecules. As shown in Figure 10.4b, the value of n can be read off from a log-log plot of the binding curve. Figure 10.9 shows such

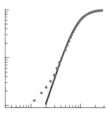

Fig. 10.4b (page 326)

---

[25]GMP and AMP are "nucleotides," that is, DNA bases joined to a sugar and a phosphate group.

[26] $\boxed{T_2}$ Also, the *local* cGMP concentration change exceeds 1% close to a disk that catches a single photon, because diffusion does not spread that change throughout the outer segment instantly.

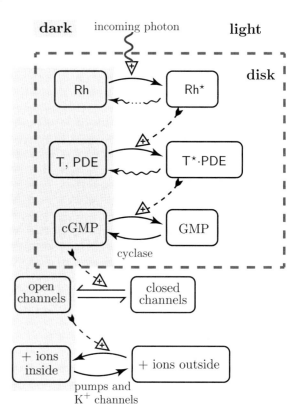

Figure 10.10: [Network diagram.] **Summary of events in the rod cell outer segment occurring in response to absorption of a single photon**—a continuation of Figure 10.7. The effect of light is to favor the states on the right. The output of the part of the diagram enclosed in the *dashed box* affects a neighborhood of 40–80 disks near the site of photon absorption.

a **dose-response relation** for the ion conductance of the cGMP-gated channels as a function of concentration. The steep slope corresponds to a cooperativity parameter value (Hill coefficient) n that, in various experiments, is found to lie between 2 and 3. The virtue of cooperativity is that it sharpens the channel's response to small changes in cGMP concentration near the resting (dark) value.

Section 10.5 has added some steps to the visual signaling transduction cascade. Figure 10.10 summarizes the steps described so far.

10.6 EVENTS AT THE SYNAPTIC TERMINAL

10.6.1 Step 6: hyperpolarization of the plasma membrane

The preceding sections explained how a light flash leads to the closing of some ion channels in the rod outer segment. The ensuing loss of inward ion flux drives the electric potential inside the cell to a value even more negative than that maintained in the dark: The cell becomes hyperpolarized.

Hyperpolarization quickly spreads throughout the entire rod cell. This claim may be clearer in the language of first-year physics: Imagine two long conductors separated along their lengths by an insulator. (The conductors represent the extracellular fluid and the interior cytosol, respectively; the insulator is the cell's plasma membrane.) If we remove charges from the end of one conductor, and place them on the same end of the other one, the conductors rapidly develop a potential difference all along their

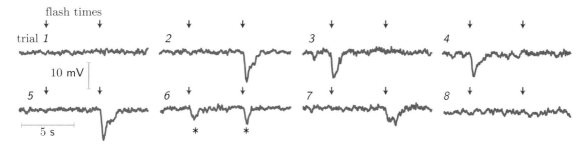

Figure 10.11: [Experimental data.] **Membrane potential responses of a mouse rod cell to dim flashes of light.** *Arrows* at top show the instants at which fixed-strength flashes of light were delivered in eight separate trials. At this flash strength, more than half of flashes elicited no response. Nonzero responses were a mixture of smaller events, interpreted as single productive photon absorptions (for example, those marked by *asterisks*), and larger ones, interpreted as double absorptions. In this experiment, the cells were maintained at room temperature; at body temperature, the response amplitude is smaller. [Data courtesy Lorenzo Cangiano; see Cangiano et al., 2012.]

lengths, because charge can move along each one. Similarly, a current imbalance gives rise to a local potential change inside the rod outer segment, which rapidly spreads to create hyperpolarization everywhere along the rod cell's plasma membrane, even at the distant synaptic terminal. The cell's interior membrane potential, relative to the exterior, falls from a steady $-40\,\mathrm{mV}$ prior to illumination to as low as $-70\,\mathrm{mV}$ for a bright flash. For a single photon absorption by a dark-adapted rod cell, the hyperpolarization is smaller, typically about $2.4\,\mathrm{mV}$ (asterisks in Figure 10.11).

Is hyperpolarization the important event in signaling, or is it some insignificant side effect? To find out, D. Baylor and R. Fettiplace artificially hyperpolarized a rod cell by using an intracellular electrode; they found that subsequent neurons in the pathway (the ganglion cells serving that rod) responded as if a real flash of light had been presented. Moreover, by artificially *depolarizing* the cell (driving its interior potential less negative than usual), they found that they could abolish signaling from the rod to its bipolar cell in response to light.

10.6.2 Step 7: modulation of neurotransmitter release into the synaptic cleft

Voltage-gated channels

So indeed cell hyperpolarization is a key step on the transduction pathway. To see how it leads to signaling, we now note that hyperpolarization spreads rapidly to the synaptic terminal and affects another class of ion channels located there. These channels are called **voltage-gated**, because they admit ions in response to changes in membrane potential ("voltage drop").[27] They permit calcium ions to pass in the dark state, but they close upon hyperpolarization (that is, in response to light).

Ex. Changing the membrane potential by $2.4\,\mathrm{mV}$ may not seem very significant, but recall that the electric field across a capacitor is the potential drop divided by the *thickness* of the insulating layer. Estimate the change in electric field (volts per meter) when a rod cell responds to a single photon. Does it seem big enough to alter the conformation of a protein embedded in the membrane?
Solution: A cell's plasma membrane is only a few nanometers thick (Figure 10.5, page 328). So the transmembrane electric field can be enormous: $2.4\,\mathrm{mV}/(5\,\mathrm{nm}) \approx$

---

[27]Section 2.4.5 (page 68) introduced voltage-gated channels in a different context: neural signaling via action potentials.

5×10^5 V/m! But is that really big? Suppose that some domain in an ion channel carries five net proton charges and can move relative to the rest of the molecule by $1\,\text{nm}$. The total electric potential energy change when this electric field drives such a motion is then roughly

$$(0.5 \times 10^6\ \text{V/m})(1\,\text{nm})(5 \cdot 1.6 \times 10^{-19}\ \text{coul}) = 4 \times 10^{-22}\ \text{J}.$$

That *doesn't* seem so big, but compared to what? Quantities with dimensions can only be declared "large" or "small" in comparison to some other relevant number.

Each molecular degree of freedom is continually supplied with thermal energy;[28] at room temperature the scale of that energy is around $k_{\text{B}}T_{\text{r}} \approx 4.1 \times 10^{-21}$ J. So the change in membrane potential gives a contribution to molecular conformational energy about $1/10$ as large as the typical thermal energy.

The Example made a meaningful comparison, but it's disappointing. How can such a small energy change have any effect in the face of thermal noise? In order to be responsive to single-photon changes in membrane potential, a channel must be so sensitive (have such a small energy difference between conformations) that it constantly flickers open and shut!

The answer lies in the fact that there are many channels, and the cell cares about the time average of their state. Even though each one is constantly flickering, nevertheless the *total* conductance of a patch of voltage-gated channels, averaged over the rod integration time, can have a well-defined value that *doesn't* have large relative standard deviation.[29] Thus, small nonrandom effects, caused by small membrane potential changes, can stand out above the noise.

Neurotransmitter release

The preceding paragraphs described voltage-gated calcium channels in the synaptic terminal that close in response to light. Like sodium in the outer segment, calcium ions are depleted inside the synaptic terminal by ion pumps that continually export them (Figure 10.8). Thus, the light-induced closing of the channels gives rise to a momentary drop in the interior Ca^{++} concentration. In the dark, interior calcium triggers the fusion of vesicles (small membrane-enclosed bags) with the rod cell's plasma membrane.[30] Each vesicle contains the neurotransmitter molecule **glutamate**; its fusion thus releases a packet of glutamate into the extracellular region. In the dark, the rod cell releases vesicles at a mean rate of about $100\,\text{s}^{-1}$. The closure of the voltage-gated ion channels, and the resulting reduction in intracellular calcium level, interrupts this release. In short,

Fig. 10.8 (page 332)

> *The rate of glutamate release from the rod's synaptic terminal falls in response to light. Variations in this rate constitute the information output of the rod cell.*

Glutamate diffuses across the gap (the **synaptic cleft**) between the photoreceptor cell and its bipolar cell's input region, where it is sensed and leads to further events to be discussed in Section 11.2.

Section 10.6 has completed our outline of signaling transduction in photoreceptor cells. Figures 10.12b and 10.13 depict machinery located in the rod cell synapse. $\boxed{T_2}$ *Section 10.6′ (page 345) discusses clearance of glutamate from the synapse.*

---

[28]Section 0.6 (page 17) introduced thermal energy.
[29]Section 0.5.2 (page 15) related RSD to sample size.
[30]Section 2.4.7 (page 70) introduced this mechanism for synaptic transmission.

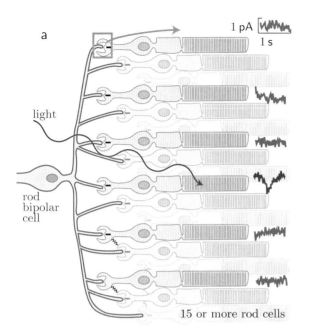
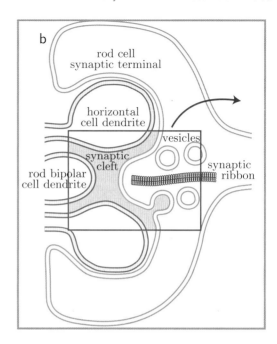

Figure 10.12: [Sketches.] **The coupling between a rod photoreceptor and its associated bipolar cells.** (a) A rod bipolar cell pools inputs from many rods. The figure depicts the situation under weak illumination: Only one rod has absorbed a photon (*red trace* at right). The remaining rods are generating continuous dark noise as usual (*blue traces*). The region in the *orange box* is magnified in (b). (b) Closeup of the synapse between a rod cell's synaptic terminal and one of its rod bipolar cells' dendrites. The region in the *red box* is shown in greater detail in Figure 10.13. Some synaptic vesicles are shown; one of them is fusing with the rod cell's plasma membrane, releasing its cargo of neurotransmitter into the synaptic cleft. For simplicity, only a few vesicles are shown; actually, a long "active zone" extends out of the page and docks over 100 vesicles. The synaptic ribbon tethers many more vesicles, readying them to move to the active zone. Two horizontal cell dendrites, not shown in panel (a), also occupy the rod synaptic terminal's cup-shaped structure. (One or more additional rod bipolar cell dendrites, displaced out of the plane of the page, are not shown.)   [(a) From Okawa & Sampath, 2007.]

10.7 SUMMARY OF THE VISUAL CASCADE

Figure 10.14 assembles all the pieces discussed so far in the rod cell's response to light. The net function of this complex network is to transduce incoming light intensity into changes in the rate of glutamate release at the synapse to the next neuron (the rod bipolar cell). In words, the information about absorption of a single photon is encoded into production of one Rh⋆, and then is transduced into a fall of cGMP→ ion channel closing → membrane hyperpolarization → calcium channel closing → calcium level drop inside the synaptic terminal → pause in vesicle release → reduction in glutamate in the synaptic cleft.

$\boxed{T_2}$ *Section 10.7′ (page 346) describes additional details of rod signaling.*

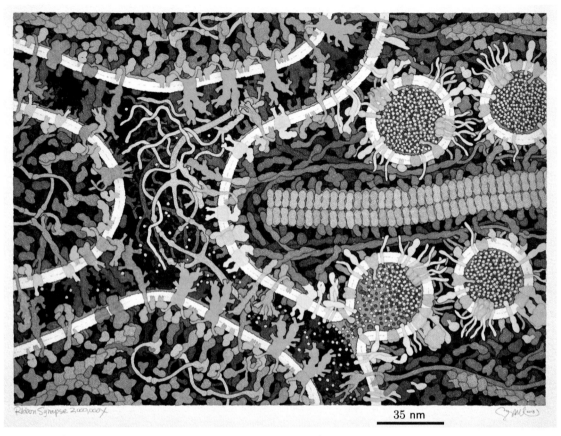

Ribbon Synapse 2,000,000X

35 nm

Figure 10.13: [Artist's reconstruction based on structural data.] **The first visual synapse.** Part of a vertebrate rod cell's synaptic terminal is shown at *right*, including four vesicles, each packed with about 2000 molecules of the neurotransmitter glutamate (*yellow dots*). One vesicle has fused its membrane with that of the rod cell, under the influence of calcium ions admitted by a voltage-gated ion channel, and is releasing its contents into the space between the cells (the synaptic cleft). Other molecular machines (*not shown*) constantly remove glutamate molecules from the synaptic cleft, creating a steady state in the dark despite continual vesicle release. (Glutamate thus recovered is eventually recycled for future use.) On the other side of the synaptic cleft, dendrites from 2–4 rod bipolar cells (one shown on *left*), and two horizontal cells (*top and bottom*), are studded with receptors (*blue and green*) that respond to changes in the concentration of glutamate. For artistic reasons, the image has been distorted somewhat: The actual distance from the rod terminal to the bipolar cells is longer than shown (10–40 times longer than the distance from rod terminal to the horizontal cells). The synaptic ribbon is shown edge-on; it extends several micrometers into and out of the page. See this book's title page or Media 21 for a key to the other structures shown. [Painting by David S Goodsell.]

THE BIG PICTURE

This chapter outlined the chain of events unleashed by a single productive photon absorption, or more precisely, the disturbance to an ongoing steady state brought about by that capture. Although the scheme is highly evolved, and hence complex, nevertheless it illustrates some unifying themes that originate all the way back with single-cell organisms. At least some of the complexity turned out to confer a huge *dynamic range* by allowing multiple points where the amplification can be regulated.

We have not yet achieved the desired synthesis between psychophysical (whole-organism) and physiological (single-cell) experiments, however. Chapter 11 will do this, after fitting one more piece into the puzzle: processing at the first synapse.

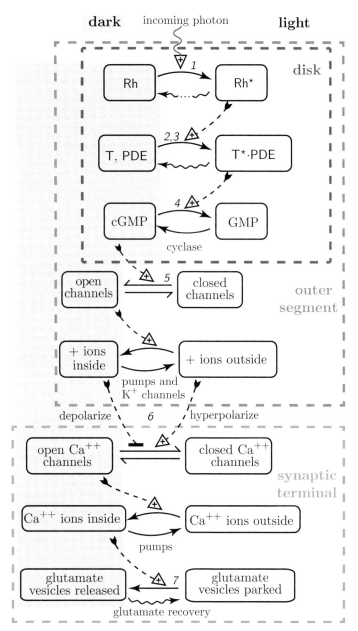

Figure 10.14: [Network diagram.] **Summary of events in a rod cell during visual transduction**, continuing Figure 10.10. In this figure, information flows downward from an initial photon capture at the top. The cascading effects of a flash of light correspond to shifts from the boxes on the left to those on the right. The seven numbered steps are discussed in Sections 10.4–10.6. *Wavy arrows* correspond to recovery events that "reset" the rod cell after a short flash (see Sections 10.7′a–c, page 346). *Dashed lines* emerging from the orange box represent membrane potential changes that spread to the synaptic terminal (*lower dashed box*). Those changes affect the release of the neurotransmitter glutamate, which is the rod cell's output signal (*bottom*). Glutamate molecules are continually scavenged from the synapse and repackaged into new vesicles (*bottom*), creating a steady state in darkness. The light response is a transient disturbance to that steady state.

KEY FORMULAS

- *Cooperative binding curve (Hill function):*

$$\mathcal{P}(\text{bound}) = \frac{1}{1 + (K_{\mathrm{d}}/c)^n}. \qquad \text{[10.3, page 326]}$$

Here K_{d} and n are called the dissociation equilibrium constant and cooperativity parameter (Hill coefficient), respectively. One measure of receptor response is the logarithmic derivative,

$$\frac{\mathrm{d}\mathcal{P}}{\mathrm{d}c}\frac{c}{\mathcal{P}}, \qquad \text{[10.4, page 327]}$$

where $\mathcal{P}$ is the probability to be in one of the receptor's activity states.

FURTHER READING

Semipopular:
Bray, 2009.

Intermediate:
Visual transduction: Sterling & Laughlin, 2015; Dowling, 2012; Rodieck, 1998; `http://webvision.med.utah.edu/book/` .
Flagellar propulsion: Nelson, 2014; Berg, 2004.
Cellular control and transduction networks: Nelson, 2015; Bialek, 2012; Phillips et al., 2012.
Cell biology and biochemistry of G protein coupled receptors and their signaling cascades: Lodish et al., 2016, chapt. 15; Alberts et al., 2015, chapt. 15; Berg et al., 2015, chapt. 33; Marks et al., 2009.
$\boxed{T_2}$ Adaptation in chemotaxis: Berg, 2004.

Technical:
General: Sterling, 2013; Sterling, 2004a, Sterling, 2004b.
Chemotaxis network: Sourjik & Wingreen, 2012; Meir et al., 2010. $\boxed{T_2}$ The adaptation model in Section 10.3.4′ originated in Barkai & Leibler, 1997.
G protein cascades: Lamb & Pugh, 2006; Arshavsky et al., 2002.
Response of ganglion cells to single photon absorptions: Barlow et al., 1971.
Spontaneous false-positive signals from rod cells: Baylor et al., 1984.
Role of hyperpolarization in photoreceptor cells: Baylor & Fettiplace, 1977; Tomita et al., 1967.
Horizontal cells: Kramer, 2014.
$\boxed{T_2}$ Light-sensitive ganglion cells: Berson, 2014; Schmidt et al., 2011.
$\boxed{T_2}$ Termination of rod response: Burns, 2010, Burns & Pugh, 2010.
$\boxed{T_2}$ Calcium feedback mechanisms in (invertebrate) photoreceptors: Pumir et al., 2008.

$\boxed{T_2}$ **Track 2**

10.2.2′ Higher light intensities

At higher light intensities, the mean waiting time between photon arrivals is shorter than the integration time of the rod cell. Even in this situation, the rod still generates a signal indicating the rate of photon arrivals. For example, the two-photon event in Figure 9.8b has twice the amplitude of the others. At even higher intensities, a photoreceptor cell's response becomes nonlinear: It reduces its amplification.[31]

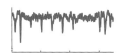
Fig. 9.8b (page 303)

$\boxed{T_2}$ **Track 2**

10.3.3′a More about two-component signaling pathways

Going beyond cell motility, a host of other bacterial cell functions, including cell division, virulence, antibiotic resistance, metabolite fixation and utilization, response to environmental stress such as osmotic shock, and sporulation are all mediated in bacteria by two-component pathways like the one described in the main text.

Homologous pathways have recently been discovered in fungi such as yeast, plants such as *Arabidopsis,* and protozoa such as *Dictyostelium,* though not yet in higher organisms.

10.3.3′b Phototaxis

Light is an important environmental variable for any organism, including unicellular ones such as bacteria and archaea. Light can be beneficial (by supplying useful energy), or it can be harmful (by causing photodamage), so cells can benefit by moving toward regions with the optimal intensity and spectrum and away from "wrong" conditions, a behavior called phototaxis.

One control network, used to implement phototaxis in halobacteria such as *Halobacterium salinarum* and *H. halobium,* is similar to the one for chemotaxis discussed in the main text (Figure 10.3b). In place of a chemoreceptor, this network substitutes an analog of the rhodopsin molecule used in human vision, called "sensory rhodopsin."[32] Sensory rhodopsins catch light by using the same retinaldehyde group as the one in our eyes. They are connected to, and influence, "transducer" molecules that are similar to the chemotaxis receptors. The transducer then activates CheA autophosphorylation as part of a standard two-component pathway.

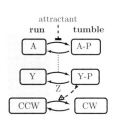
Fig. 10.3b (page 324)

$\boxed{T_2}$ **Track 2**

10.3.4′ More about adaptation in chemotaxis

The simple two-component signaling network (lower part of Figure 10.15) can de-

---

[31]See also Section 10.7′b below.

[32]Sensory rhodopsins are also similar to the halorhodopsins that are the basis for optogenetics (see Section 2.5.3, page 74).

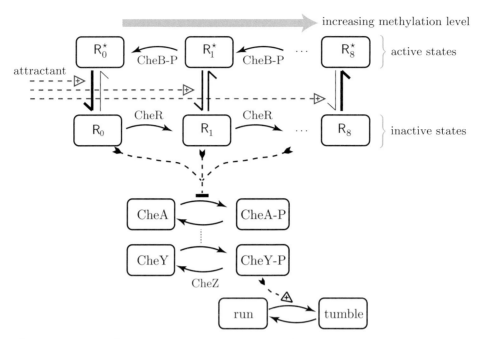

Figure 10.15: [Network diagram.] **Adaptation circuit for bacterial chemotaxis.** The lower part of this diagram is the same as Figure 10.3b (page 324). The top part represents a ladder of receptor states: R_i denotes the inactive receptors methylated i times, and similarly $R_i^\star$ for the active states. The equilibrium arrows (vertical) have unequal weights, indicating that the equilibrium between active and inactive states depends on methylation level. The *horizontal dashed* influence lines imply that, for each methylation level, increasing the concentration of chemoattractant shifts the balance toward the inactive form. (Another feedback loop, involving the interconversion of CheB and its phosphorylated version, is not shown.)

termine whether the concentration of an effector molecule (chemoattractant) near a bacterium is "high" or "low" (compared to the equilibrium constant for the effector to dissociate from the receptor). But it has limited ability to discriminate between "high" and "even higher"; that is, its **dynamic range** is limited to those concentrations over which the receptor occupancy changes from nearly always unbound to nearly always bound, typically only about a tenfold range. If this were the whole story, a cell in the "high" chemoattractant regime would be satisfied and would never seek still higher levels; and a cell in "very low" chemoattractant would lose hope and tumble constantly, making no attempt to at least find conditions that, while low, are a bit better.

In fact, real *E. coli* cells can follow chemoattractant gradients over *five orders of magnitude* in concentration. Pioneering work by H. Berg showed that they do this by estimating not the absolute concentration level, but rather its *time derivative,* as implied by Idea 10.2 (page 322). Like most other sensory systems, including our own, the cells make this estimate via **adaptation**: The network gets "accustomed" to the ambient level of chemoattractant over the past few seconds, and responds to *changes* about that level.

Figure 10.15 shows an extension of the basic circuit that can implement adaptation (a version of an idea introduced by N. Barkai and S. Leibler). Each receptor has two conformational states, called "active" and "inactive." The key point is that those states are subdivided: Each receptor has eight sites at which it can be modified by covalent attachment of a methyl group. Changes in chemoattractant concentration alter the equilibrium between the active and inactive states, but so does the methylation status, which acts as a molecular "memory." The network stores information in that memory about its recent past history.

In the model, an enzyme called a methyltransferase, or CheR, slowly adds methyl groups to the receptor, but only when the receptor is in its inactive state. Another enzyme, called a methylase, or CheB-P, slowly removes methyl groups, but only when the receptor is in its active state. Under constant conditions, the network arrives at a steady state in which there is net counterclockwise circulation in the ladder shown at the top of Figure 10.15. The model further assumes that increasing chemoattractant level biases the receptor toward its inactive state, but increasing methylation level has the opposite effect, as shown in the figure.

Suppose that the concentration of chemoattractant suddenly jumps up from a low initial level. According to the horizontal influence lines in the figure, the result is that each of the upper boxes (the "active" receptor pool) shifts population to its corresponding lower box (the "inactive" pool). The rest of the signaling cascade then responds by lengthening runs, as in the simpler model discussed in the main text.

If the increase of chemoattractant is sustained, however, something more interesting starts to happen. The increases in inactive receptor populations give more substrates for CheR to methylate. Over time, then, the entire population of inactive receptors moves to the right in the figure, to states that are more apt to jump to the active state. A new steady state is reached in which the total fraction of receptors in the inactive state is *the same* as it was in steady, low chemoattractant level.

If later the chemoattractant level jumps down, the situation is reversed: There is a transient increase in the total fraction of receptors in the active state, leading to increased tumbling, but if the low level is sustained, the active population shifts leftward, toward states more apt to jump to the inactive state, and the phosphorylation level of CheY approaches the same level as at constant high chemoattractant.

In short, under constant conditions *the total fraction of receptors in the active state is the same, regardless of chemoattractant level.* But any sudden *change* in chemoattractant level will temporarily alter the balance of active/inactive receptors, until the slow CheR and CheB have had a chance to restore the system to the steady levels. In this way, the network effectively compares the instantaneous chemoattractant level to its average over the recent past, allowing it to sense time variation over a wide range of ambient background levels.

If the cell were motionless, knowing this time derivative would not help it to navigate up a spatial gradient. But the cell is constantly swimming, so it effectively measures the dot product between its velocity vector and the spatial gradient of chemoattractant level. Thus, by estimating the time derivative, the cell can decide whether things are getting better (and so extend the current run) or worse (so terminate the run and reorient via a tumble episode), implementing the search algorithm.[33]

Y. Meir and coauthors investigated the model represented by Figure 10.15, and compared it to experiments in which *E. coli* was subjected to sudden up- and downward steps in chemoattractant concentration (Meir et al., 2010). To focus on just the adaptation mechanism, they did not measure the cell's behavior (run/tumble transitions). Instead, they used FRET to monitor the population of the phosphorylated form of CheY directly, by placing an acceptor fluorescent protein (YFP) on it and a donor (CFP) on the CheZ phosphatase, which binds only the phosphorylated form of

---

[33]Idea 10.2 (page 322) states this algorithm.

CheY.[34] The model outlined in this section, with some additional improvements, was able to reproduce both the observed adaptation and the transient responses.

$\boxed{T_2}$ **Track 2**

10.4.1′ Cone and cone bipolar cells

The main text focused on rod cells for concreteness. Phototransduction in cone cells is overall similar, with some modifications related to their function (discussed in Chapter 3).

Cones use the same chromophore as rods (11-*cis* retinal), but embed it in slightly different proteins called photopsins (instead of opsin) to form complexes called iodopsins (instead of rhodopsin). Each cone cell expresses just one of three subtly different photopsins; in this way, each tunes its retinal's absorption to one of the three spectra characteristic of L, M, or S cells (see Figure 3.11).[35] In particular, the L and M photopsins must lower the energy barrier for photoisomerization of retinal below that in rod cells, in order to be excited by the lower-energy red and green photons. This lowered barrier increases the rate of spontaneous thermal isomerization, compared to either the S cones or rod cells (Luo et al., 2008). The corresponding increase in false-positive signaling would pose a problem if cones needed to be as sensitive as rods, but they do not: Cones are optimized for vision under high light conditions, with dark-adapted sensitivity 1–4% that of rods. At still longer wavelength, however, the inevitable thermal noise would be even worse, which may explain why we do not have photoreceptors that are sensitive in the infrared.

At the opposite extreme of illumination, our eyes must also be able to function in bright daylight, under illumination levels up to about 10^{10} times the dim-light level discussed in Chapter 9 (Figure 3.10, page 121)! At these high intensities, rod photoreceptors give no useful information—their response saturates at about 10^6 times the dim-light level. Remarkably, however, cone cells never saturate. As the ambient light increases, each cone cell becomes depleted of activatable iodopsin molecules, and hence less sensitive. This depletion takes time, so we can be momentarily dazzled by a sudden increase in illumination.

The overall architecture of mammalian cone cells is also slightly different from rods: Instead of disks enclosed entirely within the outer segment (as in rods), in cone cells the disks remain attached to the outer segment membrane. Although in principle such a design could eliminate the need for a diffusing second messenger like cGMP, in fact, cones use a transduction cascade very similar to the one in rods. However, the cone response is several times faster than that of rods, a useful adaptation for high illumination conditions, in which the high photon arrival rate allows fast discrimination of changes.

The main text discussed univariance in the context of rod responses, but our analysis of color vision in Chapter 3 rested on this hypothesis being true for cones. Baylor, B. Nunn, and J. Schnapf indeed found that, for each class of cone cell, the

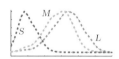

Fig. 3.11 (page 123)

---

[34]Section 2.8 introduced FRET.

[35]More precisely, *humans* have three cone pigments. Chapter 3 mentioned that most birds have four. Some mantis shrimp appear to have 12 distinct opsins, and "primitive" organisms like corals and scallops have been found to have genes encoding dozens.

entire time course of the response to a flash of light was independent of the flash's wavelength, provided only that the flash intensity was adjusted to account for wavelength dependence of the photon absorption spectrum (Baylor et al., 1987). These experiments also established the spectral sensitivity functions shown in Figure 3.11.

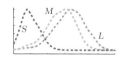

Fig. 3.11 (page 123)

 Track 2

10.4.3′ Recently discovered vertebrate photoreceptors

Chapter 3 discussed two broad classes of photoreceptors: the rod and cone cells. These cells connect to bipolar cells, which in turn connect to the eye's main outputs (the retinal ganglion cells). Axons from the ganglion cells form a bundle called the optic nerve, which sends signals into the brain's visual cortex.

Remarkably, there is another class of light-sensitive cells in our eyes: In humans, for example, about 0.2% of our retinal ganglion cells are themselves sensitive to light, in addition to receiving signals as usual from rods and cones (Berson et al., 2002; Hattar et al., 2002). These cells contain a pigment named melanopsin, related to but different from the opsins found in rod and cone cells. Although their axons are bundled with the others in the optic nerve, some of them split off into the "retinohypothalamic tract," ending up in a brain region called the suprachiasmatic nucleus. Thus, these cells do not report directly to the brain's visual cortex. Their axons give light signals to the body's daily pacemaker, keeping our diurnal cycle synchronized with local daylight. In fact, some totally blind individuals can adjust to jet lag, because this alternative pathway, which makes no use of rod or cone cells, is intact.

Other axons from light-sensitive ganglion cells end up in a different brain region, the olivary pretectal nucleus, where they control pupil constriction in response to changes in illumination; for this reason, some totally blind humans also have normal pupil reflex.

Yet another class of axons from light-sensitive ganglion cells do arrive in visual areas of the brain, where they serve visual functions that were previously thought to rely only on rods and cones (Schmidt et al., 2014; Milosavljevic et al., 2016).

 Track 2

10.6′ Glutamate removal

Glutamate molecules are constantly removed from the synaptic cleft and recycled by "excitatory amino acid transporters" in the photoreceptor membrane, and by neighboring cells called Müller glia,[36] so their concentration drops once release has been interrupted by a flash of light. Under steady illumination conditions, the glutamate level in the synapse reflects a competition between the constant rate of clearance and the variable rate of release.

---

[36]Figure 11.8 (page 363) shows a Müller glia cell.

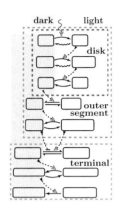

Fig. 10.14 (page 339)

Fig. 10.2c (page 321)

T_2 Track 2

10.7′a Termination of the photoreceptor response

Up to this point, we have focused on the *initiation* of the photoreceptor's response to light. The mechanism presented in the main text does result in a huge amplification of the tiny energy of one absorbed photon. But if this were the whole story, then even a single excited rhodopsin would eventually activate an unlimited number of transducins; each transducin would in turn permanently activate a phosphodiesterase, which would permanently hold cGMP to its lowest possible value, leading to a "stuck" state. That is, *the mechanism discussed so far creates a one-way switch.* That's not a useful behavior for a sensory transducer!

Instead of a one-way switch, the rod cell needs to give a *transient* response, then reset itself. Moreover, Section 9.4.5 emphasized that each response has the same form. For this to happen, each component in the cascade shown in Figure 10.14 must be reset after a well-defined time delay. Some of the resetting processes are indicated by leftward arrows in Figure 10.14:

- Rh* molecules are modified by a kinase, that is, an enzyme that tags them by adding phosphate groups.[37] Once enough of these tags have been added, an inhibitory protein (arrestin) binds to the phosphorylated Rh* and prevents it from activating any more transducins. In rod cells, the mean lifetime of an active Rh* molecule is around 40 ms. A heritable mutation in the machinery of rhodopsin inactivation leads to a form of congenital night blindness (Burns & Pugh, 2010).

- As soon as activated transducin binds to a phosphodiesterase, the transducin becomes susceptible to attack by an enzyme complex that clips the third phosphate group from its bound GTP molecule. This "GTPase activity" reverts the transducin to its GDP-bound form, which dissociates from the phosphodiesterase and eventually recombines with the other subunits to re-form the inactive transducin complex.[38] A heritable mutation of the machinery of transducin inactivation leads to a visual deficit named bradyopsia (Burns & Pugh, 2010).

- Without its transducin, the phosphodiesterase stops eliminating cGMP. The continued action of guanylate cyclase then restores the resting cGMP level.[39] The overall rate of T*·PDE deactivation is $\approx 5\,s^{-1}$, which determines the duration of the rod's single photon response.

- When the cGMP level recovers, the outer segment ion channels reopen and the rod cell depolarizes again. The voltage-gated calcium channels at the synaptic terminal respond by opening, and vesicle release is restored to its high steady-state level in darkness.

scotopic   photopic

10^{-6} 10^{8}

Fig. 3.10 (page 121)

10.7′b Negative feedback implements adaptation and standardizes signals from rod cells

Our eyes operate under an enormous range of illumination levels (Figure 3.10). Every stage of visual transduction involves processes, such as vesicle release, that have a limited range of rates. Thus, each stage must be able to adjust its sensitivity, dialing it down under high light levels, to avoid saturation. This section describes some of these **adaptation** mechanisms operating at the level of photoreceptors. The basic idea is

---

[37] Figure 10.2c (page 321) introduced such tagging.
[38] That is, transducin reverts to the trimeric form T shown in step 2 of Figure 10.6 (page 330).
[39] Feedback involving calcium ions makes the restoration faster and more precise; see (b) below.

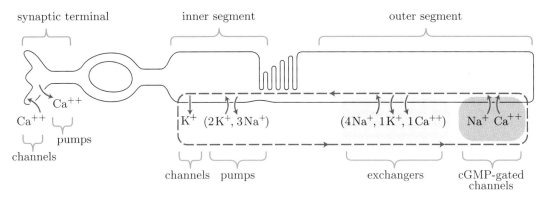

Figure 10.16: [Cartoon.] **A more detailed version of Figure 10.8** (page 332). This version shows calcium currents in the outer segment of a rod cell, and the exchangers (*green*) that drive them. Calcium dynamics in the synaptic terminal is decoupled from that in the outer segment.

that each photoreceptor incorporates slow negative feedbacks, which tune its responses based on its level of activity during the recent past.[40]

Negative feedback also addresses a second, seemingly very different, challenge that photoreceptors face. The strength and duration of a single-photon response depend on several processes, each involving one or a small number of molecules. We know to expect great variation in such processes, due to molecular randomness.[41] And yet, data such as those in Figure 9.12b show a rather tight distribution in single-photon response amplitudes; other data show that the entire time course is highly repeatable (stereotyped). An analogy to the nonliving world suggests a way out of this paradox: Negative feedback can shape transient signals like the single-photon response of a photoreceptor. For example, a fluctuation that would otherwise lead to an unusually large single-photon response triggers a correspondingly large negative correction.[42]

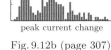

peak current change

Fig. 9.12b (page 307)

At least nine negative feedback mechanisms are now known to operate in photoreceptors. Because some of these mechanisms involve intracellular calcium ions, we must now pause to discuss their role in rod cells.

The main text focused only on sodium and potassium ions in the rod inner and outer segments, but Figure 10.16 shows that calcium (Ca^{++}) concentration is also actively managed. Thus, the inventory of calcium ions inside the outer segment amounts to another state variable for the rod cell (in addition to membrane potential, other ion concentrations, and cGMP level). Unlike membrane potential, however, the calcium levels in the outer segment and in the synaptic terminal do not affect each

---

[40] Section 11.4.1′ (page 369) describes another mechanism, which operates at a later stage of visual processing, for compressing the dynamic range of the input. Section 10.3.4′ discussed a simpler sensory adaptation mechanism used in primitive organisms.

[41] For example, the number of transducins activated by a single Rh⋆ during its lifetime is variable; see (a) above.

[42] Depending on parameter values, feedback can instead lead to oscillations, which would be undesirable in this context. We are interested in a situation called "overdamped" feedback, in which oscillations don't occur (Nelson, 2015, chapt. 9).

other; they are independent state variables.[43] Calcium in the synaptic terminal triggers vesicle release,[44] whereas our present discussion involves the calcium level in the outer segment, which is involved in adaptation and pulse shaping.

In addition to the pumps and channels mentioned earlier, the key new element is an **ion exchanger** (green blob in Figure 10.16). Unlike a molecular motor, this machine does not use any source of chemical bond energy. Instead, its working cycle involves transporting four Na^+ ions in one direction across the membrane, and one each of K^+ and Ca^{++} in the opposite direction. The electrochemical gradients of Na^+ and K^+ cause this machine to run in the direction shown in the figure, even though that entails dragging calcium ions *against* their electrochemical gradient. Thus, the exchanger constantly exports Ca^{++} from the rod cell's outer segment. The net effect of the pumps, channels, and exchangers is still to remove positive charge from the cell, and hence drive the cell's interior electric potential negative, as described in the main text.

In the dark, calcium ions constantly flow back into the rod cell through the open cGMP-gated channels (rightmost blob in Figure 10.16). Thus, besides reducing the cell's degree of polarization, these channels also limit the depletion of Ca^{++} in the cell. The light response closes some channels, leading to a drop in calcium level, which is sensed by various actors in the outer segment, including those in Figure 10.17.

Your Turn 10C

Run your finger around the feedback loop associated to one of the red influence lines in Figure 10.17, and convince yourself that it is overall negative. Repeat for the other red line.

One effect of a drop in calcium is to stimulate the production of cGMP to several times its resting rate, driving the cell back to its dark state. This feedback contributes to the observed sharp, reproducible single-photon signal and also to adaptation in dim light (see arrow *1* in Figure 10.17). A mutation of the machinery of cyclase modulation leads to inherited vision deficits called early onset Leber's congenital amaurosis and cone-rod dystrophy (Burns & Pugh, 2010). The other mechanism shown in the figure (arrow *2*) also contributes to adaptation, but more weakly; a third one, described in the caption, contributes to adaptation at higher light levels.

10.7′c Recycling of retinal

The main text described a phototransduction process that starts with a rhodopsin molecule, consisting of an opsin protein and its embedded retinal cofactor in the 11-*cis* conformation. After phototransduction, the retinal has converted to its all-*trans* form. Section (a) above described how this molecule is inactivated, terminating the neural response. But the retina must eventually *recycle* its used rhodopsins, which get taken out of service at a prodigious rate in daylight.

Inactivated rhodopsin is unstable; within about one minute it spontaneously ejects its all-*trans* retinal, which is then clipped off and separated from the opsin. A series of reactions then ensue, some of which actually take place outside the photoreceptor cell. The cell exports the used all-*trans* retinal into a neighboring layer of cells called the retinal pigment epithelium.[45] These helper cells reset and reexport 11-*cis* retinal

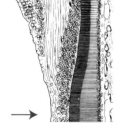

Fig. 6.6b (page 219)

---

[43]Calcium levels in the rod cell are buffered by mitochondria in the inner segment, making the outer segment and synaptic terminal effectively separate boxes as far as calcium is concerned.

[44]Section 10.6.2 (page 336) introduced this mechanism.

[45]Figures 6.6b (page 219) and 11.8 (page 363) show the retinal pigment epithelium.

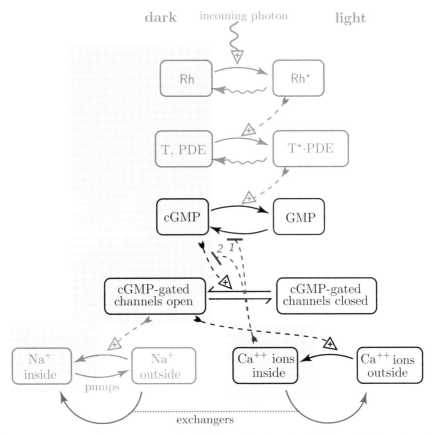

Figure 10.17: [Network diagram.] **Two negative feedback mechanisms (red) that underlie adaptation and pulse-shaping in photoreceptors.** This diagram is an elaboration of Figure 10.10 (page 334), showing further details of processes in the outer segment. *1*: Decreased calcium (Ca^{++}) concentration in the rod outer segment in response to light unblocks cGMP production, speeding the cell's recovery to its dark state. *2*: Calcium level also alters the affinity of the outer segment's ion channels to bind cGMP, and hence their ability to respond to it. *Not shown:* Calcium in the outer segment also decreases the activity of rhodopsin kinase, creating yet another negative feedback loop.

molecules back to the photoreceptors, which then recombine them with free opsin molecules, reforming light-sensitive visual pigments, and packaging them into new disks. The entire process is called the **retinoid cycle**.

PROBLEMS

10.1   *Twilight*

People say that the stars "come out" at twilight, but of course they're there in the sky all day long. Why can't we see them in the daylight, and what determines when they "appear?"

Clearly, this question involves the *contrast* between a faint source of light and a uniformly lit backdrop. Let's think about the faintest visible stars, which we only see when the sky gets quite dark. In a region of sky with no star, we receive photons in a Poisson process with some mean rate, leading to an average of μ_0 photoisomerizations per integration time. In a region with the star, we receive this background photon stream, plus an extra $\mu_\star$ from the star. What condition must be met for our eye to discriminate reliably between these two streams?

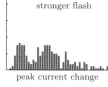

Fig. 10.4a (page 326)

10.2   *Cooperative binding*

Different plot styles reveal different things. The cooperative binding curve has sigmoidal shape in an ordinary plot (Figure 10.4a), but it's not easy to glance at that plot and read off the value of the cooperativity parameter n. The loglog plot is nice because it reveals the value of n as an ordinary slope (Figures 10.4b and 10.9), but on the other hand it obscures the existence of an inflection point (point of maximal slope in the ordinary plot).

Starting from Equation 10.3 (page 326), evaluate the derivative $d\mathcal{P}(\text{bound})/dc$ and then find the value of c at which its ordinary graph is maximally steep. Comment on the difference between the cases $n = 1$ and $n > 1$.

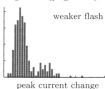

Fig. 10.4b (page 326)

10.3   *Evaluate Poisson hypothesis via single-cell recording*

Single-cell recordings actually yield richer data than just the see/no-see information used in Problem 9.3; they give the actual magnitude of every response (Figure 9.12), and so allow a more stringent test of our hypotheses. If you haven't done Problem 2.1 yet, look at it before attempting this one.

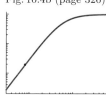

Fig. 10.9 (page 333)

a. Obtain Dataset 13 and make graphs similar to Figure 9.12. The second data series gives the responses to flashes that are four times stronger than those in the first one.

b. Partition each histogram into bumps centered near 0, 2, ... picoamperes, and count up the total number of events in each bump. Our hypothesis is that these correspond to $\ell = 0$, 1, ... productive photon absorptions in the rod after a flash of light. Let N_ℓ and N'_ℓ denote the observed numbers of instances you found for each outcome, for weaker and stronger flashes respectively.

c. Let μ be the unknown expectation of the number of productive photon absorptions from the weaker flashes, and 4μ the corresponding quantity for the stronger flashes. Find a likelihood function for μ that incorporates all the numbers you found in (b), optimize it, and evaluate μ_*.[46]

d. Express your answers to (b) as estimated probabilities, show them as two bar graphs, and superimpose the Poisson distributions you found in (c).

e. Comment on the hypothesis that rods can respond to single photons.

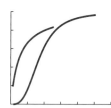
weaker flash

peak current change

Fig. 9.12a (page 307)

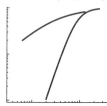
stronger flash

peak current change

Fig. 9.12b (page 307)

10.4   *Cyclic nucleotide gated channels*

Obtain Dataset 15, fit it to Hill functions with $n = 1$, 2, and 3, make the appropriate

---

[46]Section 7.2.3 (page 249) introduced likelihood maximization.

graphs, and comment.

10.5 $\boxed{T_2}$ *Predicted absorption in rods*

a. The outer segment of a human rod cell contains a stack of parallel disks with diameter $2\,\mu\mathrm{m}$. Each disk's membrane contains rhodopsin molecules with area density $\sigma \approx 25\,000\,\mu\mathrm{m}^{-2}$. The spacing of these disks is roughly one disk per $25\,\mathrm{nm}$; each disk actually contains *two* layers of rhodopsin molecules (Figure 10.1c). Calculate the average number density of rhodopsin molecules in the outer segment of a human rod cell, in m^{-3}. Multiply by the volume of the rod outer segment, roughly $100\,\mu\mathrm{m}^3$, to estimate the total number of molecules.

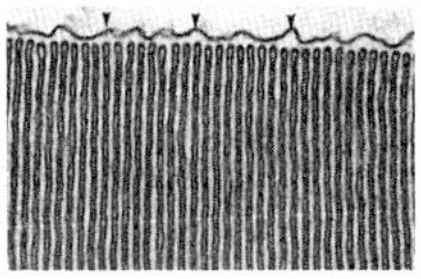

b. The extinction coefficient of rhodopsin is about $4.0 \times 10^4\,\mathrm{M}^{-1}\,\mathrm{cm}^{-1}$. Convert this figure to an absorption cross-section a_1 by using the definition in Section 9.4.2′a (page 311). Combine this with your result in (a) to estimate the quantity a_1c, and compare to the value given in Section 9.4.2′a.

Fig. 10.1c (page 319)

CHAPTER 11

The First Synapse and Beyond

> I paint, not things, but differences between things.
> — *Henri Matisse*

11.1 SIGNPOST: *FALSE POSITIVES*

Previous chapters have explored how our eyes, although tremendously impressive, are not quite perfect detectors. For example, each rhodopsin molecule has a low probability of catching an incoming photon, so each rod cell needs a very large number of rhodopsins. That large number, in turn, multiplies the tiny rate of false positive signals from thermal isomerization of rhodopsins, creating a significant background (the eigengrau). This chapter will outline some strategies that our visual system uses to avoid an unacceptable level of distracting *false positive* reports. The conclusion will be that, apart from the physically unavoidable loss and randomness in the first stage, our later visual processing of dim light signals is very efficient.

The Focus Question is

Biological question: If each rod cell can signal upon absorbing a single photon, then why do our brains impose a lower limit of several such signals before alerting the conscious mind?

Physical idea: The visual system imposes a breakpoint at the first synapse, and a threshold further downstream, to avoid the distraction of false positive signals.

11.2 TRANSMISSION AT THE FIRST SYNAPSE

11.2.1 The synapse from rod to rod bipolar cells inverts its signal via another G protein cascade

Neurons generally receive signals at their dendrites that consist of bursts of neurotransmitter, released onto them by other neurons.[1] Such signals open ion channels, depolarizing the dendrite and thus triggering a wave of depolarization (the action potential) that travels down the axon. Depolarization at the axon terminal in turn triggers a new burst of neurotransmitter release, a response of the same general form as the one that initiated the action potential.

We have seen how a flash of light, leading to the productive absorption of even just one photon, momentarily *de*creases the rate of vesicle release at the rod's synaptic terminal. However, in the dark, we are more interested in learning about the presence, not the absence, of light. Unlike a typical neuron, the next cell in the dim-light signaling pathway (rod bipolar cell) must accordingly *invert* its input, depolarizing and

---

[1] Section 2.4.7 (page 71) introduced this mechanism.

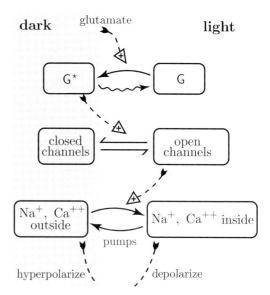

Figure 11.1: [Network diagram.] **Events involved in the response of a rod bipolar cell.** Instead of acting directly on ion channels, the neurotransmitter glutamate activates a receptor called mGluR6, which activates a G protein, which in turn closes "transient receptor potential" (TRPM1) ion channels, hyperpolarizing the rod bipolar cell in darkness. (There is no second messenger analogous to the rod cell's cGMP.)

generating a burst of neurotransmitter when glutamate momentarily *stops* appearing at its dendrite. The rod bipolar cell accomplishes this inversion by an indirect response to glutamate. Instead of being studded with ion channels that open when they bind glutamate, the rod bipolar cell expresses a G protein coupled receptor called mGluR6.[2] External glutamate activates this receptor molecule, whose intracellular part activates a G protein. The activated G protein in turn closes ion channels in the cell membrane, increasing its degree of hyperpolarization (Figure 11.1). In short,

> The rod bipolar cell is polarized in the dark, like any resting neuron. In response to light, the glutamate supply to the rod bipolar cell drops, ion channels open, and the cell depolarizes, triggering release of neurotransmitter from its own synaptic terminal.

One indication that the mGluR6 receptor is essential comes from the fact that mutations in its gene lead to complete loss of night vision (autosomal recessive congenital stationary night blindness).

11.2.2 The first synapse also rejects rod signals below a transmission breakpoint

Figure 11.2 shows that there is a fairly clear demarcation between the weaker continuous dark noise generated by the rod cell and the larger current drops generated by photon absorptions (plus occasional spontaneous isomerizations). If we were given such a signal, we might interpret it by just declaring that every event with peak current change less than, say, $0.8\,\mathrm{pA}$ is continuous dark noise, and the rest represent isomerization events. But this strategy would not give us good vision.

The problem with the discrimination strategy just outlined is that we don't just want to detect light—we want to *see things*. That is, we wish to know roughly where

---

[2]We already encountered a G protein in Section 10.4.2 (page 329). Ligand-gated ion channels are sometimes instead called "ionotropic." Receptors like mGluR6 that signal by a mechanism not directly involving ion passage are instead called "metabotropic."

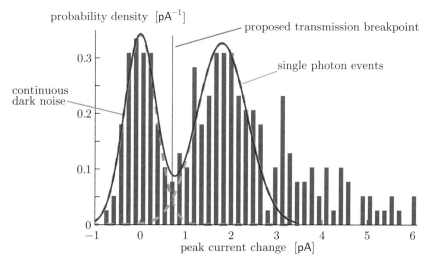

Figure 11.2: [Experimental data with idealized probability density functions.] **The distribution of rod noise overlaps that of true signals.** *Bars:* The same data as in Figure 9.12b, an observed distribution of peak amplitudes of a rod cell's extracellular current change after a flash of light. *Solid curve:* An idealized probability density function, the sum of two Gaussian distributions corresponding respectively to continuous dark noise (*left dashed curve*) and events in which a single productive photon absorption occurred (*right dashed curve*). We might declare that any event exceeding the value marked by a *vertical line* will be deemed an isomerization event, but the graph suggests that a fraction of those events really belong to the noise peak. The text explains that, although it seems small, this contamination by false positives would seriously degrade dim-light vision. [Data courtesy Greg D Field; see also Rieke, 2008.]

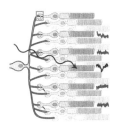

Fig. 10.12a (page 337)

photons are coming from, and when they arrive, in order to form moving images of the dimly lit world. In a world lit by starlight, photons are so scarce that each rod cell has probability less than 10^{-2} to catch a photon in a 200 ms time window. In order to get any kind of image, then, our eyes must *pool* the responses from batches consisting of many rod cells. Indeed, the retina's bipolar cells each receive input from dozens of rods (Figure 10.12a); their outputs in turn feed into ganglion cells that further combine the signals from many bipolar cells and send them on to the brain. We will call a patch of retina containing rod cells that feed into a single ganglion cell a **summation region**.

Although the current fluctuations from each individual rod cell seem to split nicely into "signal" and "noise," nevertheless Figure 11.2 shows some overlap between the two distributions. Thousands of rod cells' outputs are pooled to give the signal corresponding to a particular region of the visual field (a summation region), and every rod cell is constantly making continuous dark noise, even in darkness. If our eyes really set a breakpoint like the one imagined in Figure 11.2, then we'd misclassify a small but significant fraction of the noise events as real signals, swamping the few genuine signals in dim light. And yet, some ganglion cells do respond to single photoisomerizations, with little noise.

D. Baylor, B. Nunn, and J. Schnapf pointed out this paradox, and offered a possible resolution. They reasoned that *noise could be suppressed by sacrificing some real rod signals.* Suppose that we choose the **transmission breakpoint** to be 1.8 pA, considerably higher than the "obvious" value imagined in Figure 11.2. Then about half of the isomerization signals (second peak in Figure 11.2) will be wrongly discarded. The payoff, however, is that then practically *none* of the noise signals (first peak) will

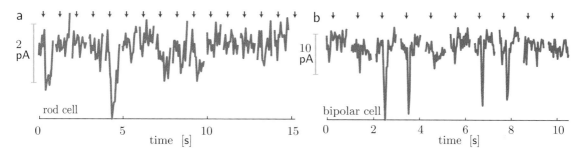

Figure 11.3: [Experimental data.] **Effect of discrimination at the first synapse.** (a) Responses to light flashes by single mouse rod cells. Circulating current is shown as a function of time. Light flashes were delivered at times marked by *arrows*. (b) Responses recorded from mouse rod bipolar cells. Imposing a transmission breakpoint at the first synapse makes the true signals stand out from noise better than in (a). [From Okawa et al., 2010.]

be wrongly passed on to the next neuron—a big improvement to the net signal-to-noise ratio, at the cost of reducing the overall quantum catch.

Baylor and coauthors noted that this discrimination operation must take place downstream from the rod cell, because continuous dark noise arises in the rod. However, it must also take place before any pooling of signals is done, because after many rod signals have been combined, it is too late to determine that some of them were noise and should be deleted. About 15–30 rod cells all combine their signals as inputs to the next level of processing (the rod bipolar cell), so the discrimination must occur at the synapse between a rod and its bipolar cell. Indeed, direct evidence for this hypothesis came years later, when techniques became available to record from the bipolar cells (see Figure 11.3).

How high should the transmission breakpoint be? A. Berntson and coauthors noted that discrimination can do nothing about the noise generated by spontaneous isomerizations of rhodopsin, because those events look *exactly* like the ones actually elicited by photons. So there is little point in raising the transmission breakpoint indefinitely: Once the rate of false positives from the continuous dark noise (no isomerizations) has been reduced to less than that from spontaneous isomerizations, increasing the transmission breakpoint further would cut the quantum catch without much improvement in the signal-to-noise ratio. These researchers estimated that the trade-off point comes when about 50% of the true photon events are sacrificed.

Section 11.2 has outlined how researchers identified an engineering trade-off imposed by the molecular character of cellular signaling and correctly predicted one surprising approach to that trade-off made by animal eyes.

$\boxed{T_2}$ *Section 11.2.2′ (page 368) discusses the first synapse in greater detail.*

11.3 SYNTHESIS OF PSYCHOPHYSICS AND SINGLE-CELL PHYSIOLOGY

The preceding section described an approach to mitigating one source of photoreceptor noise. However, another source of false positives, spontaneous isomerizations, cannot be handled at the level of individual photoreceptors because these events look exactly like true photoisomerizations.[3] Spontaneous isomerizations are an unavoid-

---

[3]Applying coincidence detection in a single photoreceptor is not a good solution; see Problems 9.1–9.2.

transduction module (parameters Q, μ_0 fixed)   decision module (parameter t flexible)

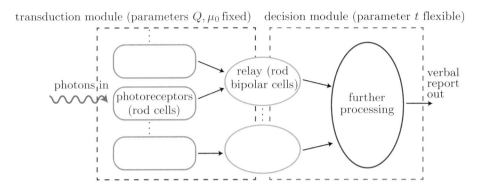

Figure 11.4: [The same schematic as Figure 9.2 (page 292).] **Barlow-type model for dim-light detection.**

able consequence of the need to have a huge number of rhodopsin molecules in each photoreceptor. Chapter 9 outlined a possible strategy to deal with them, distinct from discrimination at the first synapse. Now we are ready to test this hypothesis quantitatively.

11.3.1 Why does vision require several captured photons?

Chapter 9 showed that human subjects can have reliable behavioral responses to flashes whose mean photon content is about 150 or larger, or to events with about 15 productive photon absorptions.[4] Section 11.2.2 argued that the first synapse discards about half of the genuine rod signals in order to suppress continuous dark noise. We could register most such events by imposing a network threshold of six rod signals passing the first synapse.[5] But why six? Wouldn't *one* be better? Chapter 9 introduced H. Barlow and B. Sakitt's proposal that some level of processing downstream from the rod cells imposes a network threshold on the number of signals before it notifies the conscious mind.[6] Such a threshold would mitigate the effects of spontaneous isomerization in the rod cells (because the probability of multiple simultaneous false-positive signals is small), but would also raise the number of productive photon absorptions needed for reliable seeing. Chapter 9 postponed discussion of this proposal; now we are ready to combine the psychophysical results of Section 9.3 with the single-cell measurements of Sections 9.4 and 11.2.

Single-cell physiology sets a lower bound on the visual system's loss and random-ness, while the incredibly good dim-light vision of an entire animal sets an *upper* bound. We now ask, is there any room between these bounds, and if so, how much?

11.3.2 Review of the eigengrau hypothesis

In greater detail, Barlow and Sakitt's hypothesis was that the early, dim-light visual system takes single photon signals from rod cells, pools them by simple addition, and applies one or more network thresholds (Figure 11.4). Today, we know that even early visual processing is more complicated than this; for example, after the first

---

[4]See Section 9.3.4 (page 298).
[5]If the expectation of the number of productive absorptions is $150 \times 10\% \times 50\% = 7.5$, then the probability that the actual number is at least 6 is $\mathcal{P}_{\mathrm{pois}}(\ell \geq 6; \mu = 7.5) = 76\%$.
[6]The eigengrau hypothesis (Section 9.3.3, page 296).

synapse, there are multiple stages at which signals branch out, then are recombined with losses and nonlinearities. Accounting fully for this complexity would take us far afield, however. Instead, we will only seek to *describe* the performance of early visual processing, by seeing whether any simple physical model of this type can reproduce the psychophysical data when constrained by single-cell measurements.

Before we proceed, here is a summary of some notation and terminology introduced in Chapter 9 and the following discussion:

summation region	collection of rod cells whose signals are pooled
integration time	time window over which signals are pooled
spontaneous isomerization	generates false-positive signals indistinguishable from photon absorptions
continuous dark noise	lower-amplitude rumble superimposed on photoisomerization signals
network threshold t	applied to total photon-like signals from a summation region (an integer quantity) at some point after the first synapse
rating r	verbal response of a human subject to a flash (from a discrete list)
ℓ_{tot}	total number of rod signals that cross the first synapse
rod quantum catch Q_{rod}	the probability that a photon, traveling along the rod axis and localized to its geometric cross section, will be productively absorbed and generate a signal
$Q_{\text{rod,ret}}$	rod quantum catch for the whole <u>ret</u>ina: the probability that a photon, spread out over many rod cells, will elicit a signal from any one of them
Q_{tot}	the same as $Q_{\text{rod,ret}}$ but reduced by the probability that a signal passes the first synapse to a rod bipolar cell (Equation 9.3, page 298)
$\mu_{\text{ph,spot}}$	expectation of the number of photons in a flash of light confined to one summation region or smaller (Equation 9.3, page 298)
eigengrau parameter $\mu_{0,\text{rod}}$	mean false positive count for a single rod (Equation 9.2, page 297)
$\mu_{0,\text{sum}}$	same but for a <u>sum</u>mation region of retina (Equation 9.3, page 298)

The model includes two claims, concerning the transduction and decision modules respectively. We can now state the first claim a bit more precisely than before:[7]

> *The total number of rod signals that cross the first synapse during a brief flash, ℓ_{tot}, is Poisson distributed. Its expectation depends linearly on the flash strength:* $\langle \ell_{\text{tot}} \rangle = Q_{\text{tot}}\, \mu_{\text{ph,spot}} + \mu_{0,\text{sum}}$.   [9.3, page 298]

In this formula, the whole-retina quantum catch Q_{tot} and the eigengrau parameter $\mu_{0,\text{sum}}$ (mean rate of false-positive signals in a summation region) are fixed for any individual. The fact that some rod signals are lost at the first synapse just reduces the values of $\mu_{0,\text{sum}}$ and Q_{tot}, without altering the Poisson character of ℓ_{tot}.[8]

The second claim in Barlow and Sakitt's model is that

> *A subject reports a rating r determined by applying a set of network thresholds, $\{t_r\}$, to the rod signal count ℓ_{tot}. (The subject has unconsciously set those values during the course of prior training.) After a stimulus, the subject replies with the highest rating, r, for which ℓ_{tot} exceeds t_r.*   [9.4, page 299]

Although it is highly simplified, the model does make testable predictions for $\mathcal{P}(\text{rating}; \text{stimulus})$, which are the quantities measured in psychophysical experiments.

---

[7]The more precise feature is that now we are accounting for loss at the first synapse: From now on, "rod signals" means "rod signals that cross the first synapse."

[8]Section 3.4.1 (page 111) introduced the thinning property of Poisson processes.

11.3.3 Single-rod measurements constrain the fit parameters in the eigengrau model

We could take Q_{tot}, $\mu_{0,\text{sum}}$, and the t_r all to be fitting parameters, and seek values that seem to account for the psychophysical data. But then we would lose much of the theory's falsifiability, and all of its contact with the single-cell measurements.[9] Instead, we will use estimates of the first two parameters based on the single-cell data (Sections 9.3.4 and 11.2.2). Doing so tightly constrains the model: It's possible that *no* successful choice of assumed t_r values can then be found. (It's not easy to fit a function of two variables, such as the observed $\mathcal{P}(\text{rating}; \text{stimulus})$, if your model is wrong and has few parameters.)

The value of Q_{tot} can be estimated by combining the single-cell measurements with other known facts about the eye.[10] We can also get a rough estimate of the eigengrau parameter $\mu_{0,\text{sum}}$, because Baylor and coauthors measured the mean rate of dark events in single rod cells. A more recent measurement gives this rate, for humans, as $\approx 0.0062\,\text{s}^{-1}$. Thus, the number of spontaneous events that could be confused with real photon absorptions is roughly this rate, multiplied by the rod integration time,[11] about 200 ms, and reduced by 50% to account for loss at the first synapse. The number of rod cell signals that are effectively pooled in a summation region is not well measured, but Barlow used an indirect argument that leads to the value ≈ 1700, and hence the estimate

$$\mu_{0,\text{sum}} \approx (0.0062\,\text{s}^{-1}) \times (200\,\text{ms}) \times 0.5 \times 1700 \approx 1.05. \tag{11.1}$$

11.3.4 Processing beyond the first synapse is highly efficient

We can now ask: Is it possible to find a set of network thresholds, $\{t_r\}$, such that the probabilities of each rating r in an experiment like Sakitt's, at all the flash strength values studied, follow the eigengrau model? More precisely, the eigengrau hypothesis implies that the response ℓ_{tot} is Poisson-distributed, with expectation given by Idea 9.3 as restated above. Then Equation 9.4 (also reproduced above) says that the probability of a subject assigning a rating r to a flash of strength $\mu_{\text{ph,spot}}$ is obtained by summing $\mathcal{P}_{\text{pois}}(\ell_{\text{tot}}; \mu_{\ell_{\text{tot}}})$ over values of ℓ_{tot} in the range $t_r \leq \ell_{\text{tot}} < t_{r+1}$.

Our model implicitly includes the claim that *the quantum catch of the entire dim-light visual system is determined by that measured at the single-rod level.* Later we will see to what extent the data allow other hypotheses.

Figure 11.5 shows results from a modern recreation of Sakitt's experiment, with a maximum-likelihood fit to a physical model similar to the one just described.[12] D. Koenig and H. Hofer used four different values of flash strength, while only requesting a scale of five possible ratings ($r = 0, \ldots, 4$). This imposes a more stringent test of the model than in Sakitt's design, because there are fewer fitting parameters (the four network thresholds $t_1, \ldots t_4$) with which to match a function measured at more

---

[9]Section 9.3.4 (page 298) made this point.

[10] $\boxed{T_2}$ See Problems 9.3 or 9.7. The rod quantum catch $Q_{\text{rod,ret}}$ obtained there, times the estimate 50% for transmission loss at the first synapse from Section 11.2.2, yields Q_{tot}.

[11]Section 9.3.1 (page 294) described one way to estimate the integration time. We ignore the contribution to the false-positive rate from continuous dark noise, because as we have seen, the first synapse imposes a transmission breakpoint that eliminates most of it (Section 11.2.2).

[12]Section 7.2.3 (page 249) introduced likelihood maximization. $\boxed{T_2}$ You'll explore Sakitt's data in Problem 11.3.

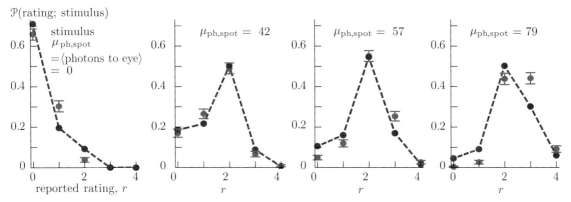

Figure 11.5: [Experimental data with fits.] **Fit to a modern version of Sakitt's experiment.** *Circles* show the estimated $\mathcal{P}(\text{rating; stimulus})$, that is, the observed probabilities of various ratings given by one human subject, for flashes of 490 nm light with four different strengths (compare Figure 9.4, page 299). *Dots* joined by lines show a fit to functions based on Ideas 9.3–9.4, with quantum catch and noise parameters fixed by single-cell measurements (see Problems 9.3 or 9.7). Thus, the only fitting parameters are the assumed network thresholds for each of the four nonzero ratings. Error bars indicate one standard deviation for a Poisson distribution with expectation equal to the number of times each rating was given (see Section 0.5.3, page 16). [Data courtesy Darren Koenig and Heidi Hofer; see also Koenig & Hofer, 2011.]

values of the stimulus strength (a total of 20 data points in the figure). Moreover, the fit parameters are restricted to be an increasing sequence of small integers.

The figure shows a maximum-likelihood fit with network threshold values $\{t_r\} = \{2, 3, 6, \text{and } 9\}$. The fit shows that the data are roughly consistent with the eigengrau hypothesis. For example, the "fairly sure" network threshold value, t_3, is consistent with the rough estimate found in Section 11.3.1. The first one, t_1, means that, for this subject, *as few as two photon signals* crossing the first synapse was enough to bias the conscious mind toward perceiving a flash. Even though there were false positives at that rating, due to the eigengrau, they were few enough to justify the subject in choosing such fine-grained perception. Not every subject in the study performed this well, but all could be fit by setting the first network threshold to either two or three rod signals.

Now that we have found one way to reconcile the whole-organism performance with the single-cell data, we can also ask if the data permit any *other* hypotheses. The model whose results appear in Figure 11.5 assumed that there was no further loss of signals after the first synapse. If present, such losses might appear as a value of the quantum catch smaller than the one we found from single-rod experiments. Repeating the fit with smaller assumed values of the quantum catch, we find that no fit is possible if Q_{tot} is less than about 80% of the value given by single-cell physiology. That is, all the complex downstream processing cannot lose more than about 20% of the signals that cross the first synapse. Similarly, if we represent additional randomness by a larger effective value of the false signaling rate $\mu_{0,\text{sum}}$, we find no fit is possible if $\mu_{0,\text{sum}}$ is taken to be 30% larger than the estimate in Equation 11.1.

Section 13.2 has completed a story begun in Chapter 9. The conclusion of this analysis is that *neural processing of the signals from faint light flashes is very reliable,* and that *the high performance of dim-light vision is understandable from a physical model,* which includes the nature of light itself, transduction mechanisms in the rods, and discrimination at the first synapse needed to mitigate the individual rods' false

Figure 11.6: [Schematic.] **The classical rod pathway through the mammalian retina.** Productive absorption of a photon by a rod cell sends signals through the cells shown, which ultimately travel to the brain via an axon of an ON ganglion cell. Some of the cells shown have other functions under higher illumination; for example, the amacrine cells perform some image processing functions, and the ON cone bipolar cell accepts input from cone cells. Moreover, rod signals can be handled by pathways other than this one, though not with single-photon sensitivity. *Dashed lines* indicate the division of the retina into layers visible under the microscope. The cells depicted here are shown in a real micrograph on the cover of this book; see also page ii and Figure 11.8. [Adapted by permission from Macmillan Publishers Ltd: Wässle, H. 2004. Parallel processing in the mammalian retina. *Nat. Rev. Neurosci.*, 5(10), 747–757, ©2004.]

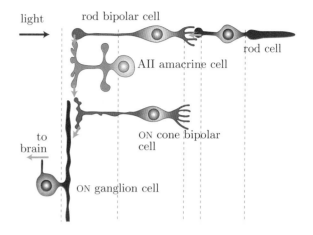

positive signals. Although future measurements will surely revise some of the details of the eigengrau hypothesis, still the general conclusion is that it describes the full system's performance well, in the important regime of dim-light vision.

<u>T2</u> *Section 11.3.4′ (page 369) describes a psychophysics experiment with single photon stimuli.*

11.4 A MULTISTEP RELAY SENDS SIGNALS ON TO THE BRAIN

11.4.1 The classical rod pathway implements the single-photon response

We have traced the eye's dim-light response from a rod cell to the rod bipolar cells that connect to it. Figure 11.6 shows the main pathway for the subsequent handling of these signals:

- The rod bipolar cell connects to a class of intermediate neurons called **amacrine cells**, located in a different layer of the retina, via another chemical synapse.
- Amacrine cells make direct electrical contact with another class of bipolar cells, called ON **cone bipolar cells**.[13] In bright light, these cells' main job is to handle signals from cone cells. In dim conditions, the cones do not respond to light, so the cone bipolar cells are available to do a second job, relaying signals from rod bipolar cells to "ON ganglion cells" located in the final retinal layer. Each amacrine cell collects inputs from many rod bipolar cells, extending the signal pooling that is already done at the previous level. Overall, each ganglion cell typically collects signals from thousands of rods.
- Up to this point, all signals have been transmitted over short distances, typically a few micrometers. The ganglion cells, however, extend into the brain, several centimeters from the retina. They therefore convert their input signals into a

---

[13]Such contacts involve tunnels joining the two cells' interiors, called "gap junctions."

form that can be carried over long distances, the "spike" (action potential) format used by most neurons throughout the body.[14]

Direct electrical recording from ganglion cells confirms that, in dim light, the "classical rod pathway" just described can send a short burst of spikes to the brain in response to a single photon absorption. Thus, the network threshold found in psychophysical experiments must have its origin downstream of the ganglion cells.

11.4.2 Other signaling pathways

Up till now, we have focused on the pathway responsible for transmitting single-photon signals at the lowest illumination. This section will widen the frame, considering other pathways and cell types present in the retina and used for other visual tasks.

G. Field, E. J. Chichilnisky, and coauthors found the connections of cone photoreceptors to retinal ganglion cells directly (Figure 11.7). The experimenters stimulated an isolated retina with various random patterns of light, with various spectra, and simultaneously monitored the action potentials generated by many RGC. They then correlated the visual input with RGC output to determine the "receptive field" of each RGC, that is, the subregion of the visual field and the spectral region that preferentially stimulated that cell to fire action potentials. These receptive fields had spatial substructure that corresponded to the known actual cone locations on the retina. Comparing the receptive fields for pairs of RGC often showed partially overlapping spatial patterns, showing that some cones are functionally connected to more than one RGC. As expected, a cone identified in the receptive fields of multiple RGC was found to have the same spectral sensitivity for each one.

Figure 11.8 shows all the players we have discussed in the retina story, together with some supporting cast. Unlike in Figure 11.7, this figure shows in a stylized way the actual wetware connections that implement the observed functional relationships. *T₂ Section 11.4.1' (page 369) discusses* ON *and* OFF *pathways, image processing steps performed by the retinal circuitry, and additional roles for amacrine cells.*

11.4.3 Optogenetic retinal prostheses

Retinitis pigmentosa refers to a group of hereditary diseases that lead to incurable blindness, affecting about two million people worldwide. Generally the patient's rod photoreceptors die, and although the cone cells survive, they become insensitive to light. Installing new phototransduction mechanisms in the retina offers a possible route to recovering vision in these patients.

One approach involves expression of archaebacterial halorhodopsin in the light-insensitive cones to substitute for the native phototransduction cascade;[15] this in-

---

[14]Section 2.4.5 (page 68) introduced action potentials.
[15]Figure 2.11 (page 75) described this and other optogenetic schemes; also see page ii and Media 6.

Figure 11.7: [Diagram.] **Locations, types, and strengths of the functional inputs to retinal ganglion cells from cone photoreceptors.** The diagram does not show physical cell-cell connections; rather, it shows the net effect of all three layers of the retina as functional relationships between individual photoreceptors (*colored dots*) and the retinal ganglion cells (RGC) that they stimulate (specifically, a class called "ON parasol" cells). The inferred locations on the retina of each cone cell are color-coded by spectral type (L, M, or S). For each RGC studied, a point was chosen near the center of the set of cone cells that contributed significantly to that ganglion cell's firing. Then *white lines* were drawn to that point to indicate inferred functional connections from cone cells. Thicker lines denote stronger connections. This class of RGC had no functional connections to S-type cones; the existence and placement of the *blue dots* were deduced though the responses of other classes of RGC. Compare the micrographs in Figure 3.14 (page 130). [Courtesy Greg D Field. Adapted by permission from Macmillan Publishers Ltd: Field et al. Functional connectivity in the retina at the resolution of photoreceptors. *Nature* (2010) vol. 467 (7316) pp. 673–677, ©2010.]

tervention has restored limited vision in mice with the disease. Another method, also applicable to other kinds of retinal degeneration, bypasses the photoreceptors altogether and adds light-sensitive channels directly in the ganglion cells.

Section 11.4 has offered a vista onto the complex processing applied to photoreceptor signals, starting in the retina itself.

11.5 EVOLUTION AND VISION

11.5.1 Darwin's difficulty, revisited

Darwin asked about how the overall anatomy of the eye could have evolved. Chapter 6 outlined his logic:[16] If a sequence of light-sensing organs can be found in contemporary animals that gradually passes from primitive to complex, then it is plausible that our own ancestors developed such organs in a similar sequence. Lacking the electron microscope, Darwin could not have posed the same question for the exquisite anatomy of a single photoreceptor cell—but we can.

---

[16]See Section 6.6 (page 230).

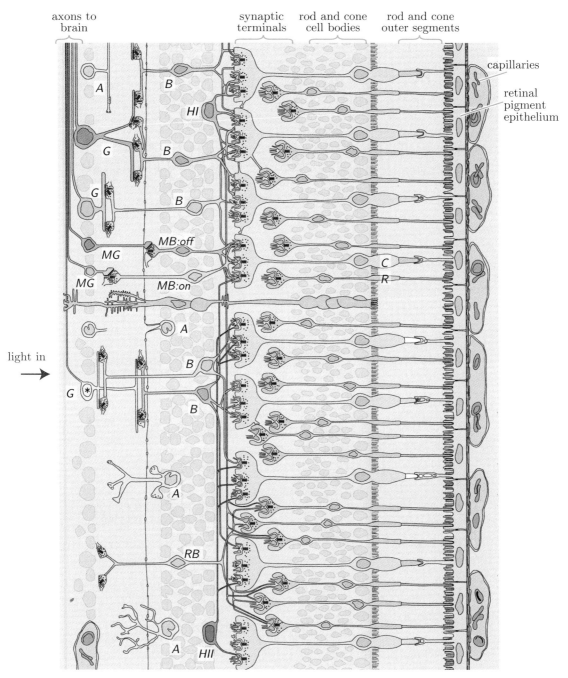

Figure 11.8: [Schematic.] **The main classes of neurons, and their connections, in primate retina.** A real micrograph of these structures appears on the front cover of this book. *C*, cone cell; *R*, rod cell; *MB*, midget bipolar cell; *RB*, rod bipolar cell; *B*, other bipolar cell types; *MG*, midget ganglion cell; *G*, other ganglion cell types; *HI* and *HII*, horizontal cells; *A*, amacrine cell. Bipolar and ganglion cells colored *orange* are OFF types; those colored *green* are ON types. One ganglion cell, indicated with an *asterisk,* subtracts signals from two neighboring but chromatically distinct cone cells. For clarity, the synaptic terminals of the photoreceptors have been enlarged relative to other objects, in order to show structural details; also, some of the cells that extend into the synaptic terminals of photoreceptors are cut off. The one *gray cell* near the center represents a class of support cells called Müller glia. Capillaries (blood vessels) at *far right* nourish a layer of cells called the retinal pigment epithelium; these in turn nourish the photoreceptors and recycle their used retinal molecules. [From Rodieck, 1998.]

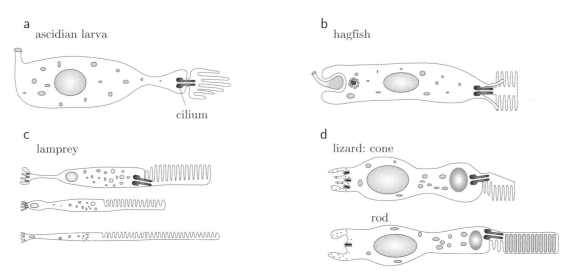

Figure 11.9: [Anatomical sketches based on electron micrographs.] **Photoreceptors of some contemporary organisms, arranged in a sequence of increasing complexity.** Note the gradual transition toward a highly organized layered structure in the outer segment, and the appearance of ribbons in the synaptic terminal. The fact that the more primitive forms still exist today lends credence to the claim that a similar sequence of structures may have evolved sequentially. The *orange blobs* in the lizard cells represent oil droplets, found in some reptiles and birds. Their general function is not known, although some are colored and hence adjust the spectral sensitivity of their photoreceptor. [Adapted by permission from Macmillan Publishers Ltd: Lamb et al. Evolution of the vertebrate eye: Opsins, photoreceptors, retina and eye cup. *Nat. Rev. Neurosci.* (2007) vol. 8 (12) pp. 960–976, ©2007.]

Figure 11.9 shows a sequence of photoreceptor morphologies from contemporary animals, displaying a gradation from disorganized to highly layered structures in photoreceptor outer segments. Each of the examples shown falls into the class called **ciliary photoreceptors**, which include our own rods and cones.

T_2 *Section 11.5.1′ (page 373) mentions a second class of photoreceptor cells.*

11.5.2 Parallels between vision, olfaction, and hormone reception

The visual cascade is complex. Even though many of its subtleties have now been understood as conferring improved performance, it may still appear to be *unnecessarily* complex.

Our first intuition might be that, if asked to design a transducer from light to nerve impulses, we might invent some neurons with light-activated ion channels, a one-step solution. In fact, evolution did already invent such channels in ancient organisms like archaea and algae.[17] Doesn't this mean that the entire visual cascade is unnecessary?

To begin to address this issue, we must think about that tricky word, "necessary." What's *necessary* in biology is that each gadget in our bodies must have evolved somehow, through a sequence of steps, none of which was too injurious to survival. Evolution is a tinkerer; it works with existing solutions and modifies them. One way to generate a visual sense might be to start with a light-sensitive ion channel. But there may be other possibilities, with better potential for high performance.

---

[17]Section 2.5 (page 72) introduced such channels. T_2 In fact, our intrinsically photosensitive ganglion cells, mentioned in Section 10.4.3′ (page 345), implement a similar idea, although they do not play a central role in vision.

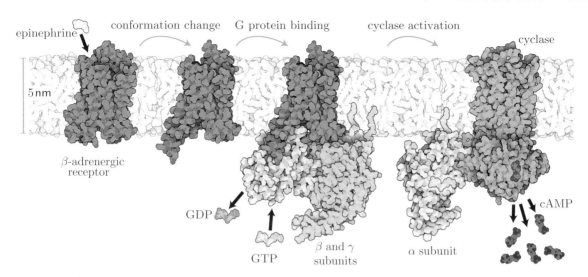

Figure 11.10: [Artist's reconstruction based on structural data.] **Hormone reception.** A sequence of steps is shown. From left to right: A receptor embedded in a cell's plasma membrane encounters and binds its effector, in this case a molecule of epinephrine (adrenaline). Binding entails an allosteric change in the receptor's shape, which allows docking of a G protein complex. When bound, the complex undergoes its own allosteric change, allowing it to eject a GDP molecule and replace it by GTP. Thus activated, the α subunit (*blue*) sheds the other subunits and wanders through the membrane until it encounters and activates a cyclase. The activated cyclase then creates the internal signaling molecule cyclic AMP (cAMP). Up to this point, the signaling cascade partly resembles the one used in vision. Instead of electrical signaling, however, the next steps involve activation of protein kinase A by the cAMP molecules; protein kinase A then phosphorylates many enzymes, reprogramming the cell's activity to prepare an animal for fight or flight. [Art by David S Goodsell.]

Our senses of smell and taste (olfaction and gustation) must, like vision, address a daunting task: We can detect exposure to just a few molecules of some odorants, so we need very high amplification. But we must not overload when exposed to high concentrations—we also need a high dynamic range of useful sensory transduction. These criteria are hard to achieve with a simple, one-step detection scheme. A multistep approach, like the one used in vision, would allow an adjustable amount of amplification at every step, and hence the possibility of a very wide range of overall sensitivities. And indeed, our chemical senses were found to work similarly to the visual cascade. In addition to olfactory and gustatory receptors, the list includes many that "smell" our own blood, detecting certain hormones, such as epinephrine (adrenaline) and dopamine, as well as other signaling molecules like histamine and serotonin.[18]

Figure 11.10 shows an example of a hormone receptor, the β-adrenergic receptor (smell and taste operate in similar ways, but are exposed to the environment outside the body). In each case, cells that sense chemicals have membranes that are studded with a protein that has the same general structure as visual opsin (seven helices spanning the plasma membrane). However, the site corresponding to where opsin holds its chromophore (retinal) is *vacant*. Binding an appropriate hormone, odorant, or tastant molecule distorts the receptor's shape, just as a light-induced conformational change in retinal distorts its opsin wrapper.

---

[18]Even fungi such as yeast have G protein based cascades that detect pheromones. (Another class of hormones, including steroids, act quite differently; they pass through the cell membrane and act in the nucleus.)

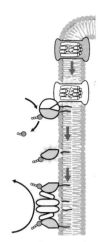

Fig. 10.6 (page 330)

Once an appropriate effector molecule binds to the receptor, a G protein signaling cascade ensues that is similar to the one used in vision. Figure 11.10 shows the first steps of this cascade (compare Figure 10.6). Instead of stimulating a phosphodiesterase that pulls down the concentration of a second messenger, in this case the activated G protein stimulates a cyclase that *raises* a concentration. But in either case, the resulting second messenger can then leave the membrane, affecting downstream signaling elements throughout the cell. In short, the key difference between chemical senses and vision is that *visual opsin carries its own effector's precursor* at all times, instead of waiting for effector molecules to waft by. Capturing a photon converts that precursor (11-*cis* retinal) to the effector (all-*trans* retinal).

A significant fraction of our entire genome is dedicated to G protein coupled receptors; over 20 000 are now known. Their significance is also underscored by the fact that receptors in this family are the targets for a significant fraction of all pharmaceutical drugs. The connection between olfactory, gustatory, hormone, and visual receptors goes beyond function: When genomic analyses became available, all of their genes were found to belong to one superfamily, originating with single-cell organisms.

Each cell type in the body chooses which chemical broadcasts it will receive by expressing appropriate receptors on its outer membrane. Each cell type can also program a different response to receiving the message, by controlling which pathways are connected to each receptor type. This flexible "plugboard" scheme permits a single message (such as elevated epinephrine) to trigger very different responses in different body tissues.

Section 11.5 has hinted at the unity of Life, not only across species but also across the diverse tasks that a single organism may need to perform.

THE BIG PICTURE

This chapter uncovered the need for a discrimination step at the first visual synapse to reduce the *false positive* signals that would otherwise degrade dim-light vision. Adding that ingredient allowed us to complete our model and confront it with psychophysical data.

We have now followed one thread of the vision story from photon to neuron. In keeping with this book's other themes, we have concentrated mainly on the dim-light limit of vision, which is critical for many animals. Other, much longer, books discuss other illumination regimes, many more features of early visual performance, and the elaborate yet efficient processing that occurs later.

The story in Chapters 9–11 illustrates the power of reductionism, the unbundling of a complex system into simpler components, each of whose behaviors we hope will be similar in isolation to how it is in the whole system. We characterize the components and summarize them quantitatively. Then we combine them mathematically, and see whether the resulting model accurately predicts the behavior of the complete system. Although we did not attempt a complete representation of the first steps in vision, more detailed physical models of this living system have been found by following procedures familiar from the case studies in this book.

Moreover, vision is emblematic of other sensory and control systems, in part because evolution constantly tinkers with the solution to one problem, as it incrementally finds the solutions to others. Lessons learned in vision, for example about adaptation, recur with variations in every sensory modality.

KEY FORMULAS

- *False positives:* We estimated the expectation of the number of false but photon-like rod signals crossing the first synapse that could be confused with the real events elicited by a short flash as

$$
\begin{aligned}
\mu_{0,\text{sum}} \quad \approx \quad & (\text{rate of rod false positives}) \times (\text{integration time}) \\
& \times (\text{fraction not rejected at first synapse}) \\
& \times (\text{number of rods in a summation region}).
\end{aligned}
$$

[11.1, page 358]

- *Sakitt experiment:* We defined ℓ_{tot} as the total number of photon-like signals passing the first synapse in a collection of rod cells corresponding to a summation region, in an integration time. Section 11.3.4 then explored the hypothesis that a subject's verbal ratings of flash strength are obtained by comparing ℓ_{tot} to a set of network threshold values.

FURTHER READING

Semipopular:
Higher levels of processing: Daw, 2012; Snowden et al., 2012.
Evolution of eyes and their pigments: Lane, 2009, chapt. 7.

Intermediate:
Transmission of information by neurons: Nelson, 2014, chapt. 12; Phillips et al., 2012, chapt. 17; Dayan & Abbott, 2001.
General neuroscience, including vision: Nicholls et al., 2012; Purves et al., 2012; Nolte, 2010.
Evolution of opsins: Zimmer & Emlen, 2013; Lamb, 2011.
Olfaction, gustation, and hormone reception: Berg et al., 2015.
Luminance versus radiance: Bohren & Clothiaux, 2006, chapt. 4.
$\boxed{T_2}$ Different design trade-offs made in invertebrate photoreceptors: Sterling & Laughlin, 2015.

Technical:
The first synapse: Sampath, 2014; Bialek, 2012; Dhingra & Vardi, 2012; Taylor & Smith, 2004; Berntson et al., 2004; van Rossum & Smith, 1998.
Optogenetic approaches to retinal degeneration: Barrett et al., 2014; Nirenberg & Pandarinath, 2012; Busskamp et al., 2012.
Higher levels of processing: Zhaoping, 2014.

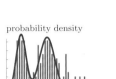
probability density

peak current change

Fig. 11.2 (page 354)

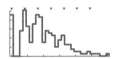

Fig. 2.8 (page 72)

$\boxed{T_2}$ **Track 2**

11.2.2′a Instrumental noise

Figure 11.2 shows electrical noise in the absence of any photoisomerization (the leftmost peak is broad). Some of this noise came from the measuring apparatus, but not all. One can saturate the cell with bright light, shutting off *all* the gated ion channels and hence eliminating that source of intrinsic rod noise. The resulting traces show that instrumental noise is significantly less than what is seen in Figure 11.2. The difference, called "continuous dark noise" in the main text, actually does come from the rod cell.

11.2.2′b Quantal release noise

There is another source of randomness created at any synapse, arising because of the discrete character of the neurotransmitter vesicles (Figure 2.8). A flash of light will momentarily reduce the mean rate of vesicle release, but the number actually released in any time window must be an integer. A stimulus whose effect is to change that number by an average of, say, 20.5 must in each instance actually change it by 20, 21, or some other whole number.

This source of randomness is often called "quantal release noise," but it has nothing to do with quantum physics! "Quantum physics" is concerned with the discreteness of light, atomic energy levels, and so on. "Quantal release noise" refers to the discrete nature of vesicle release, but the analogy goes no deeper than that.

11.2.2′c Mechanism of discrimination at the first synapse

Section 11.2.2 reviewed Baylor and coauthors' prediction of a transmission breakpoint at the synapse between rod and rod bipolar cells, and the subsequent experimental confirmation of this phenomenon. What is the origin of this breakpoint, which contrasts with the roughly linear response of the previous steps at low illumination?

M. van Rossum and R. Smith (1998) proposed that in darkness, the high rate of glutamate release is more than enough to keep some step of the mGluR6 cascade busy—that is, the rod bipolar cell's response saturates, analogously to the saturation of photosynthesis under conditions of high light intensity.[19] The system's operating point is tuned so that under a small reduction in glutamate release, characteristic of continuous dark noise (Figure 9.8b), the cascade remains saturated, and hence its ion channels do not respond at all. A true photoisomerization in the rod cell leads to a bigger reduction in glutamate; in this situation, the cascade moves farther from its dark operating point, and some of the bipolar cell's ion channels do open, creating a response.

Rod bipolar cells have another mechanism of noise rejection, not mentioned in the main text. The mGluR6 cascade acts as a low-pass frequency filter, transmitting components of the incoming glutamate signal with temporal frequencies characteristic of real photoisomerization responses, while partially blocking the continuous dark noise, which is of higher frequency. Other noise sources, including quantal release noise [(b) above] and the open/shut flickering of the discrete ion channels in the rod outer segment, are even higher-frequency than this.

To summarize, rod bipolar cells enhance vision by filtering, applying a transmission breakpoint, and only then summing the outputs of many rod cells.

---

[19]Section 2.9.3 (page 96) introduced this phenomenon.

11.2.2′d Why discrimination at the first synapse is advantageous

Rod bipolar cells discard many true photon events coming from rod cells, to avoid being swamped by noise. This statement sounds paradoxical—how can it be advantageous to discard information-bearing signals? The key point is that linearly summing multiple inputs from rod cells would *also* discard information that was originally present in their individual values.[20] The main text argued that the combined operation of discrimination followed by summing gives a more informative output than summing alone.

11.2.2′e Thresholding at later stages of processing

Additional noise rejection has been found to occur via thresholding at the level of ganglion (retinal output) cells (Ala-Laurila & Rieke, 2014).

T_2 **Track 2**

11.3.4′ Psychophysics with single photon stimuli

A human subject can be presented with stimuli containing exactly one photon, using an apparatus similar to the one discussed in Section 9.4.3 (page 305). Under ideal conditions, subjects in such an experiment in fact distinguished one-photon from zero-photon stimuli correctly in 51.6%±1.1% of trials, just slightly better than random guessing (Tinsley et al., 2016). That's qualitatively consistent with the eigengrau model, which predicts no effect on behavior if the number of absorbed photons is smaller than the first network threshold t_1; our fit to data yielded t_1 values greater than 1 for each of several subjects.

Interestingly, however, the experiment also asked its subjects to express their *confidence* in their answer to each trial. Not surprisingly, subjects usually rated their confidence as low. But when the data were restricted to only those trials with high self-reported confidence, subjects' ability to discriminate one- from zero-photon stimuli accurately rose to 60%±3%.

T_2 **Track 2**

11.4.1′a ON and OFF pathways

The main text discussed the pathway that rod signals take to the brain in low light (Figure 11.6). It involved pooling the signals of many rods, because photons are so sparsely distributed at low illumination. At the higher illumination levels to which cone cells respond, however, pooling is less important, and would degrade visual acuity. Accordingly, each cone bipolar cell connects to just a few cones (or just one, in the fovea).

A new problem arises in bright light, however: The enormous dynamic range of illumination levels greatly exceeds the range of spiking rates available to the ganglion

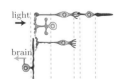

Fig. 11.6 (page 360)

---

[20]Section 7.2.5 (page 250) introduced this principle.

cells. Partly for this reason, each cone cell is connected to two distinct pathways for further signal processing:

- ON bipolar cells act similarly to rod bipolars, depolarizing in response to light via a G protein cascade.
- OFF bipolar cells do not invert their input signals; they depolarize in response to a *de*crease in illumination. Instead of the metabotropic mGluR6 receptors, they have ordinary glutamate-sensitive ion channels in their dendrites.[21]

These two classes of bipolar cells[22] in turn communicate with corresponding ON and OFF ganglion cells, effectively doubling the range of intensities that can be transmitted to the brain (Sterling & Laughlin, 2015). In the fovea, where the acuity requirement is greatest, each cone has a dedicated pair of ON and OFF bipolar cells, and corresponding dedicated ganglion cells.

11.4.1′b Image processing in the retina

The main text discussed retinal response to very dimly illuminated scenes. Under such conditions, all the retina can hope to do is to report roughly where and when individual photons arrive; the circuitry described in the main text performs that task. When photons are more plentiful, the retina must avoid completely saturating its response. After all, if at high illumination the retina reported to the brain that every part of the scene is receiving photons at or beyond the maximal rate that a retinal ganglion cell can encode, then all details would be obliterated. Moreover, the overall illumination level is generally not of great interest to us, and in fact we are almost unaware of uniform changes in illumination. What interests us much more is *variation* in illumination, either in space (edges of objects) or in time (motion or change).

Previous sections have described some strategies used to avoid overwhelming the retinal responses in high illumination:

Fig. 6.6a (page 219)

- Our eye's pupil is an adjustable aperture, which constricts in high illumination to admit less light (Figure 6.6a).
- We have two classes of photoreceptors (rods and cones), with differing sensitivities, serving different ranges of overall illumination (Chapter 3).
- Each photoreceptor adapts, reducing its sensitivity in response to higher illumination (Section 10.7′b, page 346).
- Our photoreceptors connect to separate ON and OFF pathways [see (a) above].

None of these mechanisms, however, makes use of the opportunity mentioned in the preceding paragraph: Schematizing an image by just its edges, or at least enhancing its contrasts, would allow us to compress its dynamic range still further without loss of detail.

Figure 11.11 illustrates the idea just mentioned. The upper-left panel shows an image in which the light intensities in each pixel have been binned into 256 categories, each represented by a different shade of gray. We could reduce the cost of transmitting this image by using a coarser binning scheme, with just two categories (lower-left panel). Each pixel now requires just one bit of information, instead of $\log_2 256 = 8$

---

[21]There is also a pathway from rod cells to OFF bipolar cells. Although it is not used for single-photon vision, this pathway is important for motion detection at higher illumination.
[22]Actually, there are about 10 known types of bipolar cells, but they can be grouped into these two broad classes.

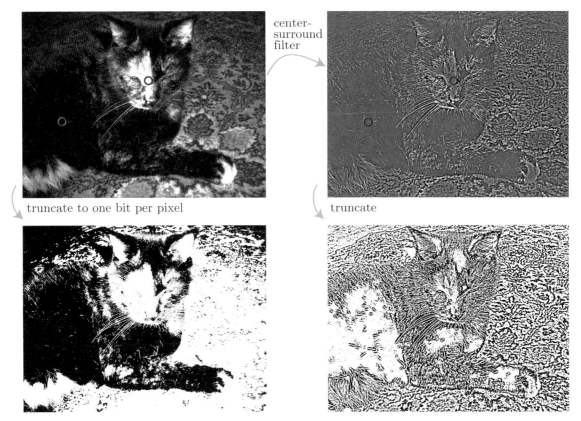

center-
surround
filter

truncate to one bit per pixel

truncate

Figure 11.11: [Photographs.] **Effect of center-surround processing.** See text. Each of these four images consists of the same number of pixels, but the two lower panels have been rendered in just two levels (black and white). Of these, the lower-right image shows more detail, because a center-surround filter was applied (*upper right*) prior to truncating to two levels. The *red circles* illustrate how uniform regions map to the same value, regardless of whether they were light or dark in the original.

bits, but clearly much detail has been lost. The image looks better, however, if we first accentuate spatial contrasts between dark and light (upper right). This time, when we threshold the pixels down to a single bit of information, we see much more detail (lower right), while maintaining the same 8-to-1 data compression as before!

The almost magical operation alluded to in the preceding paragraph involves replacing each pixel's intensity by a new value, computed by subtracting the *average* illumination of nearby pixels. An image processed in this way will not change if we add a constant to the illumination of every pixel of the original image. For example, a uniformly bright region will have every pixel replaced by *zero*, regardless of the overall illumination level. A brighter pixel against a darker background, however, yields a positive value; conversely, a darker pixel against a brighter background yields a negative value. The operation just described is called a **center-surround filter**. Mathematically, it is a convolution.[23] It is closely related to a photographic touch-up operation called "unsharp masking."

---

[23]Section 0.5.5 introduced convolution in the context of probability distributions, which must always be positive. The same formula Equation 0.48 (page 16), however, also makes sense with a filter function that combines image points with a more general weighting. You'll explore it in Problem 11.4.

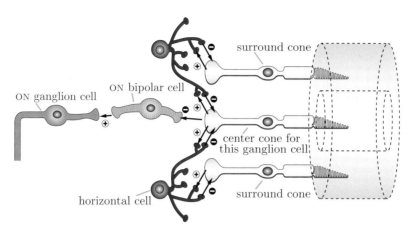

Figure 11.12: [Schematic.] **Formation of center-surround receptive fields.** The drawing shows one ON type ganglion cell (*left*) and some of the photoreceptors that influence its signaling. One photoreceptor, labeled "center cone," transmits to its bipolar cell by the inverting first synapse (*minus sign*). Light received by the center photoreceptor hyperpolarizes it, and therefore tends to depolarize the bipolar cell, increasing the mean rate of spiking from the ganglion cell. Illumination of photoreceptors also tends to hyperpolarize the horizontal cells (*plus signs*). Each horizontal cell receives signals from many photoreceptors; its membrane depolarization corresponds roughly to the average of its inputs. This "surround" signal is inverted and fed back on all of the photoreceptors in the surround field (*minus signs*). This mechanism implements one form of center-surround filter: The ganglion cell shown reports on the difference between the illumination in its field's center and the average in the surround. [After Purves et al., 2012.]

Our retinas implement center-surround processing, in part for the advantages just mentioned. Figure 11.12 shows the idea, an example of **lateral inhibition**, schematically. A generic photoreceptor is shown in the center, making a sign-inverting synapse onto its ON bipolar cell like the one for rods (Section 11.2.1, page 352). However, this photoreceptor is also connected to its neighbors by a network of **horizontal cells**, which cover a patch of nearby photoreceptors.[24] Each photoreceptor's hyperpolarization in response to light contributes a bit to hyperpolarizing the network of horizontal cells. The resulting spatially averaged signal is then fed back negatively to all the photoreceptors in the patch, depolarizing them and hence partially offsetting any hyperpolarization each may have in response to light.

In short, although illuminating a single photoreceptor stimulates (depolarizes) its corresponding ON bipolar cell, it also inhibits its neighbors. Each bipolar cell's output accordingly reflects the difference in illumination between a spot corresponding to its photoreceptor, and the average of a surrounding region, thus implementing a center-surround filter (Figure 11.13). A similar mechanism holds for the OFF bipolar cells as well.

The retina also uses center-surround processing in other ways. For example, color signals are sent to the brain not as raw intensities of stimulation in L, M, and S cone cells, but as the *differences* between a central field reporting one color and a surround reporting another one (or a combination). In Figure 11.8 (page 363), the ganglion cell marked with an asterisk is of this type.

So far, this section has emphasized the role of retinal image processing in compressing the dynamic range needed to transmit what is essentially still a pixel-by-pixel image to the brain. However, an additional layer of laterally connected cells performs

---

[24]Figure 11.8 (page 363) shows horizontal cells. They receive signals from both rods and cones.

Figure 11.13: [Diagram.] **Measurement of center-surround receptive fields.** Experimentally determined functional connectivity of individual cone photoreceptors to a ganglion cell. The figure is similar to Figure 11.7 (page 362), but only shows connections for one output retinal ganglion cell. Also, this diagram shows both positive "center" connections (*white lines*) and negative "surround" connections (*black lines*) (Figure 11.7 showed only center connections). Thicker lines denote stronger connections; the surround line thicknesses were increased relative to center line thicknesses for visibility. [Courtesy Greg D Field. Adapted by permission from Macmillan Publishers Ltd: Field et al. Functional connectivity in the retina at the resolution of photoreceptors. *Nature* (2010) vol. 467 (7316) pp. 673–677, ©2010.]

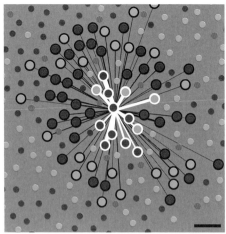

25 μm

even more sophisticated preprocessing. These amacrine cells can both receive input from the bipolar cells and also transmit signals back to them, creating the possibility of a rich feedback system; moreover, they also connect to one another, and to the ganglion cells.[25] They combine signals in such a way as to make specific classes of ganglion cells sensitive only to particular features in the visual scene, for example, motion in one particular direction. In fact, there are at least 27 different types of amacrine cell, feeding into at least 15 types of ganglion cell. Each of these ganglion cell types is repeated in a network that covers the entire visual field; each therefore sends the brain a different motion picture of the world, reporting a particular aspect of what we see (Werblin & Roska, 2007; Asari & Meister, 2012). Segmenting the world in this way prior to transmission appears to conserve the limited bandwidth available on the optic nerve, and to begin the process of making visual sense of the world. More such levels of processing then take place in our brain's visual cortex.

The retina also performs a kind of "center-surround" processing in *time*. Thus, steady illumination provokes little or no response, whereas changes in illumination over time elicit strong responses. The response characters of individual bipolar cells, and feedbacks from amacrine cells, are thought to implement this processing.

$\boxed{T_2}$ **Track 2**

11.5.1′ Rhabdomeric photoreceptors

Another class of photoreceptors, called "rhabdomeric," are found in the eyes of invertebrates such as insects, squid, and octopus. They are morphologically distinct from the "ciliary" form found in vertebrates. Unlike our rods and cones, which share the same cGMP-based signaling pathway, rhabdomeric receptors use a different internal scheme (Yau & Hardie, 2009).

---

[25]See Figure 11.8 (page 363) and Figure 11.6 (page 360).

Some researchers have concluded that a common ancestor of vertebrates and invertebrates possessed both types, and that in vertebrates, the rhabdomeric receptors evolved into retinal ganglion cells (Lamb et al., 2007). One reason for this identification is the finding that melanopsin (contained in the light-sensitive retinal ganglion cells[26]) is a member of the rhabdomeric class of opsin proteins. In addition, rhabdomeric photoreceptors and ganglion cells share similar developmental transcription cascades.

---

[26]Section 10.4.3′ (page 345) introduced these cells.

11.1 *Scotopic–mesopic transition*

Review Section 6.7.2 and Problem 6.2 before working this problem. Under very dim illumination, our cone photoreceptors do not contribute usefully to vision; this regime is called scotopic. As the illumination level increases, we enter a mesopic regime, in which both rods and cones contribute. Figure 3.10 gave the transition point between these regimes as occurring at light intensity about a thousand times greater than the minimum for human vision, but did not specify absolute numbers.

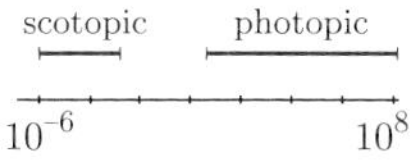

scotopic   photopic

10^{-6} 10^{8}

Fig. 3.10 (page 121)

In fact, the axis in Figure 3.10 is labeled in multiples of a unit called candelas per meter squared ($\mathsf{cd/m^2}$). This unit differs from others we have met so far: It describes luminance, a concept we have not yet encountered.

To get started, first note that most of the objects we look at are either giving off light in many directions (a computer screen) or reflecting light in many directions (a piece of paper). Even if light comes in from one direction (as from the Sun), a sheet of paper will reflect it in many directions. Few everyday objects give off light in a single direction (a laser pointer) or reflect it in a single direction (a mirror).

Thus, every small area element of an illuminated object is independently emitting light with some probability density function to leave in any direction. An appropriate way to describe the "brightness" of such an object is to state the energy per time that it gives off, per unit of its area, *per angular area;* this quantity is called the **radiance** L_{e} of the source, with typical units $\mathsf{W\,m^{-2}\,sr^{-1}}$. (The subscript stands for "energy.")

Suppose that we view such an object through a camera or eye like the simplified one in Figure 11.14, and consider a point **B** on that object. Ignoring aberration, diffraction and photon absorption, all the light from **B** that passes through the aperture (the eye's pupil) lands on a single point of the retina. The distance $d \approx 20\,\mathrm{mm}$.

a. Find a formula for the angular area Ω subtended by the aperture in terms of the circular pupil diameter (about $8\,\mathrm{mm}$ in dim light) and other quantities shown.

b. Suppose that a single photoreceptor covers area a on the retina. Find an expression for the energy per time that lands anywhere on that area in terms of the source's radiance L_{e} and geometrical parameters and evaluate for a rod cell outer segment with diameter $2\,\mu\mathrm{m}$. How does your expression depend on D? Is that reasonable?

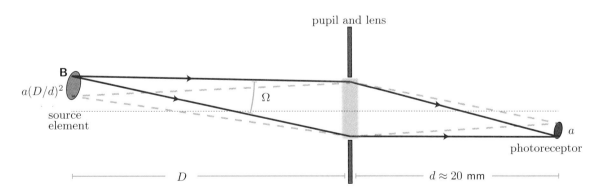

Figure 11.14: [Ray diagram.] **See Problem 11.1.** All the light emitted from point **B** into angular area Ω is assumed to arrive at a single point on the photoreceptor cell. A range of such points (*orange*) all feed light into the photoreceptor's cross-sectional area a. *Dashed lines* show rays from another source point in this range.

c. Convert your formula to express the answer in photons per second, assuming that the light is monochromatic with wavelength 507 nm (the wavelength at which rod cells are most sensitive).

Phrasing illumination in terms of radiance is a good start, but in these units every wavelength of light would have a different threshold for the crossover from scotopic to mesopic vision, because at wavelengths different from 507 nm, more energy is needed to get the same rod cell response. To give a more broadly applicable criterion, vision scientists construct a new kind of quantity called **luminous flux**, which we can think of as "ability to stimulate the sensation of brightness in the human visual system," and a corresponding unit called the lumen, abbreviated lm. In the rod-dominated (scotopic) regime, the lumen is defined as the luminous flux supplied by a light with wavelength 507 nm that delivers energy at overall rate $(1/1700)$ W. For monochromatic light at some other wavelength, the same amount of power will deliver less illumination (less luminous flux, or fewer lumens).[27]

Earlier we noted that the physical "brightness" of a visual scene can be expressed as power per area per angular area. The analogous quantity for perceptual "brightness" is called **luminance**, so its typical units are $lm\,m^{-2}\,sr^{-1}$. The **candela** is a convenient derived unit defined as $1\,cd = 1\,lm/sr$.

d. Figure 3.10 gives the threshold separating scotopic from mesopic vision as about $10^{-3}\,cd/m^2$. Still neglecting diffraction, aberration, and photon loss, convert this figure to a mean photon arrival rate for a single photoreceptor cell.

e. Actually, in a typical human eye only 45% of incoming photons with wavelength 507 nm arrive at the retina. Of the surviving photons, only about 30% are absorbed by rhodopsin and trigger rod responses. Finally, only about 50% of the resulting rod signals cross the first synapse (Section 11.2.2). Get an estimate for the mean rate of a such signals at the crossover from scotopic to mesopic illumination.

f. Repeat for the lowest value of luminance permitting useful vision, $10^{-6}\,cd/m^2$, and comment.

11.2 *Unfair die*

Figure 11.5 (page 359) shows the result of a maximum-likelihood fit to data. To get started thinking about this calculation, consider a warmup problem. Imagine a six-sided die with fixed, unequal probabilities to land on each of its faces, specified by numbers $\xi_1, \ldots, \xi_6$ whose sum equals 1. We observe M rolls of this die, and summarize them by six integers $\ell_1, \ldots, \ell_6$ whose sum is M. That is, we don't care in what order the M outcomes came; we only care about the frequencies.

Of course, if we do another M rolls, we're likely to get different frequencies; that is, the ℓ_i are a six-component random variable. Write an expression for the probability of any set of $\ell_1, \ldots, \ell_6$. [*Hint:* The expression you seek is a generalization of the Binomial distribution, which is the case of a 2-sided "die."]

11.3 $\boxed{T_2}$ *Sakitt experiment*

Chapter 9 discussed a psychophysics experiment done by B. Sakitt.[28] Dark-adapted subjects were asked to rate the strengths of flashes of light. Each flash was randomly chosen to have one of three different strengths: "strong, medium, or blank," where the last option refers to no flash at all. In response to each flash or blank, the subject

---

[27] In the cone-dominated (photopic) regime, a lumen is defined as the luminous flux supplied by light with wavelength 555 nm that delivers $(1/683)$ W. This arbitrary number arose from the original definition of the candela, which actually involved the burning of a standard kind of candle!

[28] Section 9.3.4 (page 298) described the experiment.

reported a "rating" on a scale from zero to six.

For each flash strength, Sakitt found the estimated probability distribution of the responses $r = 0, \ldots, 6$ (Figure 9.4). In this problem, you'll fit the eigengrau hypothesis outlined in the text to her data. If you haven't done Problem 11.2 yet, do it before proceeding.

a. Obtain Dataset 11 and begin by making plots of the empirical distributions like Figure 9.4.

Fig. 9.4a (page 299)

Our version of the eigengrau hypothesis supposes that some neural circuitry counts the total number ℓ_{tot} of photon (or photon-like) signals that cross the first synapse of any rod in a summation region of the retina during an integration time.[29] If there were no eigengrau, we might have conscious access to the exact number, and report $r = \ell_{\text{tot}}$. More generally, the circuitry may impose a set of network thresholds, $t_1, \ldots, t_6$, reporting to the conscious brain which of the values was met or exceeded by a particular event's count.

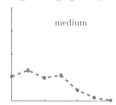

Fig. 9.4b (page 299)

b. The retinal quantum catch can be estimated from single-cell experiments. For this problem, take the single-rod quantum catch to be $Q_{\text{rod}} \approx 0.30$, and multiply by other factors discussed in Section 9.4.2′d (page 313) to obtain the retinal rod quantum catch $Q_{\text{rod,ret}}$. Finally, multiply by 50% for loss at the first synapse, obtaining an estimate of the psychophysical quantum catch Q_{tot}.[30]

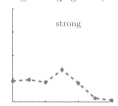

Fig. 9.4c (page 299)

c. Multiply your answer to (b) by 0, 55, or 66 photons delivered to the eye to determine the expectation of the number of true photon signals crossing the first synapse under each of the three stimulus types.

d. Now estimate the spontaneous event rate $\mu_{0,\text{sum}}$, as follows. Section 11.3.3 (page 358) stated that an individual human rod emits photon-like signals in the dark at a mean rate of about $0.0062 \, \text{s}^{-1}$. Again suppose that 50% of these pass the first synapse, and that the subsequent neural circuitry counts the number that occur in any of the ≈ 1700 rod cells in a summation region, throughout a $\approx 200 \, \text{ms}$ integration time.

e. The number of photon-like signals passing the first synapse, ℓ_{tot}, is a Poisson random variable with contributions from both spontaneous events [part (d) above] and true photoisomerizations [part (c)]. Choose a set of network thresholds, for example, $t_1 = 1, \ldots, t_6 = 6$. The model then states that the probability to elicit a ranking of 0 is the probability that $\ell_{\text{tot}} < t_1$, and so on. Evaluate these probabilities for each ranking. Then compute the likelihood for the set of t's that you chose, given Sakitt's data.[31] See if you can find another choice of t values that gives a better likelihood score than the set just proposed. What are the best values?

f. Graph your model's predictions for the probability of each ranking, for each of the three stimulus strength values, and compare to Sakitt's data.

11.4 $\boxed{T_2}$ *Center-surround image processing*

Obtain Dataset 16, or use your own black-and-white photograph.

a. Find how to get your computer to display these data as an image.

b. Make a 7×7 array called **sqMask**, each of whose entries equals $1/7^2$. Use this array to replace each pixel of the original image by the average of that pixel and 48 of its neighbors. What happened to your image?

---

[29]Section 9.3.1 (page 294) introduced the concepts of summation region and integration time.
[30]Alternatively, you may use your value for $Q_{\text{rod,ret}}$ from Problem 9.3 (page 315). Section 11.2.2 (page 353) introduced the first-synapse loss factor.
[31]Section 7.2.3 (page 249) introduced likelihood maximization.

c. Go back to the original image and repeat (b), but this time use `blip - sqMask`, where `blip` is a 7×7 array that's zero everywhere except for a 1 in the center. After this operation, your new image will not consist of numbers between 0 and 255, so rescale it to restore that property and again render it as an image.

d. Find the median value of all the pixels in the original image,[32] and replace each pixel value in the original image by 0 or 255 depending on whether its original value was less than or greater than the median. Render your image and comment.

e. Repeat (d) for your filtered image in (c) and comment.

---

[32] Any computer math package has a function that computes the median, which is the numerical value that separates the lowest 50% of pixels from the rest.

PART III

Advanced Topics

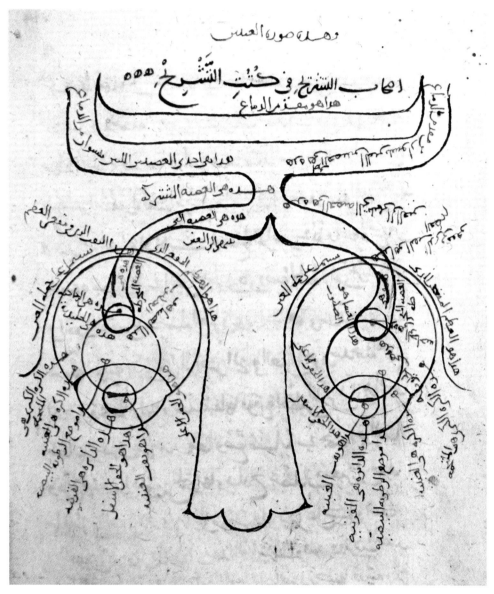

The oldest known depiction of the nervous system, drawn in eleventh-century Cairo by Ibn al-Haitham, depicts eyes on either side, and, flowing out of each eye, an optic nerve. Al-Haitham's drawing was based in part on the teachings of Galen of Pergamum, a second-century Roman physician and anatomist. [Courtesy of the Süleymaniye Library, Istanbul.]

CHAPTER 12

Electrons, Photons, and the Feynman Principle

> First we have reality, which we don't understand. We then create a complicated model. Now we have two things we don't understand.
>
> — *Peter J. Turchi*

12.1 SIGNPOST: *UNIVERSALITY*

Chapters 1 and 4 introduced the Light Hypothesis and Electron State Hypothesis as though they were independent ideas. But light and electrons have a key behavior in common:

- Light has a lumpy aspect. Its particles (photons) carry momentum p and energy E, related by $E = pc$ (Section 1.3.3′b, page 51). Light also has a wave aspect. Monochromatic light diffracts like a wave with $\lambda = 2\pi\hbar c/E$, that is, $\lambda = 2\pi\hbar/p$.
- Electrons, too, display wave-particle duality. For example, monoenergetic electrons can create diffraction patterns (see Media 11). Experimentally, those patterns obey the *same* relation $\lambda = 2\pi\hbar/p$ that we found for light.

Is this similarity a fluke? Certainly electrons and photons are also different in some ways. For example, electrons carry electric charge, which lets them bind to protons, forming atoms and molecules. Photons don't do that.[1] Still, it would be elegant if we could extend our existing photon framework to embrace electrons, without having to construct a whole new conceptual structure. Remarkably, this level of *universality* does turn out to be possible: The quantum rules may seem bizarre to our common sense, but at least the *same crazy rules* that apply to light can also be generalized to apply to electrons and all other fundamental particles. For example, each type of fundamental particle is distinguished from others by its own characteristic mass, electric charge, and so on. But all share the main aspects of quantum theory that we have studied in this book.

    This chapter will make no attempt at mathematical rigor. Nor is this treatment intended as a self-contained introduction to quantum mechanics. Instead, the goal is just to hint at the relation between this book's formulation and that found in other introductory books.

    The chapters in this Part of the book are short and independent; if you don't have the background for one of them, you may nevertheless be able to work through the others. In this chapter, Section 12.3 assumes background in the special theory of

---

[1] Electrons also differ from photons in their behavior under interchange, but in this chapter we mostly study one-electron problems, in which this behavior does not arise.

relativity (the notion of invariant interval), and in differential geometry (the notion of a metric).

This chapter's Focus Question is not explicitly biological:

Question: How can electrons be like photons? Photons don't form atoms!

Physical idea: Despite some differences in detail, in each case the Feynman principle tells us to find probability amplitudes by performing a sum over trajectories.

12.2 ELECTRONS

12.2.1 From paths to trajectories

Another big difference between electrons and photons is that electrons can move at a variety of different speeds, whereas photons always move through vacuum at speed c. Thus, when an electron moves through space, it's not enough to state the path it took, as we did for light in Chapter 4; we must instead specify its entire *trajectory* $r(t)$ through space and time. Two different trajectories could traverse the same path, and even do so with the same total elapsed time, yet still differ in their speeds at specific moments.

It is true that a *free* electron's velocity will not change, but that is a dynamical statement, which we incorporate as an equation of motion in classical physics: $\mathrm{d}^2 r / \mathrm{d}t^2 = 0$ (a special case of Newton's third law). We have come to expect that such statements are provisional, approximations to the real quantum behavior of particles; for example, neither light nor electrons travel on straight lines in diffraction experiments. So let's try to write an electron's probability amplitude as a weighted sum over *all* the ways that an electron can get between two points, as we did for light in Chapter 4. Unlike for photons, however, the preceding paragraph suggests that we must sum over trajectories (curves in space–time), not just paths (in space). The weighting factor that we seek must attribute a complex phase to each trajectory, in such a way that the classical trajectory is a stationary-phase point in the space of all trajectories. Thus, we write the weighting factor as $\exp(\mathrm{i}S[r(t)]/\hbar)$, where the square brackets emphasize that the **action functional** S depends on the entire trajectory $r(t)$. The factor of $\hbar$ implies that the numerical value of $S/\hbar$ will be large in macroscopic situations. In such situations, the stationary-phase principle says that one trajectory will dominate the sum.[2]

As in previous chapters, we will ignore the physics of polarization (in this case, electron spin). Although the system we study is highly simplified, it does share some features with real chromophores, and we will obtain an order-of-magnitude result in Section 12.2.7 about the spectrum of energy levels that fits qualitatively with observations of light absorption.

12.2.2 The action functional singles out classical
trajectories as its stationary points

To make progress, we must now make a proposal for the form of the action functional S. For free electrons, we know that our choice should meet some criteria:

- S should be unchanged if we modify the trajectory by translating (rigidly shifting) it in space, by rotating it in space, or by shifting it in time. This condition

_____

[2]Section 4.7 (page 161) introduced this principle.

guarantees that the set of trajectories at which S is stationary will also have those invariances.

- S should have dimensions $\mathbb{M}\mathbb{L}^2\mathbb{T}^{-1}$, matching those of $\hbar$, so that we can exponentiate $iS/\hbar$.
- S should be stationary on those trajectories $\boldsymbol{r}(t)$ that obey Newton's law, $\mathrm{d}^2\boldsymbol{r}/\mathrm{d}t^2 = 0$.

A simple expression that meets the first two conditions is[3]

$$S[\boldsymbol{r}(t)] = \frac{m_e}{2} \int \mathrm{d}t \left\| \frac{\mathrm{d}\boldsymbol{r}}{\mathrm{d}t} \right\|^2. \tag{12.1}$$

Your Turn 12A

Confirm that this quantity really does have the same dimensions as $\hbar$.

To check the third condition, we begin with a trajectory $\boldsymbol{r}(t)$ that starts at position $\boldsymbol{r}_i$ at time zero and ends at position $\boldsymbol{r}_f$ at time T. Any other such trajectory can be written as $\boldsymbol{r}(t) + \delta\boldsymbol{r}(t)$, where $\delta\boldsymbol{r}(0) = \delta\boldsymbol{r}(T) = 0$. Now we compute the difference in the action to lowest order in $\delta\boldsymbol{r}(t)$:[4]

$$S[\boldsymbol{r}(t) + \delta\boldsymbol{r}(t)] - S[\boldsymbol{r}(t)] = m_e \int \mathrm{d}t \, \frac{\mathrm{d}\boldsymbol{r}}{\mathrm{d}t} \cdot \frac{\mathrm{d}(\delta\boldsymbol{r})}{\mathrm{d}t}. \tag{12.2}$$

Integrating by parts lets us rephrase this first-order variation as

$$-m_e \int \mathrm{d}t \, \delta\boldsymbol{r} \cdot \frac{\mathrm{d}^2\boldsymbol{r}}{\mathrm{d}t^2}.$$

A stationary trajectory is one for which this quantity is zero for *any* $\delta\boldsymbol{r}(t)$, that is, for which $\mathrm{d}^2\boldsymbol{r}/\mathrm{d}t^2 = 0$. Thus, the proposed action functional's stationary trajectories are the ones obeying Newton's law, as desired.[5]

Your Turn 12B

a. Consider a trajectory that starts at a given point $\boldsymbol{r}_i$ at time zero, ends at given $\boldsymbol{r}_f$ at time T, but changes speed in the middle. That is, initially the speed is a constant v, but at some intermediate time T_{switch} it changes to another constant v'. First find a formula for v' in terms of $\boldsymbol{r}_i$, $\boldsymbol{r}_f$, T, T_{switch}, and v.
b. Next, consider a family of trajectories, all with the same $\boldsymbol{r}_i$, $\boldsymbol{r}_f$, T, and T_{switch}, but with various values of v. Compute the action functional for each of these trajectories and comment.

---

[3]The action functional requires modifications to handle electrons moving near the speed of light, because Newton's laws themselves become invalid in that regime.
[4]The next steps may be familiar from Problem 5.6.
[5]We could modify the action functional to account for forces acting on the electron, for example, its electrostatic attraction to an atomic nucleus.

12.2.3 The Feynman principle expresses probability amplitudes as sums over trajectories

We are exploring the idea that

> *The probability amplitude for a process is a constant times the sum of many contributions, corresponding to all possible continuous trajectories. Each contribution is of the form* $\exp(iS[\boldsymbol{r}(t)]/\hbar)$, *where $\hbar$ is the reduced Planck constant and S is the appropriate action functional on the space of all trajectories.*

Feynman principle

(12.3)

We now define the **electron propagator** $g(\boldsymbol{r}_\mathrm{f}; \boldsymbol{r}_\mathrm{i}, T)$ to be the probability amplitude for an electron initially at $\boldsymbol{r}_\mathrm{i}$ to be detected at $\boldsymbol{r}_\mathrm{f}$ after elapsed time T. Idea 12.3 states that

$$g(\boldsymbol{r}_\mathrm{f}; \boldsymbol{r}_\mathrm{i}, T) = \mathrm{const} \times \int_\mathrm{traj} [\mathrm{d}\boldsymbol{r}(t)] \mathrm{e}^{iS[\boldsymbol{r}(t)]/\hbar}, \qquad (12.4)$$

where the integral is over all trajectories with the specified start point, end point, and duration.

12.2.4 States and operators arise from partial summation over trajectories

Equation 12.4 is certainly elegant and compact, but as usual it needs interpretation before it will divulge any testable predictions. For one thing, we are very interested to know about the ground and excited states of a bound electron, but such states live a long time (forever, in the case of the ground state). What is the "initial position" of an electron in such a situation? To sidestep such questions, we will summarize the electron's prehistory by defining its **wavefunction** $\psi_0(\boldsymbol{r})$ to be a similar expression to Equation 12.4, but integrating over all trajectories from time minus infinity to 0. Rather than attempt to compute it, we will just ask how it *changes* under *further* passage of time.

To find the time evolution of the wavefunction, first note that any trajectory from the distant past to time T can be divided into the part from the past to time zero, continued by the part from time zero to T. We can classify the trajectories according to their location $\boldsymbol{r}$ at time zero. The action functional of such a trajectory (Equation 12.1) can also be split into a sum, and its exponential into a product. Then the full sum over trajectories can be broken down: For each $\boldsymbol{r}$ value, we sum over all paths from the distant past that arrive at $\boldsymbol{r}$ at time zero, over all paths from time zero to T that start from $\boldsymbol{r}$, and finally over $\boldsymbol{r}$ itself. But we have already given names to the two factors: The first is $\psi_0(\boldsymbol{r})$, and the second is g from Equation 12.4. Thus,

$$\psi'(\boldsymbol{r}') = \int \mathrm{d}^3\boldsymbol{r}\, g(\boldsymbol{r}'; \boldsymbol{r}, T)\psi_0(\boldsymbol{r}). \qquad (12.5)$$

We can think of $\psi(\boldsymbol{r})$ as a vector in the big space of all complex functions of $\boldsymbol{r}$. It's traditional to introduce the notation $|\Psi\rangle$ for that vector. Then Equation 12.5 says that $|\Psi'\rangle$ is the result of applying a *linear operator* U_T to $|\Psi_0\rangle$:

$$|\Psi'\rangle = \mathsf{U}_T|\Psi_0\rangle.$$

Later chapters will use the notation $\langle\Psi|$ to denote the vector corresponding to the complex conjugate of the function ψ, and the notation $\langle\Psi|\Psi\rangle$ for the integral of $|\psi(\boldsymbol{r})|^2$

over position, a single number. This peculiar-looking notation, introduced by P. Dirac, has a justification. Suppose that X is the operator corresponding to some physically observable quantity. That quantity will in general give a different value each time we measure it; it is a random variable. However, if we normalize our state vector so that $\langle\Psi|\Psi\rangle = 1$, then the quantity $\langle\Psi|\mathsf{X}|\Psi\rangle$ will be the expectation of the observed value of X, a notation reminiscent of the expectation in ordinary probability.

12.2.5 Stationary states are invariant under time evolution

We are interested in **stationary states**, those that, like the ground state, are unchanged by time evolution.[6] But physical observables, for example probabilities, always involve the modulus squared of a probability amplitude. So we must interpret "unchanged" to mean "except possibly for an overall phase."

Thus, a stationary state is one that has

$$\left|\Psi'\right\rangle = \mathsf{U}_T\left|\Psi_0\right\rangle = \mathrm{e}^{\mathrm{i}\varphi_T}\left|\Psi_0\right\rangle \tag{12.6}$$

for some real phase φ_T. We get a more specific characterization by noting that evolving forward by time T can be rewritten as the net effect of many little steps by Δt:

$$\left(\mathsf{U}_{\Delta t}\right)^{T/\Delta t}\left|\Psi_0\right\rangle = \mathrm{e}^{\mathrm{i}\varphi_T}\left|\Psi_0\right\rangle. \tag{12.7}$$

Now take the limit $\Delta t \to 0$. On the left side of the expression, the infinitesimal time evolution operator has a Taylor series expansion as

$$\mathsf{U}_{\Delta t} = 1 + \mathsf{L}\Delta t + \cdots,$$

for some constant operator L, because $\mathsf{U}_0 = 1$ must be the identity operator. Suppose that $\mathsf{L}\left|\Psi_0\right\rangle = -\mathrm{i}\omega\left|\Psi_0\right\rangle$. Then by the compound interest formula,[7] we find

$$\left|\Psi'\right\rangle = \mathrm{e}^{-\mathrm{i}\omega T}\left|\Psi_0\right\rangle. \tag{12.8}$$

By comparing this result to Equation 12.6, we see that *eigenvectors of the operator L correspond to stationary states*. The eigenvalue must be imaginary, so that the exponential in Equation 12.7 will have modulus equal to one. The product of $\mathrm{i}\hbar$ with the infinitesimal time evolution operator L is called the **Hamiltonian operator**.

Equation 12.8 also tells us that stationary states *oscillate in time*. We are already familiar with a connection between oscillation frequency and energy: the Einstein relation.[8] So let's tentatively identify the angular frequency ω appearing in Equation 12.8 with the energy of the stationary state, divided by $\hbar$.

To test whether a proposed state is stationary, then, we can calculate its time evolution and show that it has the form Equation 12.8 for some frequency ω. If it does, then we also get the energy of that state as $\hbar\omega$.

12.2.6 A confined-electron problem

To start getting testable consequences from the Feynman principle, we'd like to compute some energy levels for a system containing an electron localized in space. A

---

[6]Even excited states will appear stationary if we omit the coupling between electrons and the photons that they emit, as we do in this chapter. See Chapter 13.

[7]This formula appears as Equation 0.50, page 18.

[8]Idea 1.6 (page 34) states this relation. Recall the relation between angular frequency and cycle frequency: $\omega = 2\pi\nu$ (Section 6.7, page 230).

hydrogen atom is such a system, but we will consider something that is easier to handle mathematically. So instead of an inverse-square law force holding the electron to its nucleus, we will suppose that some unspecified agency confines an electron to move on a circular track of radius R. The electron moves freely along its track, but is forbidden to leave it. This example is not as fanciful as it may at first sound. Many important biomolecules, particularly chromophores specialized to interact with light, contain rings of atoms with one or a few mobile electrons nearly free to move around the ring.

Our simplified problem thus involves an electron moving in only one direction, not three, and moreover that one direction is along a circle. Thus, we replace the position coordinates $\boldsymbol{r}$ by a single quantity $R\theta$, and remember that changing the value of θ by 2π does not change our electron's location in space.

We will also simplify the problem, at first, by considering only contributions to the integral Equation 12.4 arising from stationary-phase trajectories. For a particle moving freely through space, there is only one constant-speed trajectory that starts at a specified $\boldsymbol{r}_i$, ends at specified $\boldsymbol{r}_f$, and has specified duration: The only choice is $\boldsymbol{r}(t) = \boldsymbol{r}_i + (t/T)(\boldsymbol{r}_f - \boldsymbol{r}_i)$. In our problem, however there are *infinitely many* different trajectories meeting the requirements, because of the periodicity of θ. A trajectory can move either forward or backward, and still hit the desired end point. Moreover, in each direction it can loop any number of times around the ring. In fact, for any integer j, there is a constant-speed trajectory of length $L_j = R|\theta_f - \theta_i + 2\pi j|$. The speed of this trajectory is L_j/T, so our approximate expression is

$$g(\theta_f, \theta_i; T) \approx \sum_{j=-\infty}^{\infty} \exp[i\alpha(\theta_f - \theta_i + 2\pi j)^2]. \tag{12.9}$$

Here α is an abbreviation for $m_e R^2/(2\hbar T)$.

Your Turn 12C
Convince yourself that $g(\theta_f + 2\pi, \theta_i; T) = g(\theta_f, \theta_i; T)$, as required, and that $g(\theta_i, \theta_f; T) = g(\theta_f, \theta_i; T)$.

We wish to find the stationary states of our electron on a ring, and to evaluate their energies. That is, we seek wavefunctions $\psi(\theta)$ with the property that the time-evolved wavefunction ψ' (Equation 12.5) is the same as the original, apart from an overall phase (Equation 12.8):

$$\psi'(\theta_f) = \int_0^{2\pi} d\theta_i \, g(\theta_f, \theta_i; T)\psi(\theta_i) = e^{-iET/\hbar}\psi(\theta_f). \tag{12.10}$$

It may seem hard to know where to begin looking for such functions, and in general this is indeed difficult. But for our highly symmetrical problem, it's easy to guess a promising set of periodic functions. Our ring has no particular starting point; it's unchanged if we rotate it by any angle. So we may guess that a trial wave function with the corresponding symmetry (it equals itself, up to an overall constant phase, when we shift θ) would be worth checking. The exponential functions $\psi_k(\theta_i) = e^{ik\theta_i}$ all have this property, for any integer k, so let us investigate them.

To summarize the argument so far, we wish to show that, for any integer k, the expression

$$\int_0^{2\pi} d\theta_i \sum_{j=-\infty}^{\infty} \exp\left[i\alpha(\theta_f - \theta_i + 2\pi j)^2 + ik\theta_i\right] \tag{12.11}$$

equals $e^{ik\theta_f}$ times a phase that is independent of θ_f (but that may depend on k and T). If that's true, then the discussion in Section 12.2.5 tells us to examine that phase, to find the energy of the stationary state $\left|\Psi_k\right\rangle$.

Before proceeding, we should clean up our expression a bit. First, divide it by $e^{ik\theta_f}$. Next, notice that the combined effect of the sum and the integral is the same as integrating over all real values of the quantity $u = \theta_f - \theta_i + 2\pi j$ holding θ_f fixed. So we wish to show that the following expression is independent of θ_f, and to evaluate it:

$$\int_{-\infty}^{\infty} du\, e^{i(\alpha u^2 + k(2\pi j - u))}.$$

The independence from θ_f is clear. We can also simplify by noting that $e^{i2\pi kj}$ is *always equal to 1*, because k and j are both integers. Completing the square then gives

$$\int_{-\infty}^{\infty} du\, e^{i\alpha(u - k/(2\alpha))^2} e^{-i\alpha k^2/(4\alpha^2)}.$$

After a shift of integration variable, we recognize this expression as a Fresnel integral, times the last factor.[9] Recalling that α is an abbreviation for $m_e R^2/(2\hbar T)$, the last factor equals $\exp[-ik^2\hbar T/(2m_e R^2)]$, which indeed is a phase that depends linearly on time,[10] with frequency $k^2\hbar/(2m_e R^2)$. According to Section 12.2.5, then,

> The stationary states of an electron confined to a ring of radius R have discrete energies of the form $(k\hbar/R)^2/(2m_e)$, for any integer k. (12.12)

Your Turn 12D

a. Confirm that this formula has the appropriate units to be an energy.

b. Suppose that we confine the electron to a macroscopic ring with radius $R = 0.5\,\text{m}$. Are we likely to notice the discreteness of levels in this system?

c. Suppose instead that the ring structure is that of an aromatic molecule, or a typical dye pigment, about $0.4\,\text{nm}$ in radius, and that at least one of the outer electrons in this molecule is effectively free to run around that ring. Suppose also that transitions among energy levels are accompanied by emission or absorption of a photon, in such a way that total energy is conserved. Find the difference in energy levels between the states with $k = 0$ and 1. Is there anything significant about the general order of magnitude of your answer?

12.2.7 Light absorption by ring-shaped molecules

The preceding discussion of an electron on a ring was motivated by the mathematical simplicity of this kind of confinement. However, some biologically relevant molecules

---

[9] Section 4.7.1 (page 161) analyzed the Fresnel integral.

[10] The Fresnel integral also has some dependence on T, but this just reflects the approximation made in our calculation. When all trajectories are included (not just the stationary-phase trajectories), an additional factor of the square root of T emerges and cancels this dependence, leaving only the phase discussed here (Schulman, 2005, §23.1).

really do act as though they had a few nearly free electrons confined to a ring. From the simple benzene molecule (a ring of six carbon atoms) to the more complex porphyrins and even chlorophyll, ring motifs are ubiquitous. Moreover, when the valences of the atoms involved give rise to alternating single and double bonds, then we may think of the electrons participating in the double bonds as delocalized, nearly free to wander around the ring consisting of the nuclei and the other, more tightly bound, electrons. There are two such free electrons per double bond, that is, one per carbon atom in the ring.

To get a rough approximation of the low-lying energy states of such a system, suppose that a ring of N atoms, spaced by distance d, binds N delocalized electrons. Thus, the ring's circumference is Nd. If we could remove $N-1$ of the delocalized electrons, holding the nuclei fixed, then Idea 12.12 gives the spectrum of the remaining electron's allowed energy states:

$$E_k = \frac{k^2 \hbar^2}{2m_{\mathrm{e}}(Nd/(2\pi))^2}.$$

We must now consider adding the remaining $N-1$ electrons to get the complete neutral molecule.

We have not yet considered many-electron states, but you may recall a rule from chemistry:[11]

> *Each electron state may only be occupied by 0, 1, or 2 electrons.*

exclusion principle

To find the ground state of all N electrons, then, we fill energy levels starting from the lowest ($k = 0$, two electrons), followed by $k = \pm 1$ (four electrons), and so on. To fill all levels up to k_* requires $4k_* + 2$ electrons. If we have such a configuration, then an excitation requires that some electron be promoted to level ($k_* + 1$) (the "lowest unoccupied molecular orbital"), with energy E_{k_*+1}. The least-costly way to perform that excitation is to draw the needed electron from level $\pm k_*$ (the "highest occupied molecular orbital"), with a net energy cost of[12]

$$E_{k_*+1} - E_{k_*} = \frac{(2\pi\hbar)^2}{2m_{\mathrm{e}}(Nd)^2}(2k_* + 1).$$

The preceding discussion also gives $k_* = (N-2)/4$ if the highest occupied molecular orbital is full, or

$$\Delta E \leq \frac{(2\pi\hbar)^2}{2m_{\mathrm{e}}(Nd)^2}\frac{N}{2}.$$

(The inequality reminds us that the excitation energy will be lower if the highest occupied orbital is not full.)

---

[11]Sometimes the exclusion principle is stated as "Each electron state may only be occupied by 0 or 1 electron." In that formulation, "state" refers to both orbital and spin degrees of freedom; our treatment does not explicitly treat spin.

[12]Our discussion is primitive, in that we are neglecting the electrostatic energy of interactions between the delocalized electrons, or at least assuming that it doesn't change when electrons change state.

Figure 12.1: [Photograph.] **Relation between physical size and energy levels.** Each vial contains a suspension of cadmium selenide nanocrystals, fluorescing under a common light source. The differences in emission wavelength are due only to differences in size of the nanocrystals, determined by details of their preparation. [Courtesy Marija Drndić.]

Converting from energy to wavelength gives a prediction for the wavelength of absorbed light:

$$\lambda \geq \frac{c}{\Delta E/(2\pi\hbar)} = \left(\frac{2cm_{\mathrm{e}}d}{\pi\hbar}\right)(Nd) \gg Nd. \tag{12.13}$$

This formula shows us the limitation of thinking of molecules as analogous to musical instruments. An organ pipe or violin string emits sound with wavelength comparable to its physical size (or shorter, for higher harmonics). But Equation 12.13 gives the wavelength of absorbed light as the circumference Nd times a large prefactor.

Your Turn 12E

a. Take the atomic spacing to be $d \approx 0.14\,\mathrm{nm}$ and evaluate the prefactor.
b. Evaluate the predicted wavelength for benzene ($N = 6$) and chlorophyll ($N = 20$) and comment.

Your result is not an exact explanation of the blue and red absorption of chlorophyll, but it's the right order of magnitude, and certainly far larger than the physical size of the molecule.

Figure 12.1 emphasizes the relation between size and wavelength, not for ring-shaped molecules, but for suspensions of nanocrystals differing only in size.

12.2.8 The Schrödinger equation emerges in the limit of an infinitesimal time step

Section 12.2.6 studied the simplest possible problem involving a confined electron, but other problems are mathematically more demanding than that one. To make further progress, we now return to the idea of advancing time by an infinitesimal step, as in Equation 12.7. In that limit, we can simplify Equations 12.4–12.5. For an electron on a ring, those equations give the time-evolved wavefunction as

$$\psi'(\theta_{\mathrm{f}}) = \text{const} \times \int \mathrm{d}\theta_{\mathrm{i}} \left[\int_{\mathrm{traj}} [\mathrm{d}\theta(t)]\, \mathrm{e}^{\mathrm{i}S[\theta(t)]/\hbar}\right] \psi_0(\theta_{\mathrm{i}}). \tag{12.14}$$

The integral inside the square brackets depends on θ_i, θ_f, and T, because those parameters specify the set of trajectories over which to integrate.

The simplification is that, for infinitesimal Δt, we may restrict attention to only those trajectories that go directly from θ_i to θ_f, without making a long loop (or multiple loops) around the ring. Moreover, any such trajectory can be represented by the simplest one, which makes the short trip at uniform speed.[13] Thus, we get

$$\psi'(\theta_f) = \text{const}' \times \int d\theta_i \left[e^{i m_e R^2 (\theta_f - \theta_i)^2 / (2\hbar \Delta t)} \right] \psi_0(\theta_i). \tag{12.15}$$

The large factor $1/\Delta t$ in the exponential also implies that only the region $\theta_i \approx \theta_f$ is important for the integral; the other regions' contributions cancel as in our study of the Fresnel integral.[14] Accordingly, we let $\theta_i = \theta_f + \delta\theta$ and write $\psi_0(\theta_i)$ as a Taylor series around θ_f:

$$\psi'(\theta_f) = \text{const}' \times \int d(\delta\theta)\, e^{i m_e R^2 (\delta\theta)^2 / (2\hbar \Delta t)} \left[\psi_0(\theta_f) + \delta\theta \frac{d\psi_0}{d\theta}\bigg|_{\theta_f} + \tfrac{1}{2}(\delta\theta)^2 \frac{d^2\psi_0}{d\theta^2}\bigg|_{\theta_f} + \cdots \right]. \tag{12.16}$$

Let's abbreviate the terms of this integral as I_0, I_1, I_2, and $I_{...}$ respectively.

I_0 is a Fresnel integral:

$$\int_{-\infty}^{\infty} d\xi\, e^{i\alpha\xi^2} = \sqrt{i\pi/\alpha}. \tag{12.17}$$

Recalling that $\alpha = m_e R^2 / (2\hbar \Delta t)$ thus gives

$$I_0 = \text{const}' \times \left(\frac{2 i \pi \hbar \Delta t}{m_e R^2} \right)^{1/2} \psi_0(\theta_f). \tag{12.18}$$

I_1 equals zero, because it is the integral over a symmetric range of an odd function of $\delta\theta$.

To compute I_2, first evaluate the derivatives of both sides of Equation 12.17 with respect to α:

$$i \int_{-\infty}^{\infty} d\xi\, \xi^2 e^{-i\alpha\xi^2} = -\tfrac{1}{2}\sqrt{i\pi/\alpha^3}.$$

Thus

$$I_2 = \text{const}' \times \frac{1}{2} \frac{1}{i} \left(\frac{-\sqrt{i\pi}}{2} \right) \left(\frac{2\hbar \Delta t}{m_e R^2} \right)^{3/2} \frac{d^2\psi_0}{d\theta^2}\bigg|_{\theta_f}. \tag{12.19}$$

The remaining terms, $I_{...}$, are all zero (for odd powers of $\delta\theta$), or else vanish with higher powers of Δt than the two leading terms. Moreover, I_2 has a higher power of Δt than I_0.

Up to now, we have been vague about the overall constant appearing in Equation 12.15. To fix it, note that we must have $\psi' \to \psi_0$ as $\Delta t \to 0$. To get that behavior,

---

[13]Trajectories that deviate from this one make corrections that will vanish in the limit $\Delta t \to 0$. Note that this is not an approximation; we really will take the mathematical limit. In contrast, other derivations in this book, for example restricting to stationary-phase paths in optics, are only approximately valid, only in certain parameter regimes.
[14]Section 4.7.1 (page 161) analyzed the Fresnel integral.

we must choose the constant in front of our integral to be equal to $\sqrt{m_{\mathrm{e}}R^2/(2\pi i \hbar \Delta t)}$, or

$$\psi'(\theta_{\mathrm{f}}) \to \psi_0(\theta_{\mathrm{f}}) + \frac{i\hbar}{2m_{\mathrm{e}}R^2}\Delta t \frac{\mathrm{d}^2\psi_0}{\mathrm{d}\theta^2}\bigg|_{\theta_{\mathrm{f}}}, \text{ for } \Delta t \to 0. \tag{12.20}$$

We have now found a formula for the infinitesimal time evolution of a wavefunction, to first order in Δt. Rephrasing Equation 12.20 as a derivative yields

$$\boxed{\frac{\partial \psi'}{\partial t} = \frac{i\hbar}{2m_{\mathrm{e}}R^2}\frac{\partial^2\psi_0}{\partial\theta^2}. \quad \text{Schrödinger equation}} \tag{12.21}$$

One consequence of this result is a compact criterion for a state to be stationary: Taking the derivative of Equation 12.8 (page 385) and evaluating at time zero yields

$$\omega\psi_0 = \frac{-\hbar}{2m_{\mathrm{e}}R^2}\frac{\mathrm{d}^2\psi_0}{\mathrm{d}\theta^2}. \tag{12.22}$$

If a wavefunction $\psi(\theta)$ satisfies this condition, then it represents a stationary state with energy $\hbar\omega$.

We can now compare the result with one that we previously found. The eigenfunctions of the second derivative operator are all of the form $\psi_k(\theta_{\mathrm{i}}) = e^{ik\theta_{\mathrm{i}}}$, which agree with what we guessed earlier. And their corresponding energy values are $(k\hbar/R)^2/(2m_{\mathrm{e}})$, in agreement with our earlier result (Idea 12.12, page 387).

One virtue of the Schrödinger equation approach is that we did not make the approximation of considering only stationary-phase trajectories. Also, the Schrödinger equation can be readily amended to include forces acting on the electron, and hence to find energy levels for a more realistic atomic model.

Section 12.2 has outlined a unification of two different-seeming formulations of quantum physics. We also saw in an example how the Electron State Hypothesis is not an independent assumption, but rather is a consequence of the deeper Feynman principle.[15]

12.3 PHOTONS

Chapter 4 introduced a version of the Light Hypothesis suitable for macroscopic optical phenomena involving a monochromatic point light source. Suppose that the light is emitted at spatial position $r_{\mathrm{e}} = \mathbf{0}$ and has frequency ν. We place a photon detector at position r_{d}, and turn it on briefly at time $t_{\mathrm{d}} = 0$. The procedure we adopted in

---

[15]Section 1.6.1 (page 40) introduced the Electron State Hypothesis.

earlier chapters says[16]

- *List all piecewise-straight paths from source to detector, with kinks only at barrier edges and interfaces.*
- *Each such path contributes an additive term to* Ψ.
- *Each of those terms is the product of factors for each straight segment, equal to a constant times* $r^{-1}\exp(2\pi iT\nu)$. *Here* r *is the length of the   (12.23) segment and* $T = r/c$ *is its transit time when traversed at speed* c.
- *The probability of detecting a photon at* $(\boldsymbol{r}_\mathrm{d}, t_\mathrm{d})$ *is then* $|\Psi|^2$.
- *There can also be extra factors arising from passage through, or reflection from, interfaces between media.*

Chapters 4–8 discuss how this recipe correctly predicts many observed optical phenomena. However, it is limited to the case of a monochromatic light source (one with definite frequency ν). The main text alluded to a grander synthesis, embracing other light phenomena and even other fundamental particles such as electrons. The following sections outline how we can obtain Idea 12.23 for light, starting from the same Feynman principle that we used for electrons (Idea 12.3, page 384).

12.3.1 The action functional for photon trajectories

We would like a single, unified framework for photons and electrons. But Section 12.2.1 argued that, for electrons, the proper meaning of the words "all paths" extends even beyond unusual curves in space: We must sum over all *trajectories* in space and time. For photons, this means that we must even include those traversed at speeds different from c.

To see if such a generalization is possible, we must propose a choice of action functional appropriate for photons; that is, one whose stationary points are the classical trajectories for light. Thus, $S[\mathrm{traj}]$ must have the property that its first-order variation is zero when evaluated for the trajectories used in ray optics: straight lines in space–time moving at velocity c. Then such trajectories will also be stationary-*phase* trajectories of $\exp(iS[\mathrm{traj}]/\hbar)$, and hence will dominate the total probability amplitude in the classical limit.

Before we begin, one further point needs to be made. The monochromatic light source mentioned in the Light Hypothesis, for example, a laser, is a macroscopic system, consisting of innumerable electrons and atomic nuclei. No fundamental description of such a source is practical, so we must make an idealization to connect it to the single-photon world of Idea 12.3. We do this by proposing a model of such a light source that is consistent with the Einstein relation:[17]

A monochromatic light source with frequency ν is an apparatus whose probability amplitude Ψ_emis for emitting a photon at time t_e equals a   (12.24) constant times $\exp(-i2\pi\nu t_\mathrm{e})$. That is, Ψ_emis oscillates with frequency ν.

When evaluating probability amplitudes for photons emitted by such a source, we must sum all trajectories starting from the source at *all times* t_e,[18] and ending at the

---

[16]Ideas 4.5 (page 154) and 4.6 (page 155) state the Light Hypothesis parts 2a,b. The present chapter continues to neglect phenomena involving polarization of light; see Chapter 13.

[17]Equation 1.6 (page 34) states this relation. The same link between energy and oscillation frequency arose in our study of electrons, Section 12.2.5 (page 385). Section 13.7.6 will discuss how lasers generate monochromatic light, and why it has this property.

[18]Section 4.6.3′ (page 174) introduced this idea.

detector location at the time t_d of detection. Each contribution to the sum equals the factor Ψ_emis, multiplied by the factor given in Idea 12.3 (page 384). We would like to see that performing this sum does lead to the result in Idea 12.23.

The first step is to identify a good candidate for the action functional appearing in Idea 12.3. The technical difficulty stems from the fact that photons, being massless, are intrinsically relativistic. We would like an action functional that keeps this invariance visible, that is, one that treats space and time equivalently. Thus, a "trajectory" must be specified by four functions $\underline{X}^\mu(\xi)$, where $\mu = 0, 1, 2,$ or 3. Here $\underline{X}^0$ is the product of time and the speed of light; $\underline{X}^i = \boldsymbol{r}_i$ for $i = 1, 2, 3$ are the three components of position. The dimensionless variable ξ parameterizes the trajectory; it always runs from 0 to 1.

The same physical trajectory can appear in various guises, reflecting the possibility of reparameterizing: $\xi' = f(\xi)$, where f is any increasing function with $f(0) = 0$ and $f(1) = 1$, will yield a different set of four functions, $\underline{X}'^\mu(\xi')$, describing the same photon trajectory. Surprisingly, the way to deal with this reparameterization invariance involves introducing another variable, a positive "metric" function along the parameter space, $e(\xi)$. In terms of these variables, we now propose the action functional

$$S[\underline{X}^\mu(\xi), e(\xi)] = \frac{\hbar}{2} \int \mathrm{d}\xi \, (e^{-1}\|\underline{\dot{X}}\|^2 - e\bar{m}^2). \tag{12.25}$$

Here the dot means derivative with respect to ξ, and the notation $\|\dot{X}\|^2$ means the Lorentz-invariant quantity $-(\underline{\dot{X}}^0)^2 + \|\dot{\boldsymbol{r}}\|^2$. The parameter $\bar{m}$ is the particle's mass times $c/\hbar$; for light, we are interested in the limiting case $\bar{m} \to 0$ holding c and $\hbar$ fixed.

We can now find the condition for S to be stationary about a trajectory $\{\underline{X}(\xi),\ e(\xi)\}$:

Your Turn 12F

Show that Equation 12.25 is stationary about any trajectory that obeys

$$0 = e^{-2}\|\underline{\dot{X}}\|^2 \text{ and} \tag{12.26}$$

$$e^{-1}\underline{\dot{X}} = \text{constant}. \tag{12.27}$$

Equations 12.25 and 12.26 are reparameterization-invariant if we give the function $e(\xi)$ a transformation rule appropriate for a metric. That rule in turn implies that we can always find a parameterization in which $e(\xi)$ is a constant function. Let K denote its value. Then Equation 12.27 says that the stationary-phase trajectory is a *straight line* in space–time:

$$\underline{X}^\mu(\xi) = \underline{X}_\mathrm{e}^{\ \mu} + (\Delta\underline{X}^\mu)\xi\,, \quad \text{where} \quad \Delta\underline{X}^\mu = \underline{X}_\mathrm{d}^{\ \mu} - \underline{X}_\mathrm{e}^{\ \mu}.$$

Equation 12.26 implies that $\|\Delta X\|^2 = 0$, or that $c^2(\Delta t)^2 = (\Delta\boldsymbol{r})^2$. That is, the stationary-phase trajectories are straight lines in space-time with velocity equal to c. These are the properties of the ray-optics limit of light that we wished to program into our theory, so Equation 12.25 is a suitable choice of action functional.

12.3.2 The special case of a monochromatic light source reduces to our earlier formulation

We now arrive at a difficult mathematical result that must be brought in from outside:[19] The sum over all trajectories that start at some emission point $\underline{X}^{\,\mu}_{\,\mathrm{e}}$, and end at some other detection point $\underline{X}^{\,\mu}_{\,\mathrm{d}}$, is a constant times the **photon propagator** function:

$$G(\underline{X}^{\,\mu}_{\,\mathrm{e}}, \underline{X}^{\,\mu}_{\,\mathrm{d}}) = \int_0^\infty \frac{\mathrm{d}K}{K^2} \exp\!\big[\mathrm{i}\|\Delta\underline{X}\|^2/(2K)\big]. \qquad (12.28)$$

Equation 12.28 needs careful interpretation, because the integral doesn't converge when the invariant interval $B = \|\Delta\underline{X}\|^2$ equals zero. To ensure that it is well defined, we replace B by $B + \mathrm{i}\epsilon$, where ϵ is a small positive number, and later take the limit $\epsilon \to 0$. Changing the integration variable from K to $\eta = 1/K$ and performing the integral then yields

$$G(\underline{X}^{\,\mu}_{\,\mathrm{e}}, \underline{X}^{\,\mu}_{\,\mathrm{d}}) = \frac{2\mathrm{i}}{B + \mathrm{i}\epsilon}. \qquad (12.29)$$

We wish to apply Equation 12.29 to a monochromatic point source. Thus, the place of emission is $\boldsymbol{r}_\mathrm{e} = \mathbf{0}$, but the *time* of emission t_e is not specified. Instead Idea 12.24 says that the source contributes a phase $\mathrm{e}^{-\mathrm{i}\omega t_\mathrm{e}}$, where $\omega = 2\pi\nu$ is the angular frequency of the source. The probability amplitude to detect a photon at time $t_\mathrm{d} = 0$ and position $\boldsymbol{r}_\mathrm{d}$ is then

$$\int_{-\infty}^\infty \mathrm{d}t_\mathrm{e}\,(\text{emit})(\text{propagate}) = \int_{-\infty}^\infty \mathrm{d}t_\mathrm{e}\,\mathrm{e}^{-\mathrm{i}\omega t_\mathrm{e}} \frac{2\mathrm{i}}{-(ct_\mathrm{e})^2 + r^2 + \mathrm{i}\epsilon}. \qquad (12.30)$$

This integral can be evaluated by extending it to a closed contour in the lower half complex plane, which yields an overall constant times $r^{-1}\exp[\mathrm{i}(r + \mathrm{i}\epsilon/2)\omega/c]$. Now we may safely take the limit $\epsilon \to 0$, which yields

$$\text{probability amplitude} = \text{const} \times r^{-1}\mathrm{e}^{\mathrm{i}r\omega/c}.$$

Thus, the very general Feynman principle (Idea 12.3, page 384), specialized to the case of monochromatic light, implies the Light Hypothesis in the form we have used in earlier chapters, stated in Idea 12.23.[20]

12.3.3 Vista: reflection, transmission, and the index of refraction

The goal of this section was to find an improved underpinning for the Light Hypothesis, but so far we have only touched on "Part 2a," which describes light in vacuum. What about Parts 2b and 2c, which describe the behavior of light in transparent media?[21]

Media are complicated: They may be in crystalline or disordered states; their constituent atoms or molecules themselves have internal structure; and so on. So we

---

[19] Equation 12.28 is reasonable: It's dimensionally correct and Lorentz invariant. One way to obtain it is by directly computing the sum of $\exp(\mathrm{i}S/\hbar)$ over trajectories (Kleinert, 2009, sect. 19.1). Feynman's way was to show first that the propagator must obey the "Klein-Gordon equation" and proceed from that (Feynman, 1950, eqn. 8A; Schulman, 2005, Eqn. 25.4–25.5).

[20] Earlier chapters sometimes used an approximate form, in which the slowly varying factor r^{-1} was treated as a constant.

[21] Equations 4.5 (page 154), 4.6 (page 155), and 5.1 (page 184) state the Light Hypothesis part 2a,b,c, respectively.

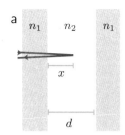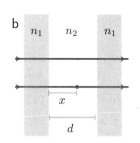

Figure 12.2: [Path diagrams.] **Interaction of light with a medium.** (a) One path contributing to reflection: The path penetrates to a depth x before it interacts with an electron and is reemitted to the source. (b) Two paths contributing to transmission.

cannot delve too deeply into this topic. Still, however, it would at least be good to get a sense for how the formulas in Ideas 4.6 and 5.1, which connect the slowing of light in media to its reflection at a boundary, arise. In particular, it does seem odd that these ideas do not explicitly invoke anything happening as light passes through a medium. All the action seems to happen at surfaces, but at the atomic level, is the "surface" even well defined? Moreover, the Light Hypothesis states that photons interact with individual electrons—not with "surfaces."

Reflection

To keep things as simple as possible, suppose that one medium is vacuum, so that $n_1 = 1$, and the other is of very low density, with $n_2 = 1 + \delta n$ where δn is small. In such a medium, it is unlikely that an incoming photon will interact with more than one electron, so we will only consider contributions to the probability amplitude involving zero or one such interaction.

We consider light paths that travel perpendicular to a flat layer of medium *2* suspended in vacuum. An incoming photon could interact with an electron anywhere within the thickness of the layer. The interaction of interest to us involves an electron absorbing the photon, then emitting another one with the same energy, leaving the material unchanged. The simplest hypothesis we could entertain about this process is that it contributes some constant factor Λ to the probability amplitude. The numerical value of Λ will depend on the nature of the medium, for example, how its electrons are bound into atoms or molecules, and also on the frequency of the photon. However, we are assuming that the medium is uniform within the layer and the illumination is monochromatic, so we take Λ to be a constant throughout the layer.

Let's write the probability amplitude to interact with an electron somewhere in a slice of thickness dx as $\kappa\,dx$, where the value of the constant κ includes Λ but also depends on the density of electrons. Then a photon path that travels to a depth x, interacts with an electron, and travels back to a detector (Figure 12.2a) will have probability amplitude given by the sum of many contributions:[22]

$$r_{1\ \text{layer}} = \int_0^d (\kappa\,dx)\mathrm{e}^{2\pi\mathrm{i}2x\nu/c}.$$

The phase involves distance traveled in the medium; overall factors involving light transit to and from the front surface are common to every contribution and have been dropped. No index of refraction enters into the phase, because in between the

---

[22] For a thin layer, the overall $1/r$ factor in the propagator does not vary much, so we have approximated it as a constant, as in Chapter 4.

electrons there's *nothing* (vacuum).[23]

The integral is easy:

$$r_{1 \text{ layer}} = \frac{\kappa c}{4\pi i \nu} \left(e^{4\pi i \nu d/c} - 1 \right). \qquad (12.31)$$

Thus, although every intermediate layer does contribute, we may summarize the total amplitude for reflection by saying "two processes are possible: light may bounce off the front surface, with a minus sign, or it may bounce off the back surface, with no minus sign."[24]

We can now compare Equation 12.31 to the Light Hypothesis,[25] which says that

$$r_{1 \text{ layer}} = \frac{n_1 - n_2}{n_1 + n_2} \left(1 - e^{4\pi i \nu d/c} \right).$$

This does agree with Equation 12.31, if we identify $\delta n = \kappa c/(2\pi i \nu)$.

To see whether we are on the right track, we must now consider the index of refraction.

Transmission

Next, consider light paths that pass through the sample (Figure 12.2b). Here again, a light path may interact with an electron at any point inside medium *2*, but this time every path has the same total length. Hence, the probability amplitude for this process is

$$e^{2\pi i \nu d/c} \left[1 + \int_0^d (\kappa \mathrm{d}x) \right] = e^{2\pi i \nu d/c} (1 + \kappa d).$$

To first order in κ, the expression in parentheses equals $e^{\kappa d}$, so we can rewrite everything as

$$\exp \left(\frac{2\pi i \nu d}{c/n_2} \right) \qquad \text{where} \quad n_2 = 1 + \frac{\kappa c}{2\pi i \nu}. \qquad (12.32)$$

Thus, our complicated probability amplitude, with many contributions, just amounts to transit with *no* interactions but with reduced speed c/n_2, as claimed in the Light Hypothesis part 2b; n_2 is its index of refraction. Moreover, n_2 differs from 1 by the same amount as what we found by considering reflection above, a nontrivial check of consistency.

For a transparent material, n_2 must be a real quantity, or in other words the constant κ characterizing the material must be purely imaginary. (If this is not the case, we see that the probability amplitude for transmission will decrease exponentially with layer thickness; that is, the real part of κ, if any, describes *absorption* of light by the medium.)

In addition to the explicit frequency dependence of Equation 12.32, the interaction parameter κ depends on the details of how electrons in the medium are bound to their atoms, in ways that may also depend on the frequency of the light. Thus, the index of

---

[23] We may neglect the protons in the medium because they turn out to interact much more weakly with light than the electrons.

[24] Slowdown of light within the medium only arises at higher orders in the interaction κ, which we are neglecting.

[25] Part 2c, Idea 5.1 (page 184).

refraction is in general frequency-dependent, a physical origin for the dispersion that is responsible for rainbows (and chromatic aberration).

Section 12.3 has established a connection between the most general formulation of quantum physics (the Feynman principle) and the simpler version used for light in earlier chapters.

THE BIG PICTURE

This chapter began by noting that for electrons, which can move at any speed less than c, we need a sum over all trajectories, not just over paths in space. We proposed a formulation by requiring that we recover classical trajectories in the macroscopic limit, then investigated the concept of steady state and obtained a special case of the Electron State Hypothesis. Along the way, we also made contact with the state space and operators introduced in Schrödinger's formulation of quantum theory.

Generalizing quantum physics in this way may seem to break its promised *universality,* so we had another look at photons. We found that their behavior, too, can be formulated in terms of a sum over trajectories. For the special case of monochromatic light, we found that the generalized form (the Feynman principle) is equivalent to the form used earlier in this book. In contrast, the Schrödinger equation, which is limited to slowly moving particles, cannot handle light.

In fact, the Feynman principle's applicability extends far beyond electrons and photons. Even the particles responsible for the weak and strong nuclear interactions, and other exotic particles such as neutrinos, are now thought to be described by instances of this all-embracing principle.

FURTHER READING

Semipopular:
Styer, 2000.
Although Feynman, 1985, is a popularization, it contains sketches of many technical points, including those discussed in this chapter.

Intermediate:
Schulman, 2005.
Feynman et al., 2010c; Müller-Kirsten, 2006.
Equivalence of the wavefunction-based formulation of quantum physics and that based on the Feynman principle: Feynman et al., 2010c; Das, 2006.

Technical:
Kleinert, 2009.

CHAPTER 13

Field Quantization, Polarization, and the Orientation of a Single Molecule

> The sages of Chelm began to argue about which was more important: the Moon or the Sun.... The reigning wise man ruled: "The Moon *must* be more important, because without its light our nights would be so dark we could not see anything. The Sun, however, shines only by day—which is when we don't need it!"
>
> — *Leo Rosten*

13.1 SIGNPOST: *FIELDS*

Earlier chapters described light as packets of energy and momentum (the Light Hypothesis, Section 1.5.1), whose arrivals at a detector or other target followed a probabilistic law (Section 4.5). These principles were compact and easy to state, and had far-reaching implications for a wealth of biophysical and other phenomena. Nevertheless, several outstanding questions remain unaddressed in this framework:

- What is the mechanism for the creation and destruction of photons?
- How can we understand the polarization of light?
- What is the connection between light, with its lumpy aspect, and electricity and magnetism, traditionally regarded as continuous *fields?* J. C. Maxwell found that radiation traveling at speed c was an automatic consequence of the laws of electrodynamics; does his classical wave model have any relevance in our quantum world?

A bit more subtly, we have almost always considered scenarios involving *just one* photon, for example, as it passes through an optical system. We argued that, despite this limitation, we could understand diffraction, because diffraction is observed even under single-photon conditions. We may still wonder, however,

- Are there new phenomena for which multiphoton states play an essential role?

To work through the formulas in this chapter, you'll need some standard background not developed in this book, concerning classical electrodynamics, the quantization of the harmonic oscillator, and the rates of transitions in perturbation theory. See "Further Reading" for some suggested sources.

The Focus Question is

Biological question: How can insects and crustaceans see the polarization state of light? Why can't we do this?

Physical idea: The probability for a molecule to absorb or emit a photon depends on its orientation relative to the photon's polarization, but a particular cellular architecture is needed to take advantage of that fact.

13.2 A SINGLE MOLECULE EMITS PHOTONS IN A DIPOLE DISTRIBUTION

Before we unleash a lot of formulas, let's first frame the issues with an experimental observation. A concrete example of what we'd like to understand is the pattern of light seen from a single immobilized fluorophore, for example in defocused orientation imaging (Figure 7.4b). The distribution of photon arrivals resembles the "dipole" radiation pattern found in classical electrodynamics, but the emission of *single photons* by a *single molecule* is as far from being classical as one can get. Is the observed agreement in radiation patterns just a coincidence? This chapter will argue that in fact, a quantum-mechanical treatment recapitulates the classical distribution of energy flow as a probability density function for photon arrivals.

Fig. 7.4b (page 260)

13.3 CLASSICAL FIELD THEORY OF LIGHT

Classical electrodynamics describes a system whose states are field configurations. But we have come to expect that Nature is described by probability amplitudes, not classical state variables. The goal of this section is therefore to recast Maxwell's eight equations for the electric and magnetic fields in a form that is suitable for quantization. Later, Section 13.4 will recover the photon concept as a *consequence* of field quantization, not as a separate hypothesis.

A key result from classical electrodynamics is that we can represent electric and magnetic fields via a scalar potential field, $\Phi(t, \boldsymbol{r})$, and a vector potential field, $\boldsymbol{A}(t, \boldsymbol{r})$:

$$\boldsymbol{E} = -\frac{\mathrm{d}}{\mathrm{d}t}\boldsymbol{A} - \boldsymbol{\nabla}\Phi; \quad \boldsymbol{B} = \boldsymbol{\nabla} \times \boldsymbol{A}. \tag{13.1}$$

In this representation, half of Maxwell's equations are identities (automatically true), so we only need to substitute Equation 13.1 into the other ones and solve them.[1]

There is some freedom in how we represent a given field configuration by potentials. To eliminate it, we can impose additional "gauge conditions" on Φ and $\boldsymbol{A}$. We will choose to work in Coulomb gauge, that is, use only vector potentials that obey $\boldsymbol{\nabla}\cdot\boldsymbol{A} = 0$. Moreover, in a world with no charged particles we can always specialize further, supplementing Coulomb gauge with the extra condition that the scalar potential $\Phi = 0$ everywhere. (Later sections will reinstate Φ when we consider coupling of the field to electrons.)

We wish to show that Maxwell's equations reduce to a set of simple, decoupled dynamical systems. It's convenient to imagine a finite world of some very large size L, which will ultimately be taken to be infinity, and specifically to take that world to be a cube with periodic boundary conditions. Then the vector potential can be expanded as

$$\boldsymbol{A}(t, \boldsymbol{r}) = \tfrac{1}{2} \sum_{\boldsymbol{k}}{}' \left(\boldsymbol{A}_{\boldsymbol{k}}(t)\mathrm{e}^{\mathrm{i}\boldsymbol{k}\cdot\boldsymbol{r}} + \mathrm{c.c.}\right). \tag{13.2}$$

In this formula, each coefficient $\boldsymbol{A}_{\boldsymbol{k}}$ is a complex 3D vector depending on time. There are many such vectors, indexed by the discrete label $\boldsymbol{k}$. The abbreviation "c.c." denotes the complex conjugate of whatever precedes it; this term is needed in order to

---

[1] Faraday's law, and the magnetic Gauss law, are satisfied regardless of what we choose for Φ and $\boldsymbol{A}$; conversely, those two laws are what guarantee the existence of potentials that represent $\boldsymbol{E}$ and $\boldsymbol{B}$ via Equation 13.1.

ensure that $\boldsymbol{A}$ is real. The primed summation denotes a sum over all vectors $\boldsymbol{k}$ with components of the form $2\pi\nu_i/L$; the ν_i are integers, not all of which are zero. For each such wavevector $\boldsymbol{k}$, we exclude the redundant $-\boldsymbol{k}$ in primed summations.

The Coulomb gauge condition implies that $\boldsymbol{k} \cdot \boldsymbol{A_k} = 0$, or in other words that the component of each $\boldsymbol{A_k}$ along its $\boldsymbol{k}$ must equal zero. The other two components are unrestricted, so for each $\boldsymbol{k}$, we choose a basis of two real unit vectors perpendicular to it and to each other; we denote these **polarization basis vectors** by $\boldsymbol{\varepsilon}_{\boldsymbol{k}}^{(\alpha)}$, where the index α runs from 1 to 2. Then Equation 13.2 becomes

$$\boldsymbol{A}(t, \boldsymbol{r}) = \tfrac{1}{2} \sum_{\boldsymbol{k},\alpha} {}' \left(A_{\boldsymbol{k},\alpha}(t)\boldsymbol{\varepsilon}_{\boldsymbol{k}}^{(\alpha)}\mathrm{e}^{\mathrm{i}\boldsymbol{k}\cdot\boldsymbol{r}} + \mathrm{c.c.} \right). \tag{13.3}$$

The polarization basis vectors are not dynamical variables. The dynamical variables, whose equations of motion we wish to find and quantize, are the mode expansion coefficients $A_{\boldsymbol{k},\alpha}(t)$.

Your Turn 13A

Show that, with these definitions, Maxwell's equations in Coulomb gauge become simple:

$$\frac{\mathrm{d}^2}{\mathrm{d}t^2} A_{\boldsymbol{k},\alpha} = -(ck)^2 A_{\boldsymbol{k},\alpha}. \tag{13.4}$$

Here α runs over 1,2, $\boldsymbol{k}$ runs over the nonredundant set described earlier, and k denotes the length of the vector $\boldsymbol{k}$ (that is, $\|\boldsymbol{k}\|$).

Equation 13.4 shows that every distinct combination of polarization α and wavevector $\boldsymbol{k}$ corresponds to an independent dynamical system, decoupled from the others. To make the system more familiar, we now give separate names to the real and imaginary parts of $A_{\boldsymbol{k},\alpha}$:[2]

$$A_{\boldsymbol{k},\alpha} = (\epsilon_0 L^3/2)^{-1/2} \left(X_{\boldsymbol{k},\alpha} + \mathrm{i}Y_{\boldsymbol{k},\alpha} \right). \tag{13.5}$$

The real scalar quantities $X_{\boldsymbol{k},\alpha}$ and $Y_{\boldsymbol{k},\alpha}$ separately obey Equation 13.4, so we see that

> Maxwell's equations in vacuum are mathematically equivalent to a set of decoupled harmonic oscillators. (13.6)

The harmonic oscillator has a well known quantum-mechanical formulation, so Idea 13.6 achieves the first goal of this section.

To understand the meaning of these oscillators better, we now express the electromagnetic field energy $\mathcal{E}$ and momentum $\boldsymbol{P}$ in terms of the new variables X and Y.

---

[2]In Equation 13.5, ϵ_0 is a constant of Nature called the "permittivity of vacuum." The overall rescaling chosen in the definitions of X and Y will simplify some later formulas.

Let $\dot{\boldsymbol{A}}$ denote the time derivative $\partial \boldsymbol{A}/\partial t$. Then

$$
\begin{aligned}
\mathcal{E} &= \frac{\epsilon_0}{2} \int \mathrm{d}^3\boldsymbol{r} \left(\boldsymbol{E}^2 + c^2 \boldsymbol{B}^2\right) = \frac{\epsilon_0}{2} \int \mathrm{d}^3\boldsymbol{r} \left((-\dot{\boldsymbol{A}})^2 + c^2(\boldsymbol{\nabla} \times \boldsymbol{A})^2\right) \\
&= \frac{\epsilon_0}{2} {\sum_{\boldsymbol{k}_1,\alpha}}' {\sum_{\boldsymbol{k}_2,\beta}}' \int \mathrm{d}^3\boldsymbol{r} \Big[\tfrac{1}{2}\big(-\dot{A}_{\boldsymbol{k}_1,\alpha}\boldsymbol{\varepsilon}_{\boldsymbol{k}_1}^{(\alpha)}\mathrm{e}^{\mathrm{i}\boldsymbol{k}_1\cdot\boldsymbol{r}} + \text{c.c.}\big) \cdot \tfrac{1}{2}\big(-\dot{A}_{\boldsymbol{k}_2,\beta}\boldsymbol{\varepsilon}_{\boldsymbol{k}_2}^{(\beta)}\mathrm{e}^{\mathrm{i}\boldsymbol{k}_2\cdot\boldsymbol{r}} + \text{c.c.}\big) \\
&\qquad + c^2 \tfrac{1}{2}\big(A_{\boldsymbol{k}_1,\alpha}\mathrm{i}\boldsymbol{k}_1 \times \boldsymbol{\varepsilon}_{\boldsymbol{k}_1}^{(\alpha)}\mathrm{e}^{\mathrm{i}\boldsymbol{k}_1\cdot\boldsymbol{r}} + \text{c.c.}\big) \cdot \tfrac{1}{2}\big(A_{\boldsymbol{k}_2,\beta}\mathrm{i}\boldsymbol{k}_2 \times \boldsymbol{\varepsilon}_{\boldsymbol{k}_2}^{(\beta)}\mathrm{e}^{\mathrm{i}\boldsymbol{k}_2\cdot\boldsymbol{r}} + \text{c.c.}\big) \Big].
\end{aligned}
\tag{13.7}
$$

The integrals are easy to do, because most of them vanish: Only those cross-terms with $\boldsymbol{k}_1 = \boldsymbol{k}_2$, and hence involving $\mathrm{e}^{\mathrm{i}\boldsymbol{k}_1\cdot\boldsymbol{r}}\mathrm{e}^{-\mathrm{i}\boldsymbol{k}_1\cdot\boldsymbol{r}}$, survive. Moreover, we have $\boldsymbol{\varepsilon}_{\boldsymbol{k}}^{(\alpha)}\cdot\boldsymbol{\varepsilon}_{\boldsymbol{k}}^{(\beta)} = \delta_{\alpha\beta}$, leaving

$$
\begin{aligned}
\mathcal{E} &= \frac{\epsilon_0 L^3}{4} {\sum_{\boldsymbol{k},\alpha}}' \left(|\dot{A}_{\boldsymbol{k},\alpha}|^2 + (ck)^2|A_{\boldsymbol{k},\alpha}|^2\right) \\
&= \tfrac{1}{2} {\sum_{\boldsymbol{k},\alpha}}' \left(\dot{X}_{\boldsymbol{k},\alpha}{}^2 + (ck)^2 X_{\boldsymbol{k},\alpha}{}^2 + \dot{Y}_{\boldsymbol{k},\alpha}{}^2 + (ck)^2 Y_{\boldsymbol{k},\alpha}{}^2\right).
\end{aligned}
\tag{13.8}
$$

The field momentum is given by a similar calculation, starting with the Poynting vector:

$$
\boldsymbol{P} = \epsilon_0 \int \mathrm{d}^3\boldsymbol{r}\, \boldsymbol{E} \times \boldsymbol{B}
\tag{13.9}
$$

$$
\begin{aligned}
&= \epsilon_0 {\sum_{\boldsymbol{k}_1,\alpha}}' {\sum_{\boldsymbol{k}_2,\beta}}' \int \mathrm{d}^3\boldsymbol{r}\, \tfrac{1}{2}\big(-\dot{A}_{\boldsymbol{k}_1,\alpha}\boldsymbol{\varepsilon}_{\boldsymbol{k}_1}^{(\alpha)}\mathrm{e}^{\mathrm{i}\boldsymbol{k}_1\cdot\boldsymbol{r}} + \text{c.c.}\big) \times \left(\boldsymbol{\nabla} \times \tfrac{1}{2}\big(A_{\boldsymbol{k}_2,\beta}\boldsymbol{\varepsilon}_{\boldsymbol{k}_2}^{(\beta)}\mathrm{e}^{\mathrm{i}\boldsymbol{k}_2\cdot\boldsymbol{r}} + \text{c.c.}\big)\right) \\
&= -\frac{\epsilon_0 L^3}{4} {\sum_{\boldsymbol{k},\alpha}}' \sum_{\beta} \big(\dot{A}_{\boldsymbol{k},\alpha}A_{\boldsymbol{k},\beta}^* \boldsymbol{\varepsilon}_{\boldsymbol{k}}^{(\alpha)} \times (-\mathrm{i}\boldsymbol{k} \times \boldsymbol{\varepsilon}_{\boldsymbol{k}}^{(\beta)}) + \text{c.c.}\big) \\
&= \frac{\epsilon_0 L^3}{4} {\sum_{\boldsymbol{k},\alpha}}' \big(\mathrm{i}\boldsymbol{k}\dot{A}_{\boldsymbol{k},\alpha}A_{\boldsymbol{k},\alpha}^* + \text{c.c.}\big) \\
&= \frac{1}{2} {\sum_{\boldsymbol{k},\alpha}}' \boldsymbol{k}\big((\mathrm{i}\dot{X}_{\boldsymbol{k},\alpha} - \dot{Y}_{\boldsymbol{k},\alpha})(X_{\boldsymbol{k},\alpha} - \mathrm{i}Y_{\boldsymbol{k},\alpha}) + \text{c.c.}\big) \\
&= {\sum_{\boldsymbol{k},\alpha}}' \boldsymbol{k}\big(\dot{X}_{\boldsymbol{k},\alpha}Y_{\boldsymbol{k},\alpha} - \dot{Y}_{\boldsymbol{k},\alpha}X_{\boldsymbol{k},\alpha}\big).
\end{aligned}
\tag{13.10}
$$

Section 13.3 has obtained compact formulas for the energy and momentum of the electromagnetic field in terms of the harmonic-oscillator representation (Equation 13.8 and 13.10). The interpretation is that every mode of the field, labeled by $\boldsymbol{k}$ and α, makes an independent contribution to $\mathcal{E}$, and to each component of $\boldsymbol{P}$. Note, however, that the momentum gets mixed contributions from the X and Y oscillators. We will soon remove this remaining inconvenience.

13.4 QUANTIZATION REPLACES FIELD VARIABLES BY OPERATORS

Finding the quantum-mechanical version of a harmonic oscillator is a standard problem, whose solution we can now employ for each of our many independent oscillators. One approach to quantization is via the path-integral method, similarly to the discussion in Chapter 12, but for this particular system, we'll see that a shortcut makes explicit solution unnecessary. The shortcut involves another change of variables. To motivate the required change, we will break it down into four steps. It is worthwhile to verify each of the steps, which are straightforward if a bit tedious; ultimately the goal is to replace the X and Y variables by a set of quantum operators called Q and their Hermitian conjugates (Equation 13.22). Note that this book uses different typefaces to distinguish quantum operators from their corresponding classical dynamical variables.[3]

Step 1: Quantize

For brevity, at first consider only one pair of modes X and Y, that is, only a particular $\boldsymbol{k}, \alpha$. We introduce two Hermitian operators[4] X and U, with the property that their commutator is $[\mathsf{X}, \mathsf{U}] = i\hbar$. In the energy function, Equation 13.8, we substitute $X \to \mathsf{X}$ and $\dot{X} \to \mathsf{U}$ to obtain the Hamiltonian operator for X:

$$\mathsf{H}_X = \tfrac{1}{2}\big(\mathsf{U}^2 + (ck)^2 \mathsf{X}^2\big). \tag{13.11}$$

This operator both represents the energy of a quantum state and also determines its time evolution. For example, the time evolution of $\big|\Psi(t)\big\rangle$ is given by $\exp(-i\mathsf{H}_X t/\hbar)\big|\Psi\big\rangle$.[5] It implies that

$$\frac{\mathrm{d}^2}{\mathrm{d}t^2}\big\langle\Psi_1\big|\mathsf{X}\big|\Psi_2\big\rangle = \frac{\mathrm{d}}{\mathrm{d}t}\big\langle\Psi_1\big|\tfrac{i}{\hbar}[\mathsf{H}_X,\mathsf{X}]\big|\Psi_2\big\rangle = \frac{\mathrm{d}}{\mathrm{d}t}\big\langle\Psi_1\big|\mathsf{U}\big|\Psi_2\big\rangle = \big\langle\Psi_1\big|\tfrac{i}{\hbar}[\mathsf{H}_X,\mathsf{U}]\big|\Psi_2\big\rangle$$
$$= -(ck)^2\big\langle\Psi_1\big|\mathsf{X}\big|\Psi_2\big\rangle, \tag{13.12}$$

which is the desired equation of motion for a harmonic oscillator (Equation 13.4).

We proceed in the same way with the other oscillator family, introducing operators Y and V analogous to X and U. Then the operator corresponding to $A_{\boldsymbol{k},\alpha}$ in Equation 13.5 is

$$\mathsf{A} = (\epsilon_0 L^3/2)^{-1/2}(\mathsf{X} + i\mathsf{Y}). \tag{13.13}$$

Step 2: Diagonalize energy

We could now finish constructing the state space, for example, by writing and solving a set of decoupled Schrödinger equations for each pair of operators (X, U) and (Y, V). However, the harmonic oscillator problem has an elegant reformulation that simplifies the math. Change variables once again by defining new operators

$$\mathsf{S} = (2\hbar ck)^{-1/2}(ck\mathsf{X} + i\mathsf{U}) \text{ and } \mathsf{R} = (2\hbar ck)^{-1/2}(ck\mathsf{Y} + i\mathsf{V}). \tag{13.14}$$

Then it is straightforward to verify that

$$[\mathsf{S}, \mathsf{S}^\dagger] = 1, \qquad [\mathsf{R}, \mathsf{R}^\dagger] = 1, \qquad [\mathsf{S}, \mathsf{R}] = [\mathsf{S}, \mathsf{R}^\dagger] = 0, \tag{13.15}$$

$$\mathsf{H} = \mathsf{H}_X + \mathsf{H}_Y = \hbar ck(\mathsf{S}^\dagger\mathsf{S} + \mathsf{R}^\dagger\mathsf{R} + 1), \text{ and} \tag{13.16}$$

$$\mathbf{P} = i\hbar\boldsymbol{k}(\mathsf{S}^\dagger\mathsf{R} - \text{h.c.}). \tag{13.17}$$

---

[3]Section A.1 outlines these conventions.

[4]In the analogy to a harmonic oscillator, these represent the position and momentum respectively, but in electrodynamics they have no direct connection to physical position $\boldsymbol{r}$ or field momentum $\boldsymbol{P}$.

[5]See Section 12.2.5 (page 385).

In the last formula, "h.c." denotes the Hermitian conjugate, that is, $R^\dagger S$.

Step 3: Diagonalize momentum

The Hamiltonian operator has the nice property that S and R make independent, additive contributions to it. The momentum operator still mixes S and R, but we can diagonalize it, without spoiling H, by a unitary transformation. Define two new **lowering operators** by

$$Q = (S + iR)/\sqrt{2}, \qquad \widetilde{Q} = (S - iR)/\sqrt{2}. \qquad (13.18)$$

Your Turn 13B

Show that

$$[Q, Q^\dagger] = 1, \qquad [\widetilde{Q}, \widetilde{Q}^\dagger] = 1, \qquad [Q, \widetilde{Q}] = [Q, \widetilde{Q}^\dagger] = 0, \qquad (13.19)$$

$$H = \hbar c k (Q^\dagger Q + \widetilde{Q}^\dagger \widetilde{Q} + 1), \text{ and} \qquad (13.20)$$
$$\boldsymbol{P} = \hbar \boldsymbol{k} (Q^\dagger Q - \widetilde{Q}^\dagger \widetilde{Q}). \qquad (13.21)$$

We now have new field operators Q and $\widetilde{Q}$ that, unlike S and R, enter independently into both the field energy and momentum.

Step 4: Relabel

We now reinstate the mode indices $\boldsymbol{k}$ and α. Until now, all mode sums were over a half-space of discrete $\boldsymbol{k}$ values, but now we can simplify the notation: Define operators for *all* nonzero $\boldsymbol{k}$ by renaming $\widetilde{Q}_{\boldsymbol{k},\alpha}$ as $Q_{-\boldsymbol{k},\alpha}$. Then

$$[Q_{\boldsymbol{k}_1,\alpha}, Q^\dagger_{\boldsymbol{k}_2,\beta}] = \delta_{\alpha\beta}\delta_{\boldsymbol{k}_1,\boldsymbol{k}_2}, \quad [Q_{\boldsymbol{k}_1,\alpha}, Q_{\boldsymbol{k}_2,\beta}] = 0, \quad \text{for all nonzero } \boldsymbol{k}_1 \text{ and } \boldsymbol{k}_2. \qquad (13.22)$$

Our final formulas then become unrestricted sums:

$$H = \sum_{\boldsymbol{k},\alpha} \hbar c k \big(Q^\dagger_{\boldsymbol{k},\alpha} Q_{\boldsymbol{k},\alpha} + \tfrac{1}{2} \big), \text{ and} \qquad (13.23)$$

$$\boldsymbol{P} = \sum_{\boldsymbol{k},\alpha} \hbar \boldsymbol{k} \big(Q^\dagger_{\boldsymbol{k},\alpha} Q_{\boldsymbol{k},\alpha} \big). \qquad (13.24)$$

Section 13.4 has constructed a set of operators in terms of which the energy and momentum of light will have simple interpretations.

13.5 PHOTON STATES

13.5.1 Basis states can be formed by applying creation operators to the vacuum state

Previous sections pointed out that although Maxwell's equations do not account for the particle-like character of light, nevertheless they account for phenomena not yet included in our model, such as the emission of light by a charged object. To address such phenomena we have started over, creating a model in which field-like *operators* obey Maxwell-like equations. We must now connect up this picture to the one we used in earlier chapters.

Your Turn 13C
Show that

$$[\mathsf{H}, \mathsf{Q}_{\boldsymbol{k},\alpha}] = -\hbar c k \mathsf{Q}_{\boldsymbol{k},\alpha} \quad \text{and} \quad [\mathbf{P}, \mathsf{Q}_{\boldsymbol{k},\alpha}] = -\hbar \boldsymbol{k} \mathsf{Q}_{\boldsymbol{k},\alpha}. \qquad (13.25)$$

Equations 13.25 justify the term "lowering operator":

> *Applying the lowering operator $\mathsf{Q}_{\boldsymbol{k},\alpha}$ to a state lowers its energy by $\hbar c k$, and changes its momentum by $-\hbar \boldsymbol{k}$. Conversely, applying the* **raising operator** $\mathsf{Q}_{\boldsymbol{k},\alpha}^{\dagger}$ *has the opposite effects.* (13.26)

Next, note that both of the terms in the classical electromagnetic energy function (Equation 13.7) are nonnegative. So it must not be possible to lower that energy indefinitely; there must be a state for which any lowering operator yields *zero*. We'll denote that **photon ground state** by the symbol $|0\rangle$. Any other state is obtained from this one by the actions of the various raising operators, each of which may be applied any number of times, always raising the energy by $\hbar c k$ and changing the momentum by $\hbar \boldsymbol{k}$. When a raising operator acts n times, we can obtain a normalized state as follows:[6]

$$|n_{\boldsymbol{k},\alpha}\rangle = \sqrt{\frac{1}{n!}} (\mathsf{Q}_{\boldsymbol{k},\alpha}^{\dagger})^{n} |0\rangle. \qquad (13.27)$$

More generally, we can define $|n_{\boldsymbol{k}_1,\alpha_1}; n_{\boldsymbol{k}_2,\alpha_2}, \ldots\rangle$ as a state obtained by applying several different raising operators to the ground state, each multiple times. States of this form with different sets of "occupation numbers" are all linearly independent and orthogonal. In fact,

> *The quantum states of light form a linear space spanned by basis vectors of this form, which act like states of noninteracting particles ("photons").* (13.28)

That is, each one-photon basis state is labeled by a wavevector and a polarization, and carries energy and momentum related by Equation 13.25:

$$E_{\boldsymbol{k},\alpha} = \hbar c k; \qquad \boldsymbol{p}_{\boldsymbol{k},\alpha} = \hbar \boldsymbol{k}; \qquad \text{so} \quad E_{\boldsymbol{k},\alpha} = c\|\boldsymbol{p}_{\boldsymbol{k},\alpha}\|, \qquad (13.29)$$

as claimed in the Einstein relation (Equation 1.6, page 34) and in Section 1.3.3′b (page 51). For multiphoton states, we add the corresponding quantities, just as we would do with any noninteracting particles.[7]

The interpretation of the quantum basis states as containing particles motivates another commonly used set of terms for the raising and lowering operators: Because they can be interpreted as raising and lowering the *number of photons* in a state, they are also called **creation and destruction operators**; $|0\rangle$ is also called the **vacuum state**.

---

[6]Section 12.2.4 (page 384) introduced the idea of normalization of states.

[7]The facts that the energy levels are discrete, and that each mode's state is characterized solely by its occupation numbers, agree with assumptions that were needed to obtain the thermal energy spectrum in Section 1.3.3′c (page 52). That is, two or more individual photons with the same wavevector and polarization must be regarded as in principle indistinguishable.

We may guess that these concepts will be key to understanding how a fluorescent molecule in its excited state can create photons from "nothing" (and how other processes can make photons disappear).

Equation 13.29 may be surprising, because planets and baseballs instead obey the relation $E = \frac{1}{2}mv^2 = vp/2$; substituting $v \to c$ gives a formula that differs by a factor of 2 from Equation 13.29. But there is precedent for our result, from classical electrodynamics. Comparing the pressure exerted by a beam of light (its momentum per time per area) to its energy per time per area shows that they are indeed related by a factor of c, not $c/2$. Textbooks on relativity theory reconcile the two conflicting formulas by showing that they are in fact two extreme limiting cases of a more general formula for objects moving at any speed.[8] Planets and baseballs correspond to the limit where $p \ll mc$, whereas photons correspond to the opposite limit, because $m = 0$ but $p \neq 0$.

> *Photons may be regarded as particles, but any photon's mass must be equal to zero, regardless of frequency.*

13.5.2 Coherent states mimic classical states in the limit of large occupation numbers

The states we have called "one-photon" are far from being classical. Indeed, no state with a definite number of photons can be an eigenvector of the field operators corresponding to the classical electric and magnetic field, because $\mathbf{A}(\mathbf{r})$ involves both raising and lowering operators:

Your Turn 13D

Use Equations 13.3, 13.13, 13.14, and 13.18 to show that

$$\mathbf{A}(\mathbf{r}) = \sum_{\mathbf{k},\alpha} \sqrt{\frac{\hbar}{2L^3 \epsilon_0 ck}}\, \boldsymbol{\varepsilon}_{\mathbf{k}}^{(\alpha)} \left(\mathsf{Q}_{\mathbf{k},\alpha}\, \mathrm{e}^{\mathrm{i}\mathbf{k}\cdot\mathbf{r}} + \text{h.c.} \right). \qquad (13.30)$$

However, we can find eigenvectors of $\mathsf{Q}_{\mathbf{k},\alpha}$, called **coherent states**: For any complex number $u_{\mathbf{k},\alpha}$, define

$$\left| u_{\mathbf{k},\alpha} \right\rangle = \exp(-\tfrac{1}{2}|u_{\mathbf{k},\alpha}|^2) \sum_{n=0}^{\infty} (n!)^{-1/2} (u_{\mathbf{k},\alpha})^n \left| n_{\mathbf{k},\alpha} \right\rangle. \qquad (13.31)$$

Your Turn 13E

a. Show that the states $\left| u_{\mathbf{k},\alpha} \right\rangle$ just defined are all properly normalized for any complex number $u_{\mathbf{k},\alpha}$.
b. Show that $\mathsf{Q}_{\mathbf{k},\alpha} \left| u_{\mathbf{k},\alpha} \right\rangle = u_{\mathbf{k},\alpha} \left| u_{\mathbf{k},\alpha} \right\rangle$, and hence also $\left\langle u_{\mathbf{k},\alpha} \right| \mathsf{Q}_{\mathbf{k},\alpha}^\dagger = u_{\mathbf{k},\alpha}^* \left\langle u_{\mathbf{k},\alpha} \right|$.
c. Then show that Equation 13.30 implies

$$\left\langle u_{\mathbf{k},\alpha} \right| \mathbf{A}(\mathbf{r}) \left| u_{\mathbf{k},\alpha} \right\rangle = (2L^3 \epsilon_0 ck/\hbar)^{-1/2} \boldsymbol{\varepsilon}_{\mathbf{k}}^{(\alpha)} u_{\mathbf{k},\alpha}\, \mathrm{e}^{\mathrm{i}\mathbf{k}\cdot\mathbf{r}} + \text{c.c.}$$

Your results show that the coherent state based on a particular wavevector and polarization is the quantum analog of a classical single-mode state (Equation 13.3, page

---

[8] Section 1.3.3′b (page 51) gave the general formula.

400). Moreover, as the amplitude $|u|$ becomes large (and hence also the expectation of the photon number), the relative standard deviation of the electric field in this state goes to zero, leading to classical behavior. In this limit, the coherent states correspond to classical states of the electromagnetic field, for example the radiation emitted by a radio broadcast antenna.[9]

Your Turn 13F

The coherent states are superpositions of states with different numbers of photons. Find the length-squared of the individual terms of Equation 13.31 to get the probabilities of getting exactly ℓ photons in a measurement on that state. Is this a distribution you have seen previously?

Section 13.5 has established contact between the field quantization procedure in this chapter, the particle picture from earlier chapters, and Maxwell's original classical fields.

13.6 INTERACTION WITH ELECTRONS

13.6.1 Classical interactions involve adding source terms to the field equations

If we wish to study the creation of light by a molecule, then we must acknowledge that the light field interacts with that molecule's electrons. In the presence of charged matter, we can no longer find a gauge transformation that eliminates the scalar potential Φ, though we can still impose $\boldsymbol{\nabla} \cdot \boldsymbol{A} = 0$. Gauss's law then says

$$\boldsymbol{\nabla} \cdot \boldsymbol{E} = -\nabla^2 \Phi = \rho_{\mathrm{q}}/\epsilon_0, \tag{13.32}$$

where ρ_{q} is the charge density. This formula looks just like the corresponding equation in electrostatics, and it leads to the usual potential that binds the molecule's electrons to its nuclei.

Ampère's law also involves charges, via the electric current density $\boldsymbol{j}(t, \boldsymbol{r})$:[10]

$$\boldsymbol{\nabla} \times \boldsymbol{B} = \mu_0 \boldsymbol{j} + \mu_0 \epsilon_0 \frac{\mathrm{d}}{\mathrm{d}t} \boldsymbol{E}. \tag{13.33}$$

Casting everything into plane wave mode expansions as before gives the full Maxwell equations as

$$k^2 \Phi_{\boldsymbol{k}} = \frac{1}{\epsilon_0} \rho_{\mathrm{q},\boldsymbol{k}} \quad \text{and} \tag{13.34}$$

$$\frac{\mathrm{d}^2}{\mathrm{d}t^2} \boldsymbol{A}_{\boldsymbol{k}} + (ck)^2 \boldsymbol{A}_{\boldsymbol{k}} = -\mathrm{i}\boldsymbol{k} \frac{\mathrm{d}\Phi_{\boldsymbol{k}}}{\mathrm{d}t} + \frac{1}{\epsilon_0} \boldsymbol{j}_{\boldsymbol{k}}, \tag{13.35}$$

where $c = (\mu_0 \epsilon_0)^{-1/2}$ and $\Phi_{\boldsymbol{k}}, \rho_{\mathrm{q},\boldsymbol{k}}$, and $\boldsymbol{j}_{\boldsymbol{k}}$ are the plane-wave components of Φ, ρ_{q}, and $\boldsymbol{j}$, respectively. We now take the dot product of both sides of Equation 13.35 with the

---

[9]Books on quantum optics show that the light created by a single-mode laser, operated well above threshold, is also a coherent state (Loudon, 2000, chapt. 7).

[10]In this formula, μ_0 is a constant of Nature called the "permeability of vacuum." It is related to ϵ_0 by $\mu_0 \epsilon_0 = 1/c^2$.

two transverse basis vectors $\varepsilon_{\boldsymbol{k}}^{(\alpha)}$ to find the desired generalization of Equation 13.4:

$$\frac{\mathrm{d}^2}{\mathrm{d}t^2} A_{\boldsymbol{k},\alpha} = -(ck)^2 A_{\boldsymbol{k},\alpha} + \frac{1}{\epsilon_0} \boldsymbol{j}_{\boldsymbol{k}} \cdot \varepsilon_{\boldsymbol{k}}^{(\alpha)} \quad \text{for each } \boldsymbol{k}, \alpha. \tag{13.36}$$

The scalar potential Φ has dropped out of this equation of motion.

13.6.2 Electromagnetic interactions can be treated perturbatively

There is no need to quantize the scalar potential Φ, because Equation 13.32 shows that in Coulomb gauge, it is not an independent dynamical variable: It just tracks whatever the charge density is doing.

The last term of Equation 13.36 describes the interaction of the vector potential with charges. To discuss the radiation of a molecule, we treat this term as a perturbation. That is, we set up an "unperturbed" Hamiltonian operator describing the quantum mechanics of the electrons making up the molecule, with their Coulomb attraction to the nuclei mediated by the scalar potential Φ as usual. There is another term describing the free electromagnetic field (Equation 13.23). To these terms we then add the perturbation

$$-\int \mathrm{d}^3 \boldsymbol{r}\, \mathbf{j}(\boldsymbol{r}) \cdot \mathbf{A}(\boldsymbol{r}), \tag{13.37}$$

where $\mathbf{j}(\boldsymbol{r})$ is the operator version of the current density and $\mathbf{A}(\boldsymbol{r})$ is given by Equation 13.30. This term modifies the quantum equations of motion, introducing the last part of Equation 13.36.

Each electron in the atom or molecule of interest contributes a delta function to $\mathbf{j}$ that is localized at the electron's position $\boldsymbol{r}_{\mathrm{e}}$, with strength equal to its charge, $-e$, times its velocity, $\mathbf{p}_{\mathrm{e}}/m_{\mathrm{e}}$. Thus, each electron makes a contribution to the integral in Equation 13.37 equal to

$$-\sum_{\boldsymbol{k},\alpha} \sqrt{\frac{\hbar}{2L^3 \epsilon_0 ck}} \varepsilon_{\boldsymbol{k}}^{(\alpha)} \cdot (-e\mathbf{p}_{\mathrm{e}}/m_{\mathrm{e}})\left(\mathsf{Q}_{\boldsymbol{k},\alpha} \mathrm{e}^{\mathrm{i}\boldsymbol{k}\cdot\boldsymbol{r}_{\mathrm{e}}} + \mathrm{h.c.}\right). \tag{13.38}$$

The effect of this perturbation is to allow transitions between eigenstates of the unperturbed Hamiltonian operator, that is, between states that would be stationary were it not for the perturbation term. For example, the transitions that interest us are those from a molecule with initially excited electron state and no photons present, to a deexcited electron state and one photon present. To find the probability per unit time that this transition will occur, we need to compute the modulus squared of Equation 13.38 sandwiched between the molecule's initial and final states.[11] The Hermitian conjugate term, involving $\mathsf{Q}_{\boldsymbol{k},\alpha}^{\dagger}$, can create the photon, so we want the matrix element of the remaining factors of this term sandwiched between the molecular states.

To make progress, notice that for transitions in the visible spectrum, $k \approx 10^{-2}\,\mathrm{nm}^{-1}$. But r_{e} cannot exceed the size of the atom or molecule, typically $\approx 1\,\mathrm{nm}$, so $\boldsymbol{k} \cdot \boldsymbol{r}_{\mathrm{e}}$ is a small dimensionless number. Accordingly, we will approximate $\exp(\mathrm{i}\boldsymbol{k} \cdot \boldsymbol{r}_{\mathrm{e}})$ by its leading-order Taylor series term, which is 1.

---

[11] Quantum mechanics textbooks call this scheme the "Golden Rule" of time-dependent perturbation theory.

13.6.3 The dipole emission pattern

We now ask for the probability that the emitted photon will be observed to be traveling in a particular direction with a particular energy and polarization. Dropping overall constant factors, the answer is proportional to

$$\left| \left\langle \text{ground}; \boldsymbol{k}, \alpha \right| \sum_{\boldsymbol{k}',\beta} \mathsf{Q}^{\dagger}_{\boldsymbol{k}',\beta} \varepsilon^{(\beta)}_{\boldsymbol{k}'} \cdot \mathbf{p}_{\text{e}} \left| \text{excited} \right\rangle \right|^2$$

$$= \left| \left\langle \text{ground} \right| \mathbf{p}_{\text{e}} \left| \text{excited} \right\rangle \cdot \varepsilon^{(\alpha)}_{\boldsymbol{k}} \right|^2. \tag{13.39}$$

One further transformation helps to clarify the meaning of this quantity. The electron momentum operator, whose matrix element we need, can be rephrased in terms of the electron *position* operator, as the commutator

$$[\mathsf{H}_{\text{e}}, \mathbf{r}_{\text{e}}] = \frac{-\mathrm{i}\hbar}{m} \mathbf{p}_{\text{e}}.$$

Sandwich this relation between the ground and excited states to find

$$\left\langle \text{ground} \right| (E_0 \mathbf{r}_{\text{e}} - \mathbf{r}_{\text{e}} E_*) \left| \text{excited} \right\rangle = \frac{-\mathrm{i}\hbar}{m} \left\langle \text{ground} \right| \mathbf{p}_{\text{e}} \left| \text{excited} \right\rangle.$$

The right-hand side of this formula is a constant times the quantity needed in Equation 13.39. The left-hand side is can be written in terms of the electric dipole moment operator, $\mathbf{d} = -e\mathbf{r}_{\text{e}}$, so we find that the probability of photon emission involves the matrix element of the dipole moment, a vector called the molecule's **transition dipole**. This is encouraging news: In classical electrodynamics the rate of energy radiation is also proportional to the dipole moment squared.

If the molecular states are such that the transition dipole is nonzero, then we can choose a coordinate system in which it points along the z axis:

$$\left\langle \text{ground} \right| \mathbf{d} \left| \text{excited} \right\rangle = D_{\text{e}} \hat{\boldsymbol{z}}. \tag{13.40}$$

Suppose that, as is the case in many experiments, we record every photon received regardless of its polarization. The sum of Equation 13.39 over α includes the factor

$$\sum_{\alpha} \hat{\boldsymbol{z}} \cdot \varepsilon^{(\alpha)}_{\boldsymbol{k}} \varepsilon^{(\alpha)}_{\boldsymbol{k}} \cdot \hat{\boldsymbol{z}}. \tag{13.41}$$

We can simplify this expression by realizing that it involves the *projection* of $\hat{\boldsymbol{z}}$ onto the plane perpendicular to $\boldsymbol{k}$. Another expression for that projection operator is $1 - \hat{\boldsymbol{k}}\hat{\boldsymbol{k}}$, so we get

$$\hat{\boldsymbol{z}} \cdot (1 - \hat{\boldsymbol{k}}\hat{\boldsymbol{k}}) \cdot \hat{\boldsymbol{z}} = \hat{\boldsymbol{z}} \cdot \hat{\boldsymbol{z}} - (\hat{\boldsymbol{z}} \cdot \hat{\boldsymbol{k}})^2 = 1 - \cos^2 \theta = \sin^2 \theta, \tag{13.42}$$

where θ is the polar angle between the direction of observation, $\hat{\boldsymbol{k}}$, and the transition dipole.

Equation 13.42 shows that the probability density function for the angles at which photons are emitted has a "doughnut" or **dipole pattern**: No photons are emitted along $\pm\hat{\boldsymbol{z}}$; instead, they are preferentially emitted in an equatorial belt, the circle

$\theta = \pi/2$. A similar argument shows that the probability to *absorb* light also follows a dipole pattern.

The mean rate at which photons are emitted is determined by the transition dipole D_e defined by Equation 13.40, which itself is essentially the matrix element of the molecule's electric dipole moment operator.

> *If the matrix element of the dipole moment operator is nonzero, then the dominant mechanism of energy loss by a molecule is the one just described, with its characteristic angular distribution $\wp(\theta, \phi) \propto \sin^2 \theta$.*   (13.43)

Section 13.6 has resolved the puzzle posed at the start of this chapter: The pattern of photon emission observed in defocused orientation imaging (Figure 7.4b) agrees with the dipole radiation pattern in classical electrodynamics because the same angular factors enter each calculation.

Fig. 7.4b (page 260)

13.6.4 Electrons and positrons can also be created and destroyed

Chapter 12 announced the goal of unifying electrons and photons, and yet the present chapter has treated them quite differently: Photons were treated as excitations of a quantized field, subject to creation and destruction, whereas each electron was treated as a permanent entity and assigned its own position and momentum operator.

Section 1.5.1′a (page 54) mentioned, however, that some kinds of radioactivity involve ejecting an electron (or its antiparticle the positron) from an atomic nucleus. This observation puzzled scientists at first—where was the electron just prior to its emission? According to the Uncertainty Principle, confining an electron to a region the size of the nucleus would come with a prohibitive energy cost. Enrico Fermi broke this impasse by creating a theory in which electrons and positrons were also assigned creation and destruction operators, paralleling the situation with photons. Thus, the electron or positron *did not exist* prior to emission from the nucleus. Field quantization also explains how an electron and positron can mutually annihilate, the key process underlying positron emission tomography. In fact, *all* fundamental particles can be created and destroyed.

Electrons do differ from photons in some key respects. For example, photons are massless, and hence can be created from arbitrarily small amounts of energy. In contrast, an electron must carry a minimal amount of energy regardless of its momentum (Equation 1.18, page 52). This minimum energy is available in nuclear reactions, but is about a million times larger than the energy scale characteristic of chemical reactions. Thus, the creation and destruction of electrons can be neglected when we discuss fluorescence. Proceeding in that way obscures the unity of electrons and photons, but it also leads to simpler formulas.

13.7 VISTAS

13.7.1 Connection to the approach used in earlier chapters

Instead of computing the probability that a photon will emerge from a molecule in some direction, we can apply the same approach to compute the probability amplitude for the process where a photon is created at some particular $(t_e, \boldsymbol{r}_e)$ and then absorbed at a different $(t_a, \boldsymbol{r}_a)$. The answer agrees with Equation 12.29 (page 394), establishing the connection between the quantized-field approach and the one used elsewhere in this book.

13.7.2 Some invertebrates can detect the polarization of light

Many species of insects, including honeybees, ants, crickets, flies, and beetles, have the ability to detect and act on the polarization of light.[12] The first firm evidence for this sense was obtained by K. von Frisch in his studies of honeybees in 1948. Von Frisch knew that when returning from a successful foraging trip, a worker bee uses the location of the Sun in the sky to determine its own orientation. With this information, the bee can effectively integrate its instantaneous velocity to get an overall vector indicating the displacement to the source of food. Upon its return to the hive, the bee must communicate this information to others, via a "dance."

The dance includes a segment of straight walking with a tail-wagging motion, indicating direction to the food by the direction of this straight segment.[13] Because the bee only knows the direction relative to that of the Sun, it must again determine the Sun's location before it can know which way to walk, and the others watching it must in turn remember that direction relative to the Sun if they are to follow that course. But remarkably, von Frisch found that the returning bee could successfully communicate even when he blocked the view of the Sun at the hive: As long as a small patch of blue sky was visible, communication was accurate.[14] Von Frisch discussed the phenomenon with physicist H. Benndorf, who pointed out that the polarization pattern of the blue sky is related to the location of the Sun.

Von Frisch therefore hypothesized that bees could discern and act on the polarization of light. To test his hypothesis, he filtered the sky light visible to the bees through polarizers, modifying this one aspect of the bee's environment while holding everything else unchanged. When the polarizer's axis aligned with that of the sky's polarization, then it had no effect other than to enhance the degree of polarization, and the bee's dance was unaltered. When the polarizer was rotated, however, altering the apparent direction of polarization of the blue sky, then the bee's dance changed, inaccurately reporting the location of the food.

13.7.3 Invertebrate photoreceptors have a different morphology from vertebrates'

Before describing how a polarization sense is possible, let's first see why most vertebrates *don't* have it. Photoreceptor cells in the vertebrate eye contain stacks of membrane layers oriented perpendicular to the incoming light (Figures 6.6b, 10.1c). Embedded rhodopsin molecules hold their retinal cofactors parallel to the membrane (Figure 10.5a), but within that plane the orientation is random. Thus, each chromophore's transition dipole also points randomly in the plane perpendicular to the incoming photons' direction of motion. Regardless of whether the incoming photons are polarized, their polarization vectors make random, Uniformly distributed angles with the transition dipoles they encounter. Thus, although the photon's probability to be absorbed by any particular chromophore depends on polarization, the *overall* absorption probability does not.[15]

Insect and crustacean eyes have a different morphology (Figure 13.1). For example, bees have compound eyes each consisting of about 5000 individual units called

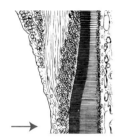

Fig. 6.6b (page 219)

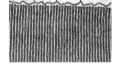

Fig. 10.1c (page 319)

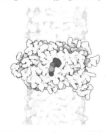

Fig. 10.5a (page 328)

---

[12]Many other invertebrates, including spiders, crustaceans, and cephalopods, have this ability as well.

[13]Distance to the food is encoded in other characteristics of the dance such as tempo. To repeat the dance, the bee must loop around to the start of the straight segment; it does this loop without the tail-wagging motion, which indicates to others that this segment of the dance is to be ignored.

[14]If no blue sky was visible, either by design or on an overcast day, then communication failed.

[15]Actually some fish, for example the northern anchovy *Engraulis mordax,* have cone cells with layers oriented parallel to the incoming light, enabling polarization vision (Horváth, 2014, Chapt. 9).

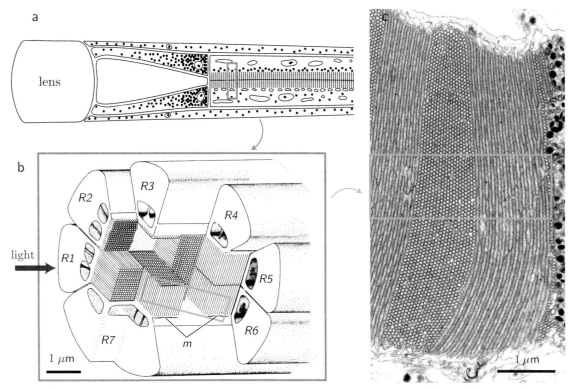

Figure 13.1: [Sketches; electron micrograph.] **Invertebrate photoreceptors.** (a) Cross-section through one facet (ommatidium) of a generic insect eye. Light passes through the lens (*left*) to the rhabdom (*right*). (b) Enlargement of the *orange box* in (a). Several long photoreceptor cells (*R1–R7*) run lengthwise along the ommatidium (in this example, from a crab). Each has many parallel, hairlike projections (microvilli, *m*) that together form the rhabdom along the central axis of the ommatidium. Each photoreceptor cell has many microvilli all oriented parallel to each other, but differently from those of neighboring cells. (c) Enlargement of the *green box* in (b). Electron micrograph of the microvilli confirming this arrangement in the mantis shrimp *Gonodactylus oerstedii*. Alternating layers of microvilli are seen either end-on or sideways. [(b) From Stowe, 1980. (c) Courtesy Christina A King-Smith.]

ommatidia, about 150 of which are specialized for polarization vision (those in the "dorsal rim area"). Each ommatidium has its own rudimentary lens serving several photoreceptor cells. Each photoreceptor is long, in order to present many light-sensitive molecules to incoming photons that traverse its length; each guides light down its length much like an optical fiber.[16] Unlike the stack of disks bearing rhodopsin in the vertebrate photoreceptor, however, these **rhabdomeric** receptor cells embed their chromophores in an array of parallel, tubular projections called the **rhabdomere**. Each of these tubes (**microvilli**) is oriented with its long axis perpendicular to the incoming light; each microvillus in turn carries chromophores with their transition dipoles predominantly parallel to its axis.[17]

The discussion in Section 13.6.2 averaged light emission probability over all polarizations. Had we not taken this step, we would have found that the probability to emit (or absorb) light depends on the light's polarization relative to the molecule's

---

[16]One way to understand this guidance is via total internal reflection (Section 5.3.4, page 194).

[17]The light-sensitive pigment is different from the ones found in vertebrate photoreceptor cells; see Section 11.5.1′ (page 373). However, the embedded cofactor, retinal, is the same.

transition dipole. Thus, the effect of the insect's ommatidial arrangement is that *each photoreceptor cell has an overall preference for catching photons of a particular polarization.* By comparing the outputs of different photoreceptors viewing the same patch of sky through the same lens, the bee can determine the orientation of polarization relative to that of its head, and from that information deduce its orientation relative to the Sun.

13.7.4 Polarized light must be used for single photoreceptor measurements

Section 9.4 described how, in their measurements on single photoreceptor cells, Baylor and coauthors presented light to the cell from the side. Of the two allowed polarizations (perpendicular to the direction of propagation), one lay in the plane of the membrane and therefore made random, Uniformly distributed angles with the chromophores' transition dipole. The other polarization, however, was directed perpendicular to the membrane, and so *always* made a right angle to every transition dipole. The experimenters eliminated the second polarization, that is, they used light polarized in the plane of the membrane to stimulate their photoreceptor cells, to ensure that the quantum catch they found would be related in a simple way to that of photoreceptors presented with light traveling axially (Section 9.4.2′b, page 313).

13.7.5 Some transitions are far more probable than others

Section 13.6 focused on the relative mean rates to emit photons in different directions. To find the absolute rates, we need various other factors provided by the "Golden Rule" of time-dependent perturbation theory. The derivation of the rule also shows why energy must be conserved in photon emission and absorption, or more precisely, it must be conserved to within a tolerance set by the uncertainty relation.

For simplicity, we chose to expand the vector potential A in a basis of linearly polarized, plane wave states. Other bases may be better adapted to the problem at hand, for example, a basis of circularly polarized plane waves. Also, a basis of outgoing *spherical* waves, centered on the emitting object, is better suited to study light emitted by a very small object and traveling out to infinity. That basis can be chosen such that each element carries definite angular momentum away from the emitter. When we do this, we find that certain kinds of photons cannot be emitted at all by certain kinds of transitions, because doing so would violate the conservation of angular momentum. Other transitions appear impossible when we make the approximation $\exp(\mathrm{i}\boldsymbol{k}\cdot\boldsymbol{r}_{\mathrm{e}}) \approx 1$, as was done above, but not when we retain higher terms in the Taylor series. Such transitions are called "forbidden," but more precisely their rates are just suppressed by powers of the small factor $(kr_{\mathrm{e}})^2$.

The statement that some transitions are "forbidden" is an example of a **selection rule**. Another class of selection rules arises from considerations of electron spin in multi-electron atoms or molecules. It is possible for a molecule to get trapped in an excited state, from which transitions to the ground state are suppressed by a spin selection rule. Such an excited state can eventually make its transition, but with mean rate far slower than most fluorescence transitions, leading to the phenomenon of **phosphorescence** (ultra-slow fluorescence). Spin selection rules also ensure very slow exit from the dark states of some fluorophores, which is useful for localization microscopy (Section 7.4, page 256).

13.7.6 Lasers exploit a preference for emission into an already occupied state

Earlier sections restricted attention to the case in which a photon is emitted into a world originally containing *no* photons. Although photons do not interact in the usual sense of colliding, nevertheless a very important new phenomenon arises when we consider *adding* a photon to a state that is already occupied. If a mode initially contains n photons, Equation 13.27 (page 404) implies

$$\langle n+1|Q^\dagger|n\rangle = \langle 0|\frac{1}{\sqrt{(n+1)!}}Q^{n+1}(Q^\dagger)^{n+1}\frac{1}{\sqrt{n!}}|0\rangle = \langle 0|\sqrt{\frac{(n+1)!}{n!}}|0\rangle = \sqrt{n+1}.$$

This factor gets squared when it enters into the rate for photon emission into this mode. Because this matrix element depends on n, we conclude that

> When an atom or molecule emits a photon, it preferentially chooses a mode that is already occupied. (13.44)

If we have a population of many excited atoms or molecules, then this result implies that there can be an avalanche-type effect, in which one particular mode gets the vast majority of all emitted photons. This mechanism for obtaining nearly single-mode light is called l̲ight a̲mplification by s̲timulated e̲mission of r̲adiation—the **laser**.

13.7.7 Fluorescence polarization anisotropy

The preceding sections focused mostly on deexcitation by photon emission. The inverse process (excitation via photon absorption) is also probabilistic, and also depends on the orientation of the molecule, polarization of the incoming light, and transition dipole. That is, certain combinations of molecular orientation and light polarization make excitation more probable. If we illuminate a collection of molecules with fixed but random orientations, using polarized light, then only the subset with favorable orientation will become excited. When those molecules in turn fluoresce, the distribution of directions and polarizations of the outgoing light will reflect the nonrandom character of their orientations, giving a **fluorescence polarization anisotropy**.

Fluorescence polarization anisotropy becomes more interesting when the molecules are *not* immobilized: As before, there is preferential excitation in the subpopulation whose momentary orientation is favorable. But later, when the molecules fluoresce, their orientations will have partly or completely changed, degrading the polarization signature. Thus, measuring the time-dependent loss of fluorescence polarization anisotropy gives a measurement of the angular *mobility* of the fluorophores in question, and hence also of any larger macromolecules to which they are attached.

An analogous technique is even more powerful when applied to single molecules. The ability of a single molecule to absorb photons of specific incoming polarizations, and then emit photons of specific outgoing polarizations, can be measured. Although each absorption and each emission is probabilistic, nevertheless a likelihood function[18] can be constructed for the overall probability as a function of molecular orientation. Maximizing this likelihood then gives a real-time determination of the orientation, as long as enough photons can be collected between successive conformational changes. This p̲olarization method can be combined with t̲otal i̲nternal r̲eflection f̲luorescence excitation (to reduce background noise), yielding a technique called "polTIRF."

---

[18]Likelihood maximization was introduced in Section 7.2 (page 248).

THE BIG PICTURE

Returning to the questions posed in Section 13.1,

- Creation (emission) and destruction (absorption): The mechanism is that the time evolution operator contains raising and lowering (creation and destruction) operators coupled to electron degrees of freedom.
- Polarization is an attribute of the classical electromagnetic *field* that carries over to the quantum states.
- Light/electricity/magnetism: In our Coulomb gauge formulation, there is a contribution to the electric field equal to minus the gradient of the electric potential Φ created by charged particles, as usual (Equations 13.1 and 13.32). Macroscopic magnetic fields are less simple, as they involve coherent states, but in the nonrelativistic world of atoms and molecules electrostatic interactions usually dominate the magnetic ones.
- Multiphoton effects: More complex interactions, including light-with-light, can occur by compounding multiple simple ones. Such interactions lie outside our scope, however, as they do not occur at lowest order in the perturbation theory that we have used here.

We have had to work hard for our results, but as often happens in physics, they are far more general than the problem that motivated them. For example, the same framework that we used to discuss visible light emission from molecules also applies to gamma-ray emission from atomic nuclei. Suitably generalized, it even describes the creation and destruction of electrons and positrons.

FURTHER READING

Semipopular:
Walmsley, 2015.

Intermediate:
For a concise introduction to classical electrodynamics in SI units, see, for example, Fleisch, 2008. For more details, see, for example, Garg, 2012.
Quantum mechanics and the radiation field: Feynman et al., 2010c, chapt. 9.
Specifically on the quantum theory of light: Lipson et al., 2011; Leonhardt, 2010; Loudon, 2000.
Radiation; forbidden transitions: van der Straten & Metcalf, 2016.
Use of light to interrogate biomacromolecules: van Holde et al., 2006; Cantor & Schimmel, 1980.
Polarization vision in insects: Cronin et al., 2014; Johnsen, 2012, chapt. 8; Smith, 2007; Warrant & Nilsson, 2006.

Technical:
General: Berman & Malinovsky, 2011; Mandel & Wolf, 1995.
Defocused orientation imaging: Toprak et al., 2006; Böhmer & Enderlein, 2003; Bartko & Dickson, 1999a; Bartko & Dickson, 1999b.
Polarization vision in invertebrates: Homberg & el Jundhi, 2014; Horváth, 2014; Sweeney et al., 2003; Wehner, 2003.
Polarized total internal reflection microscopy for the determination of molecular motions: Forkey et al., 2005; Rosenberg et al., 2005.

CHAPTER 14

Quantum-Mechanical Theory of FRET

> Magic is imagination working together with dexterity to persuade experience how limited its experience really is, the heart working with the fingers to remind the head how little it knows.
>
> — *Adam Gopnik*

14.1 SIGNPOST: *DECOHERENCE*

Chapter 2 outlined the phenomenon of FRET (the process depicted by horizontal dashed lines in Figure 2.21) without detailed justification. This short chapter gives a quantum-mechanical calculation.

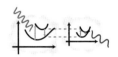

Fig. 2.21 (page 86)

The analysis requires concepts not developed in this book, including the traditional states-and-operators formulation of quantum physics and the density operator, which is discussed in many textbooks.[1] Working through the calculations also points out some physical approximations that are made along the way. The same approach can be adapted to handle other situations, when those approximations are not accurate. The Focus Question is

Biological question: Why do FRET, and related energy transfers like the ones in photosynthesis, obey first-order kinetics? (Other radiationless two-state transitions, such as the ammonia molecule's inversions, obey a different rule.)

Physical idea: This and other features arise from quantum *decoherence,* which is strong in the parameter regime appropriate for FRET.

14.2 TWO-STATE SYSTEMS

14.2.1 FRET displays both classical and quantum aspects

Section 2.8.1 described some remarkable features of FRET, but lurking behind some of these features is a deeper surprise. You were asked to imagine the excitation *state* of a fluorophore as a discrete *thing* that could be transferred intact, as basketball players pass the ball. The discussion never mentioned the possibility of quantum-mechanical superposition states, in which an excitation is delocalized (simultaneously located on two different fluorophores). Also, we assumed that the transfer could be described by a fixed probability per unit time.[2] Other kinds of transitions do not behave this way (see Section 14.2.2 below).

We are accustomed to classical behavior from macroscopic objects (micrometer scale and larger). After studying quantum mechanics, we become accustomed to

---

[1] See, for example, Schumacher & Westmoreland, 2010, chapt. 3.
[2] That is, the transfer follows a "rate equation" with "first-order kinetics."

quantum behavior from microscopic objects (such as a single hydrogen atom, with diameter about 0.1 nm). In between these extremes, we might expect intermediate behavior for mesoscopic objects (like individual fluorophores). The surprise is that FRET seems to involve fluorophores simultaneously displaying *strongly* quantum behavior (discrete energy levels), but also *strongly* classical behavior (no superpositions, localized excitations, rate equations). *How could anything like that possibly happen at all?*

As we address such questions, we will not discuss the process that excites the donor, nor the eventual fluorescence of the acceptor, instead concentrating on the transfer of the excitation from one to the other. We will also make some simplifying assumptions:

- We suppose that only two electronic states of the donor are relevant: the ground state $|D_0\rangle$ and one excited state $|D_\star\rangle$. Similarly, we consider only two acceptor states $|A_0\rangle$ and $|A_\star\rangle$. We are particularly interested in transitions between joint states of the form

$$|1\rangle = |D_\star A_0\rangle, \quad |2\rangle = |D_0 A_\star\rangle, \tag{14.1}$$

  whose energies are nearly equal (the vertical arrows joined by dashed lines in Figure 2.21 are of nearly equal length). Direct transitions between those two states, without any photon emission, are at least compatible with energy conservation.
- Later, we will define a "decoherence time" T, and assume that it is much shorter than the hopping time ($T \ll \Omega^{-1}$ below). We will also assume that T is much shorter than the mean waiting time before loss processes other than FRET deexcite the donor ($T \ll \tau$ below).

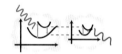

Fig. 2.21 (page 86)

14.2.2 An isolated two-state system oscillates in time

The transition between the states in Equation 14.1 would be easy to describe in a world containing only two atoms. Suppose for a moment that the two states of interest have exactly the same electronic-state energy in isolation (they are exactly resonant). Make the convenient convention that these energy values are $E_1 = E_2 = 0$. When the two atoms are brought near each other, they will have a coupling giving rise to a Hamiltonian operator with an off-diagonal entry in the $1, 2$ basis. Suppose that that entry, V, is real:

$$\mathsf{H} = \begin{pmatrix} 0 & V \\ V & 0 \end{pmatrix}. \tag{14.2}$$

The system's evolving state can then be expanded as

$$|\Psi(t)\rangle = a(t)|1\rangle + b(t)|2\rangle, \tag{14.3}$$

where the coefficient functions obey the Schrödinger equation:[3]

$$i\hbar \begin{bmatrix} da/dt \\ db/dt \end{bmatrix} = V \begin{bmatrix} b \\ a \end{bmatrix}.$$

We can now find a solution with the initial state $|\Psi(0)\rangle = |1\rangle$; at later times, we find that $|b(t)|^2 = \sin^2(\Omega t/2)$, where $\Omega = 2V/\hbar$. Interpreting this quantity as the

---

[3]To obtain this formula, substitute Equation 14.2 for the operator on the right-hand side of Equation 12.21 (page 391).

probability to find the system in state *2*, we conclude that the probability initially increases with time as t^2. But this means that the *initial* growth rate of the probability is zero, contrary to the observed first-order kinetics. Moreover, at almost every time the state is a quantum superposition of $|1\rangle$ and $|2\rangle$, in contrast to the "basketball" picture of resonance energy transfer alluded to above. Finally, the solution just found is oscillatory: The system periodically reverts to being completely in state *1*, in contrast to the one-way transfer characteristic of FRET.

14.2.3 Environmental effects modify the behavior of a two-state system in solution

To see where we have gone astray, we must remember that our two fluorophores are hardly alone: They are just a *subsystem* of the entire world. Each constantly suffers collisions with surrounding water molecules, as well as less obvious influences involving fluctuating electric fields in its neighborhood and so on. A good fluorophore is robust to such disturbances, in the sense that they are not strong enough to knock it into a different electronic state. Nevertheless, environmental influences can affect the quantum-mechanical *phase* of a fluorophore's state, by momentarily perturbing its energy levels during each collision. The colliding particle in turn also undergoes a phase change; we say that the system of interest rapidly becomes **entangled** (quantum-mechanically correlated) with its environment.

Entanglement sounds like a disaster for our goal of predicting the behavior of a s̲ubsystem 𝕤 embedded in an e̲nvironment 𝕖 consisting of dense matter. But as often happens in physics, the enormous complexity of the environment can lead to a simplifying net effect on the subsystem of interest.

Suppose that we can separate the total state space into the product of a two-dimensional state space $\mathcal{H}_\mathfrak{s}$ for the subsystem and a large space $\mathcal{H}_\mathfrak{e}$ for the environment. Then some states of the complete system can be written as simple products $|\psi\rangle_\mathfrak{s} \otimes |\phi\rangle_\mathfrak{e}$; we call these **pure states**. Most states, however, cannot be represented in this way—they are entangled.[4]

14.2.4 The density operator summarizes the effect of the environment

We are interested in state measurements that could in principle be made internally to the subsystem. That is, we wish to study observables O that act only on $\mathcal{H}_\mathfrak{s}$. Given a pure state, we can express the measured value of such an observable without needing to know anything about the environment:

$$\langle \mathsf{O} \rangle = {}_\mathfrak{s}\langle \psi | \mathsf{O} | \psi \rangle_\mathfrak{s} \qquad \text{for a pure state } |\Psi\rangle = |\psi\rangle_\mathfrak{s} \otimes |\phi\rangle_\mathfrak{e}. \tag{14.4}$$

(To get this formula, use the fact that $|\phi\rangle_\mathfrak{e}$ is normalized.)

Unfortunately, even if we could prepare a pure initial state, it would quickly evolve into an entangled state due to the interactions between 𝕤 and 𝕖. However, we can still compactly summarize the effect of the environment on the measured values of observables that, like O, refer only to the subsystem. To do this, we introduce a Hermitian operator on $\mathcal{H}_\mathfrak{s}$ called the **density operator** ρ, defined by constructing the

---

[4]An entangled state can, however, be expressed as a *sum* of pure states.

dyad[5] $|\Psi\rangle\langle\Psi|$ and taking the trace over the environment state space:

$$\rho = \mathrm{Tr}_{\mathfrak{e}}\left(|\Psi\rangle\langle\Psi|\right). \tag{14.5}$$

In our problem, ρ can be represented by a two-dimensional matrix with respect to the basis $|1\rangle$, $|2\rangle$. If we know ρ, then the measured value of any subsystem observable can be expressed as

$$\langle\mathsf{O}\rangle = \mathrm{Tr}_{\mathfrak{s}}\left(\rho\mathsf{O}\right) \qquad \text{for any state, represented by } \rho. \tag{14.6}$$

In order for this formulation to be useful, we need to be able to compute ρ, at least approximately. This is not difficult when $\mathfrak{s}$ is perfectly isolated from its environment, because in that case a pure (unentangled) state remains pure:

$$\left|\Psi(t)\right\rangle = \left|\psi(t)\right\rangle_{\mathfrak{s}} \otimes \left|\phi(t)\right\rangle_{\mathfrak{e}} \qquad \text{for isolated subsystem.} \tag{14.7}$$

Here $\left|\psi(t)\right\rangle_{\mathfrak{s}}$ denotes the time development of the subsystem under its Hamiltonian, independent of that of the environment, $\left|\phi(t)\right\rangle_{\mathfrak{e}}$.

Your Turn 14A

a. Show that Equation 14.6 reduces to the more familiar form Equation 14.4 in the case of an isolated subsystem.
b. Use the Schrödinger equation to show that in this case, the time development of ρ is the solution to

$$\frac{d\rho}{dt} = \frac{1}{i\hbar}[\mathsf{H}_{\mathfrak{s}}, \rho]. \qquad \text{for isolated subsystem} \tag{14.8}$$

Still considering the isolated case, notice that Equations 14.3, 14.5, and 14.7 give

$$\rho(t) = \left|\psi(t)\right\rangle_{\mathfrak{s}\,\mathfrak{s}}\langle\psi(t)| \quad \text{so} \quad \rho_{ij} = \begin{pmatrix} |a(t)|^2 & a(t)b(t)^* \\ a(t)^*b(t) & |b(t)|^2 \end{pmatrix}_{ij}. \tag{14.9}$$

This formula shows that the diagonal elements of ρ reflect the respective probabilities to be in the two states. Unlike the off-diagonal elements, they are unaffected if we change basis states to new versions differing by phases from the old ones, for example, $|1'\rangle = e^{i\theta}|1\rangle$.

14.2.5 Time development of the density operator

As mentioned earlier, interactions with the environment $\mathfrak{e}$ will destroy the simple form of Equation 14.7, converting an initially pure state to one that is entangled with the environment. Although these interactions are complicated, Section 14.2.3 above suggested that they could be summarized by saying that the subsystem's *phase* is altered by the many environmental particles that interact with it. When we perform the trace operation in Equation 14.5, the entanglement leads to the sum of many

---

[5]Some authors call this construction the "outer product." The analogous construction in ordinary 3-space takes a column vector $\boldsymbol{v}$ and builds the symmetric 3×3 matrix $\boldsymbol{v}\boldsymbol{v}^{\mathrm{t}}$, whose ij entry is the product $v_i v_j$. The trace of this matrix is just the dot product, or $\|\boldsymbol{v}\|^2$. The extension to the complex state space of quantum mechanics yields the Hermitian matrix in Equation 14.5.

random phase factors in the off-diagonal elements of ρ, effectively suppressing them within some **decoherence time scale** T (Schlosshauer, 2007, chap. 3). The diagonal terms are unaffected, however.

We must also extend the simplified discussion above (Equation 14.2) by allowing for the possibility that the energies of $|1\rangle$ and $|2\rangle$ may not be exactly equal. Thus, let $\mathsf{H} = \mathsf{H}_0 + \mathsf{V}$, where H_0 is diagonal with eigenvalues E_1 and E_2 and V is the off-diagonal interaction operator appearing in Equation 14.2. Equation 14.8 then becomes

$$\frac{d\rho_{22}}{dt} = \frac{1}{i\hbar}[\mathsf{V}, \rho]_{22} \qquad (14.10)$$

$$\frac{d\rho_{ij}}{dt} = \frac{1}{i\hbar}\left([\mathsf{V}, \rho]_{ij} + (E_i - E_j)\rho_{ij}\right) - \frac{1}{T}\rho_{ij} \quad \text{for } i \neq j. \qquad (14.11)$$

The environment enters via the last term above, which would simply lead to exponential decay if the other terms were equal to zero. This term contains the decoherence time scale T.

Another effect of the environment is that it can directly mediate deexcitation of the donor, without transfer of energy to the acceptor. We approximate this effect as a decay term in the equation for ρ_{11}:[6]

$$\frac{d\rho_{11}}{dt} = \frac{1}{i\hbar}[\mathsf{V}, \rho]_{11} - \frac{1}{\tau}\rho_{11}. \qquad (14.12)$$

Section 14.2 began by pointing out that an isolated two-state system's behavior conflicts with what we observe in FRET, then upgraded our description to account for the relevant effects of environment.

14.3 FRET

14.3.1 The weakly coupled, strongly incoherent limit displays first-order kinetics

T. Förster studied the situation in which the decoherence rate, $1/T$, is much faster than either the transition rate, $\Omega = 2V/\hbar$, or the donor deexcitation rate, $1/\tau$ (the "fast decoherence" limit).[7] In this case, we may find the relevant solution by perturbation in Ω, as follows.

Let $S = (E_1 - E_2)/\hbar$. Change variables to the four real quantities $U = \rho_{11}$, $W = \rho_{22}$, $X = (\rho_{12} - \rho_{21})/i$, and $Y = \rho_{12} + \rho_{21}$. Then the dynamical equations take the real form

$$\begin{aligned}
dU/dt &= -\tfrac{1}{2}\Omega X - U/\tau \\
dW/dt &= \tfrac{1}{2}\Omega X \\
dX/dt &= \Omega(U - W) - X/T - SY \\
dY/dt &= -Y/T + SX.
\end{aligned} \qquad (14.13)$$

This is a set of coupled linear differential equations with constant coefficients, so its solutions will combinations of exponentials.

---

[6]Equation 14.10 neglects the analogous effect. That is, we are simplifying by also assuming that the decay rate for acceptor fluorescence is much smaller than that for donor fluorescence.

[7]Typical numbers for chromophores in solution are $1/T \approx 10^{14}\,\mathrm{s}^{-1}$ (Gilmore & McKenzie, 2008), compared with $\tau^{-1} \approx \Omega \approx 10^8\,\mathrm{s}^{-1}$.

Let $\boldsymbol{Z}(t)$ be the 4-component vector with entries $U(t)$, $W(t)$, $X(t)$, and $Y(t)$, so that Equation 14.13 can be written symbolically as $\mathrm{d}\boldsymbol{Z}/\mathrm{d}t = \mathsf{M}\boldsymbol{Z}$, where M is a 4×4 matrix whose entries don't depend on time. When the coupling $\Omega = 0$, we easily find one solution:

$$\boldsymbol{Z}_0(t) = \begin{bmatrix} U(t) \\ W(t) \\ X(t) \\ Y(t) \end{bmatrix} = \mathrm{e}^{-\beta_0 t} \boldsymbol{B}_0 \quad \text{where} \quad \boldsymbol{B}_0 = \begin{bmatrix} 1 \\ 0 \\ 0 \\ 0 \end{bmatrix} \quad \text{and} \quad \beta_0 = 1/\tau. \tag{14.14}$$

This solution describes spontaneous deexcitation of the donor, for example, via fluorescence.

At small but nonzero Ω, we expand all quantities in powers of $\epsilon = (T\Omega)$, for example, writing the dynamical equations (Equations 14.13) as $\mathrm{d}\boldsymbol{Z}/\mathrm{d}t = (\mathsf{M}_0 + \epsilon\mathsf{M}')\boldsymbol{Z}$, and again seek a trial solution that is exponentially changing in time. Expanding the eigenvalue β as $\beta_0 + \epsilon\beta' + \epsilon^2\beta'' + \cdots$, we can use standard perturbation theory to find $\beta' = 0$. At the next order, however,

$$\beta'' = \tfrac{1}{2} \frac{T^{-1}}{1 + (TS)^2}. \tag{14.15}$$

Altogether, we again find exponential decay for the initial state population, that is, first-order kinetics, this time with rate given by

$$\beta = \tau^{-1} + \frac{\Omega^2 T/2}{1 + (TS)^2} = \tau^{-1} + \frac{2V^2 T}{\hbar^2 + T^2(E_1 - E_2)^2}. \tag{14.16}$$

Like the unperturbed solution, the full equation has $\rho_{11}(t)$ falling exponentially with time (first-order kinetics), but with a new contribution to its rate constant β. The extra contribution reflects the rate of the excitation transfer $|1\rangle \to |2\rangle$, which is what we sought. Equation 14.16 shows that this rate is peaked as a function of the energy difference.[8]

Your Turn 14B

Derive Equation 14.15 in the fast-decoherence (small ϵ) limit. [*Hints:* Expand the eigenvector as $\boldsymbol{B}_0 + \epsilon\boldsymbol{B}' + \cdots$. Equation 14.14 gives $\boldsymbol{B}_0$. You will need to find $\boldsymbol{B}'$ as an intermediate step to finding β''.]

14.3.2 Förster's formula arises in the electric dipole approximation

To apply our result to FRET, we need to know that the interaction energy of two electric dipoles is proportional to the product of their electric dipole moments $\mathbf{d}_\mathrm{D}$ and $\mathbf{d}_\mathrm{A}$, and to the inverse cube of the distance between them.[9] Thus, V in Equation 14.16 is proportional to $r^{-3}\langle 2|\mathbf{d}_\mathrm{D} \cdot \mathbf{d}_\mathrm{A}|1\rangle$.

---

[8] In fact, it resembles the Cauchy distribution; see Section 1.6.2′b (page 57). This "resonant" behavior is similar to that of the energy transfer rate in the analogous classical derivation; see Problem 2.4.
[9] We need not account for the quantum nature of electromagnetic interactions, because Chapter 13 showed that, for nonrelativistic systems like molecules, those interactions divide into two parts. One part of the interaction is equivalent to classical electrostatics and gives rise to the dipole-dipole interaction discussed here. The other part involves photon emission and absorption; although those processes are important for the donor's initial excitation and the acceptor's eventual emission, they are not involved with the transfer studied here.

From this point, we can follow the reasoning of Problem 2.4, introducing a distribution of excited-donor energies, and a distribution of ground-state acceptor energies. The sharply peaked form of Equation 14.16 implies that the mean FRET rate will be proportional to the overlap integral of the two distributions; the V^2 dependence implies that it will be proportional to r^{-6}. The main features of FRET (Idea 2.4, page 89) then follow from these two facts. In fact, in the stated limit Förster was able to find a prediction with *no free parameters* for the FRET rate, in terms of the donor's measured emission spectrum and fluorescence rate, the acceptor's measured excitation spectrum and fluorescence cross section, the medium's index of refraction, and the distance and relative orientation between donor and acceptor.

In particular, the derivation just outlined explains the most surprising aspect of FRET, which is that there can be highly specific energy transfer between two particular molecular species, despite the multitude of other directions into which the donor could instead emit a photon, and the crush of other molecules that could instead receive the energy:

- To understand the dominance of FRET over photon emission, note that the "near fields" of a fluctuating dipole fall off with distance as r^{-3}, independent of its frequency. The "radiation fields" fall off more slowly, as r^{-1}, and they do depend on frequency. Turning these statements around, at *small* distances the near fields are stronger by a factor of $(\lambda/r)^2$, where λ is the wavelength of light corresponding to donor fluorescence. The square of this ratio can exceed 10^4.
- Turning to the other molecules, the strongly peaked form of Equation 14.16 ensures that only those with a transition resonant with the donor's emission will have significant probability per unit time to gain energy from it.

Finally, the derivation can also be modified to account for other transfer processes, such as the electron transfers that follow photon capture in photosynthesis.

14.3.3 More realistic treatment of the role of FRET in photosynthesis

Oppenheimer and Arnold made some approximations in their analysis of photosynthetic energy transfer—the same approximations as the ones made above.[10] Although these approximations are often excellent for FRET used as a lab technique, later work has shown that they, and the physical picture that they support, are not fully accurate in photosynthesis.

First, we assumed that the interaction between two molecules can be approximated as a dipole-dipole interaction. For the tightly spaced photosynthetic chromophores, this approximation is not always valid and a more detailed model must be used.

Second, the derivation of Equation 14.16 assumed that excitation transfer is much slower than quantum decoherence. In that situation, we got a simple rate law for excitation transfer, and a nice picture of localized excitations. But there is a hierarchy of substructures in the photosynthetic apparatus, and within some of them the transfers are extremely fast. In this situation, it makes more sense to regard a whole array of chromophores as a single "supermolecule," with delocalized excitations called "Frenkel excitons." The supramolecular units in turn transfer excitons among themselves via FRET-like processes. For more about intrinsically quantum behavior in these systems, see Strümpfer et al., 2012; Şener et al., 2011; Engel, 2011.

---

[10]Section 2.9 (page 93) gave the context for this work.

Section 14.3 has obtained the key features of FRET by applying the master equations, Equations 14.10–14.12, in an appropriate limiting case.

THE BIG PICTURE

We have seen in a quantitative framework how FRET connects the quantum world to classical behavior (excitations localized on single molecules and making one-way transitions with first-order kinetics). Quantum *decoherence* played a key role in the mechanism.

Once again, our hard work is repaid by the generality of the results. The mechanism of FRET was originally worked out by Oppenheimer in the context of a nuclear physics problem,[11] but was then applied to a wealth of problems related both to biological function (photosynthesis) and to instrumentation. Moreover, a similar framework is also relevant for other one-way transfer interactions involving not only energy but also electrons.

FURTHER READING

Semipopular:
Clegg, 2006.

Intermediate:
Isolated two-state systems are analyzed in Feynman et al., 2010a, chaps. 7–9. The modifications in this chapter to account for the environment follow the treatment in Agranovich & Galanin, 1982, chapt. 1.

Technical:
Jang, 2007.

---

[11]Section 2.9.3 (page 96) told this story.

Epilogue

For the rest of my life I want to reflect on what light is!
— *Attributed to Albert Einstein, around 1917*

The shortest summary

This book has explored the interplay between three areas of inquiry:

$$probability \leftrightarrow quantum\ theory \leftrightarrow imaging\ and\ vision.$$

We had to stretch our imaginations even to find an acceptable physical model for light. Only with the right model in hand could we begin to reconcile what happens in the front of our eyes (focusing, the diffraction limit, and so on) with what happens in the back (photoisomerization and the transduction cascade). That is, understanding some of Life's tricks forced us to abandon a hugely successful physical model, which scientists thought was the true theory of light up until the 20th century. The agility to discard one model, and then find a better one, is an important scientific skill.

More generally, we have seen how miraculous some of Life's mechanisms appear, even if we took them for granted before we looked carefully. Then we worked through some details of how they are implemented by steps consistent with natural law.

Follow the rabbit

The job of scientists is to see to what extent we can understand the world without positing supernatural interventions. Sometimes such insights can lead to improvements in health, sustainability, or some other big goal. The skills and frameworks emphasized in this book can help you with this search.

For example, at various points we called on concepts such as dimensional analysis, probability density functions, likelihood maximization, and biochemical reaction networks; on skills like numerical computer math and statistical inference; and on whole areas of science and medicine such as quantum optics, psychophysics, ophthalmology, and neuroscience. In science, you must run where the rabbit runs, not down the well-worn tracks that respect the boundaries of existing disciplines. I hope that the experience has left you with the impression that all of science is deeply interconnected. I personally find that inspiring.

Models

In science, we have multiple ways to describe and manipulate ideas. Each kind of discourse supports the others in a process called "modeling." Each also supplies its own kind of imagination, especially connections to seemingly unrelated phenomena. One example, illustrated in the boxes below, involves the idea that the stationary-phase approximation to a class of oscillatory integrals could underpin the reconciliation of the wave and particle aspects of light (Chapter 4):

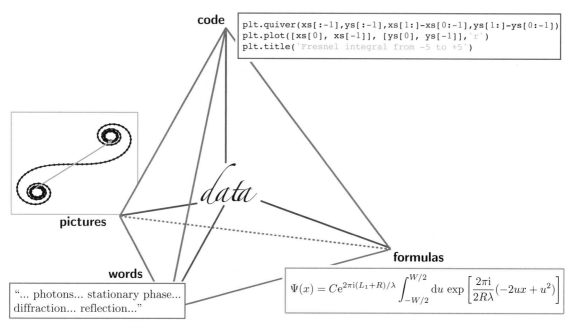

code

```
plt.quiver(xs[:-1],ys[:-1],xs[1:]-xs[0:-1],ys[1:]-ys[0:-1])
plt.plot([xs[0], xs[-1]], [ys[0], ys[-1]],'r')
plt.title('Fresnel integral from -5 to +5')
```

pictures

data

words

"... photons... stationary phase... diffraction... reflection..."

formulas

$$\Psi(x) = C e^{2\pi i (L_1 + R)/\lambda} \int_{-W/2}^{W/2} du \, \exp\left[\frac{2\pi i}{2R\lambda}(-2ux + u^2)\right]$$

We have been especially interested in *physical* models. These generally involve a few classes of actors, following a few rules, rooted in other, nonbiological phenomena. We use math to tease out the falsifiable, quantitative consequences of the hypothesis that those actors and rules generate the phenomena of interest. For example, Chapter 11 explored the idea that over one hundred billion identical rhodopsin molecules participate in a mechanism for seeing faint flashes of light, and we constructed a model that made detailed predictions about psychophysical experiments.

It's remarkable that such a chain of reasoning can actually lead to *new knowledge*. Philosophers may quibble that that knowledge was really present in the existing experimental data, but over and over we see examples in which the lens of analysis does uncover insights not previously known to any human being. In fact, the key data are rarely available in advance; preliminary analysis also helps us *discover what experiment is needed* to probe the truth of a hypothesis. The art of framing promising hypotheses, and imagining feasible experiments to test them, is an important part of your training; seeing how it has worked in the past prepares you for the future.

Wonder

Some say that the quantitative study of Nature, while useful, erodes our sense of wonder. But in many cases, we need a considerable amount of quantitative understanding even to grasp what is truly remarkable. Appreciating the limited understanding achieved so far only deepens my sense of wonder. For example, if the first steps in sensing light now seem slightly less mysterious than they did before reading this book, that just means you're ready to appreciate the harder puzzle of adaptation to a hundred million-fold range of ambient illumination levels, when the output ganglion cells have a dynamic range of only about a hundred—or the much bigger mystery of how every individual animal spontaneously wires up its own central nervous system during development.

Beauty

Some of the ideas we have encountered in this book are things that many scientists describe as "beautiful." What does that mean? There are as many definitions as there are scientists, but I think many would agree that part of the answer is that a beautiful physical idea is *surprising yet inevitable;* it may also be *simple yet unexpectedly general.* The quantum ideas do have these characteristics.

But is an idea likely to be true *because* it seems beautiful? Surely not—to think so would be to anthropomorphize Nature. Rather, the role of beauty may simply be that a scientist who is moved by a beautiful idea will follow it to the ends of the Earth, without being overwhelmed by the many red herrings that seem to say the idea contradicts some aspect of reality, nor by the myriad distractions of everyday life.

Why did evolution install this imperative in our brains? Certainly humans are programmed to figure things out, and to make connections; the pleasure we get from using these skills may be reinforcement for a behavior that enhanced our survival in difficult times. We habituate, so we need novelty to keep getting that reinforcement. In science, this means that the most powerful jolts come from unexpected connections that nevertheless carry conviction—the quality called "surprising yet inevitable" earlier. We call that beauty, both in art and in science.

Is this a depressingly mundane take on "beauty?" No: It puts its pursuit squarely in the same frame as the rest of the human project. Our species is the one that is able to transcend biological evolution, via culture. Some of that project involves making the struggle for survival less central, finding ways to be healthier, better fed, and more harmonious. That's not mundane, even if evolution has arranged for us to feel good when we use our brains for that purpose.

Last

As I write, dawn sunlight filters through trees and onto my page. Some shadows form sharp images; others do not. A moment later, the Sun has shifted and the dance is gone, replaced by something else ordinary, subtle, but understandable. For millions of years, my species got no particular evolutionary fitness advantage from our latent capacity to understand shadows, and yet in the most recent, tiny blink of our history, we obtained such an understanding and parlayed it into a cabinet of wonders, which have changed our world in ways that their inventors could never have imagined. Now it's Your Turn to put another bead on the string.

Acknowledgments

Knowing what
thou knowest not
is in a sense
omniscience.

— Piet Hein

Many teachers have taught me by example how far one must go in order to grasp—and then transmit—an idea. Some were technically *my* own students; some I never met in person; but all of them were my teachers. I have tried to remember the passion, precision, and verve that they radiated (and demanded of me). Of the many names that belong here, I can mention only Sidney Coleman and my long-time collaborators Sarina Bromberg, Ann Hermundstad, and Jesse Kinder.

The only way I can write is first to tell a story, and then later figure out what I ought to have said instead. Many classes of Penn students have been on the receiving end of such exercises, and students in those classes have fired back weekly bulletins to me about things that were obscure, wrong, and not-even-wrong. Their enthusiastic collaboration has really been the mainspring driving this project to completion. I'm also grateful to Stephanie Palmer and Aravinthan Samuel, and their students, for bravely using draft versions of this book at Chicago and Harvard, and supplying feedback.

Kevin Chen, Yaakov Cohen, Ann Hermundstad, Jesse Kinder, and Sharareh Tavaddod read everything, offering suggestions from micro to macro, from visual to verbal to conceptual. John Briguglio, Edward Cox, Chris Fang-Yen, Ethan Fein, Heidi Hofer, Xavier Michalet, Natasha Mitchell, Rob Phillips, Jason Prentice, Brian Salzberg, Jim Sethna, Kristina Simmons, Vijay Singh, and Kees Storm, also read multiple chapters and made incisive comments.

Special thanks are due to Mark Goulian, Rob Smith, Peter Sterling, and Alison Sweeney, to whom I turned countless times for judgment calls and up-to-date information.

Some of the art in this book was imagined and then executed by Sarina Bromberg, David Goodsell, and Felice Macera, as well as many scientists mentioned below. That task included a great deal of independent scientific research by David Goodsell. Carl Goodrich also contributed wizardry with ray-tracing, and Steve Nelson with photo adjustments. In the course of a decades-long collaboration, William Berner invented some of the classroom demonstrations, perfected others, and brought it all home every time.

Many reviewers generously read and commented on the book's initial plan, including Larry Abbott, David Altman, Matthew Antonik, John Bechhoefer, André Brown, Peter Dayan, Rhonda Dzakpasu, Gaute Einevoll, Nigel Goldenfeld, Ido Golding, Ryan Gutenkunst, Steve Hagen, Gus Hart, Robert Hilborn, K. C. Huang, Greg Huber, John Karkheck, Maria Kilfoil, Jan Kmetko, Alex Levine, Ponzy Lu, Anotida Madzvamuse, Jens-Christian Meiners, Ethan Minot, Simon Mochrie, Liviu Movileanu,

Daniel Needleman, Ilya Nemenman, Kerstin Nordstrom, Julio de Paula, Erwin J. G. Peterman, Rob Phillips, Thomas Powers, Thorsten Ritz, Rob de Ruyter, Aravinthan Samuel, Ronen Segev, Anirvan Sengupta, Sima Setayeshgar, John Stamm, Yujie Sun, Dan Tranchina, Joe Tranquillo, Clare Waterman, Joshua Weitz, Ned Wingreen, Eugene Wong, Jianghua Xing, Haw Yang, Daniel Zuckerman, as well as other, anonymous, referees. Several of these people also gave suggestions on drafts of the book. I couldn't include all the great suggestions for topics that I got—at least, not this time!—but still, many topics that were new to me did end up in these pages.

Many colleagues read sections, answered questions, supplied images and data, discussed their own and others' work, and more, including Ariel Amir, Robert Anderson, Jessica Anna, Clay Armstrong, Vijay Balasubramanian, Horace Barlow, Steven Bates, Denis Baylor, John Beausang, Kevin Belfield, Hubert van den Bergh, William Bialek, Craig Bohren, Nancy Bonini, Ed Boyden, David Brainard, J. Helmut Brandstätter, André Brown, Marie Burns, Lorenzo Cangiano, Sean Carroll, William Catterall, Mojca Čepič, Constance Cepko, Brian Chow, Adam Cohen, M. Fevzi Daldal, Todorka Dimitrova, Jacques Distler, Tom Dodson, Kristian Donner, Marija Drndić, Doug Durian, Greg Field, Luke Fritzky, Yuval Garini, Clayton Gearhart, Yale Goldman, Takejip Ha, Steve Haddock, Xue Han, Paul Heiney, Heidi Hofer, Zhenli Huang, David Jameson, Seogjoo Jang, Dan Janzen, Na Ji, Patrice Jichlinski, Sönke Johnsen, Randall Kamien, Pakorn Kanchanawong, Charles Kane, Christina King-Smith, Shuichi Kinoshita, Darren Koenig, Helga Kolb, Leonid Krivitsky, Joseph Lakowicz, Trevor Lamb, Robert Langridge, Simon Laughlin, Peter Lennie, Jennifer Lippincott-Schwartz, Amand Lucas, Walter Makous, Justin Marshall, Ross McKenzie, Xavier Michalet, Alan Middleton, John Mollon, Greg Moore, Jessica Morgan, Alexandre Morozov, Jeremy Nathans, Dan-Eric Nilsson, Axel Nimmerjahn, Silvania Pereira, Erwin Peterman, Shawn Pfiel, Joe Polchinski, Jason Porter, John Qi, Fred Rieke, Austin Roorda, Tiffany Schmidt, Julie Schnapf, David Schneeweis, Greg Scholes, Carl Schoonover, Udo Seifert, Paul Selvin, Gleb Shtengel, Glenn Smith, Hannah Smithson, Lubert Stryer, Nico Stuurman, Joseph Subotnik, Yujie Sun, Alison Sweeney, Lin Tian, Noga Vardi, Peter Vukusic, Georges Wagnières, Sam Wang, Sui Wang, Eric Warrant, Antoine Weis, Edith Widder, Bodo Wilts, Ned Wingreen, Ciceron Yanez, King-Wai Yau, Ahmet Yildiz, Shinya Yoshioka, Andrew Zangwill, and Xiaowei Zhuang.

In addition, my teaching assistants over the years have made countless suggestions, including solving, and even writing, many of the exercises: They are Isaac Carruthers, Tom Dodson, Stephen Hackler, Jan Homann, Asja Radja, and, ex officio, Jason Prentice.

Two opportunities to speak in unusual circumstances helped me to learn how to address more diverse audiences than the ones I find in the classroom. I'd like to thank the Aspen Center for Physics for an invitation to give its public lecture, and Fernando Moreno-Herrero for an invitation to speak at the Nicolás Cabrera Institute's summer school.

Every topic in this book has benefited from discussions with Scott Weinstein, always while walking together. Nily Dan had a strategic insight at every fork in the road. Larry Gladney unhesitatingly made the complex arrangements needed for a scholarly leave, and generally supported my efforts at every step. And when I hit the limits of my scholarly skills, David Giovacchini, Lauren Gala, and Melissa Flamson (With Permission Inc.) went beyond them.

The US National Science Foundation has steadfastly supported my ideas about education for many years (see page 471), particularly Krastan Blagoev. I regret that Kamal Shukla, who nurtured a whole generation of scientists, could not see the completion of this project; he will be deeply missed. At Penn, Dawn Bonnell and A. T. Johnson recognized this project and its precursor as activities of the Nano-Bio Interface Center, an NSF-supported center. The project also benefited from critically timed grants from Penn's University Research Fund, Research Opportunity Grant program, and Center for Teaching and Learning. Periodic visits to the Aspen Center for Physics, which is partially supported by NSF, were also indispensable for the most arduous revisions, as well as for exposing me to cutting-edge science. Karin Rabe was President of the Center during these years, and under her leadership it acted as a catalyst for rapid advances in many fields.

At Princeton University Press, Ingrid Gnerlich undertook this complex project under unusual circumstances. I'm grateful to her and to Karen Carter, Lorraine Doneker, Dimitri Karetnikov, and Arthur Werneck for their technical expertise, professionalism, and flexibility. Teresa Wilson (TDW Communications) once again provided the tenderly severe discipline that saved me (and you, reader) from uncountably many missteps.

Last

Oliver Nelson brought me photovoltaic cells to play with. He brought me manuscripts he was editing. His hands were on mine as I learned to use tools. That will never change.

Philip Nelson
Philadelphia, 2017

APPENDIX A

Global List of Symbols

> With a plethora
> Of words
> The would-be
> Explicator
> Hides himself
> Like a squid
> In his own ink.
>
> — *John M. Burns*

A.1 MATHEMATICAL NOTATION

Abbreviated words

corr correlation coefficient of two random variables (Equation 0.20, page 7).

cov covariance of two random variables (Equation 0.19, page 6).

var variance of a random variable (Equation 0.10, page 5).

$\operatorname{Re} Z$, $\operatorname{Im} Z$ real and imaginary parts of a complex number (Appendix D).

c.c. complex conjugate of preceding term(s) (Chapter 13).

h.c. Hermitian conjugate of preceding operator(s) (Chapter 13).

Operations

Both $\times$ and $\cdot$ denote ordinary multiplication when applied to numbers. When applied between vectors, they denote the scalar and cross products, respectively.

$\mathcal{P}_1 \star \mathcal{P}_2$ convolution of two distributions (Section 0.5.5, page 16).

Z^* complex conjugate of Z (Appendix D, page 449). However, a star can have other meanings (see below).

$\langle f \rangle$ expectation (Section 0.2.3, page 4). $\langle f \rangle_\alpha$, expectation in a family of distributions depending on a parameter α.

$\overline{f}$ sample mean of a random variable (Section 0.2.3, page 4). However, an overbar can have other meanings (see below).

$|Z|$ modulus (absolute value) of a complex number (Appendix D, page 449).

∇ gradient operator; $\nabla\cdot$ divergence operator; $\nabla\times$ curl operator (Equation 13.1, page 399).

Other modifiers

$\bar{c}$ dimensionless rescaled form of a variable c.

$\bar{\boldsymbol{y}}$ equilibrium value, about which a variable $\boldsymbol{y}$ fluctuates.

431

Δ is often used as a prefix: For example, Δx is a small, but finite, change of x. Sometimes this symbol is also used by itself, if the quantity being changed is clear from context.

The subscript 0 appended to a quantity can mean an initial value of a variable, or the center of a small range of specific values for a variable. Or it can indicate a variant of that quantity associated to a system's ground state, or an average value of a fluctuating variable in thermal equilibrium.

The subscript $*$ appended to a symbol can mean an optimal value (for example, the maximally likely value), an extreme value, or some other special value (for example, an inflection point or stationary-phase point).

The subscript $\star$ appended to a symbol can mean the variant associated with the excited state of a molecule (Section 1.6.3, page 42).

The superscript $*$ can indicate the complex conjugate of a number.

The superscript $\dagger$ indicates the Hermitian conjugate of an operator.

The superscript $\star$ can indicate the activated form of an enzyme or G protein (Section 10.4.2, page 329), or a quantity associated to a "target" light in a color-matching experiment (Section 3.8.7, page 127).

Vectors

Vectors are denoted by bold italic: $\boldsymbol{v}$ or by their components (v_x, v_y, v_z) (or just (v_x, v_y) if confined to a plane). The symbols $\boldsymbol{v}^2$ or $\|\boldsymbol{v}\|^2$, refer to the total length-squared of $\boldsymbol{v}$, that is, $(v_x)^2 + (v_y)^2 + (v_z)^2$. The symbol $\mathrm{d}^3\boldsymbol{r}$ is not a vector, but rather a volume element of integration.

The notation $|\Psi\rangle$ denotes a vector in a quantum-mechanical state space (Chapter 12). One representation of that state space consists of functions of position; the notation $\psi(\boldsymbol{r})$ denotes the function that represents $|\Psi\rangle$. The notation $\langle\Psi'|$ denotes a "dual" vector in state space, which can be combined with a vector to yield a single complex number, denoted $\langle\Psi'|\Psi\rangle$ (Section 12.2.4).

Matrices

A matrix is denoted by sans serif type (Section 3.8.8, page 127): $\mathsf{M} = \begin{bmatrix} M_{11} & M_{12} \\ M_{21} & M_{22} \end{bmatrix}$. Vectors can be regarded as matrices with either one row or one column. The transpose of a matrix or vector is denoted M^{t}.

Quantum operators

Operators on quantum state space are also denoted by sans serif (Chapters 12, 13, and 14), for example, Q. Vector-valued operators are denoted by bold sans serif (Chapter 13), for example, $\mathbf{P}$.

Relations

The symbol $\approx$ means "approximately equal to."
In the context of dimensional analysis, $\sim$ means "has the same dimensions as." In the context of color matching $\sim$ means "gives a perceptual match."
The symbol $\propto$ means "is proportional to."

Miscellaneous

The symbol $\left.\frac{dG}{dx}\right|_{x_0}$ refers to the derivative of G with respect to x, evaluated at the point $x = x_0$.

Inside a probability distribution, | is pronounced "given" (Section 0.2.2, page 4).

A.2 NETWORK DIAGRAMS

See Section 10.3.4, page 324, for example, Figure 10.3.

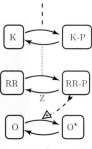

Fig. 10.3a (page 324)

- An incoming solid arrow represents production of a species, for example, by conversion from something else. Outgoing solid arrows represent loss mechanisms.

- If a process transforms one species to another, and both are of interest, then we draw a solid arrow joining the two species' boxes.

- But if a precursor species is not of interest to us, for example, because its inventory is maintained constant by some other mechanism, we can omit it, and similarly when the destruction of a particular species creates something not of interest to us. Thus, the phosphate groups that appear and disappear in Figure 10.3 come from ATP molecules in the cytosol, which are not shown. They leave as phosphate groups to be recycled by a mechanism that is also not shown.

- To describe how the population of one species affects another one's transformation, we draw a dashed "influence line" from the former to the latter's solid arrow, terminating with a symbol: A blunt end, ------∎, indicates suppression, whereas an open arrowhead, ----▷, indicates enhancement.

- A dotted line joining two processes indicates that they are coupled: For example, phosphate transfer simultaneously transforms one species from its phosphorylated to dephosphorylated state and has the opposite effect on another species.

A.3 NAMED QUANTITIES

The lists below act as a glossary for usage adopted in this book. Although symbolic names for quantities are in principle arbitrary, still it's convenient to use standard names for the ones that recur the most, and to use them as consistently as possible. But the limited number of letters in the Greek and Latin alphabets makes it inevitable that some letters must be used for more than one purpose. See Appendix B for explanation of the dimensions, and for a glossary of the corresponding units.

Latin alphabet

- a_1 absorption cross section for a single chromophore (Section 9.4.2′, page 311) [dimensions $\mathbb{L}^2$].

- $\boldsymbol{A}$ vector potential field (Equation 13.1, page 399) [dimensions $\mathbb{M}\mathbb{L}\mathbb{T}^{-1}\mathbb{Q}^{-1}$].

- **A**, **B**,... generic names for points in space, used in optics diagrams.

- $B_{i(\alpha)}$ coefficients in the color-matching equation (Equation 3.16, page 128); B, the same nine numbers regarded as a matrix.

- $\boldsymbol{B}$ magnetic field (Equation 13.1, page 399) [dimensions $\mathbb{M}\mathbb{T}^{-1}\mathbb{Q}^{-1}$].

- c speed of light [dimensions $\mathbb{L}\mathbb{T}^{-1}$].

c number density (**concentration**) of a small molecule, for example, a chromophore (Section 9.4.2′, page 311) [dimensions $\mathbb{L}^{-3}$].

d generic name for a distance, for example, a length in an optics diagram [dimensions $\mathbb{L}$]. d_{rod}, diameter of a single rod cell (Section 9.4.2′, page 311).

$\boldsymbol{d}$ electric dipole moment operator (Section 13.6.2, page 407) [dimensions $\mathbb{Q}\mathbb{L}$]. D_{e}, the length of its matrix element (the "transition dipole").

e charge on a proton, or minus the charge on an electron [dimensions $\mathbb{Q}$].

$\mathfrak{e}$ the environment surrounding a subsystem (Chapter 14) [not a quantity].

$E_{\mathrm{p}\lambda}$ spectral photon flux irradiance, expressed in terms of wavelength (Section 3.5′, page 133) [dimensions $\mathbb{L}^{-3}\mathbb{T}^{-1}$].

$\boldsymbol{E}$ electric field (Equation 13.1, page 399) [dimensions $\mathbb{M}\mathbb{L}\mathbb{T}^{-2}\mathbb{Q}$].

$\mathcal{E}_{\mathrm{FRET}}$ FRET efficiency (Equation 2.1, page 84) [dimensionless].

E generic name for an "event" in probability (Section 0.2, page 2) [not a quantity].

f generic name for a function.

f focal length of a lens or lens system (Equation 6.6, page 215) [dimensions $\mathbb{L}$].

f point spread function of an imaging system [dimensions depend on dimensionality].

$\mathcal{F}$ scaling function in the thermal radiation spectrum (Section B.4) [dimensions $\mathsf{kg\,m^{-1}\,s}$].

g electron propagator (Equation 12.4, page 384).

G photon propagator (Equation 12.28, page 394).

$\hbar$ reduced Planck constant (Sections B.4, page 442 and 1.3.3, page 33) [dimensions $\mathbb{M}\mathbb{L}^2\mathbb{T}^{-1}$].

H Hamiltonian operator (Chapter 14) [dimensions $\mathbb{M}\mathbb{L}^2\mathbb{T}^{-2}$].

i generic name for an integer quantity, for example, a subscripted index that counts items in a list [dimensionless].

I electric current (charge per time) [dimensions $\mathbb{Q}\mathbb{T}^{-1}$].

$\mathcal{I}$ spectral photon arrival rate of light, that is, the mean arrival rate of photons in a small range of wavelengths $\mathrm{d}\lambda$, divided by $\mathrm{d}\lambda$ (Section 1.5.2, page 38) [dimensions $\mathbb{T}^{-1}\mathbb{L}^{-1}$]. Some authors call this quantity "spectral photon flux" and abbreviate it as $P_{\mathrm{p}\lambda}(\lambda)$. $\mathcal{I}^{\star}$, "target" spectrum to be matched in a perceptual experiment (Section 3.6.3, page 113); $\mathcal{I}_{(\alpha)}$, spectrum of basis light α (Section 3.8.1, page 118).

j generic name for an integer quantity, for example, a subscripted index that counts items in a list. Specifically, the discrete waiting time in a Geometric distribution (Section 0.2.5, page 8) [dimensionless].

$\boldsymbol{j}$ electric current density (Equation 13.35, page 406) [dimensions $\mathbb{Q}\mathbb{L}^{-2}\mathbb{T}^{-1}$]. j, its quantum version.

k a spring constant (Problem 2.4) [dimensions $\mathbb{M}\mathbb{T}^{-2}$].

k integer label for electron states on a ring (Section 12.2.6) [dimensionless].

k_{B} Boltzmann constant; $k_{\mathrm{B}}T$, thermal energy at temperature T [dimensions $\mathbb{M}\mathbb{L}^2\mathbb{T}^{-2}$]; $k_{\mathrm{B}}T_{\mathrm{r}}$, thermal energy at room temperature (Sections 0.6, page 17 and B.4, page 442).

$\boldsymbol{k}$ wavevector (Equation 13.2, page 399) [dimensions $\mathbb{L}^{-1}$]. Its length k is sometimes called the wavenumber.

K number of slits in a diffraction setup (Section 8.3.1, page 274) [dimensionless].

K_d dissociation equilibrium constant (Section 10.3.5, page 325) [dimensions depend on the reaction].

ℓ generic name for a discrete random variable [dimensionless]. Specifically, the number of signals emitted by a rod cell in response to a flash of light (Idea 9.2, page 297). ℓ_{tot}, the number of signals emitted by all the rods in a summation region of the retina in response to a flash of light, and crossing the first synapse (Idea 9.3, page 298, and Section 11.3.2, page 356).

ℓ order of a line in a diffraction pattern, an integer (Section 8.3.1, page 274).

$\boldsymbol{\ell}(s)$ a parameterized curve in space with parameter s (Problems 5.6, page 204, and 6.14, page 245) [dimensions $\mathbb{L}$].

L generic name for a distance, for example, in an optics setup [dimensions $\mathbb{L}$].

L_e radiance (Problem 11.1) [dimensions $\mathbb{M}\mathbb{T}^{-3}$].

m mass; m_e, mass of electron [dimensions $\mathbb{M}$].

m generic name for an integer quantity [dimensionless].

M generic name for an integer quantity, or specifically, total number of coin flips summed to obtain a Binomial distribution [dimensionless].

M quantity controlling when diffractive effects will be significant (Equation 4.20, page 167) [dimensionless].

n index of refraction of a transparent medium, a real number ≥ 1 (Equation 4.6, page 155) [dimensionless]; $n_w \approx 1.33$, the specific case of water at visible-light frequencies.

n cooperativity parameter ("Hill parameter"), a real number ≥ 1 (Section 10.3.5, page 325) [dimensionless].

$n_{\boldsymbol{k},\alpha}$ photon occupation number (Equation 13.27, page 404) [dimensionless].

N generic name for an integer quantity, or specifically, the number of times a particular outcome has been measured in a random system [dimensionless].

NA numerical aperture (Section 6.8.2, page 234) [dimensionless].

p repeat distance in a crystal (Section 8.3.1, page 274), or helical pitch of DNA (Figure 8.1c, page 273) [dimensions $\mathbb{L}$].

$\wp_x(x)$ probability density function for a continuous random variable x, often abbreviated $\wp(x)$ (Section 0.4, page 10) [dimensions match those of $1/x$]. $\wp(x \mid y)$, conditional PDF [dimensions match those of $1/x$].

$\mathbf{p}_e$ electron momentum operator (Section 13.6.2, page 407) [dimensions $\mathbb{M}\mathbb{L}\mathbb{T}^{-1}$].

P power (energy per unit time) [dimensions $\mathbb{M}\mathbb{L}^2\mathbb{T}^{-3}$].

$\boldsymbol{P}$ momentum of electromagnetic field (Equation 13.9, page 401) [dimensions $\mathbb{M}\mathbb{L}\mathbb{T}^{-1}$]. $\mathbf{P}$, corresponding quantum operator.

$\mathcal{P}(\mathsf{E})$ probability of event E [dimensionless]. $\mathcal{P}_\ell(\ell)$, probability mass function for the discrete random variable ℓ, often abbreviated $\mathcal{P}(\ell)$ (Section 0.2, page 2). $\mathcal{P}(\mathsf{E}\,|\,\mathsf{E}')$, $\mathcal{P}(\ell\,|\,s)$, conditional probability (Section 0.2.2, page 4). $\mathcal{P}_{see}$, probability of seeing a dim flash of light (Section 9.3.1, page 293). $\mathcal{P}(\text{rating};\text{stimulus})$, probability of a subject giving a particular rating to flashes of light with a particular mean photon number.

$\mathcal{P}_{name}(\ell; p_1, \ldots)$ or $\wp_{name}(x; p_1, \ldots)$ mathematical function of ℓ or x, with parameter(s) $p_1, \ldots$, that specify a particular idealized distribution, for example, $\mathcal{P}_{unif}$ (Section 0.2.5, page 8), $\wp_{unif}$ (Section 0.4.2, page 12), $\mathcal{P}_{bern}$ (Section 0.2.5, page 8), $\mathcal{P}_{binom}$ (Section 0.2.5), $\mathcal{P}_{pois}$ (Equation 0.28, page 9), $\wp_{gauss}$ (Equation 0.38, page 13), $\mathcal{P}_{\mathrm{geom}}$ (Section 0.2.5, page 8), $\wp_{\mathrm{exp}}$ (Equation 1.10, page 36), $\wp_{\mathrm{cauchy}}$ (Equation 0.40, page 13).

Q_{rod} rod cell axial quantum catch: the probability that a photon, traveling along the rod axis and localized to its geometric cross section, will be productively absorbed and actually generate a signal at that rod cell's synaptic terminal (Equation 9.2, page 297) [dimensionless]. $Q_{\mathrm{rod,side}}$, similar but for a photon traveling perpendicular to the rod axis, polarized perpendicular to that axis, and localized to the rod's outer segment (Problem 9.3, page 315). $Q_{\mathrm{rod,ret}}$, rod quantum catch for the whole retina. Q_{tot}, whole-retina quantum catch: the probability that a photon, spread out over many rod cells, will induce a signal in any one of them that passes the first synapse to a rod bipolar cell (Equation 9.3, page 298).

Q, $\mathsf{Q}^{\dagger}$ lowering (destruction) and raising (creation) operators, respectively, for electromagnetic field (Equation 13.18, page 403) [dimensionless].

r reflection factor for an interface (Section 5.2.2, page 183) [dimensionless].

r discrete rating given by a subject in a psychophysical experiment (Section 9.3.4, page 298).

r_{F} Förster radius (the Example on page 88) [dimensions $\mathbb{L}$].

$\boldsymbol{r}_{\mathrm{e}}$ electron position [dimensions $\mathbb{L}$]; $\mathbf{r}_{\mathrm{e}}$, corresponding quantum operator (Equation 13.38, page 407).

R_{DNA} radius of the DNA molecule (Figure 8.1c, page 273) [dimensions $\mathbb{L}$].

R_{c} radius of curvature of cornea [dimensions $\mathbb{L}$]; R_{a} of lens, front; R_{b} of lens, rear (Figure 6.6b, page 219) [dimensions $\mathbb{L}$].

s arc length along a curve [dimensions $\mathbb{L}$].

$\mathfrak{s}$ a subsystem of interest, for example, a donor/acceptor pair (Chapter 14) [not a quantity].

S action functional (Sections 12.2.2 and 12.3.1) [dimensions $\mathbb{M}\mathbb{L}^2\mathbb{T}^{-1}$].

$\mathcal{S}_i(\lambda)$ spectral sensitivity function of a photoreceptor cell in class i (Section 3.8.5, page 124) [dimensionless]; $\bar{\mathcal{S}}_i(\lambda)$, relative (or rescaled) form.

t transmission factor for an interface (Section 5.2.2, page 183) [dimensionless].

t_{w} waiting time between events in a random process (Equation 0.41, page 14) [dimensions $\mathbb{T}$].

t_r network threshold (number of signaling events needed) to elicit a rating of r in a psychophysical experiment (Section 9.3.4, page 298) [dimensionless].

u in optics, transverse displacement within an aperture or lens (Figure 6.5, page 217) [dimensions $\mathbb{L}$].

U potential energy [dimensions $\mathbb{M}\mathbb{L}^2\mathbb{T}^{-2}$]. $U_0(\boldsymbol{y})$, net potential energy of nuclei and electrons in the ground state for given nuclear coordinates $\boldsymbol{y}$ (Figure 1.9, page 42). Similarly $U_{\star}(\boldsymbol{y})$ for the excited state.

W width or diameter of an aperture in optics [dimensions $\mathbb{L}$].

W energy barrier to remove an electron from a metal surface [dimensions $\mathbb{M}\mathbb{L}^2\mathbb{T}^{-2}$].

x in optics, transverse displacement on a projection screen [dimensions $\mathbb{L}$]; x', transverse displacement in object space; $\bar{x}$, dimensionless form (Figure 6.5, page 217).

$\boldsymbol{y}$ abstract nuclear coordinates (or nuclear coordinates of the donor in a FRET pair) (Figure 1.9, page 42). Similarly $\boldsymbol{y}'$ are nuclear coordinates for the acceptor in a FRET pair (Figure 2.21, page 86). $\bar{\boldsymbol{y}}$, nuclear coordinates of the minimal-energy configuration for a particular electronic state, for example, $\bar{\boldsymbol{y}}_0$ for ground state or $\bar{\boldsymbol{y}}_\star$ for excited state.

Z generic name for a complex number.

Greek alphabet

α generic name for an index that counts items in a list.

α parameter describing a power-law distribution. (Equation 0.43, page 14; Problem 7.6, page 269) [dimensionless].

β probability per unit time, appearing, for example, in a Poisson process or its corresponding Exponential distribution (Section 1.4.1, page 35) [dimensions $\mathbb{T}^{-1}$]. (But mean photon arrival rate is denoted by the special symbol Φ_{p}.) β_i, mean rates of productive photon absorption in each class of cone photoreceptor cells (Equation 3.9, page 125); $\boldsymbol{\beta}$, the same three numbers regarded as a single vector.

Δ amount by which some quantity changes. Usually used as a prefix: Δx denotes a small change in x.

ϵ extinction coefficient (Section 9.4.2′a, page 311) [dimensions $\mathbb{L}^2$]. ϵ_0, permittivity of vacuum (Equation 13.32, page 406) [dimensions $\mathbb{Q}^2\mathbb{T}^2\mathbb{M}^{-1}\mathbb{L}^{-3}$].

$\varepsilon_{\boldsymbol{k}}^{(\alpha)}$ basis of unit polarization vectors ($\alpha = 1, 2$) for plane waves traveling along $\boldsymbol{k}$ (Equation 13.3, page 400) [dimensionless].

ζ absorption coefficient (Section 9.4.2′a, page 311) [dimensions $\mathbb{L}^{-1}$].

$\zeta_{(\alpha)}$ relative amount of reference light α present in a mixture, where $\alpha = 1$, 2, or 3 (Equation 3.6, page 119) [dimensionless]. $\boldsymbol{\zeta}$, these quantities regarded as a single vector.

η width parameter of a Cauchy distribution (Equation 0.40, page 13) [same dimensions as its random variable].

θ polar angle in spherical polar coordinates [dimensionless].

Θ angle of the phosphate backbone helix in DNA (Section 8.4.1, page 280) [dimensionless].

κ fractional offset of two repeating elements in a diffracting object (Section 8.3.3, page 277) [dimensionless].

λ wavelength of a photon with a particular frequency in vacuum, or that a photon of the same frequency would have in vacuum (Equation 1.1, page 26) [dimensions $\mathbb{L}$]. $\lambda^\star$, wavelength of a monochromatic target light; $\lambda_{(\alpha)}$, wavelength of a monochromatic basis light.

μ parameter describing a Poisson distribution (Equation 0.28, page 9) [dimensionless]. $\mu_{0,\mathrm{rod}}$, eigengrau parameter for a single rod; $\mu_{0,\mathrm{sum}}$, eigengrau parameter for a summation region of retina (Section 9.3.3, page 296); μ_{ph}, expectation of the number of photons in a flash of light (Equations 9.2 and 9.3).

μ_{x} parameter setting the expectation of a Gaussian or Cauchy distribution in x (Equation 0.38, page 13) [same dimensions as its random variable x].

μ_0, permeability of vacuum (Equation 13.33, page 406) [dimensions $\mathbb{M}\mathbb{L}\mathbb{Q}^{-2}$].

ν frequency (cycles per unit time) [dimensions $\mathbb{T}^{-1}$]. ν_*, frequency threshold for the photoelectric effect on a particular metal (point **2** on page 31).

ξ parameter describing a Bernoulli trial ("probability to flip <u>heads</u>") (Section 0.2.5, page 8) [dimensionless]. ξ_{thin}, thinning factor applied to a Poisson process (Section 1.4.2, page 36).

ξ generic parameter describing a curve in space, not necessarily arclength.

Π projection operator (Sections 3.8.6 and 3.8.7).

ρ_{q} bulk charge density (charge per unit volume) (Equation 13.32, page 406) [dimensions $\mathbb{Q}\mathbb{L}^{-3}$].

ρ quantum density operator (Chapter 14).

σ area density of chromophores in a thin sheet (Section 9.4.2′a, page 311) [dimensions $\mathbb{L}^{-2}$].

σ variance parameter of a Gaussian distribution (Equation 0.38, page 13) [same dimensions as its random variable].

φ generic name for the phase of a complex number (Appendix D, page 449) [dimensionless].

ϕ azimuthal angle in polar coordinates.

ϕ quantum yield (Section 2.9.2, page 93) [dimensionless]. ϕ_{sig}, probability that a photon will initiate a rod cell signal, if it was absorbed (Section 9.4.2′b (page 313)).

Φ electric potential, also called scalar potential [dimensions $\mathbb{M}\mathbb{L}^2\mathbb{T}^{-2}\mathbb{Q}^{-1}$].

Φ_{p} total mean photon arrival rate (Section 1.5.2, page 38) [dimensions $\mathbb{T}^{-1}$]. (Some authors call this quantity "photon flux.") $\Phi_{\mathrm{p}}^{\star}$, rate for a target light; $\Phi_{\mathrm{p}(\alpha)}$, rate for a basis light, where $\alpha = 1, 2$, or 3.

ψ wavefunction of a stationary state (Section 12.2.4, page 384) [dimensions depend on situation].

Ψ probability amplitude (Section 4.5, page 154) [dimensions depend on situation]. $\left|\Psi\right\rangle$, vector in state space (Section 12.2.4, page 384).

$\omega = 2\pi\nu$ angular frequency (radians per unit time) (Section 6.7, page 230) [dimensions $\mathbb{T}^{-1}$].

Ω angular area, sometimes called solid angle (Section 6.7.2, page 232) [dimensionless].

APPENDIX B

Units and Dimensional Analysis

> Nothing can be more contrary to the organization of the mind, the memory, and the imagination [than SI units]....
> The new system of weights and measures will be a subject of embarrassment and difficulties for several generations....
> Thus are nations tormented about trifles.
>
> — *Napoleon Bonaparte*

Some physical quantities are naturally integers, like the number of discrete clicks made by a Geiger counter. But others are continuous, and most continuous quantities must be expressed in terms of conventional units. This book uses the Système Internationale, or **SI units**, but you'll need to be able to convert units when reading other works. Units and their conversions in turn form part of a larger framework called **dimensional analysis**.

Dimensional analysis gives a powerful method for catching algebraic errors, as well as a way to organize and classify numbers and situations, and even to guess new physical laws, as we'll see in Section B.4.

To handle units systematically, remember that

> A "unit" acts like a symbol representing an unknown quantity. Most continuous physical quantities should be regarded as the **product** of a pure number times one or more units.

(A few physical quantities, for example, those that are intrinsically integers, have no units and are called **dimensionless**.) We carry the unit symbols along throughout our calculations. They behave just like any other multiplicative factor; for example, a unit can cancel if it appears in the numerator and denominator of an expression.[1] We know relations among certain units; for example, we know that $1\,\text{inch} \approx 2.54\,\text{cm}$. Dividing both sides of this formula by the numeric part, we find $0.39\,\text{inch} \approx 1\,\text{cm}$, and so on.

B.1 BASE UNITS

The SI chooses "base" units for length, time, mass, and electric charge: Lengths are measured in meters (abbreviated m), masses in kilograms (kg), time in seconds (s), and electric charge in coulombs (which this book abbreviates as coul).[2] The system also creates related units via the prefixes giga ($=10^9$), mega ($=10^6$), kilo ($=10^3$), deci ($=10^{-1}$), centi ($=10^{-2}$), milli ($=10^{-3}$), micro ($=10^{-6}$), nano ($=10^{-9}$), pico ($=10^{-12}$), or femto ($= 10^{-15}$), abbreviated as G, M, k, d, c, m, μ, n, p, and f respectively. Thus, $1\,\text{nm}$ is a nanometer (or $10^{-9}\,\text{m}$), $1\,\mu\text{g}$ is a microgram, and so on.

A symbol like μm^2 means $(\mu\text{m})^2 = 10^{-12}\,\text{m}^2$, not "$\mu(\text{m}^2)$."

---

[1] One exception involves temperatures expressed using the Celsius and Fahrenheit scales, each of which differ from the absolute (Kelvin) scale by an offset.

[2] The standard abbreviation is C, but this risks confusion with the speed of light, a concentration or capacitance variable, or a generic constant.

B.2 DIMENSIONS VERSUS UNITS

Other quantities, such as electric current, derive their standard units from the base units. But it is useful to think about current in a way that is less strictly tied to a particular unit system. Thus, we define abstract **dimensions**, which tell us *what kind of quantity* a variable represents. For example,

- The symbol $\mathbb{L}$ denotes the *dimension* of length. The SI assigns it a base *unit* called "meters," but other units exist with the same dimension (for example, miles or centimeters). Once we have chosen a unit of length, we then also get derived units for area (m^2) and volume (m^3), which have dimensions $\mathbb{L}^2$ and $\mathbb{L}^3$, respectively.
- The symbol $\mathbb{M}$ denotes the dimension of mass. Its SI base unit is the kilogram.
- The symbol $\mathbb{T}$ denotes the dimension of time. Its SI base unit is the second.
- The symbol $\mathbb{Q}$ denotes the dimension of electric charge. Its SI base unit is the coulomb.
- Electric current has dimensions $\mathbb{Q}\mathbb{T}^{-1}$. The SI assigns it a standard unit $\mathsf{coul/s}$, also called "ampere" and abbreviated A.
- Energy has dimensions $\mathbb{M}\mathbb{L}^2\mathbb{T}^{-2}$. The SI assigns it a standard unit $\mathsf{kg\,m^2/s^2}$, also called "joule" and abbreviated J.
- Power (energy per unit time) has dimensions $\mathbb{M}\mathbb{L}^2\mathbb{T}^{-3}$. The SI assigns it a standard unit $\mathsf{kg\,m^2/s^3}$, also called "watt" and abbreviated W.

Suppose that you are asked on an exam to compute an electric current. You work hard and write down a formula made out of various given quantities. To check your work, write down the dimensions of each of the quantities in your answer, cancel whatever cancels, and make sure the result is $\mathbb{Q}\mathbb{T}^{-1}$. If it's not, you may have forgotten to copy something from one step to the next. It's easy, and it's amazing how quickly you can spot and fix errors in this way.

When you multiply or divide two quantities, the dimensions combine like numerical factors: Photon flux irradiance ($\mathbb{T}^{-1}\mathbb{L}^{-2}$) times area ($\mathbb{L}^2$) has dimensions appropriate for a rate ($\mathbb{T}^{-1}$). On the other hand, you cannot add or subtract terms with different dimensions in a valid equation, any more than you can add rupees to centimeters. Equivalently, an equation of the form $X = Y$ cannot be valid if X and Y have different dimensions. (If either X or Y equals zero, however, then we may omit its units without ambiguity.)

You *can* add dollars to yuan, with the appropriate conversion factor, and similarly cubic centimeters to fluid ounces. Cubic centimeters and fluid ounces are different units that both have the same dimensions ($\mathbb{L}^3$). We can automate unit conversions, and reduce errors, if we restate the conversion $1\,\mathsf{US\ fluid\ ounce} \approx 29.6\,\mathsf{cm}^3$ in the form

$$1 \approx \frac{\mathsf{US\ fluid\ ounce}}{29.6\,\mathsf{cm}^3}.$$

Because we can freely insert a factor of 1 into any formula, we may introduce as many factors of the above expression as we need to cancel all the ounce units in that expression. This simple prescription ("multiply or divide by 1 as needed to cancel unwanted units") eliminates confusion about whether to place the numeric factor 29.6 in the numerator or denominator.

Functions applied to dimensional quantities

If $x = 1\,\mathrm{m}$, then we understand expressions like $2\pi x$ (with dimensions $\mathbb{L}$), and even x^3 (with dimensions $\mathbb{L}^3$). But what about $\sin(x)$ or $\log_{10} x$? These expressions are meaningless;[3] more precisely, they don't transform in any simple multiplicative way when we change units, unlike say $x/26$ or x^2.

Additional SI units

   frequency: One hertz (Hz) equals one complete cycle per second, or $2\pi\,\mathsf{rad/s}$.

   temperature: One kelvin (K) can be defined by saying that the atoms of an ideal monoatomic gas have mean kinetic energy $(3/2)k_\mathrm{B}T$, where $k_\mathrm{B} = 1.38 \times 10^{-23}\,\mathsf{J\,K^{-1}}$.

   focusing power: One diopter (D) equals $1\,\mathsf{m^{-1}}$.

   electric potential: One volt (V) equals $1\,\mathsf{J/coul}$.

   luminous flux: Problem 11.1 (page 375) defines a unit called the lumen (abbreviated lm). It is not directly equivalent to other kinds of units, so it deserves its own dimension; however, we will not give that dimension any symbolic name.

   luminous intensity: The candela is defined as a lumen per steradian (see Section B.3.2 below).

Traditional but non-SI units

   time: One minute is $60\,\mathsf{s}$, and so on.

   length: One Ångstrom unit ($\mathsf{Å}$) equals $0.1\,\mathsf{nm}$.

   volume: One liter (L) equals $10^{-3}\,\mathsf{m^3}$. Thus, $1\,\mathsf{mL} = 1\,\mathsf{cm^3}$.

   number density: A $1\,\mathsf{M}$ solution has a number density of $1\,\mathsf{mole/L} = 1000\,\mathsf{mole\,m^{-3}}$, where "mole" represents the number $\approx 6.02 \times 10^{23}$.

   energy: An electron volt (eV) equals $e \times (1\,\mathsf{V}) = 1.60 \times 10^{-19}\,\mathsf{J} = 96\,\mathsf{kJ/mole}$.

B.3   ABOUT GRAPHS

When you make a graph involving a continuous quantity, state the units of that quantity in the axis label. For example, if the axis label says `waiting time [s]`, then we understand that a point aligned with the tick mark labeled `2` represents a measured waiting time that, when divided by $1\,\mathsf{s}$, yields the pure number 2.

   The same interpretation applies to logarithmic axes. If the axis label says `flash photon density [photons/`μ`m`^{2}`]`, and the tick marks are unequal, as they are on the horizontal axis in Figure 9.9c, then we understand that a point aligned with the first minor tick after the one labeled `10` represents a quantity that, when divided by the stated unit, yields the pure number 20 (in this case, 20 photons $/\mu\mathrm{m}^2$). Alternatively, we can make an ordinary graph of the logarithm of a quantity x, indicating this in the axis label, which says `log`$_{10}$ `x` or `ln x` instead of `x`. The disadvantage of the second system is that, if x carries units, then strictly speaking we must instead write something like `log`$_{10}$`(x/(1 m`^{2}`))` or `log`$_{10}$`(x [a.u.])`, because the logarithm of a quantity with dimensions has no meaning.

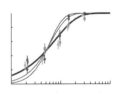

Fig. 9.9c (page 304)

---

[3]One way to see why such expressions are meaningless is to use the Taylor series expansion of $\sin(x)$, and notice that it involves adding terms with incompatible units.

B.3.1 Arbitrary units

Sometimes a quantity is given in some unknown or unstated unit. It may not be necessary to be more specific, but you should alert your reader by saying something like `emission spectrum [arbitrary units]`. Many authors abbreviate this as "[a.u.]"

B.3.2 Angles

See Section 6.7 (page 230). Angles are dimensionless: We get the angle between two intersecting rays, in the dimensionless unit radians (abbreviated `rad`), by drawing a circular arc of any radius r between them and centered on the intersection, then dividing the length of that arc (with dimensions $\mathbb{L}$) by r (with dimensions $\mathbb{L}$). Another clue is that if θ carried dimensions, then trigonometric functions like sine and cosine wouldn't be defined (see Section B.2). The angle corresponding to a complete circle is 2π `rad`. An alternative expression for this quantity is 360 `deg`.

Angular area (also called solid angle) is also dimensionless. Given a patch on the surface of a sphere, we get its angular area, in the dimensionless unit steradians (abbreviated `sr`), by finding the area of that patch and dividing by the sphere's radius squared.

B.4 PAYOFF

Dimensional analysis is more than just an excellent way to avoid, or catch, calculation errors. It can also *help you to discover new physical laws*. Here is a spectacular example from the realm of light.

Toward the end of the 19th century, it became clear that the light emitted by a hot body was of great interest. This realization came first from a theoretical result of G. Kirchoff. Kirchoff considered a hollow body, with nothing (vacuum) inside its interior cavity. As the body is heated, the cavity becomes filled with light, called thermal radiation.[4] The light carries energy, and even before the correct (quantum) description of light was known, it made sense to describe its energy density as a spectrum: The spectral energy density at temperature T, or $u_\nu(\nu, T)$, is defined as the energy per volume carried by light in a frequency range $\Delta\nu$, divided by $\Delta\nu$.

Kirchoff showed theoretically that the spectral energy density inside a cavity is universal: It depends only on the frequency ν and the absolute temperature T. (For example, the material from which the cavity walls are constructed is irrelevant.) Later, W. Wien proved an even stronger result: The temperature T enters the energy density only via the combined quantity ν/T. More precisely, Wien's "displacement law" states that there is a universal function $\mathcal{F}$ for which

$$u_\nu(\nu, T) = \nu^3 \mathcal{F}(x), \text{ where } x = \nu/(k_\mathrm{B}T). \tag{B.1}$$

We included the Boltzmann constant k_B in the definition of x because temperature enters physics only via the product $k_\mathrm{B}T$; that product has the dimensions of energy.[5]

Although Wien could not predict the form of the function $\mathcal{F}$ theoretically, his result already had a profound, and testable, implication: *All the spectral energy density functions, when divided by ν^3, should collapse onto a single curve*, the graph of $\mathcal{F}$.

---

[4]Ideal thermal radiation is sometimes called "blackbody" radiation ("Schwarzer Körper" in Figure B.1a).

[5]Section 0.6 (page 17) introduced the Boltzmann constant.

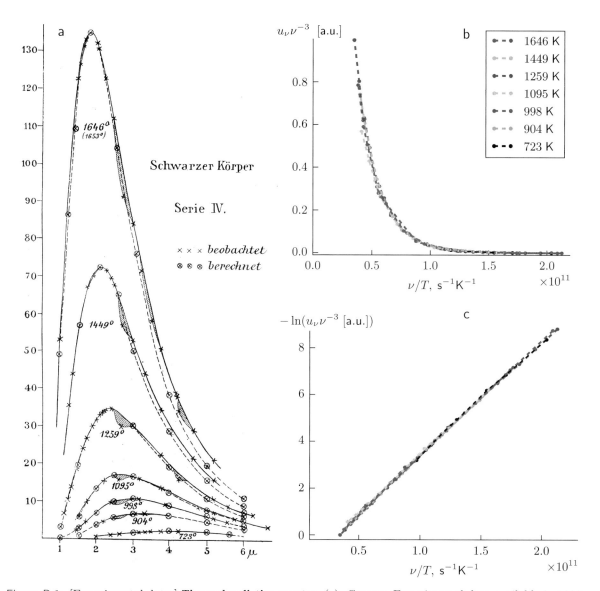

Figure B.1: [Experimental data.] **Thermal radiation spectra.** (a) *Crosses:* Experimental data available in 1899 for the spectral energy density of thermal radiation at various temperatures, given in kelvins. The horizontal axis gives wavelength in micrometers; the vertical axis gives $u_\lambda(\lambda, T)$, a quantity related to the u_ν discussed in the text by the transformation of variables $\nu = c/\lambda$ (see Problem 3.2, page 142). (*Hatched regions* are dips in the experimental data due to absorption by water vapor in the air, and do not reflect the actual thermal radiation spectrum.) [From Lummer & Pringsheim, 1899.] (b) The data in (a) have been converted to frequency spectra, divided by frequency cubed, and plotted as functions of $x = \nu/T$. All the data then collapse onto a single curve, as predicted by Equation B.1. *Dashed lines* just join successive data points; they do not reflect any theory. (c) Semilog plot of the same data shown in (b). The slight but systematic difference between experiment [*crosses* in (a)] and theoretical expectations [*circled crosses*] corresponds to a deviation from straight-line behavior at the low end of this graph, but the figure makes clear that for the data available in 1899, this deviation was not yet significant. Later experiments probed smaller values of ν/T, and disclosed larger deviations that invalidated the exponential model, leading Planck to his famous formula (Equation B.3). The text explains, however, that even prior to this work, Planck correctly realized that the 1899 data yielded a value for a new constant of Nature.

When such data later became available (Figure B.1a), they confirmed this expectation (Figure B.1b).

Max Planck considered this situation in 1899, before his discovery of the law that now bears his name. $\mathcal{F}$ is a nonlinear function of a quantity x that carries dimensions $(\mathbb{M}^{-1}\mathbb{L}^{-2}\mathbb{T})$. But we cannot directly apply a general nonlinear function to a quantity with dimensions (Section B.2)! Planck concluded that

> $\mathcal{F}(x)$ must take the form $Bf(ax)$, where B and a are constants of Nature and f is a dimensionless function of the dimensionless quantity ax. (B.2)

Planck knew that the quantities B and a must be universal constants, because Kirchoff had shown that the entire thermal radiation spectrum was universal, apart from its dependence on ν and T.

Thus, dimensional reasoning usefully subdivides the thermal radiation problem into two subproblems:

- What are the values of B and a, and what do they mean?
- What is the function f, and what does *it* mean?

Start with the constant a. It was natural to suppose that it was some combination of the constants of Nature that were already known. But in 1899, there were only two known constants with the universal character required for B and a:

- The speed of light in vacuum (c), with dimensions $\mathbb{L}\mathbb{T}^{-1}$, and
- Newton's gravitational constant (G), with dimensions $\mathbb{M}^{-1}\mathbb{L}^{3}\mathbb{T}^{-2}$.

Planck realized that *no combination of c and G could give something with the dimensions needed for a*. He then boldly asserted that therefore a must be a *new* universal constant of Nature, even though he knew of no physical mechanism that might involve it. Today, we find it more convenient to talk about the equivalent quantity $\hbar = a/(2\pi)$, the "reduced Planck constant."[6]

Turning to the other constant B in Equation B.2, we find that its dimensions are the same as those of $\hbar c^{-3}$. No further new constant of Nature is needed, therefore, to accommodate B.

In 1899, the existing data and theory seemed to give the form of the scaling function as $f(ax) = \mathrm{e}^{-ax}$ (Figure B.1c). Thus, we can estimate the value of a simply by making a semilog plot of the experimental data and finding the slope. Applying this procedure to the data in Figure B.1c yields $a \approx 6.8 \times 10^{-34}$ J s, or $\hbar \approx 1.08 \times 10^{-34}$ J s, remarkably close to today's value.[7]

Planck's achievement is all the more remarkable for being ahead of its time: At the time, the available data were not extensive enough to disclose the true form of f, which begins to deviate from the simple exponential form at smaller values of ax than those shown in Figure B.1a. Later, when the experimental failure of the exponential formula became apparent, Planck himself would guess the correct law, which is

$$f(ax) = (\mathrm{e}^{ax} - 1)^{-1}. \tag{B.3}$$

But already in 1899, Planck knew that thermal radiation must involve a new fundamental constant, and he knew its numerical value.

---

[6]Many authors abbreviate by calling $\hbar$ "the Planck constant." Others, however, instead use that phrase (and the letter h) to denote what we have called a.

[7]Planck actually used a different procedure, but arrived at a similar numerical value.

Your Turn BA

Use the value of a just given to make semilog plots of the functions e^{-ax} and $(e^{ax} - 1)^{-1}$ over the range of x values shown in Figure B.1c and comment.

Six years later, Einstein proposed that light came in discrete lumps, each with energy related to its frequency.[8] How much energy should a lump contain? To convert from frequency (s^{-1}) to energy (J) required a constant of Nature with units $J\,s$. Although the light-quantum hypothesis seemed crazy, Einstein knew that Planck had already uncovered a new constant with exactly those units, whose role in physics was not yet known. So he proposed that the energy of one lump should equal $2\pi\hbar\nu$, and showed that this hypothesis explained the thermal radiation spectrum.[9] Then he went further, making a falsifiable prediction about the photoelectric effect.

Planck made yet another prescient observation in 1899. He observed that, with the addition of his new constant, the suite of known fundamental constants of Nature (c, G, and now $\hbar$) was rich enough to create quantities with the dimensions $\mathbb{L}$, $\mathbb{M}$, and $\mathbb{T}$. Those quantities could serve as *universal units* set by Nature, unrelated to the size of a king's foot, the Earth, or any other arbitrary object. As Planck put it, they "necessarily retain their meaning for all times and for all civilizations, even extraterrestrial and non-human ones." Indeed, these "Planck units," obtained *before* the birth of quantum theory, are still considered fundamental in theoretical physics.

Your Turn BB

Using modern values for the constants $\hbar$, c, and G, work out the universal units for length, mass, and time and reexpress them in usual SI units. Why are they so different from any scale relevant to biology?

For more details on Planck's discovery see Stone, 2013, and Pais, 1982, chapt. 19.

---

[8] The Einstein relation; see Equation 1.6 (page 34).
[9] $\boxed{T_2}$ Actually, more work was needed to understand the full spectrum. What Einstein showed in 1905 was that his hypothesis correctly gave the high-frequency end of the spectrum. Section 1.3.3′c (page 52) gives the full derivation.

APPENDIX C

Numerical Values

> Nature, like a cautious testator, ties up her estate so as not to bestow it all on one generation, but has a forelooking tenderness and equal regard to the next, and the next..., and the fortieth age.
>
> — *R. W. Emerson*

(See also `http://bionumbers.hms.harvard.edu/` and Milo & Phillips, 2016.)

C.1 FUNDAMENTAL CONSTANTS

Planck constant (reduced), $\hbar = 1.05 \times 10^{-34}\,\mathrm{J\,s}$.

Proton charge, $e = 1.6 \times 10^{-19}$ coul. Electron charge is $-e$.

Electron mass, $m_\mathrm{e} = 9.1 \times 10^{-31}\,\mathrm{kg}$.

Speed of light, $c = 3.0 \times 10^8\,\mathrm{m/s}$.

Avogadro's number, $N_\mathrm{mole} = 6.02 \times 10^{23}$.

Boltzmann constant, $k_\mathrm{B} = 1.38 \times 10^{-23}\,\mathrm{J\,K^{-1}}$.

Typical thermal energy at room temperature $k_\mathrm{B}T_\mathrm{r} = 4.1\,\mathrm{pN\,nm} = 4.1 \times 10^{-21}\,\mathrm{J} = 2.5\,\mathrm{kJ\,mole^{-1}} = 0.59\,\mathrm{kcal\,mole^{-1}} = 0.025\,\mathrm{eV}$.

Permittivity of vacuum, $\epsilon_0 \approx 9 \times 10^{-12}\,\mathrm{coul^2}N^{-1}\mathrm{m^{-2}}$. Permeability of vacuum, $\mu_0 = 4\pi \cdot 10^{-7}\mathrm{m\,kg\,coul^{-2}}$.

C.2 OPTICS

C.2.1 Index of refraction for visible light

These approximate values neglect dispersion (dependence on wavelength).

Air at standard temperature and pressure: $n_\mathrm{air} = 1.0003$. This book uses the approximate value 1, except when studying the mirage phenomenon; there, we use more precise values for light of wavelength 633 nm. At 30°C: $n_\mathrm{air} = 1.00026$; at 50°C: $n_\mathrm{air} = 1.00024$.

Water: $n_\mathrm{w} = 1.33$.

Glass: 1.5–1.7. This book uses the illustrative value 1.52.

Typical oil used in oil-immersion lens microscopy: 1.52.

Human vitreous and aqueous humors, for visible light: 1.34.

Human eye lens: The index varies from 1.43 at the center to 1.32 at edge; this book approximates it as uniform with $n = 1.42$.

Fish eye lens, when approximated as uniform: 1.45. At periphery: 1.38.

Human cornea: 1.38.

C.2.2 Miscellaneous

Maximum numerical aperture with dry lens ≈ 0.95; with oil-immersion lens: 1.6.

Angular diameter of Moon viewed from Earth: 32 arcmin.

Maximum energy of solar radiation per area at Earth surface: $1.4\,\text{kW/m}^2$.

C.3 EYES

C.3.1 Geometric

Distance from pupil to central retina: 20 mm. Diameter of eye, front to back: 24 mm.

Pupil diameter in dim light: This varies considerably among individuals, for example, it decreases with age. Some humans can open to 8 mm, so this is the figure we will use. In bright light: 2 mm.

Radius of curvature of central human cornea: 7.8 mm.

Human eye lens radius of curvature: 10 mm at the front surface and 6 mm at the back.

Ocular media factor (page 313): For a 35-year-old subject viewing 507 nm light, only 45% of photons that arrive at the eye actually end up at the retina without suffering absorption or scattering along the way.

Tiling factor at 7 deg from central vision in the temporal direction (page 313): 0.56.

1 mm on a human retina corresponds to 3.7 deg of arc.

Fovea: Diameter 0.3 mm.

C.3.2 Rod cells

The sensitivity of rod vision is maximal for 507 nm light. This figure includes the spectrum of photon losses prior to the retina as well as the absorption spectrum or rhodopsin itself.

Rhodopsin absorption peaks at wavelength near 500 nm.

Absorption of light by rhodopsin in primate rod cell: Absorption cross section at peak, times concentration $= a_1 c \approx 0.044\,\mu\text{m}^{-1}$ (Section 9.4.2′a, page 311). Traditionally stated as an absorption coefficient, $\zeta = a_1 c/(\ln 10) = 0.019\,\mu\text{m}^{-1}$. The extinction coefficient (molar absorptivity) of retinal rhodopsin is $a_1/(\ln 10) = 4.0 \times 10^4\,\text{M}^{-1}\text{cm}^{-1}$.

Number of rod cells in human retina: 1.2×10^8.

Number of rhodopsin molecules in human or macaque rod cell: 1.4×10^8.

Quantum yield for rhodopsin photoisomerization: 0.67.

Rate of false photon-like signals per time per rod cell in the dark, in macaque: $0.0037\,\text{s}^{-1}$.

Single photon response, current: 1–2 pA in a typical preparation.

Rod cell membrane potential in dark: $-40\,\text{mV}$; maximum hyperpolarization $-70\,\text{mV}$; single photon response 2.4 mV.

Rod integration time: 200 ms.

Duration of rod single-photon response: 300 ms.

Rod outer segment diameter (macaque): $d_{\text{rod}}^{\text{m}} = 2\,\mu\text{m}$. For human rods, the number is similar (Figure 3.9a, page 120).

Rod cell outer segment length, human, mid-peripheral retina: $42\,\mu$m. For the macaque rod cells studied in Baylor et al., 1984: $25\,\mu$m.

Each rod cell contains about 1000 disks.

Hill coefficient (cooperativity parameter) for cyclic nucleotide gated channels in rod outer segment (Figure 10.9, page 333): 2.4.

Vesicle release rate of a rod cell in darkness: $100\,\mathrm{s}^{-1}$.

C.3.3 Cone cells

The sensitivity of cone vision is maximal for 555 nm light. This figure is a composite for the different cone types and includes the spectrum of photon losses prior to the retina as well as the absorption spectra for iodopsins themselves.

Number of cones in human retina: about 5–7×10^6.

Human cone diameter: about $6\,\mu$m (Figure 3.9, page 120a), but thinner (1.5–$3\,\mu$m) in the fovea (rod-free region, Figure 3.9b, page 120).

Cone cell photopsins I, II, and III (analogs of rod opsin) form complexes with retinal called iodopsins, with absorption spectra peaking near 426, 530, and 560 nm. The curves shown in Figure 3.11 (page 123) don't quite peak at those values, because the curves have been corrected for pre-retinal filtering by the lens and macular pigment and for self-screening.

C.3.4 Beyond photoreceptors

Fraction of one-photon rod signals rejected at the first synapse: 0.5.

The receptive fields of retinal ganglion cells serving the fovea have angular diameter about 0.1 deg.

Minimal signaling rate per rod cell for useful vision: $0.0002\,\mathrm{s}^{-1}$. The absolute threshold for detecting any light in a static scene is lower than this value.

C.4 B-FORM DNA

Fig. 8.1c (page 273)

See Figure 8.1c.

Diameter: 2.3 nm.
Helical pitch: 3.5 nm.
Base pair rise: 0.34 nm.

APPENDIX D

Complex Numbers

> I find imaginary numbers useful when computing my tax
> deductions.
>
> — *R. Shankar*

To define complex numbers, we extend the ordinary real numbers by introducing a new quantity, called i, that is not equal to any real number.[1] A complex number Z is then an arbitrary combination of 1 and i with real coefficients: $Z = a + ib$ for real a and b. The coefficients are rather fancifully named the **real and imaginary parts** of Z; that is, $a = \operatorname{Re} Z$ and $b = \operatorname{Im} Z$.

Complex numbers are similar to two-dimensional vectors in some ways, because a vector $\boldsymbol{v}$ can also be written as a combination of two basis vectors. So we can borrow the idea of vector addition to define the sum of two complex numbers, $Z = a + ib$ and $Z' = a' + ib'$, by adding corresponding components:

$$(a + ib) + (a' + ib') = (a + a') + i(b + b'). \tag{D.1}$$

As with vectors, we can visualize this definition by imagining each vector as an arrow, placing the first one's tail at the origin, and sliding the second one until its tail coincides with the first one's head. Then the geometrical meaning of Equation D.1 is that the sum is represented by the arrow that starts at the origin and ends with the shifted second vector's tip. (Figures 4.11–4.12, page 162, show examples of this construction.)

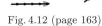

Fig. 4.12 (page 163)

Pursuing the analogy to vectors, a complex number Z has length (called its **modulus**, "absolute value," or "magnitude"), given by $\sqrt{a^2 + b^2}$ and denoted $|Z|$. Like a vector, any nonzero Z also has a direction, which we can specify by stating the angle that it makes with the positive real axis (called its **phase** or "argument"). This is the angle whose tangent equals b/a.

What is new is that we can now introduce a multiplication operation that is *not* the same as either of the familiar ones for vectors.[2] The multiplication is completely specified by declaring that $i^2 = -1$, because then we have

$$Z \times Z' = (a + ib) \times (a' + ib') = (aa' - bb') + i(ab' + a'b). \tag{D.2}$$

Unlike the cross product, this product is commutative. Unlike the dot product, it's associative. Moreover, every nonzero Z has a multiplicative inverse. That is, Equation D.2 behaves algebraically just like ordinary multiplication of real numbers, so the usual manipulations of algebra continue to work with complex numbers.

---

[1] Some engineering texts instead use the letter j to represent this quantity. Some computer math packages use the symbol I or j. Sometimes physicists use the symbol $\hat{\imath}$ to represent an unrelated concept, namely, the unit vector that this book calls $\hat{\boldsymbol{x}}$.

[2] This operation takes two complex numbers and generates a third quantity of the same sort. In contrast, the dot product of two vectors is a scalar; the cross product of two vectors in the same plane is a vector perpendicular to that plane.

To every complex number $Z = a + ib$ we also associate its **complex conjugate** $Z^* = a - ib$. The operation of complex conjugation therefore corresponds to reflection in the ab plane across the a axis.

Your Turn DA

a. Show that when we multiply two complex numbers, ZZ' has modulus equal to the product $|Z||Z'|$, and phase equal to the sum of the phases of Z and Z'. [*Hint:* The formula for the tangent of the sum of two angles may be useful.]

b. Show that

$$|Z|^2 = (Z)(Z^*) \tag{D.3}$$
$$\operatorname{Re} Z = (Z + Z^*)/2 \quad \text{and} \quad \operatorname{Im} Z = (Z - Z^*)/(2i). \tag{D.4}$$

c. Show that $(Z^*)^* = Z$ and $(Z_1 Z_2)^* = Z_1^* Z_2^*$.

The Euler formula packages these and other results in a compact way: For any real number φ,

$$e^{i\varphi} = \cos \varphi + i \sin \varphi. \qquad \text{[4.4, page 153]}$$

One way to prove this identity[3] is to define $g(\varphi)$ as the expression on the left and $f(\varphi)$ as the one on the right. Clearly $f(0) = g(0)$. Also, taking derivatives shows that $dg/d\varphi = ig$ and $df/d\varphi = if$, so the quantity $h = f - g$ obeys

$$dh/d\varphi = ih \text{ and } h(0) = 0. \tag{D.5}$$

Then the squared modulus of h obeys

$$\frac{d|h|^2}{d\varphi} = \frac{d(h^* h)}{d\varphi} = (-ih^*)h + h^*(ih) = 0. \tag{D.6}$$

Because $|h|$ starts out equal to zero, and is constant by Equation D.6, it is always zero. So h itself is always zero, and $f(\varphi) = g(\varphi)$ for all φ.

Equation 4.4 shows that the exponential function bundles together the sine and cosine functions, which generally clutter formulas in wave physics, into a single, simpler object. If we ever need sine and cosine, we can recover them by solving Equation 4.4:

$$\cos \varphi = \tfrac{1}{2}(e^{i\varphi} + e^{-i\varphi}), \quad \sin \varphi = \tfrac{1}{2i}(e^{i\varphi} - e^{-i\varphi}). \tag{D.7}$$

Notice that $e^{i\pi} = -1$; we say that the phase of -1 equals π.

The Euler formula also implies that the complex number $e^{i\varphi}$ always has modulus equal to 1, for any real φ. More generally, any complex number can be expressed as a number of this form, times a real scaling factor: $Z = re^{i\varphi}$. Note that $e^{i(\varphi + 2\pi)} = e^{i\varphi}$: The phase of a complex number, like any angle, is a periodic variable.

We can convert from the "polar representation" $re^{i\varphi}$ to the original "Cartesian representation" $a + ib$ by using Equation 4.4; to go the other way, we solve for r and φ:

$$a = r \cos \varphi, \ b = r \sin \varphi \quad \text{and} \quad r = \sqrt{a^2 + b^2}, \ \varphi = \tan^{-1}(b/a).$$

---

[3] You'll construct another proof in Problem 4.1.

The Euler formula is consistent with the usual properties of the exponential function:[4]

$$re^{i\varphi} \times r'e^{i\varphi'} = (rr')e^{i(\varphi+\varphi')}. \tag{D.8}$$

Two special cases of Equation D.8 are particularly important:

- Multiplying a complex number Z by a real number changes its length without rotating it.
- Multiplication by a pure phase $e^{i\varphi}$ rotates Z about the origin without changing its length.

Finally, you should prove that

$$(re^{i\varphi})^* = re^{-i\varphi}. \tag{D.9}$$

Many more geometrical aspects of complex numbers can be found in Needham, 1997.

---

[4]You'll work this out in Problem 4.2.

Bibliography

> As with the fires in our hearths, so also with books: You take the fire from your neighbor, you light your own, you share it with others, and it belongs to everyone.
>
> — *Voltaire*

Many of the articles listed below are published in high-impact scientific journals. It is important to know that frequently such an article is only the tip of an iceberg: Many of the technical details (generally including specification of any physical model used) are relegated to a separate document called Supplementary Information, or something similar. The online version of the article will generally contain a link to that supplement.

ABRAHAM, A V, RAM, S, CHAO, J, WARD, E S, & OBER, E J. 2010. Comparison of estimation algorithms in single-molecule localization. *Proc. SPIE Int. Soc. Opt. Eng.*, **7570**, 757004.

AGRANOVICH, V M, & GALANIN, M D. 1982. *Electronic excitation energy transfer in condensed matter.* New York: Elsevier North-Holland.

AHLBORN, B. 2004. *Zoological physics.* New York: Springer.

ALA-LAURILA, P, & RIEKE, F. 2014. Coincidence detection of single-photon responses in the inner retina at the sensitivity limit of vision. *Curr. Biol.*, **24**, 1–11.

ALBERTS, B, BRAY, D, HOPKIN, K, JOHNSON, A, LEWIS, J, RAFF, M, ROBERTS, K, & WALTER, P. 2014. *Essential cell biology.* 4th ed. New York: Garland Science.

ALBERTS, B, JOHNSON, A, LEWIS, J, MORGAN, D, RAFF, M, ROBERTS, K, & WALTER, P. 2015. *Molecular biology of the cell.* 6th ed. Garland Science.

ALLEN, L J S. 2011. *An introduction to stochastic processes with applications to biology.* 2d ed. Upper Saddle River NJ: Pearson.

AMADOR KANE, S. 2009. *Introduction to physics in modern medicine.* 2d ed. Boca Raton FL: CRC Press.

AMIR, A, & VUKUSIC, P. 2013. Elucidating the stop bands of structurally colored systems through recursion. *Am. J. Phys.*, **81**(4), 253–257.

APPLEYARD, D C, VANDERMEULEN, K Y, LEE, H, & LANG, M J. 2007. Optical trapping for undergraduates. *Am. J. Phys.*, **75**(1), 5–14.

ARMSTRONG, C M, & HILLE, B. 1998. Voltage-gated ion channels and electrical excitability. *Neuron*, **20**, 371–380.

ARNOLD, W, & OPPENHEIMER, J R. 1950. Internal conversion in the photosynthetic mechanism of blue-green algae. *J. Gen. Physiol.*, **33**(4), 423–435.

ARSHAVSKY, V Y, LAMB, T D, & PUGH, JR., E N. 2002. G proteins and phototransduction. *Annu. Rev. Physiol.*, **64**, 153–187.

ASARI, H, & MEISTER, M. 2012. Divergence of visual channels in the inner retina. *Nat. Neurosci.*, **15**(11), 1581–1589.

ASPDEN, R S, PADGETT, M J, & SPALDING, G C. 2016. Video recording true single-photon double-slit interference. *Am. J. Phys.*, **84**(9), 671–677.

ATKINS, P W, & DE PAULA, J. 2011. *Physical chemistry for the life sciences.* 2d ed. Oxford UK: Oxford Univ. Press.

ATKINS, P W, & FRIEDMAN, R. 2011. *Molecular quantum mechanics.* 5th ed. Oxford UK: Oxford Univ. Press.

BACKLUND, M P, LEW, M D, BACKER, A S, SAHL, S J, GROVER, G, AGRAWAL, A, PIESTUN, R, & MOERNER, W E. 2012. Simultaneous, accurate measurement of the 3D position and orientation of single molecules. *Proc. Natl. Acad. Sci. USA*, **109**(47), 19087–19092.

BADURA, A, SUN, X R, GIOVANNUCCI, A, LYNCH, L A, & WANG, S S-H. 2014. Fast calcium sensor proteins for monitoring neural activity. *Neurophoton.*, **1**(2), 025008.

BARKAI, N, & LEIBLER, S. 1997. Robustness in simple biochemical networks. *Nature*, **387**(6636), 913–917.

BARLOW, H B. 1956. Retinal noise and absolute threshold. *J. Opt. Soc. Am.*, **46**, 634–639.

BARLOW, H B, LEVICK, W R, & YOON, M. 1971. Responses to single quanta of light in retinal ganglion cells of the cat. *Vision Res. Supplement*, **3**, 87–101.

BARRETT, J M, BERLINGUER-PALMINI, R, & DEGENAAR, P. 2014. Optogenetic approaches to retinal prosthesis. *Vis. Neurosci.*, **31**(4-5), 345–354.

BARTKO, A, & DICKSON, R. 1999a. Imaging three-dimensional single molecule orientations. *J. Phys. Chem. B*, **103**, 11237–11241.

BARTKO, A, & DICKSON, R. 1999b. Three-dimensional orientations of polymer-bound single molecules. *J. Phys. Chem. B*, **103**, 3053–3056.

BATES, M, HUANG, B, & ZHUANG, X. 2008. Super-resolution microscopy by nanoscale localization of photo-switchable fluorescent probes. *Curr. Opin. Chem. Biol.*, **12**(5), 505–514.

BATES, M, JONES, S A, & ZHUANG, X. 2011. Stochastic optical reconstruction microscopy (STORM). *Chap. 35 of:* YUSTE, R (Ed.), *Imaging: A laboratory manual.* Cold Spring Harbor NY: Cold Spring Harbor Laboratory Press.

BATES, M, JONES, S A, & ZHUANG, X. 2013. Stochastic optical reconstruction microscopy (STORM): A method for superresolution fluorescence imaging. *Cold Spring Harb. Protoc.*, **2013**(6), 498–520.

BAYLOR, D A, & FETTIPLACE, R. 1977. Transmission from photoreceptors to ganglion cells in turtle retina. *J. Physiol. (Lond.)*, **271**(2), 391–424.

BAYLOR, D A, LAMB, T D, & YAU, K W. 1979. The membrane current of single rod outer segments. *J. Physiol. (Lond.)*, **288**, 589–611.

BAYLOR, D A, NUNN, B J, & SCHNAPF, J L. 1984. The photocurrent, noise and spectral sensitivity of rods of the monkey *Macaca fascicularis*. *J. Physiol. (Lond.)*, **357**, 575–607.

BAYLOR, D A, NUNN, B J, & SCHNAPF, J L. 1987. Spectral sensitivity of cones of the monkey *Macaca fascicularis*. *J. Physiol. (Lond.)*, **390**, 145–160.

BENEDEK, G B, & VILLARS, F M H. 2000. *Physics with illustrative examples from medicine and biology.* 2nd ed. Vol. 3. New York: AIP Press.

BERG, H C. 2004. *E. coli in motion.* New York: Springer.

BERG, J M, TYMOCZKO, J L, GATTO, JR., G J, & STRYER, L. 2015. *Biochemistry.* 8th ed. New York: WH Freeman and Co.

BERMAN, H M, WESTBROOK, J, FENG, Z, GILLILAND, G, BHAT, T N, WEISSIG, H, SHINDYALOV, I N, & BOURNE, P E. 2000. The Protein Data Bank. *Nucl. Acids Res.*, **28**, 235–242.

BERMAN, P R, & MALINOVSKY, V S. 2011. *Principles of laser spectroscopy and quantum optics.* Princeton NJ: Princeton Univ. Press.

BERNTSON, A, SMITH, R G, & TAYLOR, W R. 2004. Transmission of single photon signals through a binary synapse in the mammalian retina. *Vis. Neurosci.*, **21**(5), 693–702.

BERSON, D M. 2014. Intrinsically photosensitive retinal ganglion cells. *Chap. 14 of:* CHALUPA, L M, & WERNER, J S (Eds.), *The new visual neurosciences.* Cambridge MA: MIT Press.

BERSON, D M, DUNN, F A, & TAKAO, M. 2002. Phototransduction by retinal ganglion cells that set the circadian clock. *Science*, **295**(5557), 1070–1073.

BETZIG, E. 1995. Proposed method for molecular optical imaging. *Opt. Lett.*, **20**(3), 237–239.

BETZIG, E, PATTERSON, G H, SOUGRAT, R, LINDWASSER, O W, OLENYCH, S, BONIFACINO, J S, DAVIDSON, M W, LIPPINCOTT-SCHWARTZ, J, & HESS, H F. 2006. Imaging intracellular fluorescent proteins at nanometer resolution. *Science*, **313**(5793), 1642–1645.

BIALEK, W. 2012. *Biophysics: Searching for principles.* Princeton NJ: Princeton Univ. Press.

BLITZSTEIN, J K, & HWANG, J. 2015. *Introduction to probability.* Boca Raton FL: CRC Press.

BOAL, D. 2012. *Mechanics of the cell.* 2d ed. Cambridge UK: Cambridge University Press.

BOBROFF, N. 1986. Position measurement with a resolution and noise-limited instrument. *Rev. Sci. Instrum.*, **57**, 1152–1157.

BODINE, E N, LENHART, S, & GROSS, L J. 2014. *Mathematics for the life sciences.* Princeton NJ: Princeton Univ. Press.

BÖHMER, M, & ENDERLEIN, J. 2003. Orientation imaging of single molecules by wide-field epifluorescence microscopy. *J. Opt. Soc. Am. B,* **20**, 554–559.

BOHREN, C F, & CLOTHIAUX, E E. 2006. *Fundamentals of atmospheric radiation.* Weinheim: Wiley-VCH.

BOYD, I A, & MARTIN, A R. 1956. The end-plate potential in mammalian muscle. *J. Physiol. (Lond.),* **132**(1), 74–91.

BOYDEN, E S, ZHANG, F, BAMBERG, E, NAGEL, G, & DEISSEROTH, K. 2005. Millisecond-timescale, genetically targeted optical control of neural activity. *Nat. Neurosci.,* **8**(9), 1263–1268.

BRAINARD, D H, & STOCKMAN, A. 2010. Colorimetry. *Chap. 10 of:* BASS, M, ENOCH, J M, & LAKSHMI-NARAYANAN, V (Eds.), *Handbook of optics,* 3rd ed., Vol. 3. New York: McGraw-Hill.

BRANCHINI, B R, BEHNEY, C E, SOUTHWORTH, T L, FONTAINE, D M, GULICK, A M, VINYARD, D J, & BRUDVIG, G W. 2015. Experimental support for a single electron-transfer oxidation mechanism in firefly bioluminescence. *J. Am. Chem. Soc.,* **137**(24), 7592–7595.

BRAY, D. 2009. *Wetware: A computer in every living cell.* New Haven: Yale Univ. Press.

BRESLIN, A, & MONTWILL, A. 2013. *Let there be light: The story of light from atoms to galaxies.* 2d ed. London: Imperial College Press.

BROUSSARD, G J, LIANG, R, & TIAN, L. 2014. Monitoring activity in neural circuits with genetically encoded indicators. *Front. Mol. Neurosci.,* **7**, art. no. 97.

BUHBUT, S, ITZHAKOV, S, TAUBER, E, SHALOM, M, HOD, I, GEIGER, T, GARINI, Y, ORON, D, & ZABAN, A. 2010. Built-in quantum dot antennas in dye-sensitized solar cells. *ACS Nano,* **4**(3), 1293–1298.

BUKS, E, SCHUSTER, R, HEIBLUM, M, MAHALU, D, & UMANSKY, V. 1998. Dephasing in electron interference by a 'which-path' detector. *Nature,* **391**, 871–874.

BURNS, M E. 2010. Deactivation mechanisms of rod phototransduction: The Cogan lecture. *Invest. Ophthalmol. Vis. Sci.,* **51**(3), 1282–1288.

BURNS, M E, & PUGH, JR., E N. 2010. Lessons from photoreceptors: Turning off G-protein signaling in living cells. *Physiology (Bethesda),* **25**(2), 72–84.

BUSSKAMP, V, PICAUD, S, SAHEL, J A, & ROSKA, B. 2012. Optogenetic therapy for retinitis pigmentosa. *Gene Ther.,* **19**(2), 169–175.

BYRNE, J H, HEIDELBERGER, R, & WAXHAM, M N (Eds.). 2014. *From molecules to networks: An introduction to cellular and molecular neuroscience.* 3d ed. Amsterdam: Academic Press.

CAGNET, M, FRANCON, M, & THRIERR, J C. 1962. *Atlas optischer Erscheinungen. Atlas de phénomènes d'optique. Atlas of optical phenomena.* Berlin: Springer.

CANGIANO, L, ASTERITI, S, CERVETTO, L, & GARGINI, C. 2012. The photovoltage of rods and cones in the dark-adapted mouse retina. *J. Physiol. (Lond.),* **590**(Pt 16), 3841–3855.

CANTOR, C R, & SCHIMMEL, P R. 1980. *Biophysical chemistry part II: Techniques for the study of biological structure and function.* San Francisco: W. H. Freeman and Co.

CARROLL, S B. 2006. *The making of the fittest: DNA and the ultimate forensic record of evolution.* New York: W.W. Norton and Co.

CHANDLER, D E, & ROBERSON, R W. 2009. *Bioimaging: Current concepts in light and electron microscopy.* Sudbury MA: Jones and Bartlett.

CHEEZUM, M K, WALKER, W F, & GUILFORD, W H. 2001. Quantitative comparison of algorithms for tracking single fluorescent particles. *Biophys. J.,* **81**(4), 2378–2388.

CHEN, T-W, WARDILL, T J, SUN, Y, PULVER, S R, RENNINGER, S L, BAOHAN, A, SCHREITER, E R, KERR, R A, ORGER, M B, JAYARAMAN, V, LOOGER, L L, SVOBODA, K, & KIM, D S. 2013. Ultrasensitive fluorescent proteins for imaging neuronal activity. *Nature,* **499**(7458), 295–300.

CHOW, B Y, & BOYDEN, E S. 2013. Optogenetics and translational medicine. *Sci. Transl. Med.,* **5**(177), art. no. 177ps5.

CLEGG, R M. 2006. The History of FRET. *Pages 1–45 of:* GEDDES, C, & LAKOWICZ, J (Eds.), *Reviews in Fluorescence 2006.* Reviews in Fluorescence, Vol. 2006. Springer US.

COLE, K C, MCLAUGHLIN, H W, & JOHNSON, D I. 2007. Use of bimolecular fluorescence complementation to study in vivo interactions between Cdc42p and Rdi1p of *Saccharomyces cerevisiae*. *Eukaryotic Cell*, **6**(3), 378–387.

CONN, H W. 1900. *The method of evolution: A review of the present attitude of science toward the question of the laws and forces which have brought about the origin of species.* New York: G.P. Putnam's Sons.

COX, G (Ed.). 2012. *Optical imaging techniques in cell biology.* 2d ed. Boca Raton FL: CRC Press.

CRONIN, T W, JOHNSEN, S, MARSHALL, N J, & WARRANT, E J. 2014. *Visual ecology.* Princeton NJ: Princeton Univ. Press.

CURCIO, C A, SLOAN, K R, KALINA, R E, & HENDRICKSON, A E. 1990. Human photoreceptor topography. *J. Comp. Neurol.*, **292**(4), 497–523.

DAS, A. 2006. *Field theory: A path integral approach.* 2d ed. Hackensack NJ: World Scientific.

DAW, N. 2012. *How vision works: The physiological mechanisms behind what we see.* Oxford UK: Oxford Univ. Press.

DAWKINS, R. 1996. *The blind watchmaker.* New York: Norton.

DAYAN, P, & ABBOTT, L F. 2001. *Theoretical neuroscience.* Cambridge MA: MIT Press.

DENK, W, STRICKLER, J H, & WEBB, W W. 1990. Two-photon laser scanning fluorescence microscopy. *Science*, **248**(4951), 73–76.

DHINGRA, A, & VARDI, N. 2012. mGlu receptors in the retina. *WIREs Membr. Transp. Signal.*, **1**(5), 641–653.

DICKSON, R M, CUBITT, A B, TSIEN, R Y, & MOERNER, W E. 1997. On/off blinking and switching behaviour of single molecules of green fluorescent protein. *Nature*, **388**(6640), 355–358.

DILL, K A, & BROMBERG, S. 2010. *Molecular driving forces: Statistical thermodynamics in biology, chemistry, physics, and nanoscience.* 2d ed. New York: Garland Science.

DIMITROVA, T L, & WEIS, A. 2008. The wave-particle duality of light: A demonstration experiment. *Am. J. Phys.*, **76**(2), 137–142.

DOMBECK, D, & TANK, D. 2011. Two-photon imaging of neural activity in awake mobile mice. *Chap. 76 of:* HELMCHEN, F, & KONNERTH, A (Eds.), *Imaging in neuroscience: A laboratory manual.* Cold Spring Harbor NY: Cold Spring Harbor Laboratory Press.

DONG, B, ALMASSALHA, L M, STYPULA-CYRUS, Y, URBAN, B E, CHANDLER, J E, NGUYEN, T-Q, SUN, C, ZHANG, H F, & BACKMAN, V. 2016. Superresolution intrinsic fluorescence imaging of chromatin utilizing native, unmodified nucleic acids for contrast. *Proc. Natl. Acad. Sci. USA*, **113**(35), 9716–9721.

DONNER, K. 1992. Noise and the absolute thresholds of cone and rod vision. *Vision Res.*, **32**(5), 853–866.

DOWLING, J E. 2012. *The retina: An approachable part of the brain.* Revised ed. Cambridge MA: Harvard Univ. Press.

DROBIZHEV, M, MAKAROV, N S, TILLO, S E, HUGHES, T E, & REBANE, A. 2011. Two-photon absorption properties of fluorescent proteins. *Nat. Methods*, **8**(5), 393–399.

EIBENBERGER, S, GERLICH, S, ARNDT, M, MAYOR, M, & TÜXEN, J. 2013. Matter–wave interference of particles selected from a molecular library with masses exceeding 10 000 amu. *Phys. Chem. Chem. Phys.*, **15**(35), 14696.

EINSTEIN, A. 1905. Über einen die Erzeugung und Verwandlung des Lichtes betreffenden heuristischen Gesichtspunkt. *Ann. Physik*, **17**, 132–148. English translation in J Stachel, ed. 2009. *The collected papers of Albert Einstein*, vol. 2. Princeton NJ: Princeton Univ. Press.

EMERSON, R, & LEWIS, C M. 1942. The photosynthetic efficiency of phycocyanin in chroococcus, and the problem of carotenoid participation in photosynthesis. *J. Gen. Physiol.*, **25**(4), 579–595.

ENGEL, G S. 2011. Quantum coherence in photosynthesis. *Procedia Chemistry*, **3**(1), 222–231.

FANG-YEN, C, ALKEMA, M J, & SAMUEL, A D T. 2015. Illuminating neural circuits and behaviour in *Caenorhabditis elegans* with optogenetics. *Philos. Trans. R. Soc. Lond. B Biol. Sci.*, **370**(1677).

FAUTH, C, & SPEICHER, M R. 2001. Classifying by colors: FISH-based genome analysis. *Cytogenet. Cell Genet.*, **93**(1-2), 1–10.

FELDER, G N, & FELDER, K M. 2016. *Mathematical methods in engineering and physics.* New York: Wiley.

FEYNMAN, R P. 1950. Mathematical formulation of the quantum theory of electromagnetic interaction. *Phys. Rev.*, **80**(3), 440–457.

FEYNMAN, R P. 1967. *The character of physical law.* Cambridge MA: MIT Press.

FEYNMAN, R P. 1985. *QED: The strange theory of light and matter.* Princeton NJ: Princeton Univ. Press.

FEYNMAN, R P, LEIGHTON, R, & SANDS, M. 2010a. *The Feynman lectures on physics.* New milennium ed. Vol. 3. New York: Basic Books. Free online: `http://www.feynmanlectures.caltech.edu/`.

FEYNMAN, R P, LEIGHTON, R, & SANDS, M. 2010b. *The Feynman lectures on physics.* New milennium ed. Vol. 1. New York: Basic Books. Free online: `http://www.feynmanlectures.caltech.edu/`.

FEYNMAN, R P, HIBBS, A R, & STYER, D F. 2010c. *Quantum mechanics and path integrals.* Emended ed. Mineola NY: Dover.

FINE, A. 2011. Confocal microscopy: Principles and practice. *Chap. 5 of:* YUSTE, R (Ed.), *Imaging: A laboratory manual.* Cold Spring Harbor NY: Cold Spring Harbor Laboratory Press.

FISHER, J A N, BARCHI, J R, WELLE, C G, KIM, G-H, KOSTERIN, P, OBAID, A L, YODH, A G, CONTRERAS, D, & SALZBERG, B M. 2008. Two-photon excitation of potentiometric probes enables optical recording of action potentials from mammalian nerve terminals in situ. *J. Neurophysiol.*, **99**(3), 1545–1553.

FLEISCH, D. 2008. *A student's guide to Maxwell's equations.* Cambridge Univ. Press.

FORKEY, J N, QUINLAN, M E, & GOLDMAN, Y E. 2005. Measurement of single macromolecule orientation by total internal reflection fluorescence polarization microscopy. *Biophys. J.*, **89**(2), 1261–1271.

FRANKE, T, & RHODE, S. 2012. Two-photon microscopy for deep tissue imaging of living specimens. *Microscopy Today*, **20**, 12–16.

FRANKLIN, K, MUIR, P, SCOTT, T, WILCOCKS, L, & YATES, P. 2010. *Introduction to biological physics for the health and life sciences.* Chichester UK: John Wiley and Sons.

FRANZE, K, GROSCHE, J, SKATCHKOV, S N, SCHINKINGER, S, FOJA, C, SCHILD, D, UCKERMANN, O, TRAVIS, K, REICHENBACH, A, & GUCK, J. 2007. Muller cells are living optical fibers in the vertebrate retina. *Proc. Natl. Acad. Sci. USA*, **104**(20), 8287–8292.

FRITZKY, L, & LAGUNOFF, D. 2013. Advanced methods in fluorescence microscopy. *Anal. Cell Pathol. (Amst.)*, **36**(1-2), 5–17.

GABRECHT, T, GLANZMANN, T, FREITAG, L, WEBER, B-C, VAN DEN BERGH, H, & WAGNIÈRES, G. 2007. Optimized autofluorescence bronchoscopy using additional backscattered red light. *J. Biomed. Opt.*, **12**(6), 064016.

GARG, A. 2012. *Classical electromagnetism in a nutshell.* Princeton NJ: Princeton Univ. Press.

GARINI, Y, YOUNG, I T, & MCNAMARA, G. 2006. Spectral imaging: Principles and applications. *Cytometry A*, **69**(8), 735–747.

GATES, F L. 1930. A study of the bactericidal action of ultra violet light: III. The absorption of ultra violet light by bacteria. *J. Gen. Physiol.*, **14**(1), 31–42.

GELLES, J, SCHNAPP, B J, & SHEETZ, M P. 1988. Tracking kinesin-driven movements with nanometre-scale precision. *Nature*, **331**(6155), 450–453.

GILMORE, J, & MCKENZIE, R H. 2008. Quantum dynamics of electronic excitations in biomolecular chromophores: Role of the protein environment and solvent. *J. Phys. Chem. A*, **112**(11), 2162–2176.

GRADINARU, V, THOMPSON, K R, ZHANG, F, MOGRI, M, KAY, K, SCHNEIDER, M B, & DEISSEROTH, K. 2007. Targeting and readout strategies for fast optical neural control in vitro and in vivo. *J. Neurosci.*, **27**(52), 14231–14238.

GRANGIER, P, ROGER, G, & ASPECT, A. 1986. Experimental evidence for a photon anticorrelation effect on a beamsplitter. *Europhys. Lett.*, **1**, 173–179.

GREENSTEIN, G, & ZAJONC, A G. 2006. *The quantum challenge: Modern research on the foundations of quantum mechanics.* 2d ed. Sudbury MA: Jones and Bartlett.

HADDOCK, S H D, & DUNN, C W. 2011. *Practical computing for biologists.* Sunderland MA: Sinauer Associates.

HADDOCK, S H D, MOLINE, M A, & CASE, J F. 2010. Bioluminescence in the sea. *Annu. Rev. Marine Sci.*, **2**(1), 443–493.

HAGINS, W A, PENN, R D, & YOSHIKAMI, S. 1970. Dark current and photocurrent in retinal rods. *Biophys. J.*, **10**(5), 380–412.

HAN, X, & BOYDEN, E S. 2007. Multiple-color optical activation, silencing, and desynchronization of neural activity, with single-spike temporal resolution. *PLoS ONE*, **2**(3), e299.

HARVEY, C D, COEN, P, & TANK, D W. 2012. Choice-specific sequences in parietal cortex during a virtual-navigation decision task. *Nature*, **484**(7392), 62–68.

HATTAR, S, LIAO, H W, TAKAO, M, BERSON, D M, & YAU, K W. 2002. Melanopsin-containing retinal ganglion cells: architecture, projections, and intrinsic photosensitivity. *Science*, **295**(5557), 1065–1070.

HECHT, E. 2002. *Optics*. 4th ed. Reading, MA: Addison-Wesley.

HECHT, S, SHLAER, S, & PIRENNE, M H. 1942. Energy, quanta, and vision. *J. Gen. Physiol.*, **25**, 819–840.

HEGEMANN, P. 2008. Algal sensory photoreceptors. *Annu. Rev. Plant Biol.*, **59**, 167–189.

HELL, S W. 2007. Far-field optical nanoscopy. *Science*, **316**(5828), 1153–1158.

HELL, S W. 2009. Microscopy and its focal switch. *Nat. Methods*, **6**(1), 24–32.

HENSHAW, J M. 2012. *A tour of the senses: How your brain interprets the world*. Baltimore MD: Johns Hopkins University Press.

HERMAN, I P. 2016. *Physics of the human body: A physical view of physiology*. 2d ed. New York: Springer.

HERTZ, H. 1893. *Electric waves, being researches on the propagation of electric action with finite velocity through space*. London: Macmillan. Translated by D E Jones.

HESS, S T, GIRIRAJAN, T P K, & MASON, M D. 2006. Ultra-high resolution imaging by fluorescence photoactivation localization microscopy. *Biophys. J.*, **91**(11), 4258–4272.

HILL, C. 2015. *Learning scientific programming with Python*. Cambridge UK: Cambridge Univ. Press.

HINTERDORFER, P, & VAN OIJEN, A (Eds.). 2009. *Handbook of single-molecule biophysics*. New York: Springer.

HOBBIE, R K, & ROTH, B J. 2015. *Intermediate physics for medicine and biology*. 5th ed. New York: Springer.

HOCHBAUM, D R, ZHAO, Y, FARHI, S L, KLAPOETKE, N, WERLEY, C A, KAPOOR, V, ZOU, P, KRALJ, J M, MACLAURIN, D, SMEDEMARK-MARGULIES, N, SAULNIER, J L, BOULTING, G L, STRAUB, C, CHO, Y K, MELKONIAN, M, WONG, G K-S, HARRISON, D J, MURTHY, V N, SABATINI, B L, BOYDEN, E S, CAMPBELL, R E, & COHEN, A E. 2014. All-optical electrophysiology in mammalian neurons using engineered microbial rhodopsins. *Nat. Methods*, **11**(8), 825–833.

HOFER, H, & WILLIAMS, D R. 2014. Color vision and the retinal mosaic. *Chap. 33 of:* CHALUPA, L M, & WERNER, J S (Eds.), *The new visual neurosciences*. Cambridge MA: MIT Press.

HOFER, H, CARROLL, J, NEITZ, J, NEITZ, M, & WILLIAMS, D R. 2005. Organization of the human trichromatic cone mosaic. *J. Neurosci.*, **25**(42), 9669–9679.

HOLT, A, VAHIDINIA, S, GAGNON, Y L, MORSE, D E, & SWEENEY, A M. 2014. Photosymbiotic giant clams are transformers of solar flux. *J. R. Soc. Interface*, **11**, art. 20140678.

HOMBERG, U, & EL JUNDHI, B. 2014. Polarization vision in arthropods. *Chap. 84 of:* CHALUPA, L M, & WERNER, J S (Eds.), *The new visual neurosciences*. Cambridge MA: MIT Press.

HORVÁTH, G (Ed.). 2014. *Polarized light and polarization vision in animal sciences*. 2nd ed. New York: Springer.

HUANG, B, BATES, M, & ZHUANG, X. 2009. Super-resolution fluorescence microscopy. *Annu. Rev. Biochem.*, **78**, 993–1016.

HUBEL, D H. 1995. *Eye, brain, and vision*. New York: Scientific American Press. Available at `http://hubel.med.harvard.edu`.

IQBAL, A, ARSLAN, S, OKUMUS, B, WILSON, T J, GIRAUD, G, NORMAN, D G, HA, T, & LILLEY, D M J. 2008. Orientation dependence in fluorescent energy transfer between Cy3 and Cy5 terminally attached to double-stranded nucleic acids. *Proc. Natl. Acad. Sci. USA*, **105**(32), 11176–11181.

ISHIZUKA, T, KAKUDA, M, ARAKI, R, & YAWO, H. 2006. Kinetic evaluation of photosensitivity in genetically engineered neurons expressing green algae light-gated channels. *Neurosci. Res.*, **54**(2), 85–94.

JACOBS, G H, WILLIAMS, G A, CAHILL, H, & NATHANS, J. 2007. Emergence of novel color vision in mice engineered to express a human cone photopigment. *Science*, **315**(5819), 1723–1725.

JAMES, J F. 2014. *An introduction to practical laboratory optics*. Cambridge UK: Cambridge Univ. Press.

JAMESON, D M. 2014. *Introduction to fluorescence*. Boca Raton FL: Taylor and Francis.

JANG, S. 2007. Generalization of the Förster resonance energy transfer theory for quantum mechanical modulation of the donor-acceptor coupling. *J. Chem. Phys.*, **127**, 174710.

JAYNES, E T, & BRETTHORST, G L. 2003. *Probability theory: The logic of science*. Cambridge UK: Cambridge Univ. Press.

JOHNSEN, S. 2012. *The optics of Life: A biologist's guide to light in nature.* Princeton NJ: Princeton Univ. Press.

JOHNSON, D A, LEATHERS, V L, MARTINEZ, A M, WALSH, D A, & FLETCHER, W H. 1993. Fluorescence resonance energy transfer within a heterochromatic cAMP-dependent protein kinase holoenzyme under equilibrium conditions: New insights into the conformational changes that result in cAMP-dependent activation. *Biochemistry,* **32**(25), 6402–6410.

JOHNSON, G. 2008. *The ten most beautiful experiments.* New York: Alfred A. Knopf.

JORDAN, G, DEEB, S S, BOSTEN, J M, & MOLLON, J D. 2010. The dimensionality of color vision in carriers of anomalous trichromacy. *J. Vis.,* **10**(8), art. no. 12.

JUDSON, H F. 1996. *The eighth day of creation: The makers of the revolution in biology.* Commemorative ed. Cold Spring Harbor NY: Cold Spring Harbor Laboratory Press.

KAMBE, M, ZHU, D, & KINOSHITA, S. 2011. Origin of retroreflection from a wing of the *Morpho* butterfly. *J. Phys. Soc. Jpn.,* **80**, art. no. 054801.

KARP, G, IWASA, J, & MARSHALL, W. 2016. *Karp's cell and molecular biology: Concepts and experiments.* 8th ed. Hoboken NJ: John Wiley and Sons.

KIM, YI RANG, KIM, SEONGHOON, CHOI, JIN WOO, CHOI, SUNG YONG, LEE, SANG-HEE, KIM, HOMIN, HAHN, SEI KWANG, KOH, GOU YOUNG, & YUN, S H. 2015. Bioluminescence-activated deep-tissue photodynamic therapy of cancer. *Theranostics,* **5**(8), 805–17.

KINDER, J M, & NELSON, P. 2015. *A student's guide to Python for physical modeling.* Princeton NJ: Princeton Univ. Press.

KINOSHITA, S. 2008. *Structural colors in the realm of nature.* Singapore: World Scientific.

KINOSHITA, S, YOSHIOKA, S, FUJII, Y, & OKAMOTO, N. 2002. Photophysics of structural color in the *Morpho* butterflies. *Forma,* **17**, 103–121.

KLAPOETKE, N C, MURATA, Y, KIM, S S, PULVER, S R, BIRDSEY-BENSON, A, CHO, Y K, MORIMOTO, T K, CHUONG, A S, CARPENTER, E J, TIAN, Z, WANG, J, XIE, Y, YAN, Z, ZHANG, Y, CHOW, B Y, SUREK, B, MELKONIAN, M, JAYARAMAN, V, CONSTANTINE-PATON, M, WONG, G K-S, & BOYDEN, E S. 2014. Independent optical excitation of distinct neural populations. *Nat. Methods,* **11**(3), 338–346.

KLEINERT, H. 2009. *Path integrals in quantum mechanics, statistics, polymer physics, and financial markets.* 5th ed. Hackensack NJ: World Scientific.

KOENIG, D, & HOFER, H. 2011. The absolute threshold of cone vision. *J. Vis.,* **11**(1), art. no. 21.

KRALJ, J M, HOCHBAUM, D R, DOUGLASS, A D, & COHEN, A E. 2011. Electrical spiking in *Escherichia coli* probed with a fluorescent voltage-indicating protein. *Science,* **333**(6040), 345–348.

KRALJ, J M, DOUGLASS, A D, HOCHBAUM, D R, MACLAURIN, D, & COHEN, A E. 2012. Optical recording of action potentials in mammalian neurons using a microbial rhodopsin. *Nat. Methods,* **9**(1), 90–95.

KRAMER, R H. 2014. Horizontal cells: Lateral interactions at the first synapse in the retina. *Chap. 11 of:* CHALUPA, L M, & WERNER, J S (Eds.), *The new visual neurosciences.* Cambridge MA: MIT Press.

LACOSTE, T D, MICHALET, X, PINAUD, F, CHEMLA, D S, ALIVISATOS, A P, & WEISS, S. 2000. Ultrahigh-resolution multicolor colocalization of single fluorescent probes. *Proc. Natl. Acad. Sci. USA,* **97**(17), 9461–9466.

LAKSHMINARAYANAN, V, & ENOCH, J M. 2010. Biological waveguides. *Chap. 8 of:* BASS, M, ENOCH, J M, & LAKSHMINARAYANAN, V (Eds.), *Handbook of optics,* 3rd ed., Vol. 3. New York: McGraw-Hill.

LAMB, T D. 2011. Evolution of the eye. *Sci. Am.,* **305**(1), 64–69.

LAMB, T D. 2016. Why rods and cones? *Eye (Lond.),* **30**(2), 179–185.

LAMB, T D, & PUGH, JR., E N. 2006. Phototransduction, dark adaptation, and rhodopsin regeneration: The proctor lecture. *Invest. Ophthalmol. Vis. Sci.,* **47**(12), 5138–5152.

LAMB, T D, COLLIN, S P, & PUGH, JR., E N. 2007. Evolution of the vertebrate eye: Opsins, photoreceptors, retina and eye cup. *Nat. Rev. Neurosci.,* **8**(12), 960–976.

LAND, M F, & NILSSON, D-E. 2006. General-purpose and special-purpose visual systems. *Chap. 5 of:* WARRANT, E J, & NILSSON, D-E (Eds.), *Invertebrate vision.* Cambridge MA: Cambridge Univ. Press.

LAND, M F, & NILSSON, D-E. 2012. *Animal eyes.* 2d ed. Oxford UK: Oxford Univ. Press.

LANDAU, R H, PÁEZ, M J, & BORDEIANU, C C. 2015. *Computational physics: Problem solving with computers.* 3rd ed. New York: Wiley-VCH. http://physics.oregonstate.edu/~rubin/Books/CPbook/index.html.

LANE, N. 2009. *Life Ascending: The ten great inventions of evolution.* New York: Norton.

LANGRIDGE, R, SEEDS, W E, WILSON, H R, HOOPER, C W, WILKINS, H F, & HAMILTON, L D. 1957. Molecular structure of deoxyribonucleic acid (DNA). *J. Biophys. Biochem. Cytol.*, **3**(5), 767–778.

LANNI, F, & KELLER, H E. 2011. Microscopy princples and optical systems. *Chap. 1 of:* YUSTE, R (Ed.), *Imaging: A laboratory manual.* Cold Spring Harbor NY: Cold Spring Harbor Laboratory Press.

LEE, J Y K, THAWANI, J P, PIERCE, J, ZEH, R, MARTINEZ-LAGE, M, CHANIN, M, VENEGAS, O, NIMS, S, LEARNED, K, KEATING, J, & SINGHAL, S. 2016. Intraoperative near-infrared optical imaging can localize gadolinium-enhancing gliomas during surgery. *Neurosurgery*, **79**(6), 856–871.

LEE, N K, KAPANIDIS, A N, WANG, Y, MICHALET, X, MUKHOPADHYAY, J, EBRIGHT, R H, & WEISS, S. 2005. Accurate FRET measurements within single diffusing biomolecules using alternating-laser excitation. *Biophys. J.*, **88**(4), 2939–2953.

LEONHARDT, U. 2010. *Essential quantum optics.* Cambridge UK: Cambridge Univ. Press.

LI, X, GUTIERREZ, D V, HANSON, M GARTZ, HAN, J, MARK, M D, CHIEL, H, HEGEMANN, P, LANDMESSER, L T, & HERLITZE, S. 2005. Fast noninvasive activation and inhibition of neural and network activity by vertebrate rhodopsin and green algae channelrhodopsin. *Proc. Natl. Acad. Sci. USA*, **102**(49), 17816–17821.

LILLYWHITE, P G. 1977. Single photon signals and transduction in an insect eye. *J. Comp. Physiol.*, **122**(2), 189–200.

LIN, D, BOYLE, M P, DOLLAR, P, LEE, H, LEIN, E S, PERONA, P, & ANDERSON, D J. 2011. Functional identification of an aggression locus in the mouse hypothalamus. *Nature*, **470**(7333), 221–226.

LINDEN, W VON DER, DOSE, V, & TOUSSAINT, U VON. 2014. *Bayesian probability theory: Applications in the physical sciences.* Cambridge UK: Cambridge Univ. Press.

LINDNER, M, SHOTAN, Z, & GARINI, Y. 2016. Rapid microscopy measurement of very large spectral images. *Optics Express*, **24**(9), 9511–9527.

LIPPINCOTT-SCHWARTZ, J. 2015. Profile of Eric Betzig, Stefan Hell, and W. E. Moerner, 2014 Nobel Laureates in Chemistry. *Proc. Natl. Acad. Sci. USA*, **112**(9), 2630–2632.

LIPPINCOTT-SCHWARTZ, J, & PATTERSON, G H. 2003. Development and use of fluorescent protein markers in living cells. *Science*, **300**(5616), 87–91.

LIPSON, A, LIPSON, S G, & LIPSON, H. 2011. *Optical physics.* 4th ed. Cambridge UK: Cambridge Univ. Press.

LIU, X, RAMIREZ, S, PANG, P T, PURYEAR, C B, GOVINDARAJAN, A, DEISSEROTH, K, & TONEGAWA, S. 2012. Optogenetic stimulation of a hippocampal engram activates fear memory recall. *Nature*, **484**(7394), 381–385.

LIVINGSTONE, M. 2002. *Vision and art: The biology of seeing.* New York: Harry N. Abrams.

LODISH, H, BERK, A, KAISER, C A, KRIEGER, M, BRETSCHER, A, PLOEGH, H, AMON, A, SCOTT, M P, & MARTIN, K. 2016. *Molecular cell biology.* 8th ed. New York: W H Freeman and Co.

LOUDON, R. 2000. *The quantum theory of light.* 3d ed. Oxford UK: Oxford Univ. Press.

LUCAS, A A, & LAMBIN, P. 2005. Diffraction by DNA, carbon nanotubes and other helical nanostructures. *Rep. Prog. Phys.*, **68**, 1–69.

LUCAS, A A, LAMBIN, P, MAIRESSE, R, & MATHOT, M. 1999. Revealing the backbone structure of B-DNA from laser optical simulations of its X-ray diffraction diagram. *J. Chem. Educ.*, **76**, 378–383.

LUMMER, O, & PRINGSHEIM, E. 1899. 1. Die Vertheilung der Energie im Spectrum des schwarzen Körpers und des blanken Platins; 2. Temperaturbestimmung fester glühender Körper. *Verhandlung der Deutschen Physikalischen Gesellschaft*, **1**(12 "Vgl. oben S. 214"), 215–235.

LUO, D-G, XUE, T, & YAU, K-W. 2008. How vision begins: An odyssey. *Proc. Natl. Acad. Sci. USA*, **105**(29), 9855–9862.

MAHON, B. 2003. *The man who changed everything: The life of James Clerk Maxwell.* Chichester UK: Wiley.

MAISELS, M J, & MCDONAGH, A F. 2008. Phototherapy for neonatal jaundice. *N. Engl. J. Med.*, **358**(9), 920–928.

MANDEL, L, & WOLF, E. 1995. *Optical coherence and quantum optics.* Cambridge UK: Cambridge Univ. Press.

MARKS, F, KLINGMÜLLER, U, & MÜLLER-DECKER, K. 2009. *Cellular signal processing: An introduction to the molecular mechanisms of signal transduction.* New York: Garland Science.

MASLAND, R H. 1986. The functional architecture of the retina. *Sci. Am.*, **255**(6), 102–111.

MCCALL, R P. 2010. *Physics of the human body.* Baltimore MD: Johns Hopkins Univ. Press.

MEIR, Y, JAKOVLJEVIC, V, OLEKSIUK, O, SOURJIK, V, & WINGREEN, N S. 2010. Precision and kinetics of adaptation in bacterial chemotaxis. *Biophys. J.*, **99**(9), 2766–2774.

MERTZ, J. 2010. *Introduction to optical microscopy.* Greenwood Village, CO: Roberts and Co.

MILLS, F C, JOHNSON, M L, & ACKERS, G K. 1976. Oxygenation-linked subunit interactions in human hemoglobin. *Biochemistry*, **15**, 5350–5362.

MILO, R, & PHILLIPS, R. 2016. *Cell Biology by the numbers.* New York: Garland Science.

MILOSAVLJEVIC, N, CEHAJIC-KAPETANOVIC, J, PROCYK, C A, & LUCAS, R J. 2016. Chemogenetic activation of melanopsin retinal ganglion cells induces signatures of arousal and/or anxiety in mice. *Curr. Biol.*, **26**(17), 2358–2363.

MIYAWAKI, A, NAGAI, T, & MIZUNO, H. 2011. Genetic calcium indicators: Fast measurements using yellow cameleons. *Chap. 26 of:* YUSTE, R (Ed.), *Imaging: A laboratory manual.* Cold Spring Harbor NY: Cold Spring Harbor Laboratory Press.

MORTENSEN, K I, CHURCHMAN, L S, SPUDICH, J A, & FLYVBJERG, H. 2010. Optimized localization analysis for single-molecule tracking and super-resolution microscopy. *Nat. Methods*, **7**(5), 377–381.

MORTON, O. 2008. *Eating the sun: How plants power the planet.* New York: Harper.

MÜLLER-KIRSTEN, H J W. 2006. *Introduction to quantum mechanics: Schrödinger equation and path integral.* Hackensack NJ: World Scientific.

MURPHY, D B, & DAVIDSON, M W. 2013. *Fundamentals of light microscopy and electronic imaging.* 2d ed. Wiley-Blackwell.

NADEAU, J. 2012. *Introduction to experimental biophysics.* Boca Raton FL: CRC Press.

NAGEL, G, SZELLAS, T, HUHN, W, KATERIYA, S, ADEISHVILI, N, BERTHOLD, P, OLLIG, D, HEGEMANN, P, & BAMBERG, E. 2003. Channelrhodopsin-2, a directly light-gated cation-selective membrane channel. *Proc. Natl. Acad. Sci. USA*, **100**(24), 13940–13945.

NAGEL, G, BRAUNER, M, LIEWALD, J F, ADEISHVILI, N, BAMBERG, E, & GOTTSCHALK, A. 2005. Light activation of channelrhodopsin-2 in excitable cells of *Caenorhabditis elegans* triggers rapid behavioral responses. *Curr. Biol.*, **15**(24), 2279–2284.

NAKATANI, K, & YAU, K W. 1988. Guanosine 3',5'-cyclic monophosphate-activated conductance studied in a truncated rod outer segment of the toad. *J. Physiol. (Lond.)*, **395**, 731–753.

NASSAU, K. 2003. The physics and chemistry of color: The 15 mechanisms. *Pages 247–280 of:* SHEVELL, S K (Ed.), *The science of color*, 2d ed. Amsterdam: Elsevier.

NEEDHAM, T. 1997. *Visual complex analysis.* Oxford UK: Oxford Univ. Press.

NELSON, P. 2014. *Biological physics: Energy, information, life—With new art by David Goodsell.* New York: W. H. Freeman and Co.

NELSON, P. 2015. *Physical models of living systems.* New York: W. H. Freeman and Co.

NELSON, P, & DODSON, T. 2015. *Student's guide to* MATLAB *for physical modeling.* `https://github.com/NelsonUpenn/PMLS-MATLAB-Guide` .

NEWMAN, M. 2013. *Computational physics.* Rev. and expanded ed. CreateSpace Publishing.

NEWMAN, R H, FOSBRINK, M D, & ZHANG, J. 2011. Genetically encodable fluorescent biosensors for tracking signaling dynamics in living cells. *Chem. Rev.*, **111**(5), 3614–3666.

NICHOLLS, J G, MARTIN, A R, FUCHS, P A, BROWN, D A, DIAMOND, M E, & WEISBLAT, D A. 2012. *From neuron to brain.* 5th ed. Sunderland MA: Sinauer Associates.

NILSSON, D-E, WARRANT, E J, JOHNSEN, S, HANLON, R T, & SHASHAR, N. 2012. A unique advantage for giant eyes in giant squid. *Curr. Biol.*, **22**, 1–6.

NIMMERJAHN, A. 2011. Two-photon imaging of microglia in the mouse cortex in vivo. *Chap. 87, pages 961–979 of:* HELMCHEN, F, & KONNERTH, A (Eds.), *Imaging in neuroscience: A laboratory manual.* Cold Spring Harbor NY: Cold Spring Harbor Laboratory Press.

NIRENBERG, S, & PANDARINATH, C. 2012. Retinal prosthetic strategy with the capacity to restore normal vision. *Proc. Natl. Acad. Sci. USA*, **109**(37), 15012–15017.

NOLTE, J. 2010. *Essentials of the human brain.* Philadelphia PA: Mosby.

NOLTING, B. 2009. *Methods in modern biophysics.* 3d ed. New York: Springer.

NORDLUND, T. 2011. *Quantitative understanding of biosystems: An introduction to biophysics.* Boca Raton FL: CRC Press.

OBER, R J, RAM, S, & WARD, E S. 2004. Localization accuracy in single-molecule microscopy. *Biophys. J.*, **86**(2), 1185–1200.

OKAWA, H, & SAMPATH, A P. 2007. Optimization of single-photon response transmission at the rod-to-rod bipolar synapse. *Physiology (Bethesda)*, **22**, 279–286.

OKAWA, H, MIYAGISHIMA, K J, ARMAN, A C, HURLEY, J B, FIELD, G D, & SAMPATH, A P. 2010. Optimal processing of photoreceptor signals is required to maximize behavioural sensitivity. *J. Physiol. (Lond.)*, **588**(11), 1947–1960.

OPPENHEIMER, J R. 1941. Internal conversion in photosynthesis. *Phys. Rev.*, **60**(2), 158.

OTTO, S P, & DAY, T. 2007. *Biologist's guide to mathematical modeling in ecology and evolution.* Princeton NJ: Princeton Univ. Press.

PACKER, O, & WILLIAMS, D R. 2003. Light, the retinal image, and photoreceptors. *Pages 41–102 of:* SHEVELL, S K (Ed.), *The science of color*, 2d ed. Amsterdam: Elsevier.

PAIS, A. 1982. *Subtle is the lord: The science and the life of Albert Einstein.* Oxford, UK: Oxford University Press.

PALCZEWSKA, G, DONG, Z, GOLCZAK, M, HUNTER, J J, WILLIAMS, D R, ALEXANDER, N S, & PALCZEWSKI, K. 2014. Noninvasive two-photon microscopy imaging of mouse retina and retinal pigment epithelium through the pupil of the eye. *Nat. Med.*, **20**(7), 785–789.

PATTERSON, G H, & LIPPINCOTT-SCHWARTZ, J. 2002. A photoactivatable GFP for selective photolabeling of proteins and cells. *Science*, **297**(5588), 1873–1877.

PAWLEY, J B (Ed.). 2006. *Handbook of biological confocal microscopy.* 3d ed. New York: Springer.

PEARSON, B J, & JACKSON, D P. 2010. A hands-on introduction to single photons and quantum mechanics for undergraduates. *Am. J. Phys.*, **78**(5), 471–484.

PEATROSS, J, & WARE, M. 2015. *Physics of light and optics.* Available at `http://optics.byu.edu`.

PEDROTTI, F L, PEDROTTI, L S, & PEDROTTI, L M. 2007. *Introduction to optics.* 3d ed. San Francisco CA: Pearson.

PHAN, N, CHENG, M F, BESSARAB, D A, & KRIVITSKY, L A. 2014. Interaction of fixed number of photons with retinal rod cells. *Phys. Rev. Lett.*, **112**(21), 213601.

PHILLIPS, R, KONDEV, J, THERIOT, J, & GARCIA, H. 2012. *Physical biology of the cell.* 2d ed. New York: Garland Science.

PIERSCIONEK, B K. 2010. Gradient index optics in the eye. *Chap. 19 of:* BASS, M, ENOCH, J M, & LAKSHMI-NARAYANAN, V (Eds.), *Handbook of optics*, 3rd ed., Vol. 3. New York: McGraw-Hill.

POLYAK, S. 1957. *The vertebrate visual system.* Chicago IL: Univ. of Chicago Press. Ed. H Klüver.

PUMIR, A, GRAVES, J, RANGANATHAN, R, & SHRAIMAN, B I. 2008. Systems analysis of the single photon response in invertebrate photoreceptors. *Proc. Natl. Acad. Sci. USA*, **105**(30), 10354–10359.

PURVES, D, AUGUSTINE, G J, FITZPATRICK, D, HALL, W C, LAMANTIA, A-S, & WHITE, L E (Eds.). 2012. *Neuroscience.* 5th ed. Sunderland MA: Sinauer Associates.

RHODES, G. 2006. *Crystallography made crystal clear: A guide for users of macromolecular models.* 3d ed. Boston MA: Elsevier/Academic Press.

RIEKE, F. 2008. Seeing in the dark: Retinal processing and absolute visual threshold. *Pages 393–412 of:* MASLAND, R H, & ALBRIGHT, T (Eds.), *The senses: A comprehensive reference*, Vol. 1. San Diego CA: Academic Press.

RODIECK, R W. 1998. *The first steps in seeing.* Sunderland MA: Sinauer Associates.

ROORDA, A, & WILLIAMS, D R. 1999. The arrangement of the three cone classes in the living human eye. *Nature*, **397**(6719), 520–522.

ROSE, A. 1953. Quantum and noise limitations of the visual process. *J. Opt. Soc. Am.*, **43**, 715–716.

ROSENBERG, S A, QUINLAN, M E, FORKEY, J N, & GOLDMAN, Y E. 2005. Rotational motions of macro-molecules by single-molecule fluorescence microscopy. *Acc. Chem. Res.*, **38**(7), 583–593.

ROSSI-FANELLI, A, & ANTONINI, E. 1958. Studies on the oxygen and carbon monoxide equilibria of human myoglobin. *Arch. Biochem. Biophys.*, **77**, 478–492.

Roy, R, Hohng, S, & Ha, T. 2008. A practical guide to single-molecule FRET. *Nat. Methods,* **5**(6), 507–516.

Ruby, S L, & Bolef, D I. 1960. Acoustically modulated γ rays from Fe^{57}. *Phys. Rev. Lett.,* **5**, 5–7.

Rupp, B. 2010. *Biomolecular crystallography: Principles, practice, and application to structural biology.* New York: Garland Science.

Rust, M J, Bates, M, & Zhuang, X. 2006. Sub-diffraction-limit imaging by stochastic optical reconstruction microscopy (STORM). *Nat. Methods,* **3**(10), 793–795.

Sakitt, B. 1972. Counting every quantum. *J. Physiol. (Lond.),* **223**(1), 131–150.

Salzberg, B M, Davila, H V, & Cohen, L B. 1973. Optical recording of impulses in individual neurones of an invertebrate central nervous system. *Nature,* **246**(5434), 508–509.

Sampath, A P. 2014. Information transfer at the rod-to-rod bipolar cell synapse. *Chap. 5 of:* Chalupa, L M, & Werner, J S (Eds.), *The new visual neurosciences.* Cambridge MA: MIT Press.

Saxby, G. 2002. *The science of imaging: An introduction.* Bristol, UK: Institute of Physics Pub.

Schlosshauer, M A. 2007. *Decoherence and the quantum-to-classical transition.* New York: Springer.

Schmidt, T M, Chen, S-K, & Hattar, S. 2011. Intrinsically photosensitive retinal ganglion cells: many subtypes, diverse functions. *Trends Neurosci.,* **34**(11), 572–580.

Schmidt, T M, Alam, N M, Chen, Shan, Kofuji, P, Li, W, Prusky, G T, & Hattar, S. 2014. A role for melanopsin in alpha retinal ganglion cells and contrast detection. *Neuron,* **82**(4), 781–788.

Schoonover, C. 2010. *Portraits of the mind.* New York: Abrams.

Schröck, E, du Manoir, S, Veldman, T, Schoell, B, Wienberg, J, Ferguson-Smith, M A, Ning, Y, Ledbetter, D H, Bar-Am, I, Soenksen, D, Garini, Y, & Ried, T. 1996. Multicolor spectral karyotyping of human chromosomes. *Science,* **273**(5274), 494–497.

Schulman, L S. 2005. *Techniques and applications of path integration.* Mineola NY: Dover Publications.

Schumacher, B, & Westmoreland, M D. 2010. *Quantum processes, systems, and information.* Cambridge UK: Cambridge Univ. Press.

Schwab, I R. 2012. *Evolution's witness: How eyes evolved.* New York: Oxford Univ. Press.

Selvin, P R, & Ha, T (Eds.). 2008. *Single-molecule techniques: A laboratory manual.* Cold Spring Harbor NY: Cold Spring Harbor Laboratory Press.

Selvin, P R, Lougheed, T, Tonks Hoffman, M, Park, H, Balci, H, Blehm, B H, & Toprak, E. 2008. *In vitro* and *in vivo* FIONA and other acronyms for watching molecular motors walk. *Pages 37–72 of:* Selvin, P R, & Ha, T (Eds.), *Single-molecule techniques: A laboratory manual.* Cold Spring Harbor NY: Cold Spring Harbor Laboratory Press.

Şener, M, Strümpfer, J, Hsin, J, Chandler, D, Scheuring, S, Hunter, C N, & Schulten, K. 2011. Förster energy transfer theory as reflected in the structures of photosynthetic light-harvesting systems. *ChemPhysChem,* **12**(3), 518–531.

Shankar, R. 1995. *Basic training in mathematics: A fitness program for science students.* New York: Plenum.

Sharonov, A, & Hochstrasser, R M. 2006. Wide-field subdiffraction imaging by accumulated binding of diffusing probes. *Proc. Natl. Acad. Sci. USA,* **103**(50), 18911–18916.

Shevell, S K (Ed.). 2003. *The science of color.* 2d ed. Elsevier.

Shevtsova, E, Hansson, C, Janzen, D H, & Kjærandsen, J. 2011. Stable structural color patterns displayed on transparent insect wings. *Proc. Natl. Acad. Sci. USA,* **108**(2), 668–673.

Shtengel, G, Galbraith, J A, Galbraith, C G, Lippincott-Schwartz, J, Gillette, J M, Manley, S, Sougrat, R, Waterman-Storer, C M, Kanchanawong, P, Davidson, M W, Fetter, R D, & Hess, H F. 2009. Interferometric fluorescent super-resolution microscopy resolves 3D cellular ultrastructure. *Proc. Natl. Acad. Sci. USA,* **106**(9), 3125–3130.

Shtengel, G, Wang, Y, Zhang, Z, Goh, W I, Hess, H F, & Kanchanawong, P. 2014. Imaging cellular ultrastructure by PALM, iPALM, and correlative iPALM-EM. *Meth. Cell Biol.,* **123**, 273–294.

Shubin, N. 2008. *Your inner fish: A journey into the 3.5-billion-year history of the human body.* New York: Pantheon Books.

Siegel, M S, & Isacoff, E Y. 1997. A genetically encoded optical probe of membrane voltage. *Neuron,* **19**(4), 735–741.

SILVER, N. 2012. *The signal and the noise.* London: Penguin.

SIMONSON, P D, & SELVIN, P R. 2011. FIONA: Nanometer fluorescence imaging. *Chap. 33 of:* YUSTE, R (Ed.), *Imaging: A laboratory manual.* Cold Spring Harbor NY: Cold Spring Harbor Laboratory Press.

SINDBERT, S, KALININ, S, NGUYEN, H, KIENZLER, A, CLIMA, L, BANNWARTH, W, APPEL, B, MÜLLER, S, & SEIDEL, C A M. 2011. Accurate distance determination of nucleic acids via Förster resonance energy transfer: implications of dye linker length and rigidity. *J. Am. Chem. Soc.*, **133**(8), 2463–2480.

SMALL, A R, & PARTHASARATHY, R. 2014. Superresolution localization methods. *Annu. Rev. Phys. Chem.*, **65**, 107–125.

SMITH, G S. 2005. Human color vision and the unsaturated blue color of the daytime sky. *Am. J. Phys.*, **73**(7), 590–597.

SMITH, G S. 2007. The polarization of skylight: An example from nature. *Am. J. Phys.*, **75**(1), 25–35.

SMITH, G S. 2009. Structural color of *Morpho* butterflies. *Am. J. Phys.*, **77**(11), 1010–1019.

SNOWDEN, R J, THOMPSON, P, & TROSCIANKO, T. 2012. *Basic vision: An introduction to visual perception.* Revised ed. Oxford UK: Oxford Univ. Press.

SOURJIK, V, & WINGREEN, N S. 2012. Responding to chemical gradients: Bacterial chemotaxis. *Curr. Opin. Cell Biol.*, **24**(2), 262–268.

ST-PIERRE, F, MARSHALL, J D, YANG, Y, GONG, Y, SCHNITZER, M J, & LIN, M Z. 2014. High-fidelity optical reporting of neuronal electrical activity with an ultrafast fluorescent voltage sensor. *Nat. Neurosci.*, **17**(6), 884–889.

STACHEL, J (Ed.). 1998. *Einstein's miraculous year: Five papers that changed the face of physics.* Princeton NJ: Princeton Univ. Press.

STEFANI, F, HOOGENBOOM, J P, & BARKAI, E. 2009. Beyond quantum jumps: Blinking nanoscale light emitters. *Phys. Today*, **62**(2), 34–39.

STERLING, P. 2004a. How retinal circuits optimize the transfer of visual information. *Chap. 17, pages 234–259 of:* CHALUPA, L M, & WERNER, J S (Eds.), *The visual neurosciences*, Vol. 1. Cambridge MA: MIT Press.

STERLING, P. 2004b. Retina. *In:* SHEPHERD, G M (Ed.), *The synaptic organization of the brain*, 5th ed. Oxford UK: Oxford Univ. Press.

STERLING, P. 2013. Some principles of retinal design: The Proctor lecture. *Invest. Ophthalmol. Vis. Sci.*, **54**(3), 2267–2275.

STERLING, P, & LAUGHLIN, S. 2015. *Principles of neural design.* Cambridge MA: MIT Press.

STEVEN, A C, BAUMEISTER, W, JOHNSON, L N, & PERHAM, R N. 2016. *Molecular biology of assemblies and machines.* New York NY: Garland Science.

STILES, W, & BURCH, J. 1959. NPL colour-matching investigation: Final report (1958). *Opt. Acta*, **6**(1), 1–26.

STONE, A D. 2013. *Einstein and the quantum: The quest of the valiant Swabian.* Princeton NJ: Princeton Univ. Press.

STOWE, S. 1980. Rapid synthesis of photoreceptor membrane and assembly of new microvilli in a crab at dusk. *Cell Tissue Res.*, **211**(3), 419–440.

STRÜMPFER, J, SENER, M, & SCHULTEN, K. 2012. How quantum coherence assists photosynthetic light harvesting. *J. Phys. Chem. Lett.*, **3**(4), 536–542.

STYER, D F. 2000. *The strange world of quantum mechanics.* Cambridge UK: Cambridge Univ. Press.

SWEENEY, A, JIGGINS, C, & JOHNSEN, S. 2003. Insect communication: Polarized light as a butterfly mating signal. *Nature*, **423**(6935), 31–32.

TAYLOR, W R, & SMITH, R G. 2004. Transmission of scotopic signals from the rod to rod-bipolar cell in the mammalian retina. *Vision Res.*, **44**(28), 3269–3276.

THOMPSON, R E, LARSON, D R, & WEBB, W W. 2002. Precise nanometer localization analysis for individual fluorescent probes. *Biophys. J.*, **82**(5), 2775–2783.

TIAN, L, HIRES, S A, & LOOGER, L L. 2011. Imaging neuronal activity with genetically encoded calcium indicators. *Chap. 8 of:* HELMCHEN, F, & KONNERTH, A (Eds.), *Imaging in neuroscience: A laboratory manual.* Cold Spring Harbor NY: Cold Spring Harbor Laboratory Press.

TINSLEY, J N, MOLODTSOV, M I, PREVEDEL, R, WARTMANN, D, ESPIGULÉ-PONS, J, LAUWERS, M, & VAZIRI, A. 2016. Direct detection of a single photon by humans. *Nat. Commun.*, **7**, art. no. 12172.

TOMITA, T, KANEKO, A, MURAKAMI, M, & PAUTLER, E. 1967. Spectral response curves of single cones in the carp. *Vision Res.*, **7**, 519–531.

TOPRAK, E, ENDERLEIN, J, SYED, S, MCKINNEY, S A, PETSCHEK, R G, HA, T, GOLDMAN, Y E, & SELVIN, P R. 2006. Defocused orientation and position imaging (DOPI) of myosin V. *Proc. Natl. Acad. Sci. USA*, **103**(17), 6495–6499.

TOPRAK, E, KURAL, C, & SELVIN, P R. 2010. Super-accuracy and super-resolution: Getting around the diffraction limit. *Meth. Enzymol.*, **475**, 1–26.

TOWNES-ANDERSON, E, MACLEISH, P R, & RAVIOLA, E. 1985. Rod cells dissociated from mature salamander retina: Ultrastructure and uptake of horseradish peroxidase. *J. Cell Biol.*, **100**(1), 175–188.

TOWNES-ANDERSON, E, DACHEUX, R F, & RAVIOLA, E. 1988. Rod photoreceptors dissociated from the adult rabbit retina. *J. Neurosci.*, **8**(1), 320–331.

TOWNSEND, J S. 2010. *Quantum physics: A fundamental approach to modern physics.* Sausalito CA: University Science Books.

TRUONG, K, SAWANO, A, MIYAWAKI, A, & IKURA, M. 2007. Calcium indicators based on calmodulin-fluorescent protein fusions. *Meth. Mol. Biol.*, **352**, 71–82.

USTIONE, A, & PISTON, D W. 2011. A simple introduction to multiphoton microscopy. *J. Microsc.*, **243**(3), 221–226.

VAFABAKHSH, R, & HA, T. 2012. Extreme bendability of DNA less than 100 base pairs long revealed by single-molecule cyclization. *Science*, **337**(6098), 1097–1101.

VAN DE KRAATS, J, & VAN NORREN, D. 2007. Optical density of the aging human ocular media in the visible and the UV. *J. Opt. Soc. Am. A*, **24**(7), 1842–1857.

VAN DER MEER, B W, COKER III, G, & CHEN, S-Y. 1994. *Resonance energy transfer: Theory and data.* New York: VCH.

VAN DER STRATEN, P, & METCALF, H. 2016. *Atoms and molecules interacting with light: Atomic physics for the laser era.* Cambridge UK: Cambridge Univ. Press.

VAN DER VELDEN, H A. 1944. Over het aantal lichtquanta dat nodig is voor een lichtprikkel bij het menselijk oog. *Physica*, **11**, 179–189.

VAN HOLDE, K E, JOHNSON, W C, & HO, P S. 2006. *Principles of physical biochemistry.* 2d ed. Upper Saddle River NJ: Prentice Hall.

VAN MAMEREN, J, WUITE, G J L, & HELLER, I. 2011. Introduction to optical tweezers: Background, system designs, and commercial solutions. *Meth. Mol. Biol.*, **783**, 1–20.

VAN ROSSUM, M C, & SMITH, R G. 1998. Noise removal at the rod synapse of mammalian retina. *Vis. Neurosci.*, **15**(5), 809–821.

WAGNIÈRES, G, MCWILLIAMS, A, & LAM, S. 2003. Lung cancer imaging by autofluorescence bronchoscopy. *Pages 361–396 of:* MYCEK, M-A, & POGUE, B (Eds.), *Handbook of biomedical fluorescence.* New York: Marcel Dekker.

WAGNIÈRES, G, JICHLINSKI, P, LANGE, N, KUCERA, P, & VAN DEN BERGH, H. 2014. Detection of bladder cancer by fluorescence cystoscopy: From bench to bedside—The HEXVIX story. *Chap. 36, pages 411–425 of:* HAMBLIN, M R, & HUANG, Y-Y (Eds.), *Handbook of photomedicine.* Boca Raton FL: CRC Press.

WALMSLEY, I. 2015. *Light: A very short introduction.* Oxford UK: Oxford Univ. Press.

WANG, S, SENGEL, C, EMERSON, M M, & CEPKO, C L. 2014. A gene regulatory network controls the binary fate decision of rod and bipolar cells in the vertebrate retina. *Dev. Cell*, **30**(5), 513–527.

WARRANT, E J, & NILSSON, D-E (Eds.). 2006. *Invertebrate vision.* Cambridge UK: Cambridge Univ. Press.

WEHNER, R. 2003. Desert ant navigation: How miniature brains solve complex tasks. *J. Comp. Physiol. A Neuroethol. Sens. Neural Behav. Physiol.*, **189**(8), 579–588.

WERBLIN, F, & ROSKA, B. 2007. The movies in our eyes. *Sci. Am.*, **296**(4), 72–79.

WILSON, T, & HASTINGS, J W. 2013. *Bioluminescence: Living lights, lights for living.* Cambridge MA: Harvard Univ. Press.

WOODWORTH, G G. 2004. *Biostatistics: A Bayesian introduction.* Hoboken NJ: Wiley-Interscience.

XU, K, ZHONG, G, & ZHUANG, X. 2013. Actin, spectrin, and associated proteins form a periodic cytoskeletal structure in axons. *Science*, **339**(6118), 452–456.

YANG, H H, ST-PIERRE, F, SUN, X, DING, X, LIN, M Z, & CLANDININ, T R. 2016. Subcellular imaging of voltage and calcium signals reveals neural processing in vivo. *Cell*, **166**(1), 245–257.

YAU, K-W, & HARDIE, R C. 2009. Phototransduction motifs and variations. *Cell*, **139**(2), 246–264.

YAU, K W, MATTHEWS, G, & BAYLOR, D A. 1979. Thermal activation of the visual transduction mechanism in retinal rods. *Nature*, **279**(5716), 806–807.

YILDIZ, A, & VALE, R D. 2011. Total internal reflection fluorescence microscopy. *Chap. 38 of:* YUSTE, R (Ed.), *Imaging: A laboratory manual.* Cold Spring Harbor NY: Cold Spring Harbor Laboratory Press.

YILDIZ, A, FORKEY, J N, MCKINNEY, S A, HA, T, GOLDMAN, Y E, & SELVIN, P R. 2003. Myosin V walks hand-over-hand: Single fluorophore imaging with 1.5-nm localization. *Science*, **300**(5628), 2061–2065.

YUN, S H, & KWOK, S J J. 2017. Light in diagnosis, therapy and surgery. *Nat. Biomed. Eng.*, **1**(1), art. no. 0008 (16pp).

ZANGWILL, A. 2013. *Modern electrodynamics.* Cambridge UK: Cambridge Univ. Press.

ZELLWEGER, M, GROSJEAN, P, GOUJON, D, MONNIER, P, VAN DEN BERGH, H, & WAGNIÈRES, G. 2001. In vivo autofluorescence spectroscopy of human bronchial tissue to optimize the detection and imaging of early cancers. *J. Biomed. Opt.*, **6**(1), 41–51.

ZEMELMAN, B V, LEE, G A, NG, M, & MIESENBÖCK, G. 2002. Selective photostimulation of genetically chARGed neurons. *Neuron*, **33**(1), 15–22.

ZHANG, F, TSAI, H-C, AIRAN, R D, STUBER, G D, ADAMANTIDIS, A R, DE LECEA, L, BONCI, A, & DEISSEROTH, K. 2011. Optogenetics in freely-moving mammals: Dopamine and reward. *Chap. 80 of:* HELMCHEN, F, & KONNERTH, A (Eds.), *Imaging in neuroscience: A laboratory manual.* Cold Spring Harbor NY: Cold Spring Harbor Laboratory Press.

ZHAOPING, L. 2014. *Understanding vision: Theory, models, and data.* Oxford UK: Oxford University Press.

ZHONG, H. 2011. Photoactivated localization microscopy (PALM). *Chap. 34 of:* YUSTE, R (Ed.), *Imaging: A laboratory manual.* Cold Spring Harbor NY: Cold Spring Harbor Laboratory Press.

ZHOU, X X, PAN, M, & LIN, M Z. 2015. Investigating neuronal function with optically controllable proteins. *Front. Mol. Neurosci.*, **8**, art. no. 37.

ZIMMER, C, & EMLEN, D J. 2013. *Evolution: Making sense of life.* Greenwood Village CO: Roberts and Company.

Credits

Figures and quotations

Several images in this book are based on data obtained from the RCSB Protein Data Bank (Berman et al., 2000; `http://www.rcsb.org/`), which is managed by two members of the RCSB (Rutgers University and UCSD) and funded by NSF, NIGMS, DOE, NLM, NCI, NINDS, and NIDDK. The corresponding entries below include the PDB ID code.

Courtesy Constance Cepko, Harvard Medical School Department of Genetics and Ophthalmology and Howard Hughes Medical Institute, and Sui Wang, Stanford Medical School Department of Ophthalmology.

Frontispiece: Art by David S Goodsell. Used by permission.

Epigraph to the book: The lines from "For the Conjunction of Two Planets." Copyright ©2016 by the Adrienne Rich literary Trust. Copyright ©1951 by Adrienne Rich, from *COLLECTED POEMS: 1950–2012* by Adrienne Rich. Used by permission of W. W. Norton & Company, Inc.

Epigraph to To the Student: Galen of Pergamum, second century, *De alimentorum facultatibus*, 1.1, 6.480K. Konrad Koch, et al., Corpus Medicorum Graecorum 5.4.2 (Liepzig: Teubner, 1923).

Epigraph to Prologue: From *Wonderful life* by Stephen Jay Gould (WW Norton and Co, 1989). Used by permission.

Part I opener: Courtesy Richard H Masland, Harvard University; used by permission.

Epigraph to Chapter 1: June 1913 recommendation by Max Planck, Walther Nernst, Heinrich Rubens and Emil Warburg nominating Einstein to the Prussian Academy. From *Collected papers of Albert Einstein*, vol. 5 (Princeton Univ. Press, 1993) Martin J Klein, A J Kox, and R Schulmann, eds., document 445. Used by permission.

Fig. 1.2: Redrawn with permission from Nelson. Old and new results about single-photon sensitivity in human vision. *Physical Biology* (2016) vol. 13 (2) art. 025001.

Fig. 1.3: From Eric Sloane, *Diary of an early American boy* (Dover Publications, Mineola NY, ©2004); used by permission.

Fig. 1.4: From Rose. Quantum and noise limitations of the visual process. *J. Opt. Soc. Am.* (1953) vol. 43 pp. 715-716.

Fig. 1.7: PDB `1ttd` (DOI: `10.1006/jmbi.1998.2062`). Art by David S Goodsell. Used by permission.

Fig. 1.11: (a) ©Sierra Blakely; (b) Courtesy of Steven H D Haddock, Monterey Bay Aquarium Research Institute; used by permission.

Fig. 2.3: Courtesy Prof. P. Jichlinski, CHUV University Hospital; used by permission.

Fig. 2.5: Image courtesy Nico Stuurman, HHMI/University of California at San Francisco. Used by permission.

Fig. 2.6: (b) From *Biological Physics*, 1/e, by Philip Nelson, ©2013 by W.H. Freeman and Company. Used with permission of the publisher.

Fig. 2.7: PDB `3tad, 1auv, 1zbd, 1d5t, 2nwl, 2yd5, 1lar, 2id5, 3tbd, 1biw, 3sph, 4gnk, 1mt5`. Used by permission of David S Goodsell and Timothy Herman.

Fig. 2.9: Redrawn similar to Fig 2a, top trace only, page 1264 of: Boyden et al. Millisecond-timescale, genetically targeted optical control of neural activity. *Nat. Neurosci.* (2005) vol. 8 (9) pp. 1263–1268.

Fig. 2.10: Adapted from Han & Boyden, 2007 ©2007 Han, Boyden, as found at `http://dx.doi.org/10.1371/journal.pone.0000299`. Used and available under the Creative Commons Attribution 4.0 International license (`http://creativecommons.org/licenses/by/4.0/`).

Fig. 2.14: PDB `3wld`. (a) Redrawn from Fig. 1a, page 3 of: Broussard, G J, Liang, R, & Tian, L. 2014. Monitoring activity in neural circuits with genetically encoded indicators. *Front. Mol. Neurosci.*, 7, art. no. 97, as found

Fig. 2.15: Photograph by the Belfield Research Group, University of Central Florida and published with the permission of Prof. Kevin D Belfield.

Fig. 2.17: Reprinted by permission of Cold Spring Harbor Laboratory Press.

Fig. 2.19: Molecules were drawn from data at `https://www.ebi.ac.uk/chebi/searchId.do?chebiId=51247` and `http://www.ebi.ac.uk/pdbe-srv/pdbechem/chemicalCompound/show/FLU`.

Fig. 2.23: PDB *lac*: `1lbh`/`1efa` (PubMed: `8638105` and `10700279`); nucleosome: `1zbb` (DOI: `10.1038/nature03686`); bacteriophage phiX174: `1cd3` (DOI: `10.1006/jmbi.1999.2699`). Art by David S Goodsell. Used by permission.

Fig. 2.27: ©1942 Emerson and Lewis. *Journal of General Physiology.* 25:579–595. DOI: `10.1085/jgp.25.4.579`. Used by permission.

Fig. 2.28: Courtesy iBio, Inc.; used by permission.

Fig. 2.29: PDB `1jb0` (DOI: `10.1038/35082000`). Art by David S Goodsell. Used by permission.

Fig. 3.3: (b) ©2016 by Nature Connect Pty. Ltd. `http://www.steveparish-natureconnect.com.au`. Used by permission. (c) Adapted with permission.

Fig. 3.4: `http://www.bealecorner.org/best/measure/cf-spectrum/Fraunhofer_Lines_Jan3-07.jpg`.

Fig. 3.6: Courtesy Mojca Čepič, University of Ljubljana, Slovenia. Used by permission.

Fig. 3.9: (a,b) Curcio et al. Human photoreceptor topography. *J. Comp. Neurol.* (1990) vol. 292 (4) pp. 497-523. Copyright ©1990 Wiley-Liss, Inc. Used with permission of John Wiley & Sons via Copyright Clearance Center.

Fig. 3.14: Left panel: Fig. 4, upper left panel labeled HS, page 9673 of: Hofer, H, Carroll, J, Neitz, J, Neitz, M, & Williams, D R. 2005. Organization of the human trichromatic cone mosaic. *J. Neurosci.*, 25(42), 9669–9679. Used with permission.. Middle and right panels: John S Werner, and Leo M Chalupa, eds., The New Visual Neurosciences, Figure 33.1, panels (e) and (q), ©2013 Massachusetts Institute of Technology, used by permission of The MIT Press.

Fig. 3.15: From Tanja Gabrecht, Thomas Glanzmann, Lutz Freitag, Bernd-Claus Weber, Hubert van den Bergh, and Georges Wagnières, Optimized autofluorescence bronchoscopy using additional backscattered red light, *J. Biomedical Optics*, vol. 12 (6), 2007, art. no. 0642016. Used with permission from SPIE and Georges Wagnières.

Fig. 3.16: (a–c) Courtesy Yuval Garini, Bar Ilan University; used by permission. (d) From Schröck et al., 1996: Readers may view, browse, and/or download material for temporary copying purposes only, provided these uses are for noncommercial personal purposes. Except as provided by law, this material may not be further reproduced, distributed, transmitted, modified, adapted, performed, displayed, published, or sold in whole or in part, without prior written permission from the publisher.

Epigraph to Chapter 4: From *Disturbing the universe* by Freeman J. Dyson (Harper & Row, New York, 1979). Used by permission of Freeman Dyson.

Fig. 4.1: Adapted from Larry Gonick, *The cartoon guide to the computer* (Harper Perennial, 1983, page 89); used with permission.

Fig. 4.2: (c) Image courtesy Antoine Weis and Todorka L. Dimitrova. Reproduced with permission of American Institute of Physics from Dimitrova and Weis. The wave-particle duality of light: A demonstration experiment. *Am. J. Phys.* (2008) vol. 76 (2) pp. 137–142 (`http://dx.doi.org/10.1119/1.2815364`). Copyright 2008, American Association of Physics Teachers.

Fig. 4.4: Courtesy Antoine Weis; used by permission.

Fig. 4.7: (b) © Robert D Anderson; used by permission.

Fig. 4.15: (a) From Cagnet et al., 1962, *Atlas optischer Erscheinungen. Atlas de phénomènes d'optique. Atlas of optical phenomena*: (a) "Chapter III. Diffraction at infinity," page 17; ©Springer-Verlag Berlin Heidelberg, 1962. With permission of Springer. (b) From Cagnet et al., 1962, *Atlas optischer Erscheinungen. Atlas de phénomènes d'optique. Atlas of optical phenomena*: (b) "Chapter IV. Diffraction at a finite distance," page 34. ©Springer-Verlag Berlin Heidelberg, 1962. With permission of Springer.

Epigraph to Chapter 5: Eugene Delacroix.

Fig. 5.1: Reproduced with permission from Smith. Structural color of *Morpho* butterflies. *Am. J. Phys.* (2009) vol. 77 (11) pp. 1010-1019 (`dx.doi.org/10.1119/1.3192768`). ©2009, American Association of Physics Teachers.

Fig. 5.2: (a,b): Used by permission of the Physical Society of Japan. (c) Fig. 3b from p. 107 of: Kinoshita et al. Photophysics of structural color in the *Morpho* butterflies. *Forma* (2002) vol. 17 pp. 103–121. Used by permission of FORMA.

Fig. 5.4: (d) ©Alex Wild, `http://www.myrmecos.net`. Used by permission.

Fig. 5.6: Courtesy Alison M Sweeney; used by permission.

Fig. 5.10: Redrawn similar to K Franklin, P Muir, T Scott, L Wilcocks and P Yates, *Introduction to biological physics for the health and life sciences* (John Wiley and Sons 2010).

Fig. 5.12: Images courtesy Nico Stuurman, HHMI/University of California at San Francisco. Used by permission.

Part II opener: Courtesy of Penn Museum, image #152647. Used by permission.

Epigraph to Chapter 6: Hermann von Helmholtz, *Popular Scientific Lectures,* trans. E. Atkinson. D. Appleton, 1883.

Fig. 6.1: (a) Unknown artist, 17th C., US Library of Congress. Restored by Durova, `http://en.wikipedia.org/wiki/File:Camera_obscura2.jpg`.

Fig. 6.2: © Hans Hillewaert, as found at `http://en.wikipedia.org/wiki/File:Nautilus_pompilius_(head).jpg`. Used under the Creative Commons Attribution-Share Alike 4.0 International license (`http://creativecommons.org/licenses/by-sa/4.0/`).

Fig. 6.6: (b) Polyak, S. 1957. *The vertebrate visual system,* p. 276. Chicago IL: Univ. of Chicago Press. Ed. H Kluver. ©1957 by the University of Chicago. ©1955 under International Copyright Union. Used by permission of University of Chicago Press.

Fig. 6.10: From Cagnet et al., 1962, *Atlas optischer Erscheinungen. Atlas de phénomènes d'optique. Atlas of optical phenomena*: (b) "Chapter IV. Diffraction at a finite distance," page 34. ©Springer-Verlag Berlin Heidelberg, 1962. With permission of Springer.

Fig. 6.12: Courtesy Austin Roorda; used by permission.

Fig. 6.14: From Fritzky & Lagunoff, 2013. ©2013. Distributed under the Creative Commons Attribution License, `https://creativecommons.org/licenses/by/3.0/legalcode`.

Fig. 6.17: Redrawn from Fig. 5.6F page 190 of: Land, M L, & Nilsson, D-E. 2006. General-purpose and special-purpose visual systems. Chap. 5 of: Warrant, E J, & Nilsson, D-E (Eds.), *Invertebrate vision.* Cambridge MA: Cambridge Univ. Press. 2006.

Fig. 7.3: From Xu et al., 2013: Readers may view, browse, and/or download material for temporary copying purposes only, provided these uses are for noncommercial personal purposes. Except as provided by law, this material may not be further reproduced, distributed, transmitted, modified, adapted, performed, displayed, published, or sold in whole or in part, without prior written permission from the publisher.

Fig. 7.4: Adapted from Fig. 1B, p. 6496 from Toprak et al. Defocused orientation and position imaging (DOPI) of myosin V. *Proceedings of the National Academy of Sciences USA* (2006) vol. 103 (17) pp. 6495–9. ©(2006) National Academy of Sciences, U.S.A. Used with permission.

Fig. 7.5: (a,b) Adapted from Figs. 1A,B from Shtengel et al. Interferometric fluorescent super-resolution microscopy resolves 3D cellular ultrastructure. *Proceedings of the National Academy of Sciences USA* (2009) vol. 106 (9) pp. 3125–3130. Used with permission.

Fig. 7.6: Adapted from Fig. 5A,B,C from Shtengel et al. Interferometric fluorescent super-resolution microscopy resolves 3D cellular ultrastructure. *Proceedings of the National Academy of Sciences USA* (2009) vol. 106 (9) pp. 3125–3130. Used with permission.

Epigraph to Chapter 8: Jacob. Evolution and tinkering. *Science* (1977) vol. 196 (4295) pp. 1161–1166.

Fig. 8.1: (b) ©1957 Langridge et al. *J. Biophys. Biochem. Cytol.* 3:767–778. DOI: `10.1083/jcb.3.5.767`. (c) Art by David S Goodsell. Used by permission.

Fig. 8.6: (c,d): Art by David S Goodsell. Used by permission.

Epigraph to Chapter 9: From *Vision: Human and Electronic* by Albert Rose, pp. vii–viii. c1973 by Plenum Press.

Fig. 9.2: Redrawn with permission from Nelson. Old and new results about single-photon sensitivity in human vision. *Physical Biology* (2016) vol. 13 (2) art. 025001.

Fig. 9.3: Redrawn with permission from Nelson. Old and new results about single-photon sensitivity in human vision. *Physical Biology* (2016) vol. 13 (2) art. 025001.

Fig. 9.4: Redrawn with permission from Nelson. Old and new results about single-photon sensitivity in human vision. *Physical Biology* (2016) vol. 13 (2) art. 025001.

Fig. 9.5: Redrawn with permission from Nelson. Old and new results about single-photon sensitivity in human vision. *Physical Biology* (2016) vol. 13 (2) art. 025001.

Fig. 9.6: Lisa R. Wright/Virginia Living Museum, used by permission.

Fig. 9.7: Photo courtesy K.-W. Yau, Johns Hopkins University School of Medicine and D. Baylor, Stanford University. Used by permission.

Fig. 9.9: Redrawn with permission from Nelson. Old and new results about single-photon sensitivity in human vision. *Physical Biology* (2016) vol. 13 (2) art. 025001.

Fig. 10.1: (a)©1985 Townes-Anderson et al. *J. Cell Biol.* 100:175–188. DOI: `10.1083/jcb.100.1.175`. (c) Fig. 2, p. 323 from Townes-Anderson et al. Rod photoreceptors dissociated from the adult rabbit retina. *J. Neurosci.* (1988) vol. 8 (1) pp. 320–331. Used with permission.

Fig. 10.2: Adapted from *Physical Models of Living Systems*, 1/e, by Philip Nelson, ©2015 by W.H. Freeman and Company. Used with permission of the publisher.

Fig. 10.5: PDB `1f88` (PubMed: `10926528`). Art by David S Goodsell. Used by permission.

Fig. 10.6: Republished with permission of Association for Research in Vision and Ophthalmology, from Lamb and Pugh. Phototransduction, dark adaptation, and rhodopsin regeneration: The proctor lecture. *Invest. Ophthalmol. Vis. Sci.,* vol. 47 copyright 2006; permission conveyed through Copyright Clearance Center, Inc. Courtesy Trevor D Lamb, Dept of Neuroscience and Vision Centre, JCSMR, Bldg 131, ANU, Australia.

Fig. 10.12: (a) Figure 3A from "Optimization of Single-Photon Response Transmission at the Rod-to-Rod Bipolar Synapse." H. Okawa, A. P. Sampath. Physiology Vol. 22 no. 4: 279–286, 2007. DOI: `10.1152/physiol.00007.2007`. Used with permission.

Fig. 10.13: Painting by David S Goodsell. Used by permission.

Epigraph to Chapter 11: Henri Matisse Dessins; Matisse to Louis Aragon, 1942, published in *Themes et variations* (Paris 1943), 37.

Fig. 11.2: Redrawn with permission from Nelson. Old and new results about single-photon sensitivity in human vision. *Physical Biology* (2016) vol. 13 (2) art. 025001.

Fig. 11.3: Redrawn with permission from Nelson. Old and new results about single-photon sensitivity in human vision. *Physical Biology* (2016) vol. 13 (2) art. 025001.

Fig. 11.4: Redrawn with permission from Nelson. Old and new results about single-photon sensitivity in human vision. *Physical Biology* (2016) vol. 13 (2) art. 025001.

Fig. 11.5: Redrawn with permission from Nelson. Old and new results about single-photon sensitivity in human vision. *Physical Biology* (2016) vol. 13 (2) art. 025001.

Fig. 11.6: Wässle, 2004. Adapted from Demb, J. B. & Pugh, E. N. Jr. Connexin36 forms synapses essential for night vision. *Neuron* 36, 551–553 (copyright 2002), with permission from Elsevier.

Fig. 11.8: ©1998 by Sinauer Associates. Used by permission.

Fig. 11.10: PDB `2rh1` (β_2-adrenergic receptor); `3sn6` (β_2-adrenergic receptor + G_s protein); `1cul` ($G_s\alpha$ + catalytic domains of adenylyl cyclase). Art by David S Goodsell. Used by permission.

Fig. 11.12: ©2012 by Sinauer Associates. Used by permission.

Part III opener: From Ibn al-Haitham, *Kitab al-Manazir*, ca. 1027. Courtesy of the Süleymaniye Library, Istanbul.

Epigraph to Chapter 12: Used by permission of Peter J. Turchi.

Fig. 12.1: Courtesy Marija Drndic. Used by permission.

Epigraph to Chapter 13: From *The joys of Yiddish* by Leo Rosten (Pocket Books, 1970). Used by permission of Rosten LLC.

Fig. 13.1: (b) *Cell and Tissue Research,* Rapid synthesis of photoreceptor membrane and assembly of new microvilli in a crab at dusk, vol. 211, 1980, p. 420, Sally Stowe, (©Springer-Verlag 1980). With permission of Springer. Original caption: "Semi-schematic diagram of a crab ommatidium illustrating the arrangement of the microvilli in the rhabdom. R1-7 the seven main retinula cells; Pal palisade; B bridge across palisade; D desmosomes; Rh rhabdomere of one cell; m microvillus." (c) Courtesy Christina King-Smith. Used by permission.

Epigraph to Chapter 14: Adam Gopnik, "The Real Work: Modern magic and the meaning of life," *New Yorker Magazine,* March 17, 2008 p. 62. Used with permission of the author.

Epigraph to Epilog: Quoted in W. Pauli, "Albert Einstein in der Entwicklung der Physik," *Phyikalische Blätter,* vol. 15, issue 6, p. 244 (June 1959). Used with permission of the Albert Einstein Archives.

Epigraph to Acknowledgments: ©Piet Hein, from *Grooks:* OMNISCIENCE page 6. Reprinted with kind permission from Piet Hein a/s, DK-5500 Middelfart, Denmark.

Epigraph to Appendix A: From *Biograffiti* by John M. Burns (New York: Norton, 1981), p. 85. Used by permission.

Epigraph to Appendix B: Napoleon I, from *Memoirs of the History of France During the Reign of Napoleon,* Volume 4, dictated by the emperor to Gen. Gourgaud (London: H. Colburn and Company, 1824).

Epigraph to Appendix D: From *Fundamentals of Physics* by R. Shankar (Yale Univ. Press, New Haven CT, 2014). Used by permission.

Software

This book was built with the help of several pieces of freeware and shareware, including TeXShop, TeX Live, LaTeXiT, BibDesk, and DataThief, as well as the Anaconda distribution of the Python language, iPython interpreter, and Spyder IDE (Continuum Analytics Inc.).

Grant support

This book is partially based on work supported by the United States National Science Foundation under Grants PHY–1601894, EF–0928048, and DMR–0832802. The Aspen Center for Physics, which is partially supported by NSF under Grant PHY–1066293, also helped immeasurably with the conception, writing, and production of this book. Any opinions, findings, conclusions, silliness, or recommendations expressed in this book are those of the author and do not necessarily reflect the views of the National Science Foundation.

The University of Pennsylvania Research Foundation and Research Opportunity Grant program provided additional support for this project.

Trademarks

MATLAB is a registered trademark of The MathWorks, Inc. *Mathematica* is a registered trademark of Wolfram Research, Inc.

Index

Bold references indicate the main or defining instance of a key term. Symbol names and mathematical notation are defined in Appendix A.